Claiming the High Ground

Claiming the High Ground

**Sherpas, Subsistence, and
Environmental Change in the
Highest Himalaya**

Stanley F. Stevens

UNIVERSITY OF CALIFORNIA PRESS

Berkeley / Los Angeles / Oxford

Photographs by Stanley F. Stevens

University of California Press
Berkeley and Los Angeles, California

University of California Press
Oxford, England

Copyright © 1993 by The Regents of the University of California

Library of Congress Cataloging-in-Publication Data

Stevens, Stanley F.
    Claiming the high ground : Sherpas, subsistence, and environmental
change in the highest Himalaya / Stanley F. Stevens.
        p.   cm.
    Includes bibliographical references and index.
    ISBN 0-520-07699-0 (cloth : alk. paper)
    1. Human ecology—Nepal—Khumbu Valley.   2. Sherpa (Nepalese
people)—Economic conditions.   3. Landscape changes—Nepal—
Khumbu Valley.   4. Khumbu Valley (Nepal)—Economic conditions.
I. Title.
GF696.N35S74   1993
304.2'095496—dc20   92-4808
                                                              CIP

Printed in the United States of America

1 2 3 4 5 6 7 8 9

The paper used in this publication meets the minimum requirements of American
National Standard for Information Sciences—Permanence of Paper for Printed Library
Materials, ANSI Z39.48-1984 ⊗

*To Sonam Hishi of Nauje and
Konchok Chombi of Khumjung*

Contents

Maps

Figures

Tables

Preface

The Sherpas of Khumbu are a celebrated people who live in one of the most famous corners of the world, the Mount Everest region of Nepal. Their homeland on the border of Tibet lies in the heart of the greatest concentration of high mountains in all the Himalaya, amidst some of the most lofty country that has ever been settled and farmed. Even the lowest Khumbu settlements are located at an altitude of nearly 3,500 meters (12,000 feet). Sherpas harvest potatoes and barley and herd yak at altitudes comparable to the highest summits of the Rocky Mountains, the Sierra Nevada, or the Swiss Alps.

This book is about that way of life. It explores how Sherpas not only survive but thrive in a difficult environment—how they cope with the constraints of life on the roof of the world and exploit the unique opportunities of their mountain realm.

It is also a book about cultural persistence. In recent years the Sherpas' homeland has become a major international tourist attraction. More than 8,000 mountaineers and trekking tourists per year now visit the region. They bring with them new opportunities for local development from which the Sherpas have been quick to profit as well as new environmental and cultural impacts. Along with the increased international visibility of the region and the threats of adverse impact from tourism has also come increasing central government and international attention. Khumbu has become a national park, Sagarmatha (Mount Everest) National Park, a World Heritage Site, and the focus of considerable international aid efforts and new resource-use regulations. These too have had significant effects on Sherpa life. One key theme of this book is the exploration of how Sherpas have coped with tourism and the national park and have continued to maintain their identity, cultural values, and subsistence practices despite major economic change.

The heart of this book is a chronicle of the ingenuity, knowledge, and cultural traditions through which Sherpas shape their crop production, pastoralism, and forest use to high mountain conditions and maintain a much-valued way of life. This represents a four-century-long process of adaptation and innovation that continues today.

One aspect of Sherpa adaptation has been the development of a body of indigenous cultural knowledge about local microenvironments, climate, soil, vegetation, and the fitness and performance of particular crop and livestock varieties in the full range of environmental conditions found in the Everest region. Another facet has been a shared cultural repertoire of practices aimed at averting the considerable risks of agriculture and pastoralism at these altitudes and achieving good harvests within the contexts of shared cultural values about lifestyles and the organization of work, local beliefs about the cultivation of luck and supernatural assistance, and efforts to maintain yields on a long-term basis. And yet another key element has been the local resource-management institutions which Khumbu Sherpas have developed to regulate some aspects of individual household land use. Here the Sherpas developed some of the most sophisticated community systems for managing pastoralism, forest use, and crop production which have been reported in the Himalaya.

At the same time that this is a book about indigenous knowledge, adaptability, and cultural persistence it is also a book about environmental change and the role of both Sherpas and tourism in the increasing environmental degradation of the Mount Everest region. One concern here is deforestation. Himalayan deforestation, and particularly deforestation and forest degradation in Nepal, is second perhaps only to the plight of the tropical rain forests in terms of the amount of international attention it has received. Deforestation in the Mount Everest area has been widely highlighted as an outstanding example of the adverse impact of tourism. But deforestation and other vegetation change in these high valleys is much more complex and has a greater historical context than has been realized and the Sherpas themselves have a greater role in it than has been appreciated.

I hope with this book to communicate a fuller understanding of the complexity of Sherpa life and the richness of their achievements. This requires drawing on Sherpa perspectives about land-use practices, resource-management institutions, local environmental knowledge and belief, and historical economic and environmental change to a degree which has not been attempted in earlier works. At times this leads me to contradict conventional wisdom about Sherpas. Looking at Khumbu through Sherpa perspectives, for example, means setting aside some of the romanticism of earlier outsiders' accounts and acknowledging the environmental shortcomings of historical Sherpa land use and management. Generations of Sherpas, for all their intimate familiarity with Khumbu geography, have left the marks of their settlement and subsistence in the local landscape. This book is consequently a study both of adaptation and impact, of land use, resource management, and environmental transformation. It links the in-depth exploration of indigenous environmental knowledge and the operation of traditional local resource-management institutions with the analysis of environmental change during both the era before the arrival of tourism and the establishment of Sagarmatha National Park and the subsequent period. I believe that the resulting portrait is true to the Sherpas' own sense of their cultural, economic, and environmental past and present, although it may clash with stereotypes of Sherpa life and environmental conditions in Khumbu.

The book has been very nearly a decade in the making. During the last nine years I made seven research expeditions to Nepal and spent a total of three-and-a-half years in the Nepal Himalaya. Three years of that was spent in the villages of Khumbu conducting cultural ecological fieldwork, spread over a period from the winter of 1982–1983 to the spring of 1991. The book is up-to-date as of May 1991 on all major topics. The writing of it has been an epic journey itself. It began with a dissertation drafted during five months on the road in China in 1988 and polished during five months in Berkeley in 1989. There followed another half a year of fieldwork in Nepal and another year of thought, reevaluation, and writing in California, Nepal, and Hawaii to create the book out of the preliminary sketch of my dissertation.

Neither the research in the Mount Everest region nor the writing would have been possible without the help of a number of people and institutions both in the U.S. and Nepal. At the University of California, Berkeley, faculty in both geography and anthropology have provided much appreciated encouragement and advice during the course of the several research projects which led to this book. I am especially grateful for the help and friendship of Bernard Nietschmann, Robert Reed, Nelson Graburn, James Parsons, and Gerald Berreman. I would like to especially thank Bernard Nietschmann for a great deal of good advice over the years.

Many people in Nepal and particularly in Khumbu contributed in invaluable ways to my research. Foremost were the many Sherpas whose willingness to share their knowledge and experience made the study possible. Above all others was Konchok Chombi of Khumjung, a man whose knowledge of Khumbu's past is unexcelled and who took the time—measured in hundreds of hours—to share this with me in the hope, above all, that it would be passed on. Lhakpa Dorje of Nauje (Namche Bazar), Kapa Pasang of Khumjung, Mingma Tsiring and Sun Tenzing of Phurtse (Phortse), and Dawa Tenzing and Tak Nuri of Thami Teng also were special sources of information and insight into past and present Khumbu, along with many others whom I will not try to list here.

I would like to especially thank Sonam Hishi of Nauje, research assistant, guide, and friend. Few Sherpas know Khumbu as well as Sonam does or are able to communicate their knowledge of their land and people as effectively. He possesses a rare love of exploration and learning and these two shared motivations sustained us through a great many months of work which took us not only up and down the mountain-sides of Khumbu but over a great deal of eastern Nepal. Sonam's dedication, enthusiasm, and patience were essential to our being able to accomplish as much as we did.

I was also fortunate in my other assistants in Khumbu, for a number of young Sherpas took an interest in the work and lent their talents to several special projects. These Sherpa friends included Ang Nima, Tenzing Gyazu, Dawa Futi, Anu, and Natong of Nauje, Nima Dorje of Pangboche, Pasang Temba of Kunde, and Apa of Thami Og. I was especially fortunate to have Ang Nima's assistance during my work in Khumbu during the autumn of 1990 and the spring of 1991.

The interest and efforts of these many Sherpas and their continuing friendship have greatly enriched my life and work over these past years. They have made of this book something more, I hope, than a book *about* Sherpas. It is a book which was written with the Sherpas as partners in the process. This went beyond the research itself, for I took the evolving draft of the book back to Khumbu in both autumn 1990 and the spring of 1991 to share and discuss with local Sherpas. Sonam Hishi especially took the time to go over the manuscript with me in detail.

Others who deserve thanks for support, assistance, and friendship in Nepal include Elizabeth Hawley, Penny Walker, Ann Lewis, Shyam Shrestha, Rita Lama, and Sundar Joshi of the U.S. Educational Foundation in Nepal; Mingma Norbu Sherpa, Lhakpa Norbu Sherpa, and Surya Bahadur Pandey, chief administrators of Sagarmatha National Park during much of my time there; Bishwa Nath Upreti, the Director General of the Department of National Parks and Wildlife Conservation; Mangal Siddhi Manandhar, chair of the Department of Geography at Tribhuvan University, and Barbara Brower, Berkeley graduate student and geographer with whom I shared a household and early Khumbu experiences during 1983 and part of 1984.

Special appreciation is due to the people of Nauje, my home for two-

and-a-half years, for all the shared months of day-to-day village life and the ongoing warmth and welcome. There are so many friends from those days: Lhakpa Dorje, Lhakpa Dikki and their family; Sonam Hishi, Tenzing, and Chin Dikki; Pasang Kami, Namdu, and Dawa Futi; Ang Purba; Anu and Mingma Lemu; Natong; Dakki and Ang Kinnu; Ang Kami; Ang Norbu; and Tenzing Chopel and Ang Lemi. Special thanks are due to Tenzing Gyazu, cook, weather observer, friend, and integral part of my 1986 through 1987 Nauje household. And most of all I would like to thank Ang Futi for her friendship over the past eight years.

The research would not have been possible without institutional, financial, and technical support from a number of sources. Especially vital was the ongoing support of the University of California, Berkeley, for my work. The University of California, Berkeley, funded my first Khumbu field work as a undergraduate with a President's Undergraduate Fellowship. Berkeley-administered National Resource Fellowship funds subsidized language training at the University of Wisconsin, Madison, for two summers. A Berkeley Fellowship for Graduate Study provided three years of very generous support for my graduate career, and this was administered in a way which allowed me ample opportunity to make continued field journeys to Nepal. A Berkeley Graduate Humanities Fellowship funded summer dissertation-research reconnaissance in Nepal in 1985 and a 1988–89 Berkeley Chancellor's Dissertation Year Fellowship enabled me to devote a full nine months of undistracted attention to writing the dissertation. I do not imagine that any other university in the country could have been so supportive of my efforts or enabled me to acquire an education so well suited to my objectives.

Dissertation research was funded by two fellowships, a Department of Education Fulbright-Hays Doctoral Dissertation Abroad Fellowship and a Social Science Research Council Dissertation-Research Fellowship. I would like to thank also Nancy Plunkett and Joan Hammes of the University of California, Berkeley, Graduate Fellowship Office for all their work in coordinating these grants and other funding on my behalf.

I returned again to Nepal in 1990 and 1991 on a Fulbright Postdoctoral Fellowship that enabled me to update a number of important facets of the book and carry out further work on tourism development and national park management. I greatly appreciated Anita Caplan's help as South Asian program officer at the Council for International Exchange of Scholars in making necessary arrangements for the project in Washington.

Much of the final writing was done in Hawaii, where I was a Fellow at the Environment and Policy Institute of the East-West Center in 1990–1991. I am grateful to Larry Hamilton at EAPI for making it possible for me to make use of the excellent facilities in Honolulu.

At the University of California Press I would like to especially thank Stanley Holwitz, assistant director of the press, for his interest in the book and help in shepherding it into print. I also thank Rebecca Frazier, managing editor, and Eran Fraenkel, copy editor, for their efforts.

There are a number of other people who helped out along the way. Dr. John B. Garver, Jr., chief cartographer of the National Geographic Society furnished detailed maps of the eastern part of the Mount Everest region. Robert Gwaltney volunteered his time and expertise to print the black and white photographs. James Fisher and an anonymous reviewer provided valuable comments on earlier drafts of the book. P. H. C. (Bing) Lucas shared with me his experiences in Khumbu in 1974 with laying the groundwork for the establishment of Sagarmatha National Park. My father-in-law, Don Gaubatz, and his friends Skip Stevens and Del Fausey designed and constructed a solar-powered system which enabled me to use a portable computer and printer in Khumbu, a remarkable effort which added considerably to the fieldwork. The funds for computer, printer, and solar equipment were a gift from my grandmothers, Mildred Burlie and Charlotte Stevens, and from my father, Charles Stevens. I thank them not only for making it possible to work and write more effectively in Nepal, on the road in western China, and back in California, but also for a major vote of confidence. I am very grateful also to my parents, Charles and Mary Stevens, for their supportiveness and encouragement through the years.

And above all others I would like to especially thank my wife, Piper Gaubatz. Her cartographic skills were crucial to the completion of the project. Beyond that her understanding, supportiveness, and assistance, both in Nepal and during the months fashioning the dissertation and the book in China, California, and Hawaii, immeasurably enriched the entire process of research and writing.

Introduction

The mountaineering exploits, unique culture, and high-altitude life of the Sherpas of Nepal have caught the world's imagination and made them one of the most famous of the multitude of peoples who give the geography of the Himalaya a cultural complexity to match the intricate contours of its terrain. Sherpas have been the subject of scores of books, articles, films, and television documentaries and they have attracted so much academic interest that Sherpa studies have become a significant *genre* in Himalayan anthropological and geographical literature.[1]

The most renowned and well-studied of all the many Sherpa groups are the Sherpas of Khumbu, the high valleys at the foot of Mount Everest. Sir Edmund Hillary has called Khumbu "the most surveyed, examined, blood-taken, anthropologically dissected area in the world" (Rowell 1980:61), and he made these remarks before the profusion of further study of the region during the past decade.[2]

Yet many important facets of Khumbu Sherpa life remain little explored. One of these is adaptation. The broad outlines of Sherpa subsistence practices have been documented for more than thirty years. Researchers noted what kinds of crops Sherpas grew and what livestock they herded. But much less has been learned about how Sherpas combine local knowledge of environment, household land-use decisions, and community resource-management institutions into adaptive responses to the environmental challenges of Khumbu life. Even less has been known of how and when these practices were historically developed, the degree to which they may also represent responses to factors other than physical environment, and whether they were environmentally sustainable in the long term. These perspectives, however, are important in understanding Sherpa life and appreciating and evaluating the knowledge and ingenuity on which it is based.

Understanding the past and present character of Sherpa subsistence and the relation of their way of life to environmental change in Khumbu is also vital to efforts to develop effective conservation efforts. Khumbu today is a place of world importance. The entire region has been made both a national park, Sagarmatha (Mount Everest) National Park, and a World Heritage Site. In the past twenty years there has been increasing international concern about environmental change in the region. One major concern has been that tourism development and changing Sherpa lifestyles and land use have resulted in a deforestation crisis (Begley and Moreau 1987; Bjønness 1980b:124, 126; 1983:270; Byers 1986, 1987b, 1987c; Hinrichsen et al. 1983:274–281; Ives and Messerli 1989; Thompson and Warburton 1985). According to some conservation planners, "Sagarmatha has suffered more deforestation during the past two decades than in the preceeding 200 years" (Hinrichsen et al. 1983:204).

Such conclusions can have enormous implications for local resource policy and for Sherpa life, particularly, as in this case, when they have been reached by men who were responsible for planning and administering Sagarmatha National Park. Yet the conclusion that massive recent deforestation has taken place in Khumbu was made not only before any comprehensive study had been conducted of the extent and causes of forest change in recent years but also before there had been any research into either historical change in Khumbu forest patterns or the history of regional Sherpa forest use and management. Such research would have established that the crisis has, in fact, been misinterpreted. Major deforestation has not taken place in recent years in Khumbu and is not occurring today. Old deforestation has been mistaken for recent change. Assumptions about the former effectiveness of traditional Sherpa forest use and management have had to be revised. But so too has the idea that the current crisis has demonstrated that Sherpa institutions are incapable of coping with changing conditions and are not an appropriate basis for regulating resource use in the national park. The result has been that Sagarmatha National Park policies are now being revised to relax crisis regulations and restore to Sherpas more control of their own land, resources, and development.

This book explores four centuries of Sherpa settlement and subsistence in Mount Everest country from a cultural ecological perspective. I examine the past and present subsistence strategies, patterns of settlement and seasonal movement, land-use practices and techniques, local environmental knowledge, social institutions, and household decisions through which Sherpas have prospered in their high-country homeland. This approach offers new insights into past and present Sherpa agriculture, pastoralism, and forest use. It highlights how the local economy, at both the household and community levels, has been rooted in an abiding concern with the risks and rewards of particular subsistence practices in the context of Khumbu ecology. At the same time it illuminates the ways in which families and communities have adjusted their settlement patterns and land use in accord with sociocultural values, political economic conditions, and situations of varying household social and economic status. The result is a more dynamic portrait of Sherpas and their way of life which testifies to their ingenuity, knowledge, sense of civic responsibility in resource management, and ability to cope with outside interventions ranging from the repercussions of wars, mass refugee movements, and tourism to government land and resource policies and the establishment of Sagarmatha National Park. It is a portrait that recognizes also the great degree to which, in coping with recent changes, Sherpas have been able to retain their sense of identity and maintain a long-valued way of life based on local environmental knowledge and characteristic land-use strategies and techniques. The Sherpas' ability to cope with the array of new ideas and economic opportunities and the increasing incorporation of their once-remote region into Nepal's national life and the global economy has often been underestimated. Too little attention has been given to the persistence of land use based on local environmental knowledge and the efforts of some Khumbu communities to maintain traditional resource-management regulation despite all institutional, economic, and sociocultural pressures to discontinue them. Sherpas have responded to new ideas and economic opportunities, especially those of tourism, but they have largely incorporated them into their culture and way of life on their own terms. Much continuity underlies the apparent transformations of recent years.

This is a book, however, which also questions some of the excess romanticism of earlier accounts. The resourceful patterns of adaptation which Khumbu Sherpas developed over the course of many generations have sustained them in a difficult region, but they have also had an impact on Khumbu environments. Through the use of Sherpa oral traditions and history I inquire not only into past patterns of local resource use and management but also into environmental change. I find that during four centuries of settlement Sherpas have refashioned the landscapes around them as a result of their specific land-use techniques, technology, and cultural institutions. Environmental change is not something which came to Khumbu only with the arrival of mass tourism. In different forms and degrees of intensity it has been a corollary of Sherpa settlement and traditional land use for many generations.

Cultural Ecology

The study of the interaction of people and environment is a fundamental focus of geography. In the Western geographical tradition concern with the linkages between physical environmental conditions and human patterns of settlement and land use goes back to the ancient Greeks. So too does the attempt to explore the possible relationships between physical environment and culture. Both of these themes also figured in the writings of the great Arab geographers more than a millennium later, and were later central to the development of both European and U.S. geography in the nineteenth century. The debate between environmental determinism and possibilism, the concept that while the environment may limit the range of human activity it does not determine it since people choose from among a number of different opportunities, was a vital issue in the early part of this century. The subsequent prominence of the Berkeley school of geography under the leadership of Carl Sauer (1925, 1956) and his successors highlighted the commitment of geographers to people-environment studies. Sauerian geography emphasized the fieldwork-based study of local land-use practices and stressed the understanding of their basis in local environmental knowledge as well as in cultural beliefs and values. Through the concept of the "cultural landscape" Sauer also stressed the study of the historical links between settlement and land-use patterns and environmental change (1925:53). Here he was concerned with the transformation through time of a "natural" landscape into a human one (C. Sauer, "Recent Developments in Cultural Geography," cited in James 1972:400), which reflects in its features the values and practices of the people who have lived in it and used it. The concept of cultural landscape has since been superseded by other approaches to examining environmental change and the processes which underlie it. But the study of settlement, land use, and the historical human transformation of the earth remains important in geography today.

Cultural ecology is the study of peoples and their ways of life as parts of ecosystems. The development since the 1960s of cultural ecology as a major geographical subdiscipline has thus focused more attention on subsistence land-use practices and environmental change and has also brought to geography (as to ecological anthropology) a stronger grounding in the principles of ecology and systems theory. The growth of the field in both geography and anthropology has also supported a proliferation of different conceptual, theoretical, and methodological approaches—a diversity which has also led to considerable contention over defining the most profitable theoretical grounds and methodologies for research and analysis.[3] There have been major differences in the degree to which practitioners have drawn, for example, from ecological science, political economy, and decision-making theory and major contrasts in the scale of analysis (household, community, class, ethnic group, microregion, country), the degree of cross-cultural comparison, temporal framework (synchronic, diachronic, or longitudinal), and emphasis on processual or neofunctional perspectives.

My own approach to cultural ecology centers on the dynamics of environmental perception, land-use decisions, resource management, and environmental change. I take an essentially geographic approach to cultural ecology rather than an anthropological one. Ever since Julian Steward's (1936) pioneering studies, much anthropological work has emphasized the study of the ways in which environmentally adapted subsistence systems may shape population and settlement density, social structure and institutions, belief systems, and other aspects of culture. I stress instead the study of the interaction of people and environment in specific places and, in particular, the detailed study of subsistence systems and resource-use practices, their basis in local environmental knowledge and belief, and their role in maintaining and transforming local ecosystems.[4] Like much of the ecological anthropological work of the past twenty years, however, the orientation I adopt emphasizes the study of process (Orlove 1980)—an approach which the Australian geographer Harold Brookfield advocated as early as the 1960s (1964, 1973).[5]

Much cultural ecology work has focused on adaptation to environment and has devoted little attention to the study of the processes by which societies transform the ecosystems within which and from which they live (Moran 1984; Netting 1984; Orlove 1980). The concept of homeostasis with the environment (Rappaport 1968) was once widely adopted. This has been much challenged, however, and it is now widely agreed on in cultural ecological circles that such a "dynamic balance" with the environment, much less a deeper, primordial "harmony with nature," cannot be assumed even for societies which are living "traditional" ways of life with little involvement in the global economy (Blaikie and Brookfield 1987), much less for indigenous peoples who have such links (Grossman 1984; Nietschmann 1973).

I take as a starting point the idea that the study of environmental adaptation must look to local environmental knowledge as the cultural pivot of adaptation, the point where a society's shared set of perceptions and beliefs about resources, risks, and opportunities provides a basis for individual and collective decisions which adjust land-use practice to environmental conditions. Individual and group decisions provide a basis for adaptation in that they make adjustment and change possible (Bennett 1976). But, to the extent that these decisions are adaptive to environment, they must be grounded first in individual environmental perception and local environmental knowledge and, when widely shared across generations and communities, ultimately also in socially transmitted and culturally shared assumptions, knowledge, belief, and customs.

While I emphasize farmers' and other land users' and managers' efforts to attune land use to their perception of environmental conditions and risks, I am well aware that other processes and conditions also influence the decisions and practices of individuals, households, communities, and societies. These include cultural values and beliefs, social pressure, economic differentiation of wealth, political economy, and the policies and performance of government institutions. To consider any one of these factors in isolation invites distortions that may hinder understanding the dynamics of settlement, land use, and environmental change. Cultural ecology thus must not neglect to consider either the role of culture in settlement and land use or the role of social and economic factors. Shared local knowledge and belief, social organization of labor, management and distribution of resources, marriage and procreation customs, religious food prohibitions, lifestyle preferences, and other factors can significantly affect resource use. Peoples occupying highly similar environments may, as a result, differ enormously in land-use patterns (Barth 1956; Cole and Wolf 1974). Individual households within a given community, groups within societies, and regions may also vary greatly in their political economic circumstances; hence the importance also of microlevel and economic and political economic analysis. In this book I emphasize the commonalities of environmental and agronomic knowledge, land-use practices, and subsistence strategies among Khumbu Sherpas. But I also note the significance of differences of wealth and status in subsistence.

I address the study of Sherpa cultural ecology at several different scales. I begin with the perspective that the study of land use and environmental change must be rooted in understanding of individual land users' and land managers' perceptions and decisions and the factors that influence them. By land managers I mean all people and institutions who make decisions about land use, from individuals to households, communities, development agencies, and local, regional, national, and international governments and institutions (Blaikie and Brookfield 1987). Analysis must focus on the interplay of these decisions on land use and environment at the local level, within the context of household and community (and in some cases also ethnic, class, and kinship) resource-use patterns, and the relationship of these with the local ecosystems. But local decisions often reflect considerations that result from the linkages of the local economy and politics to broader regional networks, processes, and institutions. Cultural variation can also be significant between different groups of a single people who inhabit different areas. Geographers have accordingly emphasized regional context beyond the study of single villages, from the study of a microregion of a valley, watershed, or local culturally defined territory (W. Clarke 1971; Brookfield 1964) to a larger regional (B. Bishop 1990; Zurick 1988), country, continental, and global scale.[6] Understanding the way of life of Khumbu Sherpas, for example, requires learning about their role in far-reaching, multicultural trade networks that have historically integrated the economies of the peoples of the Dudh Kosi valley with those of the inhabitants of adjacent areas of Nepal, India, and Tibet. And analysis of the contemporary Khumbu economy cannot begin without acknowledgment of the role of tourism. It is dangerous to make generalizations about a society or a cultural group from a pattern identified in a particular community. There are significant differences, for example, among Khumbu villages in past and present patterns of land use and management which make it vital to discuss the region as a whole before venturing generalizations about "Khumbu Sherpa" cultural ecology, much less "Sherpa" cultural ecology. Beyond this there is considerable variation in many aspects of culture, including economic practices, between the Sherpas of Khumbu and the Sherpas in other parts of Nepal. Thus a somewhat larger regional perspective encompassing the entire span of Sherpa-inhabited northeastern Nepal is required before much can safely be said about the degree to which Khumbu Sherpa cultural ecology is or is not characteristically Sherpa. Indeed, it is valuable to expand the regional and cultural focus in this regard still further, for Sherpas are one of a number of Himalayan peoples inhabiting the higher mountain regions who have origins in Tibet and it is useful to consider contrasts and comparisons within this larger group.

This concern with context can be carried still further to compare the way of life of the Sherpas of the Mount Everest region more broadly with those of other peoples of highland Asia and mountain regions elsewhere in the world. This scale of spatial and cultural context has long been a characteristic of cultural ecology, where cross-cultural comparisons of the ways of life of different peoples inhabiting similar environments goes back to the early work of Julian Steward (1936). Mountain regions, like such other difficult-to-subsist-in or distinctive environments as deserts, the Arctic, rain forests, grasslands, and coasts have attracted a considerable amount of research (Moran 1979). They have also inspired an unusual amount of multiregional and multicultural comparison (Allan, Knapp, and Stadel 1988; Brush 1976, 1977; Guillet 1983; Orlove and Guillet 1985; Rhoades and Thompson 1975). These discussions of similarities in the cultural ecology of the peoples of the Himalaya, Andes, and Alps have highlighted comparable similarities in strategy and tactics used by indigenous peoples in high mountain regions to cope with environmental difficulties and risks and to exploit seasonal and altitudinal land-use opportunities.

Scale is important in temporal as well as spatial dimensions. Cultural ecological analysis must be diachronic. History is vital to the understanding of both adaptation (Barth 1956; Cole and Wolf 1974; MacFarlane 1976; Netting 1981) and environmental transformation (Blaikie and Brookfield 1987; Sauer 1925, 1956). A focus solely on the observable present precludes any possibility of comprehending current patterns of land use as the creative achievements of local innovation, local response to shifting social, political, political-economic, and environmental conditions, or the historical impacts on local life of national, international, and global economies. This is especially so when historical perspective is reduced to merely mistakenly consigning the past to a supposedly static "traditional" period. Historical perspective is also critical to the understanding of environmental change, for often such changes only become apparent over the course of a considerable time span. A baseline in the past also provides a vital context for the interpretation of present environmental conditions.

Earlier portraits of Khumbu Sherpas' way of life and land use have lacked many of these cultural ecological perspectives. The first accounts of Khumbu agriculture, pastoralism, and forest use were general ethnographic sketches (Fürer-Haimendorf 1964, 1975). These provided a basic overview of subsistence practices and the organization of local resource-management institutions. Subsequent work, primarily by geographers and foresters, provided more detail on some facts of Khumbu land use, particularly on herding (Bjønness 1980a; Brower 1987) and forests (Byers 1987b; Hardie et al. 1987; Naylor 1970; Sherpa 1979).[7] It also in some cases included observations of the environmental impacts of current resource-use practices. But there has been no prior attempt to examine the complexity of local crop growing and pastoral practices, the role of adaptation in land-use practices, the operation of local agropastoral and forest-management systems, or links between changes in land use and pre-1960 environmental change.

History, Oral Traditions, and Oral History

Khumbu, like many other areas of the world, is a region without an extensive tradition of written history. As far as is known the Sherpas of this area have never written clan histories, memoirs, or journals, and very few historical documents have been locally preserved. This means that the study of regional history must largely be based on the interpretation of oral traditions and oral history, with only occasional documentary substantiation. By oral traditions I mean legends and hearsay (such as stories told by one's parents or grandparents) which have been passed down through generations. Oral history, by contrast, is personal reminiscence about the events of one's own lifetime and experience (Vansina 1985). Both present enormous challenges to evaluate. The legends and stories which comprise oral traditions have been reworked over generations and it is often difficult to be sure how much is historical fact. Details can vary significantly between recitations by different individuals. Chronology is often exceedingly vague and geography may be difficult to reconstruct. Supernatural and fantastic beings and events may play prominent roles. Legends may be politically and socially charged—justifications of customs or the *status quo*—and the historical base of these oral traditions may be suspect. With such challenges it is often not possible to do more with oral traditions than to use them as a base for speculation. Oral history testimonies also pose challenges to interpretation and evaluation. Oral history accounts consist of eye-witness accounts of past events and reminiscences about change. Such types of testimony are obviously subjective and prone to distortion not simply from faulty memory but from both deliberate and unconscious misinterpretation and reinterpretation. Oral history testimonies are individuals' interpretations of the past and provide opportunities to edit, expand, excuse, and accuse. But oral history nonetheless can provide a firmer basis for historical research than oral traditions. Testimony can be evaluated with the potential for distortion in mind. Accounts can be analyzed for their degree of internal consistency. Research agendas and interview formats can be structured to include cross-examination of key points as well as repeated questioning to check the consistency with which an interviewee maintains the same story and details over a period of weeks or months. Other individuals can be interviewed to establish corroboration. Even in regions where little cultural emphasis is placed on chronological consciousness some rough chronology can often be established by pegging events with reference to certain benchmark events for which dates are known through references to life cycle events or through estimates of elapsed time.

While the methodological difficulties of working with oral history testimonies must thus not be underemphasized, it is possible to analyze them with some degree of rigor. This is worth the effort, for oral history can be an enormous resource. It can reveal a great detail about aspects of local history that would otherwise often be lost even in literate societies. Details about such things as the circumstances surrounding the introduction of a new variety of potato or a change in winter herd movements, for example, might well escape documentation in the archives of many world regions yet be subjects of vivid local recollection.

Oral history, moreover, is history as viewed by insiders, the past as experienced from indigenous perspective. Such accounts can shed light not only on events but on motives, perceptions, and decisions. This makes oral history a particularly valuable tool for historical study of processes of adaptation and change.

The study of history was once one of the major gaps in Khumbu Sherpa studies. Fürer-Haimendorf (1964) recorded a few oral traditions, mostly ones which concerned regional social, religious, and political change. But little more had been done with Khumbu oral history and traditions until recently.[8] Sherry Ortner (1989) has now considerably advanced the study of Khumbu Sherpa history through the use of oral traditions and oral history testimony. She has used this material to illuminate possible links between long-standing Sherpa values and ethos, political economic conditions in the nineteenth century and early twentieth century, and the founding of the region's first Buddhist monasteries. This remarkable theoretical analysis demonstrates the value of local oral traditions for historical reconstruction. It also establishes for the first time a detailed, even if in places still largely conjectural, chronology of Khumbu history.

All of these inquiries into the Khumbu past promote a highly important ethnohistorical reevaluation of Sherpa history from the Sherpa perspective. They begin to break down the image of Khumbu Sherpa history as consisting of two major eras: traditional times before the 1950s and the period of major change since then sparked by the arrival of Western tourists and the increasing incorporation of the region into national and international political and economic networks. Increasingly we begin to see that Sherpa life before the 1950s was not simply a long, ahistorical, static, "traditional" prelude to the major changes of the past three decades, but rather a dynamic period which James Fisher has called "a tradition of change" (1990:64).

Until now this venture into ethnohistory has largely focused on political, institutional, and religious history.[9] The history which I work with in this book is another kind history. Here I am interested primarily in historical change in resource use and environment. These aspects of Khumbu history have attracted little attention. There has been little discussion, for example, of historical change in subsistence knowledge, beliefs, technology, practices, or institutions other than controversial remarks about the significance of the adoption of the potato. Subsistence styles and strategies, resource-management institutions, and the social organization of subsistence have instead been widely assumed to have been in practice for generations, as if the way of life observed by Fürer-Haimendorf and other visitors in the 1950s was at least a century old and had changed relatively little since. A similar attitude and set of assumptions long shaped thinking about regional environmental change. Following Fürer-Haimendorf (1964), the "traditional" methods of land use and the local resource-management institutions, or at least those which regulated forest use and which were operating in the 1950s, have been widely considered to be ancient and relatively effective as conservation techniques. On the basis of Khumbu oral traditions and oral history, however, I will argue that Khumbu Sherpas continually adjusted their land use and management during the century prior to the 1950s and that in many respects the patterns of resource use observed in the 1950s were less than half a century old. I find evidence also that these "traditional" land-use practices did have an impact on the local environment, particularly on forests.

Tourism and "Westernization" are often credited with having transformed Sherpa life across a range of dimensions from agriculture and pastoralism to religious practice (Bjønness 1980a, 1980b, 1983; Fürer-Haimendorf 1984). Sherpas are conscious instead of considerable continuity underlying largely superficial change and have difficulty in recognizing themselves and their homeland in many of the now-stereotypical outsiders' images of them. Restoring historical context to discussions of Khumbu culture, land use, and subsistence strategies does testify to continuity as well as to change. This is as true in the sphere of subsistence practice as it is in other aspects of culture. Historical perspective counteracts the mistaken impression that recent change has shattered older patterns of Khumbu life. Despite the dramatic economic changes of the past thirty years, Khumbu Sherpas have also maintained characteristic central facets of their subsistence strategies and practices.

Overview of the Book

The book consists of two parts. The first five chapters examine the contemporary Khumbu Sherpa economy from a cultural ecological perspective. Chapter 1 introduces the Sherpas and Khumbu and provides a brief outline of their settlement pattern and history. Chapter 2 begins with an overview of the adaptive subsistence strategies through which Himalayan peoples have coped with the challenges and exploited the opportunities of their mountain homelands, then examines the Khumbu Sherpa economy and land use as environmentally adaptive strategies. It also calls attention to variation in subsistence strategies within Khumbu, especially among households of different degrees of wealth and demographic structure. The following three chapters take up in turn the three main components of Khumbu Sherpa land use and resource management: agriculture, pastoralism, and forest use. Chapter 3 discusses Khumbu agricultural practices as adaptive responses to Khumbu micro-

processes that have so greatly changed the Sherpa way of life since the early nineteenth century; the degree to which Sherpa subsistence has been and is environmentally adaptive, in the sense of being environmentally sustainable in the long run; and how the new insights that oral history gives of changes in Sherpa lifestyles over a period of several generations contributes to issues of continuity, identity, and direction in Sherpa society today.

Life and Work in Khumbu

This book is based on three and a half years of research in Nepal conducted between December 1982 and May 1991. I spent three years of that time in Khumbu during seven separate fieldwork sessions that ranged in length from one month to fourteen months. These research expeditions began with a 1983 senior honors research study (Stevens 1983), continued with dissertation research through 1984, winter and summer 1985, autumn 1986, and all of 1987 (Stevens 1989), and culminated with postdoctoral fieldwork in autumn 1990 and spring 1991. My familiarity with the Himalaya, Nepal, and the Mount Everest region goes back a few more years to 1979 when I devoted seven months of a nine-month trip to Asia to travel in the Himalayan and Karakorum ranges. The book also benefits from what comparative perspective I have gained from fieldwork in the Annapurna range of Nepal in 1984, 1987, and 1991, and from my extensive travels in Tibet and highland Yunnan in 1986 and travels in 1988 in the Tian Shan of Xinjiang and on the northeastern part of the Qinghai-Xizang (Tibetan) plateau in Qinghai.

When I first visited Khumbu in October and November 1979 I was struck by the place, its people, and what seemed to be an imminent threat to both from uncontrolled tourism development. I returned in December 1982 for four months to study the impacts of tourism (Stevens 1983) and expanded on this with another ten months of Khumbu fieldwork in 1984 and early 1985. In 1984 I began to examine the local economy and economic change more broadly. This soon led me to discover that there was much new to learn about Sherpa local environmental knowledge, land use, and resource management. I also began to realize that widespread assumptions about "traditional" Sherpa life were too simplistic and romantic. In 1985 I devoted three months during the summer monsoon to a reconnaissance for a cultural ecology– and oral history–based dissertation study. I found that there was great local enthusiasm about the project, and rich material. I accordingly returned in autumn 1986 for fourteen more months of research. This phase of my work was finished in November 1987. I have since had the opportunity to return to Nepal in autumn 1990 and spring 1991 as a Fulbright postdoctoral research fellow to study further the impacts of tourism in Khumbu and the Annapurna range. While doing this research I was also able to update the cultural ecology material.

From the outset of my work in Khumbu I was concerned with developing a region-wide perspective rather than focusing on a single village. The Mount Everest region is small enough that it is not difficult to work in all four valleys and all eight major villages. I found that it worked well to maintain a base in the famous village of Namche Bazar (known as Nauje to Sherpas) while making a series of one-to-three-week forays to other settlements. There are no roads in the region (the nearest road is some 120 kilometers by trail), which meant that I had to devote a good deal of time to traversing back and forth across Khumbu on foot. The days spent hiking in such magnificent mountain country were one of the great rewards of the work.

In Nauje at different times I shared a geographer colleague's house, lived with a Sherpa family, and rented my own house in the village. Nauje served as a home where I regrouped, consulted my notes and library, and used the computer which I ran from solar panels. But most of the time I spent away from home. While working in the other villages and herding settlements I often stayed in villagers' homes, eating local food and sleeping Sherpa-style on the floor or window benches. I also made use of the modest local inns which Sherpas now run during the tourist season. But I preferred to stay with families, particularly with people whom I knew to be particularly knowledgeable about local history and land use. Staying with these families meant that the research filled my waking hours and often some of the best information and insights came during the long nighttime conversations sitting around the household hearth.

The fieldwork revolved around several different projects: observation (and some measurements) of land-use activities; interview data from people questioned while at work in the fields and pastures and from in-depth interviews conducted in people's own homes; a village survey of Nauje carried out by Sherpa assistants; a survey of Khumbu lodges and shops; data on the identity and involvement in tourism of every household in the region (both in 1985 and 1990–1991); a survey of tourists carried out with the help of Sagarmatha National Park officials; and sketch mapping of the distribution of crop varieties, herding patterns, pastoral management zones, protected forests, forest-use patterns, and tourism development.[11]

Detailed interviews were the core of the research. I interviewed more than 200 Sherpas and worked more closely with a group of more than forty. The central themes of these discussions included local geography, environmental knowledge, agricultural, pastoral, and forest-use practices and decisions, local resource-management institutions, perception and response to environmental risks, and evaluation of current processes of environmental change. I also emphasized oral history, particularly the history of trade and tourism, regional settlement, land use and management, and local environmental change.

Some parts of this work I carried out on my own, without assistants, relying on my knowledge of Nepali, which most Khumbu Sherpas now speak as a second language. During most periods of the work, however, I also had the dedicated and capable help of local assistants for collecting village-survey data, taking meteorological readings, measuring crop yields, and interpreting from Sherpa to Nepali or from Sherpa to English. I was especially fortunate to have the gifted assistance of Sonam Hishi of Nauje, who worked with me in 1984, 1985, and 1987. During the many months we worked together, Sonam accompanied me on most interviews and acted as an interpreter as well as a research assistant. His remarkably good (and self-taught) English made the use of Nepali as a *lingua franca* unnecessary and both his linguistic skills and his knowledge about Khumbu added immensely to the work. During interviews when Sonam was translating I relied on making periodic summaries and asking questions in Nepali in order to participate directly in the conversations, verify the accuracy of the interpretation, clarify points, and draw out more detail.

Interviews were intensive, seldom lasting less than two hours and often considerably longer. Interviews with Konchok Chombi of Khumjung, the most knowledgeable of all Khumbu residents about local history, totaled more than 250 hours. I found virtually all Sherpas remarkably generous with their time and appreciated the enthusiasm which several of them developed for the project. The high degree of local cooperation enabled me to make multiple interviews standard practice. These allowed me to obtain further detail on particular topics, introduce new lines of questioning, corroborate other people's accounts, and verify whether key testimony was repeated without discrepancy after a period of weeks or months had elapsed. I worked at building layers of corroboration by cross-checking key testimony extensively.

Like other researchers who have worked in Khumbu I found that some kinds of research met with a better reception than others. Like Ortner I found that people readily appreciated the importance of investigating Sherpa history. Many people were less interested initially in talking about agricultural or pastoral change (which some did not consider history at all) than about settlement history, the legends of local heroes and villains, the building and renovation of local temples, shrines, and monasteries, or half a century or more of village gossip. But on the whole oral history research was very well received. Most cultural ecology work was also appreciated, and especially my efforts to learn about local environmental knowledge and land-use techniques. Many people were extremely uncomfortable, on the other hand, with questions about their household wealth, be it the number of livestock they owned, how many fields they cultivated, how much harvest they stored, or the amount of income they had from tourism or trade. Efforts to measure fields and harvests or to count livestock were regarded with special suspicion. Many people could not be relied on to give truthful replies to direct questions about household wealth and the idea of conducting land or crop measurements had to be approached gingerly even with friends. It proved extremely difficult to carry out the detailed work on household economy and on economic differentiation which I had originally intended. I was only able to make some limited headway in this effort in Nauje, where I was best known and trusted.

The 1980s turned out, however, to be an opportune time to be asking other kinds of questions. A significant number of Sherpas have become aware of discrepancies between their view of themselves, their culture, history, and present problems and the way in which these have been depicted in academic and popular writing and other media. People were interested in correcting misimpressions, presenting personal perspectives, and in countering stereotypes. Equally as important, by the 1980s a number of elderly Sherpas were concerned that local oral traditions and history were no longer being passed on effectively and were highly interested in preserving their knowledge for future generations by having it recorded in writing. Other Sherpas were enormously concerned that I tell the world (and the Nepalese government) about the impact on Sherpa life of the nationalization of forests, the establishment of Sagarmatha (Mount Everest) national park, the tourism boom, and international development efforts.

Finally, I would like readers to bear in mind a few other points about the research. I was able to obtain better data in some communities than others. The material is strongest for Nauje and Khumjung and stronger for Phurtse (called Phortse by outsiders), Pangboche, and Thami Teng (also called Thamite) than for the other villages. There may also be some gender bias in my perspectives in that the majority of Sherpas with whom I worked most closely were men. I was able to interview a number of women at length, however, as well as to benefit from the insights of a great many more during family discussions in their homes. Much of the detailed information about agricultural and environmental knowledge and crop-growing practices that I discuss is based on women's knowledge and experience.

Part One

Sherpa Cultural Ecology

1

Sherpa Country

The Himalaya rises regally above the subtropical lowlands of the Ganges valley, a vast sweep of mountains crowning the Indian subcontinent with a 2,400-kilometer tiara of the highest peaks on earth.[1] Nepal is its centerpiece. Here the Himalaya crests against the edge of Tibet in a rampart of peaks which average 6,000 meters in height and includes most of the world's highest mountains.[2] From the Tarai, the narrow Nepalese share of the Ganges valley in the south of the country, where the altitude is below 200 meters and the climate subtropical, the highlands ripple toward the Tibetan plateau in a staircase series of ranges. Three parallel ranges sweep northwest to southeast across the country. First come the low foothill ridges of the Siwalik range (also known as the sub-Himalaya and the Churia hills), which rise abruptly from the plains to attain elevations of 1,000–1,800 meters. Just to the north, sometimes so close as to merge with the foothills and otherwise only separated from them by the narrow longitudinal valleys of the Inner Tarai, are the higher ridges of the Lesser Himalaya. Here the crest of the Mahabharat Lekh, the leading edge of the Himalaya proper, sometimes surmounts 3,000 meters in altitude. Beyond it is a 70–110-kilometer-wide band of hill country known in Nepali as the *pahar,* the hills, and which is often also called the middle hills or the Nepal midlands. This hill country is the geographic, historical, and demographic heartland of Nepal. Here a complex contortion of ridges and valleys extends across most of the breadth of the country to the gleaming snow peaks of the Great Himalaya. To the north, beyond the verdant, terraced slopes of the Nepal midlands, towers the massive upthrust of the Great Himalaya, a crystal wall of snowy summits which rise with such relief that by comparison the rugged topography of the rest of Nepal seems only low, rolling foothills. In Nepali this high range is called the *himal,* the mountains, and the word has come to signify a realm distinct from the rest of the country in its height, ruggedness, climate, vegetation, and populace. The main range trends northwest to southeast across the length of Nepal and from it massive high ridges lead perpendicularly to the south into the midlands. These great ridges shape the lie of the midlands' valleys and drainage except where a few major rivers wind longitudinally along a span of the Mahabharat Lekh or the Siwalik range before finding a passage through these final barriers to the plains. In eastern Nepal the Himalaya forms the border with Tibet whereas in western and central Nepal trans-Himalayan or inner-Himalayan valleys lie beyond the Great Himalaya and south of the Tibetan border ranges.[3]

The Great Himalaya attains its greatest heights in central and eastern Nepal around the headwaters of the Dudh Kosi and Barun rivers. Here a vast line of glaciated peaks soars high above the green ridges of the midlands. The range reaches its pinnacle in the country around Mount Everest.[4] Four of the world's fourteen 8,000-meter peaks (app. 1) stand along a 90-kilometer span of the Great Himalaya known as the Mahalangur Himal.[5] Three of these, Cho Oyu, Lhotse, and Mount Everest are part of the Khumbu Himal, a wall of great peaks aligned northwest to southeast which also includes Gyachung Kang and Nuptse.[6] In this region the Great Himalaya consists of two parallel ranges, for less than twenty kilometers to the south of the main Mahalangur Himal is the Numbur-Kantega range that includes such famous Khumbu peaks as Kwangde, Tamserku, and Kantega.[7] Still more high mountains punctuate the four ridges that traverse north to south between the two ranges, among them Pumori, Tsola Tse, Tawache, and Khumbila. The scores of big mountains make this the densest concentration of high peaks in the entire vast arc of high country from the Indus to the Brahmaputra.

In the midst of this fierce verticality is the country that Sherpas call Khumbu, an 1,100-square-kilometer region bounded on the north by the Khumbu Himal, on the south by the Numbur-Kantega range, and on the east and west by high ridges that link the two and ring the region with a nearly unbroken wall of mountains (map 1). Four high valleys thread between the vertical expanses of rock and ice: the valleys of the Dudh Kosi (the "Milk River") and the Bhote Kosi (the "Tibetan River"), and the two forks of the Imja Khola, the northern Lobuche Khola and the eastern Imja Khola proper.[8] These valleys are the products of the same tectonic forces that raised up the ranges and ridges which flank them and their positions reflect the strikes of prominent faults.[9] They comprise the headwaters of the Dudh Kosi, a tributary of the Sun Kosi and one of the major rivers of eastern Nepal. The Dudh Kosi is born in the glacial snows and lakes at the foot of Cho Oyu and Gyachung Kang, fed by the melting of the Ngozumpa glacier and the waters of several high lakes. It flows south through the center of Khumbu, picking up first the tributary waters of the Imja Khola, which drains eastern Khumbu, and then those of the Bhote Kosi of western Khumbu before breaching a gap in the Numbur-Kantega range and flowing on into the Nepal midlands.[10] Relief between the floor of the Dudh Kosi valley and its tributaries and the adjacent peaks ranges from 2,000 to 4,000 meters (Vuichard 1986:48).

The upper reaches of all four valleys are wide-floored, U-shaped troughs that testify to the erosive power of ancient glacial advances. At the head of each valley there are still large valley glaciers, and the braided streams of the upper valleys flow cloudy with the silt of the glacial sources. Some of these valley glaciers are more than ten kilometers in length and the eighteen-kilometer-long Ngozumpa is one of the longest in Nepal.[11] From these high sources the Khumbu rivers lilt among alpine meadows before descending into the fir, birch, rhododendron, and juniper subalpine forests which finger up into the valleys as high as 4,000 meters.[12] Below 3,800 meters the river courses steepen and plummet into precipitous river-carved canyons. These narrow gorges are among the deepest in the Himalaya. On their north-facing slopes they are thickly wooded with temperate forests of fir, birch, and rhododendron. The sunnier south- and west-facing slopes are primarily grassland, shrubland, open temperate woodlands of juniper and fir, and settlement areas. In these reaches the Dudh Kosi meets first the Imja Khola and then the Bhote Kosi before storming out of Khumbu and on towards the Ganges through a narrow gorge between the peaks of Tamserku and Kwangde. Only in these lowest reaches of the canyons does the altitude ever dip below 3,000 meters. It reaches its lowest point, Lartsa Doban, at the 2,800-meter confluence of the Dudh Kosi and the Bhote Kosi. Only in these gorges is the altitude low enough for montane pine forests to survive and even then pine is primarily found only on sunny slopes. Most of Khumbu is too high in altitude for any forests whatsoever. More than 95 percent of the region is above 4,000 meters in altitude.

Khumbu climate ranges from temperate to arctic depending on altitude and aspect. In lower Khumbu the climate is generally milder than might be expected at such altitudes. The mean daily temperature in January, the coldest month, is −0.4°C in Nauje (3,440m), while the warmest month mean daily temperature (July) is 12°C. The mean minimum in January in Nauje is −7.9°C (Joshi 1982:399–400).[13] The lack of bitter winter temperatures in lower Khumbu, even at altitudes well above 3,000 meters, reflects the region's relatively low latitude (at 28° north it is farther south than Cairo or New Orleans) and the shielding

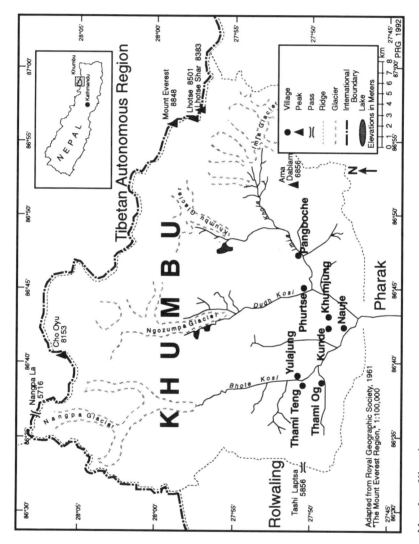

Map 1. Khumbu

effect of the wall of ridges which encircles Khumbu and shelters it from much of the impact both of the bitter northern, winter winds that sweep the Tibetan plateau and the torrential downpours of the summer monsoon that rake the area just to the south. Khumbu is one of a number of valleys in the Great Himalaya which are similarly situated in partial rainshadow conditions and which are sometimes referred to as inner Himalayan valleys.[14] The annual precipitation in even the wettest parts of Khumbu is scarcely half of the more than 2,000 millimeters recorded less than ten kilometers to the south. Precipitation is highly seasonal. Nearly 80 percent of annual precipitation falls during the four months (June through September) of the summer southwest monsoon season.[15] Rainfall varies regionally with altitude, local orographic effects, and aspect. Khumbu precipitation generally decreases with altitude and at Lhajung (4,420m, 518mm) is half or less of what Nauje (3,440m, 1,076mm) and Tengboche monastery (3,867m, 1,127mm) receive (Byers 1987b:35; Inoue 1976). Local rain shadow and orographic effects also cause rainfall to vary from valley to valley, even within valleys at sites of similar altitude. This climatic variation is of great importance for the conduct of agriculture and pastoralism. Snowfall is not heavy and below 5,000 meters most precipitation falls as rain. South- and west-facing slopes below 4,000 meters are normally not blanketed with snow for more than a few weeks each winter and most snowfalls melt within three or four days. North- and east-facing slopes retain snow much longer, often for months on end. The relatively low snowfalls of most winters are due in part to Khumbu's easterly location, for at 86° east longitude the region is far enough east to be beyond the reach of the brunt of the westerly winter storms that drop so much snow on the westernmost Himalayan regions. The heavier snowfall of some years, however, poses dangers of avalanche in some areas and jeopardizes yak herds by depriving them of opportunities for browsing senescent grass.[16]

Eight major villages nestle in the lower reaches of the region between 3,400 and 4,000 meters on the rare bits of gentle ground in a largely vertical country. These villages are Nauje, Khumjung, Kunde, Pangboche, Phurtse, Thami Og, Thami Teng, and Yulajung.[17] Their settlement sites include alluvial terraces, hanging valleys, and amphitheaterlike slumps. In population they range from 14 to 135 households in size. Each village is a congregation of one- and two-story white stone houses usually roofed with slate or fir shakes (and occasionally now with corrugated iron). Each house is surrounded by terraced fields. Large expanses of terraces, pastures, and woodlands cloak surrounding slopes beyond the cluster of houses. Many villagers also own fields and herding huts high up in the upper valleys and a few cultivate secondary fields below the main villages on the steep slopes of the lower gorges. The

forests and pastures of the lower valleys and even the high alpine meadows and tundra of the glacial trough upper valleys are sprinkled with
scores of secondary settlements that Sherpas use as seasonal bases to
maximize their use of the resources of the far corners and diverse microenvironments of the region.

Khumbu is bordered on the north by Tibet and on the west, south,
and east by Nepalese regions known as Rolwaling, Pharak, and the
Arun river region (map 2). These borders follow very clear topographical lines as well as cultural ones. On the west the border between
Khumbu and Rolwaling is a line of peaks and ridges which can be
traversed only by the dangerous glacier crossing of the 5,755-meter Tashi
Laptsa. Rolwaling is a high valley, environmentally similar to Khumbu
in many respects, which was settled in the mid-nineteenth century by
Khumbu Sherpa emigrants from the Bhote Kosi valley (Sacherer 1975,
1981). On the south Sherpas consider Khumbu to end at Lartsa Doban,
the confluence of the Dudh Kosi and the Bhote Kosi.[18] Farther south is a
lower-altitude, wetter country, the Pharak region of the Dudh Kosi
valley, inhabited by Sherpas who differ from those of Khumbu in several
respects. One of the most evident is their reliance on a rather different
system of agropastoralism based on year-round crop production and
different varieties of crops and livestock. Just south of Pharak is an area
inhabited by Sherpas and Magars and beyond that the lower Dudh Kosi
valley and its tributaries are the home of Sherpas, Rais, Gurungs, and
Hindu hill castes. In the east the upper Imja Khola valley is walled off
from the Barun and the Hongu basin regions by high ridges crossed only
by three passes, Amphu Laptsa, Sherpani Col, and Mingbo La (Mingbo
Pass), which require technical mountaineering skills to traverse.[19] The
upper Barun and Arun country on the far side is inhabited by several
peoples who appear to have Tibetan origins as well as by some Sherpa
families who migrated here in the nineteenth century from areas south
of Khumbu. Rais inhabit the valleys south of the Hongu basin and they
and Gurungs from the Ra Khola region farther to the south use some of
the high country south of Khumbu for summer pasture. To the north of
Khumbu the main crest of the Himalaya forms both a cultural and a
topographic border with Tibet. A single pass, the Nangpa La (5,716m),
leads north. Seasonally it can be negotiated by yak as well as by foot
travellers and the actual crossing is rather gentle, although its crevasses
and storms are hazards to be reckoned with. In former times a great deal
of trade was carried out over it. From Nauje Sherpa traders can reach
the pass in as little as four days with a fully loaded caravan. On the far
side lie the rain-shadow plains and arid hills of the Tingri district of
Tibet, the vast treeless expanses dotted with the black yak-hair tents of
nomadic pastoralists and villages of flat-roofed, courtyard houses. Only

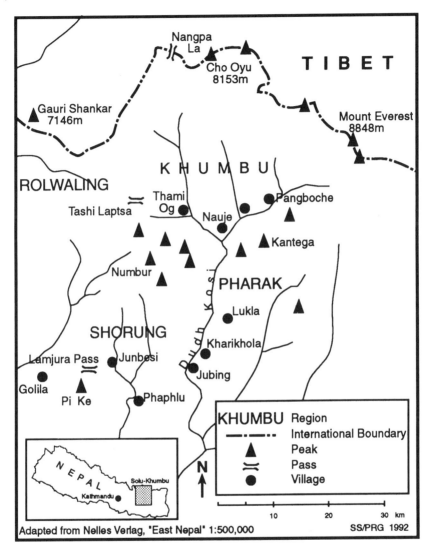

Map 2.   The Solu-Khumbu Region

six days with yak north of the Nangpa La is the village of Ganggar (also known as Tingri), which was a major Sherpa trading center until the late 1960s and where several Khumbu families once owned houses that they used as seasonal trading bases. Just to the east, at the northern foot of Mount Everest, are the ruins of Rumbu (Rongbuk) monastery, once the highest in the world and a place that long has been important to Sherpas (see Ortner 1989:130–131). West of Tingri is Rongshar, a Tibetan valley used occasionally by Khumbu Sherpas as a secondary, year-round trade route between Tibet and Nepal and where one prominent Khumbu Sherpa trader once owned a large yak herd that was cared for by hired Tibetan herders.

Politically Khumbu is today part of the Solu-Khumbu district. Khumbu forms the northernmost part of the administrative district, an area only slightly more than double the size of Khumbu but whose population is nearly thirty times larger and includes more Rais, Magars, Gurungs, and Hindu hill-caste inhabitants than Sherpas. The district center, Salleri, is a recently developed new town in a Sherpa-inhabited region of Solu (an area Sherpas call Shorung) in the southern part of the district four days walk from Khumbu. Kathmandu is yet more remote and until the 1960s to reach the national capital required two weeks of hard walking and the crossing of six passes. The Chinese-built Arniko highway which was completed during the late 1960s and the Swiss-built Jiri road that was finished in 1985 have now bridged half the distance between Kathmandu and Khumbu and may one day reach still closer.[20] Khumbu today is, however, better integrated with the rest of the country by air than by road. Except during the summer months, when the weather and the condition of dirt airstrips usually force the closing of the airstrips, nineteen-seat Twin Otter aircraft land at the STOL (Short Takeoff and Landing) airstrip at Lukla (a day's walk south of Khumbu in the Pharak region), which was constructed in 1964 by Sir Edmund Hillary. Recently a second STOL airstrip just above Nauje at Shyangboche has become active again. This was built in 1971–2 by a Japanese-led consortium as part of a luxury hotel development. It is only suitable for Pilatus Porter single prop planes that carry a maximum of seven passengers.[21]

Sherpas have their own geography of Khumbu, one much richer than that of outsiders in its intimate familiarity with every corner of the region. The Sherpa conceptual map of Khumbu is far more complex than even the finest Western maps. Western maps name little more than a few of the outstanding topographic features and settlements, often with names borrowed from Sherpa usage but so garbled in the process that the most recent editions of the best foreign maps continue to mistake the names of important herding settlements and landscape features. The Sherpa geography of the area recognizes another level of features altogether, naming a multitude of pastures, forest areas, tributary valleys, slope regions, and settlements that do not appear on any foreign maps. The awareness of local geography reflected by place names, moreover, is only the surface of Sherpa geographic knowledge of Khumbu, a working vocabulary by which to organize their enormous understanding of distinct, local microenvironments. For Sherpas the landscapes of Khumbu are also a continual evocation of their individual and collective past, the history which took place in these places. Khumbu is also for them a world alive with supernatural forces—gods, spirits, and ghosts who are also associated with certain places. The mountain Khumbila, for example, is the main residence of the Khumbu Yul Lha, the great god of the region, and there are other peaks and places that are the homes of clan gods. Springs, boulders, caves, forests, and individual trees are the domain of particular spirits and demons. This religious geography is not separate from ordinary life, for it is believed to have great bearing on the luck and well-being of people and communities and is an important consideration in the choice of settlement and house sites, use of water and attitudes towards its pollution, and forest use and protection.[22]

Sherpas

The Sherpas are one of a number of peoples of Tibetan origin and cultural affinities who have migrated south to settle the northern regions of the Himalaya. In Nepal there are several such peoples who inhabit the Great Himalaya, the trans-Himalayan valleys of western and central Nepal that lie between the Himalaya and the Tibetan plateau, and some parts of the midlands. These peoples are part of a much larger population of more or less ethnically Tibetan peoples who number more than six million and who occupy a territory of more than two million square kilometers stretching from the Chinese provinces of Gansu, Sichuan, and northern Yunnan across Tibet to Ladakh, from China's Qinghai province south to the southern slopes of the Himalaya, and along the full length of the Himalaya from Arunachal Pradesh to Kashmir. Across this Tibetan culture region (map 3) there are scores of peoples whose languages, religions, systems of social organization, economies, land-use practices, architecture, and styles of dress and ornamentation share strong resemblances. These similarities include Tibeto-Burman languages closely related to Tibetan; adherence to sects of Tibetan Buddhism; domestic and religious architecture featuring white, stone-walled structures with characteristic window patterns and other distinctive features; settlement sites at altitudes of more than 2,000 meters; cultural preferences for raising yak, sheep, and barley; distinctive traditional woolen dress consisting of a black or brown, extremely long-sleeved cloak for men and a dark, woolen dress with a striped apron for women; and special value put on certain types of jewelry, including silver amulet boxes and necklaces featuring coral, turquoise, *zi* stones, and silver.[23]

In Nepal peoples who have Tibetan-like cultures inhabit more than a quarter of the land area of the country. They comprise only a very small part, however, of Nepal's population of eighteen million.[24] Those who

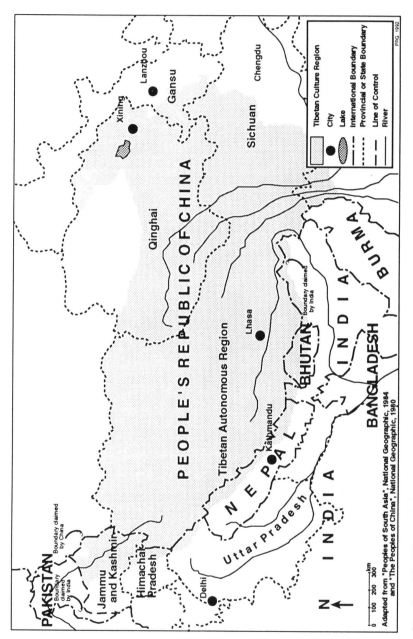

Map 3. *The Tibetan Culture Region*

Tibetan Culture Region
City
Lake
International Boundary
Provincial or State Boundary
Line of Control
River

PRG 1992

PEOPLE'S REPUBLIC OF CHINA

Lanzhou
Gansu
Xining
Chengdu
Sichuan
Qinghai

Tibetan Autonomous Region

Lhasa

Boundary claimed by India

BHUTAN

INDIA

BURMA

BANGLADESH

Kathmandu

N E P A L

Uttar Pradesh

Delhi

I N D I A

Himachal Pradesh

Jammu and Kashmir

Boundary claimed by China

Boundary claimed by India

PAKISTAN

N

0 100 200 300 km

Adapted from "Peoples of South Asia", National Geographic, 1984
and "The Peoples of China", National Geographic, 1980

live closest to the Tibetan border and most closely resemble the Tibetans in culture constitute less than 1 percent of Nepal's population. These peoples, among them the Sherpas, are sometimes referred to as Bhotes or Bhotias.[25]

People who call themselves Sherpas inhabit much of northeastern Nepal between the Sun Kosi and Arun rivers and form a part of the population of several other areas as far west as the Helambu (Yelmo) region northeast of Kathmandu and perhaps as far east as the Sikkim border.[26] The heartland of Sherpa country is the upper Dudh Kosi valley and the valley of one of its major tributaries, the Solu Khola. Often known as Solu-Khumbu, this region according to legend was the first area settled by the ancestors of the Sherpas after their migration from Tibet some nineteen generations ago. It remains the main population concentration of Sherpas today, with approximately 17,000 of what may be a total of 20–25,000 Sherpas in Nepal and as many as 30,0000 in the Himalaya as a whole.[27] This core population inhabits several geographically distinct territories that also vary to some degree in culture. The three most important of these in terms of area and population are Khumbu, Pharak, and Shorung, all of which are located in the Solu-Khumbu district of eastern Nepal. Despite their close proximity these three regions differ considerably in altitude and topography, spanning the full altitudinal range of Sherpa settlement. The Sherpas of Khumbu inhabit the rugged Great Himalayan highland valleys in the headwater region of the Dudh Kosi valley, their lands virtually all above 3,000 meters. Pharak is south of Khumbu in the Nepal midlands reaches of the Dudh Kosi valley. Here villages are situated between 2,300 and 2,800 meters. Shorung is located west of Khumbu and Pharak and reaches from the hill country of the Nepal midlands to the Great Himalaya. According to some Sherpas' reckoning it includes the Manidingma region of the west bank of the Dudh Kosi as well as the valley of the Solu Khola and the interfluvial uplands. Shorung villages vary in altitude from approximately 2,000 to 3,000 meters. Differences in the topography, climate, and vegetation of these three areas have supported very different economies based on differences in seasonal, altitudinal land use and associated variations in agricultural and pastoral practices and forest exploitation.

The Solu-Khumbu Sherpas, and particularly the Sherpas of Khumbu, are the Sherpas who have won worldwide admiration for their strength, endurance, courage, indomitable good spirits and joy in life, deep faith in Buddhism, and their relatively egalitarian and open society. Khumbu Sherpas initially earned their world renown through reports of their mountaineering exploits during the first half of this century. Their reputation was built especially on their performance on the series of British expeditions to Everest. Sherpas had become legendary in mountaineering circles by the 1920s when the British began awarding the finest climbers among them the title of Tigers. But relatively little was known about their homeland and culture even as late as 1953 when Tenzing Norgay and Sir Edmund Hillary reached the summit of Everest, an event which made *Sherpa* a household word around much of the planet.[28] Foreigners had been banned from traveling to Khumbu until 1950, when the Nepal government began to relax the policies that had previously made Lhasa more accessible to foreigners than the highlands of Nepal.[29] It was only in autumn 1950 that the first small group of British and American mountaineers was permitted to make the long journey on foot to the southern foot of Mount Everest and experienced for the first time the alpine grandeur of the Khumbu valleys and the generous hospitality of the Sherpa villages. Other mountaineers soon followed in a steady stream of expeditions and were joined by journalists, scientists, cartographers, the Hillary schoolhouse and hospital building teams, and finally in the 1960s by ordinary tourists. Sherpas also attracted considerable anthropological attention. One of the most distinguished anthropologists of South Asia, Christoph von Fürer-Haimendorf, conducted the first ethnographic studies of Khumbu Sherpas in 1953 and 1957. Fürer-Haimendorf's classic monograph, *The Sherpas of Nepal* (1964) and his later, more detailed treatment of the Sherpa economy and its changes between 1957 and 1971 (1975) focused primarily on Khumbu Sherpas and provided a firm ethnographic base for subsequent, more specialized anthropological and geographical Sherpa studies. Anthropologists have examined Khumbu Sherpa population dynamics (Pawson, Stanford, and Adams 1984), ethnohistory (Ortner 1989), religion (ibid.), shamanism (Adams 1989), festivals (Jerstad 1969), and the regional socio-economic and cultural changes that have accompanied tourism development and the establishment of such new institutions as schools and Sagarmatha National Park (J. Fisher 1990).[30] Geographers have also contributed to the increasing body of writing on Khumbu land and life with accounts of exploration and travel (Gurung 1980; Jackson 1955) and studies of geomorphology (Byers 1986, 1987b), natural hazards (Zimmerman, Bichsel, and Keinholz 1986), vegetation change (Brower 1987; Byers 1987b; Stevens 1986b, 1989), erosion (Byers 1987b), pastoralism (Bjønness 1980a; Brower 1987; Palmieri 1976; Stevens 1989), tourism impacts (Bjønness 1980b, 1983; Stevens 1988b, 1989), and cultural ecology (Stevens 1989).[31] Although most anthropological and geographical research has focused on the Sherpas of Khumbu, there have also been a number of studies of the Shorung Sherpas (including Funke 1969; Kunwar 1989; March 1977; Oppitz 1968, 1974; Ortner 1978, 1989; Paul 1979, 1982) and some work on other groups including the Rolwaling Sherpa (Kunwar 1989; Sacherer 1975, 1981), the Chyangma Sherpa (Limberg 1982), the Sherpas of the Bigu region (Fürer-Haimendorf 1984; Kunwar 1989) and the multiethnic valleys of the Helambu region (N. Bishop 1989; G. Clarke 1980a, 1980b).

From an anthropological perspective Sherpas have been widely characterized in terms of such fundamental features as language, clan structure, religion, and shared history. They have been classified as racially Mongoloid, linguistically Tibeto-Burman, and culturally "Bhotia" or "Tibetan" in terms of general orientation and origins.[32] On the basis of these basic traits the Sherpas of Solu-Khumbu can indeed be readily distinguished from neighboring non-Sherpa Tibetan, Rai, Gurung, Tamang, Sunwar, and Magar groups: Solu-Khumbu Sherpas speak Sherpa, a language distinct from the languages of these other peoples and from the national language, Nepali, which is a member of the Indo-European language family. They belong to one or another of a certain set of exogamous patrilineal clans. They believe in Tibetan Buddhism and specifically are adherants of the Nyingmapa sect. And they are descended (or at least some core of their clans trace their descent) from a small set of immigrant families who came to Nepal from Tibet more than four centuries ago.

Khumbu Sherpas distinguish themselves from the neighboring peoples they call Dongbu (Rai and Gurung), Rongba (Nepali), Tamang, and Pürba (Tibetan).[33] Usually they draw these contrasts on the grounds of territorial homeland, language, religion, and social structure.[34] While these attributes are the ones most often raised as central facets of "Sherpaness," a number of other characteristics are also often mentioned. Many Khumbu Sherpas talk of Sherpas (by which they actually usually mean specifically Khumbu Sherpas) being distinguished from other peoples by such things as their houses (including details such as materials used, styles of wall, window, and roof construction, and the placement and design of the roof, window, and house-pole prayer flags), clothes (including the cut of women's dresses and the style of their aprons), regional, village, and household festivals and rites (*Dumje* is especially cited, while *Losar, Yerchang,* and *Pangyi* are also mentioned), community offices through which temples, festivals, and natural resources are managed, and styles of livelihood and land use.[35]

Sherpas are undoubtedly most closely culturally related to their neighbors to the north, the Tibetans, with whom they share many cultural features. Their languages are closely related. While Sherpa, unlike Tibetan, is an unwritten language (and hence the considerable confusion over its orthography), it has a similar structure of Tibetan and according to one estimate may have 50 percent cognates with central Tibetan (Hutt 1986:17).[36] Sherpa social structure, like that of Tibet, includes both nuclear and stem families and accepts polygamous as well as monogamous marriage.[37] Like the people of at least some parts of Tibet, Sherpas trace their lineage and choose their spouses with reference to their affiliation in exogamous, patrilineal clans.[38] They follow a Mahayana form of Buddhism developed in Tibet (Vajrayana or "Tibetan" Buddhism) and more specifically adhere to the Nyingmapa or "Red Hat" sect developed in Tibet in the eighth century.[39] As in Tibet there is respect also for divinities and powers in the natural landscape such as *yul lha* (regional mountain gods) and *lu* (spirits of springs and trees). Sherpa domestic and religious architecture are closely related to Tibetan styles.[40] In some though not all Sherpa regions, as in Tibet, there is a preference for raising yak and barley. Sherpas also share with Tibetans some taste in clothing and personal ornamentation as well as many other customs and beliefs.

Such similarities are not surprising considering the relatively recent migration of Sherpas to Nepal and the way in which—in Khumbu at least—continuing Tibet immigration has constituted a major component of nineteenth- and twentieth-century population growth. Close cultural contact between Khumbu and Tibet has also been maintained by trade, pilgrimage, the custom of devout Sherpas going to Tibet for religious instruction, and Khumbu's long appeal to Tibetan monks as a place for meditative retreat or careers as village lamas. Together this high level of interaction may have contributed to the greater "Tibetaness" that some other Sherpas find characteristic of the Khumbu Sherpas. There may also have been political links at one time between Khumbu and Tibet. There are oral traditions of taxes paid to Tibet in the early nineteenth century. People point out the former site in Nauje of the house of one of these tax collectors, and near Nauje is a ruin which is considered to have been a Tibetan fort (*dzong*) that is said to have been overrun by Nepalese forces during the 1855–1856 war between Nepal and Tibet.[41]

Even though all Sherpas have strong historical and cultural ties to Tibet, both they and their Tibetan neighbors consider Sherpas to be a distinct people. During nearly twenty generations of life south of the Himalaya Sherpas have combined a heritage brought from Tibet with their own inspirations, developing a culture distinctively their own. They point out a number of basic differences between themselves and Tibetans. Sherpa is a distinct language. Sherpas have their own local gods, spirits of places, clan gods, and regional gods. They celebrate distinctive religious festivals such as the major summer celebration of Dumje held in Khumbu, Shorung, Dongritenga, Golila, and Rimijung in Pharak. Sherpa clans are for the most part different from those of Kham; all but two of the more than twenty clans recognized by Solu-Khumbu Sherpas are believed by them to have been developed after their arrival in Nepal. Some life-cycle rites are different. There are differences in vernacular architecture. Sherpa and Tibetan land use (at least the land use characteristic of the adjacent Tingri region) also varies. In most Sherpa regions, including Khumbu, yak and sheep are not as important as they are in Tibet. In some Sherpa areas, again including Khumbu, yak-cattle crossbreeds are raised, contrary to the practice followed in Tingri and some other parts of Tibet where there are religious reservations against such breeding. And while Khumbu Sherpas, like Tibetans, prize barley, they give it a much less important role in their agriculture and diet and emphasize buckwheat and tubers more.

Sherpa Regions

People who identify themselves as Sherpa and are recognized as such by other Sherpas and their non-Sherpa neighbors inhabit a considerable part of the highlands between the Sun Kosi and the Arun rivers, land that today falls within the administrative districts of Solu-Khumbu, Dolakha, Ramechap, Olkadunga, Sankhuwasabha, and Khotang. There are apparently also Sherpa villages west of the Sun Kosi in the upper Balephi Khola valley (Sindhupalchok district), and people who claim descent from these (and possibly other Sherpa-inhabited areas) also form a component of the multiethnic population of the Helambu region of Sindhupalchok just to the northeast of the Kathmandu valley.[42] Sherpa settlement as far west as the Sun Kosi and as far east as the Arun river can be related to migrations from Solu-Khumbu as recently as the nineteenth century.[43] As previously mentioned, there are also Sherpas in the Darjeeling district of the Indian state of West Bengal, the descendants of Solu-Khumbu emigrants who were attracted in the late nineteenth and the first half of the twentieth centuries to economic opportunities there (particularly to mountaineering and porter work).[44] The inhabitants of a number of high-altitude settlements in the far northeast corner of Nepal near Taplejung (including the village of Gunsa) are also often referred to as Sherpa (Bremer-Kamp 1987; Sagant 1976). Fürer-Haimendorf suggests that these families migrated from Shorung (1984:3). In the late nineteenth century, while traveling through that region, Sarat Chandra Das was "told that the upper part of the valley [of Kangpachan] was first inhabited by Tibetans called Sherpas, migrants from 'Shar Khumbu' " (Pradhan 1991:70).

Khumbu Sherpas consider themselves one of as many as twenty (depending on who is counting and his or her familiarity with outlying areas) different Sherpa groups in Solu-Khumbu district and the adjacent Dolakha and Ramechap districts to the west and Khotang, Bhojpur, and Sankhuwasabha districts to the east and three groups further afield. These regions are listed and shown in map 4.

Immediately to the south of Khumbu in the middle reaches of the Dudh Kosi (as far as the village of Surke and the Chutok La beyond it) is an area considered to be the homeland of the Pharak Sherpas. South of them are a group that some Khumbu Sherpas call Katangami, people of

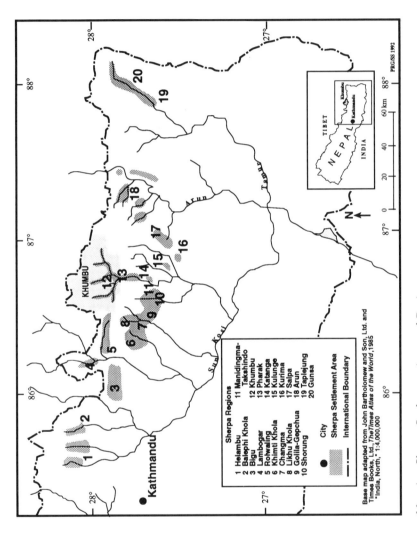

Sherpa Regions

1	Helambu	11	Manidingma-
2	Balephi Khola		Takshindo
3	Bigu	12	Khumbu
4	Lambogar	13	Pharak
5	Rolwaling	14	Katanga
6	Khimti Khola	15	Kulunge
7	Changma	16	Kurima
8	Likhu Khola	17	Salpa
9	Golila-Gepchua	18	Arun
10	Shorung	19	Taplejung
		20	Gunsa

● City

▨ Sherpa Settlement Area

–··– International Boundary

Base map adapted from John Bartholomew and Son, Ltd. and
Times Books, Ltd., *The Times Atlas of the World*, 1985
"India, North," 1:4,000,000

PRGSS 1992

Map 4.   Sherpa Settlement Area and Regions

Katanga, who live in and near the Sherpa-Magar village of Kharikhola. South of Kharikhola the eastern Dudh Kosi valley is inhabited by Dongbu (Rai in the north and Gurung in the Ra Khola tributary) and Rongba (Hindu hill castes). Farther to the east, however, there are other Sherpa settlement areas beyond the Pangkongma La in the Hinku Khola and adjacent areas including Kulunge, the Kurima area to the southeast, and Sanam, Share, and other small settlements near the Salpa Bhanjyang (pass). Still farther to the east there are Sherpas living in small settlements such as Murde, Kuwapani, and Hurhuri on the ridge-crest of the eastern slope of the Arun valley as well as at Tashigaon, Navagaon, Yangding, Shashiwa, and other places on several west-bank tributaries of the Arun and possibly at Yakua farther upstream.[45] The high country east of the Arun is inhabited primarily by peoples who settled there from Tibet in earlier times, but who are not directly related to Sherpas. The far northeast corner of the country, the Gunsa area northeast of Taplejung, however, is settled by people who may have at least some of their roots in Solu-Khumbu and who continue to make trading trips to Shorung today. West of the Dudh Kosi is more Sherpa territory. The area around Manidingma and Takshindo is sometimes referred to as Takshindo by Khumbu Sherpas who consider it different from Shorung proper. Shorung is considered to be the area from the ridge crest at Takshindo to the Lamjura pass and Pi Ke peak in the west, and from the high valleys at the foot of the sacred peaks of Numbur, Khatang, and Karyalong to the ridge south of the bend which the Solu Khola takes on its way east to the Dudh Kosi. Beyond Pi Ke and the Lamjura La there are three other Sherpa-inhabited regions east of the Khimti Khola watershed: the villages along and near the Likhu Khola, the village of Chyangma (which is jointly inhabited by Sherpas and Newars who settled there at least nine generations ago), and the villages of Golila and Gepchua. Farther west there are Sherpa villages near Jiri in the Khimti Khola region, Lambogar, Rolwaling, and the Bigu *gompa* (temple) area west of the Tamba Kosi. As already mentioned there may also be Sherpas as far west of the Sun Kosi as Helambu, although further research is needed to fully establish this. Khumbu Sherpas familiar with these various areas from trade, pilgrimage, and mountaineering and trekking work draw numerous contrasts between Khumbu culture and that of these other Sherpa groups.[46]

Khumbu Sherpas thus have a well-defined sense of themselves as distinct both from other Sherpas and non-Sherpa peoples. Yet the question of what and who are Sherpas, or even Khumbu Sherpas, remains complex. In recent years the fame, status, and economic opportunities of the Solu-Sherpas have sometimes led Tamangs (G. Clark 1980*b*; Fricke 1986:30) and Rais (Fürer-Haimendorf 1984) to claim to be Sherpas in order to obtain tourism work. In the Taplejung area some non-Sherpa high-altitude dwellers reportedly refer to themselves as Sherpas rather than as Bhotias when dealing with lower-altitude people as a way of affirming their Nepalese rather than Tibetan nationality. In the Arun region some people call themselves Sherpas who have no association with Solu-Khumbu while lower-altitude peoples in that region have begun to use the term Sherpa "as a generic term for all Tibetan-origin people" (Parker 1989:12). Within Khumbu itself there is even some confusion over who is and is not Sherpa. Khumbu Sherpas have long drawn a distinction between the older clans, the descendants of early settlers, and those who came more recently and have drawn a far sharper line between Sherpas and Khambas, the descendants of immigrants from Tibet and other Bhotia areas. Lacking membership in a patrilineal Sherpa clan, these immigrants were not only considered different from Sherpas but were treated with prejudice even though in Khumbu intermarriage with them was tolerated and took place.[47] This remains true today to a lesser degree. Khamba families intermarry freely with Sherpas, hold village offices, and are generally fully integrated into Khumbu society. Many now call themselves Sherpas and seem to be accepted as such within and outside of Khumbu. Further complexity is added by the fact that some of the central qualities by which Sherpas define themselves, as well as those by which outsiders distinguish them, have undoubtedly changed through time. Trading with Tibet, for example, was once an important aspect of Khumbu Sherpa identity, a quality which set them apart from Pharak and most Shorung Sherpas. Due to the decline in trade since the 1960s, however, this activity no longer distinguishes most Khumbu Sherpas from their neighbors.

For all the socioeconomic and cultural change of recent decades Khumbu Sherpas have not lost their sense of themselves as a people, and that to some important degree may be due to their having maintained their language, religion, social structure, and homeland. Territory is considered a fundamental aspect of identity. Khumbu Sherpas see themselves as the people who live in a particular group of four valleys on the border of Tibet. People who live beyond this region may be Sherpas, but are not Khumbu Sherpas even when, as in the case of the Pharak Sherpas, they may be culturally and socially very closely related indeed. Even the children of emigrant Khumbu Sherpas are often regarded as Darjeeling Sherpas, Shorung Sherpas, Kulunge Sherpas and so on rather than "Khumbuwa," people of Khumbu. It seems likely that the children of Khumbu Sherpas who have grown up in Kathmandu, speaking Nepali rather than Sherpa and living extremely different lives from their cousins in the mountains, will be considered to be "Kathmandu Sherpas" rather than Khumbu ones if they do not return to make their lives in Khumbu.

Finally, important facets of culture can also vary even among Khumbu villages. Local differences could be noted in such things as communal institutions and the conduct of festivals and village rites as well as in economic emphases.[48] Such variation cautions against generalizations about "Khumbu Sherpa" and "Sherpa" beliefs and practices on the sole basis of familiarity with the way things are done in a few settlements.

Population and Settlement Patterns

Most of the present population of Khumbu are Sherpas who are members of a group of "old clans" that trace their origins to the original Tibetan settlers of Solu-Khumbu. There are also substantial numbers of "new clan" Sherpas, a large percentage of Kamba families, and some families descended from unions between Sherpas and Gurungs, Tamangs, Newars, and Chetris.[49] Although they all consider themselves Sherpas there does remain some sense of social hierarchy based on these ethnic differences.[50] Besides these main components of Khumbu society there are more recent Tibetan refugees (about twenty-five families), a few households of lower-caste Hindu Kami blacksmiths (nine families) and tailors (two families), several upper-caste Brahmans and Chetri Hindu schoolteachers, two Tamang families, and a Magar family. There are also more than three hundred Hindu Nepalis stationed in the region with the police, army, post office, bank, Northern Boundary Commissioner's office, the government health post in Nauje, and the national park (Fürer-Haimendorf 1984:32). Scores of Rais, Magars, and Nepal midlands Sherpas also spend months or even years in Khumbu as household servants, agricultural laborers, fuel wood cutters, and stonecutters, and masons.[51]

Khumbu is unusual among Sherpa regions in the relatively high percentage of its population considered to be Khamba rather than of old or new clan Sherpa descent. This percentage varies regionally. Fürer-Haimendorf determined in 1957 that Khambas constituted 49 percent of the households of Khumjung, the region's largest village.[52] The percentage of Khambas in Kunde, Phurtse, and Pangboche was 33, 37.5 and 39 percent respectively (Fürer-Haimendorf 1979:26). In Nauje they were probably then a majority. For most purposes, however, there is no need to distinguish between Sherpas and Khambas and in this book I will do so only in the rare cases when historical and contemporary differences in land-use patterns and other economic activities need to be noted. Khamba families have all lived in the region now for several generations and differences between them and old-clan Sherpa families have diminished with time. Khambas intermarry with Sherpas, speak Sherpa, dress in Sherpa-style clothes, live in Sherpa-style houses, and participate fully

Table 1.   Village Household Numbers and Population

	Number of Households*		
Village	*1957*	*1991*	*Population (1979)*
Nauje	73	123	488
Khumjung	93	135	622
Kunde	45	50	237
Pangboche	58	83	347
Phurtse	63	62	287
Thamicho	192	**	707

* The 1991 Nauje total includes four Zarok households. Ninety-four of the Nauje households and 125 of the Khumjung households were Sherpa or Khamba. The 1991 Pangboche figures include five Milingo-, five Changmiteng-, and five Dingboche-based households. The Thamicho total includes all households with main dwellings in the Bhote Kosi valley other than in Nauje. No household is tallied more than once, although some own houses in more than one main settlement. Only families who were resident in the area in 1991 were counted. Monks and nuns are not included in the population totals.
** A 1991 count of Thamicho households is not available. The complex settlement pattern of Thamicho makes it easy to double-count households. Fürer-Haimendorf's 1957 count may be high.
From Fürer-Haimendorf 1964, Sherpa 1979, and fieldwork.

in community offices and religious activities. Formerly many Khambas were among the poorest families in Khumbu, but there is little distinction today between the lifestyles and land use of poor Khamba families and poor Sherpa ones or between wealthy Khambas and wealthy Sherpas. Khamba households have been among some of the very wealthiest in Khumbu since at least the 1940s when some Khambas became major traders, and the most of the conspicuously poor Khamba households of the 1950s have prospered since then as a result of involvement in tourism.

Village populations of Sherpas (including Khambas) are given in table 1 based on a 1979 census conducted by Nima Wangchu Sherpa on behalf of Sagarmatha National Park. James Fisher tallied a total of 2,474 people in a Khumbu population count in 1978 as against Nima Wangchu Sherpa's 2,836 permanent residents (including non-Sherpas) in 1979 (Sherpa 1979:7). The most recent census, conducted in 1982 by Pawson and his associates, totaled 2,524 Sherpas (Pawson, Stanford, and Adams 1984:75).[53] The collection of census data is difficult in Khumbu due to the seasonal dispersion of the population among different main and secondary settlements as well as the absence of many men on mountaineering expeditions and trekking work. An increasing number of young men and families also reside in Kathmandu for part of the year or sojourn there for an indefinite period while still maintaining a house in Khumbu and expecting to return.

The important question of the regional rate of population growth, a matter of great importance to land use and environment, is complicated by the lack of reliable census material. Ortner reports that Fricke estimates that the Sherpa population of Solu-Khumbu as a whole has doubled every sixty years since the time of original settlement (assuming that the original settlers numbered about fifty) (Ortner 1989:209, n. 1), while Oppitz earlier estimated the population doubling rate at forty-nine years (Oppitz 1968:103). These calculations, however, refer to net population growth, not natural increase. They do not distinguish between local Sherpa population growth and historical immigration and emigration, including continuing immigration since 1800 from Tibet. It is known that the population of Khumbu in 1836 consisted of at least 169 tax-paying households (Fürer-Haimendorf 1979:118), although whether or not other families went untaxed (e.g., landless immigrants) is not certain. The number of Sherpa and Khamba households in Khumbu today is slightly more than 500, meaning that the regional population has apparently tripled during the past 150 years. Much of that gain may have been in the past sixty to seventy years. Elderly residents of Kunde, Khumjung, Nauje, and Phurtse testify that the number of houses in their villages has doubled or more than doubled in their lifetimes. Even the increase in the number of households since 1957 is significant in many villages.[54]

Multialtitudinal Settlement   Khumbu Sherpas have developed a complex pattern of seasonal movement between houses in their main villages and other dwellings situated at different altitudes and as far as twenty kilometers away. Some families maintain as many as half-a-dozen houses and herding huts scattered throughout one or more valleys, living in each for a period ranging from a week to many months per year depending on their pattern of herding and the ways in which they integrate these with crop-tending requirements, social responsibilities in various places, and personal preferences. The complexity of scheduling agricultural and pastoral activities at a number of different sites often leads families to divide forces temporarily between several bases. The locations of the major villages and the scores of lower- and higher-altitude secondary settlements are shown in map 5 by settlement type, and the individual settlements of the Bhote Kosi, Dudh Kosi, and Imja Khola valleys are shown in maps 6 and 11.[55]

Villages (Yul)   Level field sites in the steep terrain below 4,000 meters are rare and prized, and the eight main villages of the area hug the most prominent bits of a relatively gentle terrain between 3,400 and 4,000

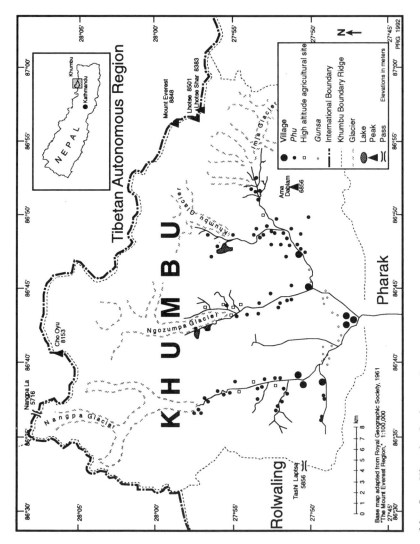

Map 5.   Khumbu Settlement Pattern

meters. Half of the total Khumbu population inhabits three villages, Nauje (3,400m), Khumjung (3,790m), and Kunde (3,820m) in southern-most Khumbu near the confluence of the Bhote Kosi and the Dudh Kosi rivers.[56] The middle reach of the Bhote Kosi valley in western Khumbu, an area known as Thamicho, is a second major area of settlement. The three main Thamicho villages, Thami Og (lower Thami), Thami Teng (upper Thami) and Yulajung, are all situated at about 3,800 meters. Thami Teng and Yulajung occupy alluvial terraces on opposite sides of the Bhote Kosi, whereas Thami Og is located on the north bank of the Thami Chu just above its confluence with the Bhote Kosi and immediately south of the prominent old lateral moraine which stands between it and Thami Teng.[57] Eastern Khumbu has two major villages, Phurtse (3,840m) in the Dudh Kosi valley just above the confluence of that river and the Imja Khola and Pangboche (4,000m) on the Imja Khola just six kilometers to the east. Besides these villages there are also several smaller settlements which serve as the main base for a majority of the families who own houses in them. These include Milingo near Pangboche on the Imja Khola, Zarok just above Nauje, and several small settlements in the vicinity of Thami Teng including Worsho, Chanekpa, and Ong. In most cases families living in these settlements join with nearby main villages for a variety of social, religious, and community land-management purposes.

Main village houses are substantial structures of unmortared stone with gabled roofs of slate, fir shakes weighted with stones, or corru-gated iron. Until recently all construction materials were locally ob-tained, including the white clay used to give houses their distinctive color. The tree-felling restrictions which the national park implemented in the late 1970s, however, have since meant that nearly all timber has had to be imported from Pharak. Sheets of corrugated iron are brought by porter from the road end at Jiri, and glass is now often flown in to Lukla.

All but the poorest families live today in two-story houses. These are constructed along very standardized designs and floor layouts and are generally about five by twelve meters in size. The lower story provides a stable for livestock and a storage area for grain, fodder, and fuel wood. The upper story is usually a single, large, open living room with front (usually south-facing) windows above a long window bench and a hearth area. The windowless back wall is lined with shelves on which the house-hold wealth of copper water-storage vessels, brass cooking pots, carpets, and other worldly goods is displayed. Wealthy families may devote a separate room on the left side of the upper floor to a family chapel, but more often a simple altar located on the left wall of the main room suffices. Often houses face in a similar direction, an orientation influ-enced not only by sunlight and slope but also by geomantic beliefs about auspicious and inauspicious directions and the danger of having a house face places such as caves inhabited by demonesses.

The eight major villages differ greatly in size and economic orientation as well as in such features as the conduct of community religious ceremonies and local resource-management institutions. Nauje is the region's second largest village.[58] Long the home of many of the region's major traders and a place where Rais, Sherpas, and Tibetans bartered salt, wool, and grain, it is today Khumbu's administrative, commercial, and tourist center and has grown into a small town with ninety-four resident Sherpa familes and a third as many non-Sherpa households.[59] Here is the region's only bank as well as a post office, government-operated clinic, elementary school, the office of the Northern Boundary Commissioner, and the first dental clinic in northeastern Nepal. On Mendelphu hill, overlooking the tiers of tightly clustered houses of the village proper, is the complex of buildings that forms the headquarters of Sagarmatha National Park. Part of the village has developed into a tourist district that in spring 1991 had twenty-one shops and eighteen lodges. Village families also run two more Sherpa lodges adjacent to national park headquarters at Chorkem. At one edge of the settlement is the site where on Saturdays the region's only periodic market is held. Here Rais and other traders sell grain and goods brought from settlements and road ends as far away as ten days' journey on foot. At night the village glows with more than 400 electric lights installed in 1983 with UNESCO aid and powered by a microhydroelectric turbine on the stream which flows from the village spring.

All the other Khumbu villages retain the atmosphere of rural settlements and their only signs of tourism development are a few modest lodges and teashops.[60] Khumjung, a sprawling village just an hour's walk from Nauje, is by far the largest of these. Its name, which may mean Khum (or Khumbila?) valley, may refer to its site in a valley at the foot of the sacred peak Khumbila.[61] With 135 Sherpa households it has the largest Sherpa population of all the Khumbu main villages. Here is the region's main elementary school, established in 1961, and its only secondary school. Both were built and maintained through the work of Sir Edmund Hillary and the Himalayan Trust organization. Adjacent to Khumjung and slightly higher in altitude is Kunde, upper Khum. Kunde has only fifty Sherpa households but is renowned as the location of the hospital established by the Himalayan Trust in 1966.[62]

Pangboche, the major village of eastern Khumbu, has two distinct sections that between them have sixty-eight households. The older, upper village (Pangboche Te Lim) occupies an amphitheater some 150 meters above the Imja Khola and directly below the spectacular summit of Tawache. This is the main settlement, with 89 percent of the village households, most of them clustered in tiers around the amphitheatre walls and the village temple. The more recently established Wa Lim occupies a large alluvial terrace just above the Imja Khola. The seventeen families here do not have land holdings in the upper village.[63] They are, however, fully integrated into a joint social and ritual life.

The village of Phurtse perches on a small shelf 400 meters above the confluence of the Dudh Kosi and the Imja Khola. The name is said to refer to Lama Sanga Dorje's miraculous feat of having flown to this spot, leaving several marks of his landing on a local boulder. Here are the dispersed houses of sixty-two families. The older section of the village occupies the center and lower parts of the slope, whereas the prominent, dense cluster of houses in the upper northeast corner is a development of the present century. The old, dense birch forest that covers a slope adjacent to the village on the north is one of several Khumbu sacred forests.

The Bhote Kosi valley villages of Thami Og, Thami Teng, and Yulajung are smaller than the other Khumbu villages. Yulajung indeed has only fourteen resident families, most of whom also have houses down valley at the *gunsa* (winterplace) settlement of Pare. Thami Teng, with twenty-two households in the village proper, is the largest of a set of closely situated communities on the west bank of the Bhote Kosi across from Yulajung. It is the home of a number of the few Sherpa families who are still trading on a small scale with Tibet and two of the largest Khumbu yak herds. Thami Og, childhood home of Tenzing Norgay, has twenty-seven households and on the slopes above it is one of Khumbu's two important monasteries. In the nineteenth century it was the home for at least three generations of the political leaders of the region. One Thami Teng Sherpa speculates that the names Thami Teng and Thami Og may refer to the flat character of the settlement sites and their relative locations. The first part he considers to derive from the Tibetan *thang,* a flat place, while the *te* and *me* suffixes refer to upper and lower sites. Yulajung may get its name from a linking of the Sherpa *yul,* "village," with *jung,* a flat valley.

Subsidiary Settlements (Gunsa, Phu)   In much of the Himalaya and Tibet herders who pasture their stock in regions far from their home villages often live in tents or in shelters largely made of bamboo mats. A small number of peoples, including the Sherpas of Khumbu, Rolwaling, and Gunsa, however, build substantial, stone herding huts. These may be located at altitudes either above or below that of the main village, and often crops are also grown at these sites. In Khumbu these huts are usually tiny, low-walled, stone structures with shake or slate roofs. Almost all are only a single story high and are seldom much larger than three by four meters with a port-hole-sized window or two, a dirt floor, and a simple hearth formed of three stones set into a tripod. Cramped and smoky at the best of times, they grow more so after the hay is cut from adjacent walled fields and the huts become not so much dwellings as hay-stores. Khumbu herders also make use of caves and *resa,* low circles of stones roofed with bamboo mats or tarps.

High-altitude herding settlements are known as *phu* and are sometimes also referred to as *chusa* (livestock place) or *yersa* (summer place). The highest are located at an altitude of 5000 meters in the alpine regions of the uppermost valleys. Gunsa (also referred to as chusa) are located below the main villages.[64] Many gunsa and some phu are not only herding bases but also have substantial areas in cropland. Houses in both gunsa and phu may be owned by families who have no interest in herding. Some of the subsidiary settlements between 4,000 and 4,400 meters are major crop-production sites, including Dingboche (4,358m) in the Imja Khola valley, Tarnga (4,050m) in the Bhote Kosi valley, and Na, Charchung, Tsom Og, and Tsom Teng (4,280–4,400m) in the Dudh Kosi valley. Houses in these places and in some crop-producing gunsa may rival a family's main village house in size. For a few families such houses are actually main residences and they may have no house in a main village. I will henceforth refer to upper valley settlements where a great deal of cultivation takes place as "secondary high-altitude agricultural sites" to distinguish them from high-altitude phu settlements that are primarily herding bases and hay-cultivation areas.

There are more than 300 high-altitude herding huts in at least eighty-six distinct settlements (not including resa). Most have associated hay fields and some also have small areas in potatoes. Some high-altitude herding settlements consist of only a single hut or resa whereas others have as many as thirty. Most are inhabited by families from several different main villages. There are twenty-three gunsa settlements in Khumbu typically situated on fragments of terraceable land deep in the gorges. Many are located as much as 500 meters below the fields of the main village. Like the high herding sites gunsa are small settlements that are often bases for families from several different villages. The largest have fewer than thirty houses. Some of the secondary high-altitude agricultural sites, by contrast, are quite large. Tarnga has more than sixty houses that belong to families from all the Thamicho villages and a few Nauje families. Dingboche has sixty-five houses, most of them owned by Pangboche villagers but some of them the property of families from Khumjung, Kunde, and Nauje. Five families have their main houses there today.

Settlement History   Khumbu Sherpas believe that their ancestors came to Nepal from Kham, a region on the eastern edge of Tibetan-inhabited territory 1,200 kilometers from Khumbu. Written clan records kept by Shorung Sherpas also claim Kham as the ancestral Sherpa homeland and in particular trace a migration from the Salmo Gang area (now in China's Sichuan province) near Derge in the upper Jinsha Jiang (Yangzi) region (Oppitz 1968; Teschke 1977). *Sharwa,* as Sherpas call themselves, means "easterners" or "people of the east" and may reflect this migration.[65] Oppitz suggests that the number of people involved was quite small, perhaps as few as twenty-five to fifty persons (Ortner 1989:26). Based on references in the texts to one of the emigrants being a student of Terton Ratna Lingba (1401–1477) Oppitz (1968:143) suggests that the original emigration from Kham "took place at the turn of the 15th to the 16th century."[66] The reasons for the departure of the small group of families from Kham can only be guessed at. Oppitz speculates that military incursions by Mongols from the Koko Nor (Qinghai Lake) region of present Qinghai province may have been a factor (ibid.:143–144) and there is mention in the texts also of dreams and visions that guided the leaders of the group (Ortner 1989:29). According to Konchok Chombi the emigrants' reluctance to take part in a coming war was also an important reason for their departure from Kham.

The migration took a period of years. Both Oppitz and Khumbu traditions agree, however, that it was accomplished by the same band of people who set out from Kham. The emigrants did not take leave of Kham with an intention to cross the Himalaya, for they clearly hoped to settle in central Tibet. They were unsuccessful, however, in their efforts to find a new homeland in south-central Tibet. According to some accounts they attempted to settle first in the Lhasa area and then in the Tinkye area, where Oppitz suggests they settled for a few decades south of Tsomo Tretung lake. He relates the subsequent abandonment of the Tinkye area to a reference in the *Ruyi* to invaders called Dohor Durkhi, whom he believes may have been the Muslim armies of Sultan Sa'id Khan of Kashgar which, under the command of Mirza Muhammad Haider, campaigned against the Buddhist "idolators" and in 1531–1533 nearly reached Lhasa (ibid.:144). Oppitz speculates that the immigrants moved rapidly from Tinkye to the Tingri region just to the north of Khumbu and then crossed the Nangpa La into Khumbu around 1533 (ibid.:144). According to Khumbu legends the ancestors of the Sherpas may have also tried to settle in the Tingri area near Langkor, but found that they were unwelcome after their yaks grazed in the barley fields of local farmers.

There are two oral traditions about the Sherpa discovery of Khumbu.

According to one the way over the Nangpa La was first found by a hunter, Kira Gombu Dorje, when he was pursuing a quarry which in some versions was a musk deer and in others a blue sheep. Some people believe that the deer was actually the embodiment of Khumbu Yul Lha. In the second a scouting party of three friends (Lama Serwa Tungel, Mijen Thakpa, and Thimi Sangbu Tashi) reconnoitered the pass and the Bhote Kosi valley beyond it. According to Konchok Chombi's reckoning of his own genealogy, eighteen generations have been born in Nepal since the ancestors of the Sherpas crossed the Nangpa La into Khumbu.[67]

The immigrants who passed across the Nangpa La into Khumbu did not all settle there permanently. Most apparently continued south after a few years, following the Dudh Kosi valley into the lower-altitude forested regions of Pharak, Shorung and Golila-Gepchua. In Khumbu it is believed that a few immigrants remained in Khumbu and adjacent Pharak and that these families constituted the nucleus of what later became the Sherpas of Khumbu. This early Khumbu population was apparently very small and remained so into the nineteenth century despite population growth from natural increase and further in-migration.[68]

The families which originally crossed into Nepal are believed to have belonged to several different clans, the ancestors of the Khumbu clans of today.[69] Oppitz (ibid.:144) refers to these as the four protoclans (Serwa, Minyagpa, Thimmi, and Chakpa) and suggests that they may originally have had separate territories, although on this point Khumbu traditions offer no confirmation. In Shorung and the area just west of the Lamjura pass early settlements were apparently established on a clan-exclusive basis. How and when the multiclan villages characteristic of Khumbu developed is not known. Oppitz believes that the Thimmi protoclan originally settled the Bhote Kosi valley (primarily the lower valley between the later settlement at Thami Og and Nauje) between 1533 and 1550 while Minyagpa families settled Phurtse and Pangboche (Fürer-Haimendorf 1984:27–28). Over the course of a few generations a number of new clans were established. Some, according to oral traditions and genealogies, were the result of the fissioning of clans into sets of exogamous brother clans following a ban on marriage between the offspring of brothers. In other cases the new clans reflected a decision by new immigrants from Tibet to take clan names based on their areas of origin.

The first settlers from Kham must have found Khumbu a rather different place than it is today. According to legend Khumbu was then in its lower reaches far more densely forested than now and richer in wildlife. Some stories describe Khumbu as a wild area and also as a "god's place" which had been set aside from the ordinary world until it was revealed to the faithful as a refuge for believers.[70] The idea that divine guardianship kept Khumbu hidden until the arrival of the Sherpas also figures in a legend of an early king who wished to settle in Khumbu but was deceived by the gods into believing that the area was uninhabitable. When he went up into the valleys of Khumbu the gods made him see only huge lakes that threatened to flood the lower country. He had established a palace on the great rock at Monjo, just south of Khumbu, but after this decided to abandon the region altogether. He is said to have eventually established his capital at Dolakha, a place more than a week's walk to the west, where he is now supposedly honored as a god.

There are a number of Khumbu oral traditions which recall that at the time of Sherpa settlement the region was already being used as summer pasture by Rai shepherds. Old stories refer to these summer visits, and certain ruins in high places in the Dudh Kosi valley are sometimes said to be remains of Rai shepherds' huts.[71] Another legend offers an explanation for the lack of permanent Rai settlement. This tale details the attempts of the Kulunge Rai culture hero Ma Pe to settle in Khumbu long ago. According to the story Ma Pe crossed into Khumbu by the now difficult Amphu Laptsa pass in eastern Khumbu and settled in the upper Imja Khola valley. He built a house and tried to cultivate fields at Dingboche. There he found his crops ill-suited to the altitude and he moved lower in Khumbu. There again the crops were poor and his subsequent attempts to farm other Khumbu sites were also discouraging. Eventually he gave up and moved farther down the Dudh Kosi valley where he found a suitable site at Bupsa near Kharikhola in Katanga. This has been, it is said, the highest and northernmost place of Rai rice cultivation ever since.[72] Other possible indications of early Rai settlement are remembered in the legend that Rais may be reincarnated in one place in Khumbu in the upper Imja Khola valley near the Tengboche monastery. There are also a number of place-names in Khumbu and Pharak which are said to date from Dongbu times (these may also refer, however, to nineteenth-century herding by Gurungs, who, like Rais, are called Dongbus by Sherpas). These names, such as Dusa and Ralha, refer to sheep- and goat-herding activities. There may even be an old recognition of Rai association with Khumbu in the name Khumbu itelf. The Dudh Kosi valley is known to Rais as a part of Majh Kirat called Khambuan, "land of the Khambu." And Rais in this part of Nepal have long referred to themselves as Khambu, tracing their lineage back to an ancestor named Khambuho (McDougal 1979:3). Perhaps Sherpas adapted a slightly modified pronunciation of this local regional name.

Rais would have found Khumbu an uncongenial place for the rice and millet that were presumably their main crops in that era before maize was a common crop in Nepal. But the possibility of Rai summer transhumance into Khumbu is plausible given current Rai sheep-herding patterns in nearby areas. Summer pastoralists may also have augmented their food supplies by planting buckwheat, barley, or tubers. The ancestors of the Rais or even earlier predecessors in the Dudh Kosi valley might very well have used swidden cultivation in the then presumably well-forested lower Khumbu valleys or set fire to woodlands in order to improve grazing. Such practices might account for the very early habitation suggested by Byers on the basis of pollen studies which revealed cerealia grains which may date back 1,500 years (Byers 1987b:199, 204).[73]

Khumbu oral traditions recall early conflict between the immigrants from Tibet and the indigenous Rais in both Khumbu and Shorung. According to one legend the first settlers had already planted some crops when they were accosted by Rais who told them they could not grow crops or settle there because the land belonged to the Rais. The Sherpas then asked permission, pointing out that they had no food and could not return to Kham. The Rais relented.[74] Another story set a few generations later describes an armed conflict between Rais and Sherpas in southern Khumbu. Sherpas are said to thereafter have had to pay an annual tax to the victorious Rais, presumably until Prithivi Narayan Shah consolidated his control over the Dudh Kosi region in the late eighteenth century and made it part of the new Shah kingdom of Nepal.[75] Ortner (1989:84) suggests a date of 1717 for the Sherpa defeat. There is also a tradition that an early division of territory between Sherpas and Rais took place in the region of present southern Solu-Khumbu and Olkadunga districts.[76] Documents from the reign of King Rajendravikram Shah (1816–1847) note that parts of Solu-Khumbu and its pasture regions were formerly under Rai control (Pradhan 1991:62).

Early Settlement   A rich body of oral traditions concerning early events in Khumbu depicts the Bhote Kosi valley as the center of population and political administration.[77] The rest of Khumbu in the early years of Sherpa habitation appears to have been a very sparsely settled, primarily wild country in which a few lamas had established religious retreats and temples.

Several legends are set within the Bhote Kosi valley during the first few generations of Sherpa habitation. A number concern Dzongnangpa, an autocratic administrator who persecuted Sherpa lamas and was ultimately assassinated. According to some accounts Dzongnangpa was originally a Tibetan official in Shekar (east of Tingri). After being disgraced there he crossed the Himalaya to Khumbu where he eventually obtained considerable power, began to rule in a style that was extremely high-handed, arbitrary, and offensive by Sherpa standards, and became one of the classic villains of Sherpa oral traditions. The degree of his misrule is most vividly illustrated in his oppression of local religious figures. He is said to have been responsible for the death of one Sherpa lama at Zamde (near Mingbo in a tributary valley north of Thami Teng) and to have persecuted a second, Lama Sanga Dorje.[78]

The Tarnga area especially figures in the legends as an early settlement site. Tarnga today is a secondary high-altitude (4,050m) agricultural site famous for growing the best-tasting potatoes in Khumbu. In the early accounts it figures as the seat of Dzongnangpa's power. Some elderly Bhote Kosi valley villagers remember a large rubble pile said to have been the remains of Dzongnangpa's palace. It was also at Tarnga that yeti are said to have caused so much damage to Sherpa crops that the Sherpas staged a devious trap which very nearly caused the extinction of yeti in the region.

Tarnga would have been an appealing early settlement site. It is situated only slightly higher in altitude than today's main villages, has one of Khumbu's largest expanses of relatively level, arable land, and is one of the two places in the region in which irrigated barley can be grown on a large scale. One shortcoming is the absence of any adjacent woodland, but in earlier times Tarnga may have had better access to timber and fuel wood than at present. The highest patches of forest remaining today are located near Mingbo and Chosero, just south of Tarnga, and there may once have been shrub juniper cover or even open woodland nearer to Tarnga itself. If Tarnga was the early focus of settlement in the Bhote Kosi it is unclear why and when the center of settlement shifted to the present main villages of the valley. This certainly took place well over 150 years ago, for through much of the nineteenth century Thami Og was the political center of Khumbu.[79]

Some clues to the settlement of eastern Khumbu can be gleaned from the extensive set of traditions concerning Lama Sanga Dorje, the great spiritual hero of Khumbu history. Besides recounting Lama Sanga Dorje's magical feats these legends tell of his travels through Khumbu in search of auspicious places to mediate and to found temples, and his persecution by Dzongnangpa who was jealous of his powers and popularity among Sherpas.[80] Although this is not the place to discuss the details of the many stories of Lama Sanga Dorje's birth, life, and death, his role in the establishment of Pangboche is pertinent. After deciding to leave his earlier hermitages in the Bhote Kosi valley because of the escalating tension with Dzongnangpa, Lama Sanga Dorje is said to have explored eastern Khumbu for a site to establish a temple. He decided against building a temple at what later became Phurtse and also rejected the ridge where the Tengboche monastery was later built. Ultimately he built a small temple at Pangboche. There may have been earlier temples in Khumbu, although what are called temples in the stories of the earliest Khumbu years may have been only the hermitages of lamas. Pangboche is clearly described as a temple and, more than that, as having originally been the place of worship for a community of celibate monks. Ortner speculates that it could have been founded between 1667 and 1672 (1989:49).[81] The site of the new community of Lama Sanga Dorje's spiritual followers was apparently intended to be in a remote place where they would not be harassed as they had been in the Bhote Kosi valley. According to one story there was both a community of monks and a nearby community of nuns.[82] After Lama Sanga Dorje left the area for Tibet, where he eventually died, these followers stayed on. Intermarriages between nuns and monks subsequently took place and a settlement grew up. The compact clustering of houses in the upper village, one villager noted, reflects its origin in the tightly grouped monks' houses that usually surround monastic temples.

It is unclear when Khumjung and Kunde were founded, but it seems likely that this took place after the immigrants had established a foothold in the Bhote Kosi valley and may have also come after the development of Pangboche. The establishment of the village of Kunde a short distance up valley from Khumjung may not have taken place at the same time as Khumjung's settlement and is assumed by some Sherpas to have come later, or at least to have developed more slowly. I know of no oral traditions concerning the establishment of the two villages, nor are they mentioned in the oral traditions of the early generations of Khumbu settlement. Khumjung was certainly settled long ago, for some families there trace eleven generations born there and believe that they were not the original settlers. Two early temples or religious retreats that do figure in the oral traditions, Gormuche and Lhonang, are said to have been located not far from Khumjung, but this casts no light on whether Khumjung had already been founded.[83] One of these hermitages belonged to Lama Sanga Dorje's father. There is an oral tradition that the valley where Khumjung later grew was a lake or marsh during the early period of Sherpa settlement. The first houses in the area are said to have been built on the slopes below Khumbila rather than on the flat valley floor.[84] Khumjung and Kunde, whenever they were established, might long have been relatively small settlements. This is suggested by the tradition that while temples were established in the sixteenth or seventeenth century in Thami Og and Pangboche there was no village temple in either Khumjung or Kunde until the 1830 establishment of Khumjung temple. Before this villagers journeyed either to Thami Teng or Pangboche for important events such as the Dumje rites (Khumjung villagers in this era went to Thami Teng whereas Kunde villagers, like Phurtse families, went to Pangboche).[85] Another suggestion of the rather late settlement of the Khumjung-Kunde region is that Khumbu traditions relate that Khumjung and Kunde were first settled by people from the Paldorje, Chusherwa, and Jongdomba (Dzongnangpa?) clans (Fürer-Haimendorf 1984:28). The latter two clans are members of the seven "new clans" that Oppitz (1968:145) believes represent immigration from Tibet long after the early Sherpa settlement.

Phurtse was presumably founded after the establishment of Pang-boche. The first families are reported to have belonged to the Shar Dorsum, Shar Tazum, and Shar Sundokpa clans (subclans of the Sherwa). The settlement took place long enough ago for subsequent changes in the definition of these "brother clans" and for abandoning the original strict bans against intermarriage between Shar Dorsum and Shar Sundokpa after the area was also settled by Shar Penakpa, Paldorji, and Sundokpa clan families. The original settlement must have been sometime before the early nineteenth century, for nine generations ago Phurtse families were among the early settlers of Nauje.

Nauje is considered to be the youngest of the Khumbu villages, dating only to the early nineteenth century. The settlement is mentioned (as Namche Bazar) in an 1810 A.D. Kathmandu document entitled "Order to Inhabitants of Solu Regarding Trading Rights Beyond Namche" discussing trade regulations (*Regmi Research Collection*, vol. 49:209, in Schrader 1988:245), and in an 1828 trade document seen in Khumbu by Fürer-Haimendorf (1975:61). Before the establishment of a village at the present site Sherpas from Khumjung had herding huts and fields in the area.[86] Permanent settlement in Nauje is thought to have begun when five Khumjung families made the place their main base. They are said to have been attracted to the site because of the spring, one of the finest in Khumbu. Other immigrants from Khumjung and also from Phurtse soon followed. The first houses were built on the south facing slope and one of them became the stopover for Tibetan tax collectors who came to Khumbu until the Nepali victory in the 1855–56 war between Nepal and Tibet.[87]

Khumbu became part of Nepal in 1772–1773 when the Dudh Kosi region was incorporated into the Kathmandu government of the Shah kings (Limberg 1982:149). Sherpas began to pay a small land and house tax to Kathmandu, the revenue being collected in Khumbu by eight Sherpa pembu who were overseen by a Sherpa gembu, the highest local official. This office was held by Thamicho men during the nineteenth century and by the powerful Tsepal, a resident of Nauje, in the early twentieth century.[88] The eight hereditary Khumbu pembu, also men of great power, prestige, and wealth, were mainly responsible to the Kathmandu government for collecting taxes, but also filled a much larger role in Khumbu society as patrons and arbiters.[89] They may once have functioned as village headmen, but even in the early nineteenth century had ceased to do so and instead collected tax from clients who often lived in several different villages. This may have reflected migration of families from long-established settlements to new areas, with the families in such cases continuing to pay tax to the same pembu even though they now cultivated different areas. Whereas some villages had no resident pembu, others had several.

During the late nineteenth and early twentieth centuries Nepali administration of Khumbu was conducted by visiting officials and by the gembu and pembu on the behalf of Kathmandu. Only in the 1940s did the central government open up an office in Nauje, before that contenting itself with occasional visits by its officials. The region was remote from the capital and in many ways largely governed its own affairs. Yet the state did play an important role in some dimensions of Khumbu life. Government edicts about subjects as diverse as trade regulations, yak eating, and marriage regulations greatly affected Khumbu lifestyles. An 1828 royal edict that gave Khumbu Sherpas a monopoly on trade north over the Nangpa La was crucial to Khumbu Sherpa development of their distinctive subsistence strategies and historical regional economy.[90] Land taxes may never have greatly influenced the orientation of Khumbu crop production, but they may have been a hardship on poorer families in the late nineteenth and early twentieth centuries and a factor in migration out of the region. Decisions handed down by the Nepali court system could have great impact on questions such as village land boundaries and rights to resources. Khumbu Sherpas have long seen the courts as a way to settle major land disputes and civil suits that could not be resolved to everyone's agreement within Khumbu as well as a way to defend local traditions and rights against the government itself.[91] Courts were also used to defend Khumbu rights and privileges against encroachment by non-Khumbu citizens. In 1947, for example, 248 Khumbu Sherpa men successfully brought a case against three Pharak Sherpas whom they accused of trading with Tibet over the Nangpa La.

In the past forty years Khumbu has undergone enormous political, economic, and social changes. Khumbu began to be much more closely integrated politically with the rest of Nepal following the 1950–1951 revolution that restored the Shah line to greater power after more than a century of rule by the Ranas, a family who had seized effective power in 1846 and ruled as hereditary prime ministers. Local administration was reorganized after 1960 with the introduction of a new, national partyless system of local government, the *panchayat* system. Until 1968 all Khumbu constituted one government unit, along with all of Pharak as far south as Surke. Thereafter it was divided into two village panchayats, one comprised of Nauje and Thamicho and the other of Khumjung,

Kunde, Phurtse, and Pangboche. Each elected a *pradhan pancha,* an assistant (*uppa pradhan pancha*) and a set of lesser officials (*adekshe*) representing the nine wards in each village panchayat.[92] The pradhan pancha were to replace the pembu, who lost their tax-collecting privileges and their command of annual corvée labor (*wulok*) from their clients. Henceforth the panchayat was to be responsible for tax collection, although in practice most Khumbu pembu continued to collect the tax from their clients and then turned it over to panchayat officials.[93] The central government established a branch district office in Nauje in 1965, charged with regulating the Tibet trade and carrying out other duties that for some years included administering a new national system of regulated forest use. Officials of this office also oversaw the extension of the national land-registration system (*bhumi sudar*) to Khumbu in 1965–1966. Other government facilities were gradually added. Most were established in Nauje, which became the administrative center for Khumbu. Perhaps the most significant institutional introduction was the establishment of Sagarmatha National Park in 1976. The Khumbu economy also underwent an important change with the decline of trade with Tibet in the 1960s, the establishment in 1965 of a Saturday market at Nauje, and the rise of the tourist industry.[94] In 1979 Nauje's increasing growth as a commercial and tourist center was marked by the opening of the region's first bank. Other important developments were the building of the Hillary schools, beginning with the Khumjung school in 1961, the construction of the Lukla airstrip in 1964, the establishment of Kunde hospital in 1966, the arrival of electricity in the region in 1983 with the completion of the Nauje hydroelectric project, and the abandonment of the panchayat system throughout Nepal in spring 1990 and its replacement in Khumbu by two village development committees corresponding to the former village panchayats.

2

A High-Altitude Economy

The distinctive way of life of the Khumbu Sherpas in many respects reflects adaptation to the environmental conditions of their high-altitude homeland. In some ways this way of life is unique and Khumbu Sherpas are well aware of how their subsistence system differs even from that of other Sherpa groups. But Khumbu subsistence practices also share many features with those of other Himalayan peoples and even with mountain peoples in other parts of the world. This chapter introduces the Khumbu economy within the larger context of the cultural ecology of mountain peoples. The chapter begins with a brief overview of the special conditions of life in mountain regions and especially in high-altitude areas. I then introduce several concepts through which the ways of life of mountain peoples can be compared and contrasted, including subsistence strategies and tactics, "verticality," and altitudinal production systems and zones. From this base I then highlight five different basic subsistence strategies that have been historically important in the Nepal Himalaya and examine in detail the two which are today the most important in the Dudh Kosi valley. This provides perspective for a closer look at the Khumbu economy, from subsistence strategies and patterns of altitudinal land use to the linkages between different sectors of the economy and types of land use and the role of differences in household wealth and demography in subsistence decisions and practices.

High-Altitude Constraints and Opportunities

High mountains are distinctive and in many ways difficult environments to live in, and at the altitudinal extremes of the higher reaches of the Himalaya they are one of the more challenging environments on earth. Peoples inhabiting the highest realms of the high country must cope with climatic and geomorphological hazards that come with the terrain and the altitude and that make high mountains places of relative unpredictability, low primary productivity, and high environmental fragility in comparison to many lowland regions. Agriculture can be a chancy endeavor here at the edge of the arable earth. Above 3,000 meters growing seasons are short and there are significant risks of killing frosts and snow in spring and early autumn. The fertility of local soils is often very low, especially where only skeletal soils have had a chance to develop since the last glacial advances and where the underlying rock that furnishes material for soil development is nutrient-poor granite or gneiss. Steep terrain can complicate crop production making it necessary to devote much time and effort to constructing and maintaining terraces in order to control erosion. High pastures may often not be suitable for winter grazing due to snow cover, making transhumance or the storage of fodder imperative. Severe temperatures may require stabling of livestock for part of the year or limit stockkeeping options to only the most hardy varieties. Settlements may be threatened by landslides, avalanches, and rockfall due to the nature of local slopes, climate, and geology, and the danger may be accented by deforestation, grazing, and agricultural expansion. At high altitudes, subalpine and alpine low temperatures and snow may make the use of fuel wood and the construction of substantial shelters important in areas where forest resources are scarce. Slow rates of tree growth make forests and woodlands particularly vulnerable to disturbances and recovery from them more difficult and uncertain. The combination of altitude, slow vegetative growth rates, and steep slopes makes many high-altitude areas particularly highly sensitive to erosion following forest and grassland degradation. Seismic activity, mountain torrents, glacial lake outbursts, volcanism, and other geomorphological hazards unique to highland regions may influence settlement and land-use patterns. And extreme altitude may also affect human physiology, influencing fertility and diet requirements in ways that also have repercussions on lifestyles and basic subsistence requirements (Moran 1979:142, 147–161).

High-altitude life may thus in a variety of ways pose significant constraints to settlement and offer relatively fewer subsistence options than many other environments. Yet mountains provide subsistence opportunities as well as challenges. Some mountain regions have volcanic soils of exceptional fertility. Mountains can be fertile, well-watered oases in arid and semi-arid regions. Sloped ground can facilitate field drainage in areas where high rainfall would otherwise inhibit crop growth and make it possible to relatively easily irrigate flights of terraces. Highlands offer rich summer pastures. Altitude may offer protection from low-altitude diseases affecting people and their livestock; the inability of the *Anopheles* mosquito to live much above 1,000 meters, for example, has undoubtedly played some role in the settlement history not only of the Andes but also of the Himalaya. Mountain lands often possess resources such as timber, minerals, and hydroelectric power which can be commercially exploited. Some regions have enormous potential for tourism development, as the Nepalese and Indians as well as the Swiss have discovered. There are sometimes even trade advantages to living in mountain lands, for mountain ranges often form environmental and political divides and those who control access and transport across them can benefit enormously from transmountain commerce. And perhaps, above all, the distinctive altitudinal and topographical variations in mountain climate and vegetation are a major resource, providing rich microenvironmental diversity in a relatively small area. Microclimatic differences within a single Himalayan or Andean valley, for example, can support crops ranging from subtropically suited rice, sugarcane, mango, and banana to temperate-thriving barley, wheat, buckwheat, and potatoes, and sustain livestock such as water buffalo, zebu (*Bos indicus*) cattle and yak that are at home in diverse environments. Even the highest-altitude regions can be ideal country for raising potatoes, barley, yak, and llamas.

Adaptive Patterns of Subsistence: Strategies and Tactics

The peoples of the Himalaya, like those of other highland regions, have developed a number of different adaptive strategies and practices for subsisting in the distinctive conditions of mountain ecosystems. Mountain subsistence systems combine strategies to exploit the characteristic microenvironmental diversity of such regions with sets of land-use practices that have been found effective in the specific conditions of particular microenvironments. Both lifestyle strategies and land-use tactics reflect an appreciation of the constraints and possibilities created by the variations in precipitation, temperature, and vegetation which are generated by altitude, aspect, rain-shadow effects, and other mountain conditions. They are thus responses to a range of microenvironmental diversity characteristic of mountain regions and not simply to altitudinal environmental variation.[1]

Cultural ecologists studying mountain peoples have analyzed adaptive patterns of mountain land use in the Andes, Alps, and Himalaya from the perspectives of both strategy and tactics.[2] A set of general characteristics has often been noted (based more on Alpine and Andean examples than on Himalayan) including the practice of mixed mountain agriculture (with grain varieties related to altitude and with root crops increasingly important at high altitudes), agropastoral transhumance, scattered (and often multialtitudinal) land holdings, systems of land tenure and resource-use decision making that combine communal management of common-property pasture and forest resources (and sometimes also community influence in crop-production decisions) with private family land and livestock ownership, and economies that combine subsistence-oriented agropastoralism with other economic ventures such as trade with other regions having different environmental conditions and natural resources (Fricke 1989; Guillet 1983; Orlove and Guillet 1985; Rhoades and Thompson 1975). In Andean work, which has thus far provided a great deal of the theoretical and case-study development in the field of mountain cultural ecology, two concepts have been emphasized during the past fifteen years: verticality and production zones. The verticality approach has focused on the examination of cultural strategies of exploiting altitudinal variation whereas the montane production zone approach has focused more closely on the specific land-use practices employed in particular altitudinal microenvironments. The verticality approach introduced by John Murra (1972) and further developed by Stephen Brush (1976, 1977) draws on a concept of the altitudinal zonation of environment and land use which was particularly developed by Carl Troll (1966, 1988). Murra noted a pre-Columbian subsistence strategy, which he called the "vertical archipelago," in parts of highland Peru that was based on direct exploitation of the crop-producing and pastoral opportunities of a number of different altitudinal environmental zones (1972, 1985*b:*16). In some cases this strategy even extended to sending out colonists to distant low-altitude areas where crops such as coca could be cultivated.[3] Brush identified several different strategies by which different Andean peoples have attempted to make use of a range of different altitudinal microenvironmental regions, including a "compressed" type of generalized strategy based on direct exploitation of these different altitudinal zones, an "extended" type of strategy in which a group specialized in agriculture at one altitudinal zone and relied on trade to obtain products from other altitudinal zones, and an archipelago strategy in which agricultural colonies were established at considerable distance from the main settlement area (Brush 1977:11).[4] The production zone concept developed by Enrique Mayer proposes a categorization based on specific land-use patterns characteristic of particular altitudinal zones. The basis of this classification system is crop production with emphasis given to both the crops grown and the specific ways in which they are grown, including the socioeconomic and political organization of production. He formally defines production zone as "a communally managed set of specific productive resources in which crops are grown in distinctive ways. It includes infra-structural features, a particular system of rationing resources such as irrigation water and natural grasses, and the existence of rule-making mechanisms that regulate how these resources are to be used" (Mayer 1985:50–51).[5] Mayer (ibid.:77, n. 1) identifies ten different production zones, for instance, in one area of Peru, including two types of rain-fed agriculture (differentiated by individual household versus community decisions about crop rotations and fallow), five types of irrigated agricultural sytems differing in types of crops raised and crop rotations, and three types of fruit production.

In my view mountain cultural ecology analysis requires both of these approaches. The concepts of verticality and production systems seem to me to address two different dimensions of sociocultural adaptation to mountain environments, and emphasizing one to the neglect of the other risks the loss of important perspectives in understanding subsistence practices. The study of verticality (in the larger sense by which I define it) brings a concern with broad economic strategies of multienvironmental resource use and management and the cultural values, environmental perception and knowledge, and social organization that support these ways of life. The study of production systems brings an understanding of tactics, of the specific land-use practices (including local knowledge, repertoires of crops and animals, technology, land tenure, labor organization and community resource-management institutions within the context of local cultural values, regional political economy, and local economic and social differentiation) which particular groups of people have developed for use in their homeland. Combining the two allows us to examine economic activities from the standpoint of both the broad strategies by which a people (or a particular group, social group, or household) exploits a range of altitudes and microenvironments and the specific tactical means by which they accomplish this in particular microenvironmental settings. This dual approach illuminates adaptation both from the standpoint of the development of techniques in agriculture and pastoralism suited to mountain conditions and the evolution of environmental perception, social organization, and lifestyle values which make possible economic strategies designed to make use of the altitudinal and topographically based complexity and diversity of mountain microenvironments. They can also illuminate the variations in settlement and land use within a group, within a given community, or, through time, in the practice of a given household. From this base it is then possible to make larger inquiries into historical change and the factors involved in it, variation in land use among villages, economic classes, and individual families, and cross-cultural comparisons with the strategies and tactics of other peoples inhabiting similar mountain microenvironments. Comparisons of economic practices among groups of a particular people who inhabit different territories can also be carried out in an attempt to highlight the degree in which some of these practices may reflect strategies and tactics of environmental adaptation whereas others may have quite different origins.

Subsistence strategies are typical of the way or ways of life followed by a people, ethnic group, social subgroup (be this defined by economic class, caste, clan, ethnicity, or other criteria) or household. Rather than speak of *the* subsistence strategy of a given group it is possible to identify a set of subsistence strategies used by different members of that group, a set which may or may not embody an underlying common set of principles. A group may, for example, tend to practice mixed agropastoralism, with virtually all households raising both crops and the livestock that furnishes manure to their fields. But the agropastoral strategies of different types of households may differ fundamentally. Some families may keep large herds for the purpose of selling dairy products, meat, and hides or wool and entrust them to herding specialists for transhumant shifts of pastures that cover considerable distances and reach areas remote from the main village. Other families may keep a single cow or a couple of goats which they care for themselves, live all year in the village, and let the livestock graze in the local environs and feed them fodder from nearby forests and fields. Households sharing a broadly similar strategy of growing a set of grains and tubers adapted to the altitude of nearby slopes may similarly vary considerably in the emphasis they place on particular crops and varieties, the degree to which they make use of fields in different microenvironmental sites, their use of irrigation, and the extent to which their crop production is influenced by market concerns.

All subsistence strategies make use of one or more production systems, and in mountain regions it is very common for multiple productions systems to be used at sites at different altitudes or other microenvironmentally different conditions. Each of these systems is a set of land-use techniques in specific environmental, social, political, and economic contexts. In a given microenvironmental site several different production systems are often in operation, sometimes implemented by different households making use of different subsistence strategies and sometimes by individual households whose subsistence strategy includes the use of multiple production systems at a given site. On Himalayan slopes below 2,000 meters, for example, it is common to find a diversity of production systems that can conveniently be characterized in terms of their cropping patterns, although each is an agroecosystem that also has

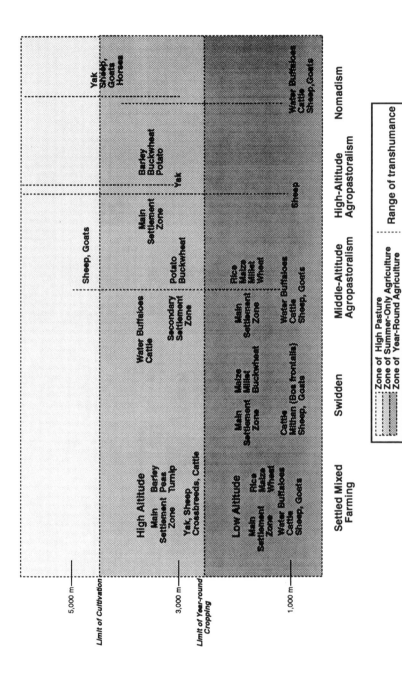

Figure 1.   Land-Use Strategies, Highland Asia

associated patterns of fertilization that may link it to particular types of local pastoralism and forest use. There is an irrigated rice (*khet*) production system—or several of them—involving single or multicrop rice cultivation, rice cultivation with winter rain-fed crop rotations, and rice cultivation with winter fallow and grazing of the terraces. There may also be production systems based on irrigated millet. At the same altitude there are also rain-fed (*bari*) production systems of various types which emphasize different crops and different types of fertilization, intercropping, relay cropping, double cropping, crop rotations, and fallow periods. There may also be swidden production systems in adjacent forests and woodlands. Some households in the community may make use of a number of these different production systems, each of which involves its own set of land, labor, and capital requirements, local knowledge, institutional resource-management arrangements, relationship to the state in the form of taxes, and involvement or lack of it with a wider market economy. Other households may confine their subsistence strategy to the use of a single production system and find their options limited by social status, wealth, family labor mobilization and talents, knowledge and experience, religious beliefs, or other factors.

Himalayan Subsistence Strategies

The environmental and cultural diversity of the Himalaya and its neighboring ranges has historically supported a diversity of different, adaptive subsistence strategies. At least five broad, traditional subsistence agricultural strategies and several common substrategies can be identified in the central and western Himalaya of Nepal and India alone: settled, mixed farming; swidden agriculture; middle-altitude agropastoralism; high-altitude agropastoralism; and pastoral nomadism (fig. 1).[6] All five have been employed for centuries and continue to be followed today.[7] Today middle- and high-altitude agropastoralism, both of which are based on mixed mountain farming and livestock herding with transhumance, are by far the most widespread adaptive strategies in the mountain regions of Nepal. Settled mixed farming is becoming more common, in part perhaps as a result of agricultural intensification following increasing population density (Boserup 1965), but also as a result of the commercialization of agriculture in response to changes in transportation (especially the expansion of road networks) and the increasing incorporation of formerly more remote areas into a developing national market economy.[8] Historically integral swidden cultivation based on forest fallow rotations was important across large areas of the country and in much of the temperate and subtropical regions remained so into the nineteenth century.[9] Supplementary swidden cultivation continues to be carried out in a few regions in central and eastern Nepal today despite government efforts to halt it out of concern for forest conditions. True pastoral nomadism, in the sense of an economy entirely based on pastoralism with no crop cultivation whatsoever, is extremely rare in the Himalaya today and is only followed by a very few groups in the western regions of Uttar Pradesh, Himachal Pradesh, and Kashmir.[10]

All of these except for the settled agricultural strategy share an emphasis on directly using resources across a range of altitudes. Although settlement patterns, crop and livestock emphases, and annual rhythms of movement up and down slopes and mountain valleys vary between the different strategies, all of these strategies except settled, mixed farming use seasonal transhumance, movement of livestock, and/or agricultural fields at multiple elevations to exploit land-use opportunities in different microenvironmental regions of the mountains. Trade is also an important mechanism used by peoples of all five patterns to gain goods from other altitudinal and microenvironmental areas of the mountains as well as from regions beyond the mountains.[11]

Middle- and high-altitude agropastoralism vary in settlement emphasis and seasonal altitudinal use. In Nepal the distinguishing characteristic of the two is that middle-altitude agropastoralism is centered on life in main villages situated below 3,000 meters, whereas people following the high-altitude strategy have villages above 3,000 meters. These two strategies differ, however, not only in their emphasis on different types of microenvironmental regions but also in their basic characteristics of crop and livestock production. Groups following the middle-altitude strategy base their crop production below 2,500 meters in country that supports the cultivation of rice, wheat, maize, and millet as staple crops and where year-round crop growing is possible. Their pastoralism tends to revolve around cattle, water buffalo, sheep, and goats that provide a critical source of manure as well as other useful products and that are supported by considerable use of crop residue and forest fodder as well as by seasonal altitudinal shifts of grazing areas, which in some cases include major long-distance shifts of residence to summer alpine pastures. Groups following the high-altitude agropastoral strategy base their agricultural production on sites between 3,000 and 4,500 meters in altitude, generally in inner valleys, partial rain-shadow valleys of the Great Himalaya, and trans-Himalayan valleys that are shielded from the full force of the southwest monsoon. Crops are grown only in the summer at these altitudes and the crop repertoire consists primarily of barley, buckwheat, and tubers. High-altitude pastoralism is based on transhumant herding of yak, cattle-yak crossbreeds, sheep, and goats. These two different altitudinal strategies are complementary. Many groups following the high-altitude pattern carry out trade to obtain grain from middle-altitude groups, offering high-altitude resources such as Tibetan salt and wool in exchange. There is some seasonal overlap in altitudinal use as well. Both strategies make use of seasonal movement between different altitudinal microenvironmental zones. High-altitude groups herd sheep and goats in lower-altitude agricultural areas in winter, in the process also providing valued manure to middle-altitude farmers' fields. Many middle-altitude groups make use of high-altitude pastures for summer herding.

Middle-altitude agropastoralism was apparently long characteristic principally of Hindu Pahari hill-caste villagers in western Nepal and adjacent western India, whose cultural origins and characteristics are closely related to the people of the Ganges plain of India. High-altitude agropastoralism, by contrast, has been practiced almost exclusively by the Buddhist peoples who live on the frontier between India and Tibet and who, as discussed earlier, in many aspects of culture and lifestyle strongly resemble their Tibetan cousins. Many of the peoples of central and eastern Nepal were long distinct from both groups in land use. Gurungs, Tamangs, Rais, and Limbus, for example, were all formerly integral swidden cultivators who relied on rotational systems of shifting agriculture with long forest or bush fallows.[12] Since the late eighteenth century, however, all of these peoples have adopted the distinctive style of transhumance and terrace agriculture of the middle-altitude agropastoral pattern, either abandoning swidden cultivation altogether or employing it only as a supplementary source of food production.[13] In some regions of mountain Nepal, particularly in lower-altitude areas, some Pahari, Tamang, and Magar villagers and communities are now abandoning the seasonal transhumance of the middle-altitude agropastoral strategy for a settled agropastoralism with year-round residence in a single village. In some areas situated near urban settlements and roads this is often associated with new patterns of crop and dairy production developed in response to new market opportunities.[14] Many peoples following the high-altitude agropastoral pattern have also shifted their adaptive strategies into less mobile patterns of pastoralism and trade during the past thirty years. These ongoing changes in lifestyle and land use highlight the need to bring to the analysis of adaptive land-use strategies not only consideration of the environmental setting and role of adaptation to environment and environmental change but also an awareness of the historical context of cultural, social, demographic, technological, and political economic changes and the ways in which these have influenced both household economic decisions and the evolution of cultural patterns of subsistence and adaptation.

Middle-Altitude Mountain
Agropastoralism

The middle-altitude agropastoral strategy combines transhumant tending of cattle, water buffalo, sheep and goats with permanent agriculture that is often carried out at multialtitudinal sites. The main focus of crop production is found in the main villages that tend to be located between 1,000 and 2,500 meters. Here terraces are carefully maintained for the cultivation of rice, maize, millet, buckwheat, mustard, vegetables, and winter wheat. Much effort is put into irrigating fields for rice production, and farmers are familiar with techniques of manuring, intercropping, double and relay cropping, and crop rotation. Surrounding forest and woodland areas supply grazing, fodder, and forest-floor litter for fertilizer as well as furnishing fuel wood and lumber. In summer those households that own substantial numbers of cattle and water buffalo often take them up to the rich pastures between 2,500 and 3,000 meters (this is sometimes handled by a single household member or hired herder rather than by the entire family) and sheep and goats may be taken as high as 5,000 meters (again often by specialists). During these weeks herders may live in simple, movable, bamboo shelters, as is common in Nepal, or more substantial herding huts such as are common in the Indian Himalaya west of Nepal (Pant 1935; Berreman 1963a). Summer herding bases sometimes provide a secondary crop-production site for the cultivation of potatoes and hardy grains such as buckwheat, wheat, and barley. In autumn livestock are led down to the main villages where they graze on field stubble and leave behind manure for the next round of field preparation. In winter the herds may be taken still farther down the valley, where herders again base in temporary shelters or herding huts.[15]

Middle-altitude agropastoral peoples often also participate in complex regional interaltitudinal and trans-Himalayan trade networks. This trade brings the people of the lower-altitude regions salt, wool, seed potatoes, and other valued products from the high valleys and Tibet in exchange for agricultural surpluses grown in the midlands. Mid-altitude-grown grains, especially rice, maize, wheat, and millet, are traded to higher-altitude regions along with some meat, dairy products, and fruits and vegetables. Much of this trade is carried out by high-altitude people who journey down valley in winter and who may make trans-Himalayan spring and autumn trade trips to Tibet. Some middle-altitude groups, however, transport their own agricultural surpluses into higher regions during the late autumn and winter and some middle-altitude farmers also trade agricultural products they have obtained from other middle- and low-altitude areas.

Figures 2 and 3 illustrate the geographic and altitudinal seasonal land-

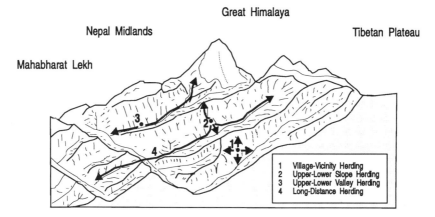

Figure 2.   Middle-Altitude Agropastoralism, Geographic Patterns (Base drawing adapted from Metz 1989)

use patterns associated with the middle-altitude mountain agropastoral strategy. In them I distinguish four different patterns of seasonal altitudinal and geographical land-use patterns. Two are relatively compressed in both altitudinal and geographical land use, making use of the resources of a single slope. Two are extended in altitudinal and geographical land use, making fuller use of the resources not only of variations in slope but also of longer-distance up and down valley movements. The first of the slope-based patterns is a village-based, relatively settled way of life such as is practiced particularly by groups living in the lower-altitude regions of the Nepal midlands. Here the entire focus is on crop production in the main village with herding and fodder collection from slope areas within easy reach of the settlement. All the other subsistence patterns require seasonal shifts of residence by at least some family members. The second pattern makes use of microenvironmental variation on a single slope, but the distances and relief involved make it necessary for herders to base seasonally in herding huts (*goths*) a day or two away from the main village. Cattle, water buffalo, sheep and goats may thus be taken on a limited-distance transhumant migration that may nonetheless cover several thousand meters of altitudinal variation. This movement of only a few kilometers takes them to summer pasture high on the ridge in the temperate forest or even subalpine zone and may also take them to winter pastures below the village in subtropical reaches of the gorge. The third pattern is a longer-distance transhumance in which flocks of sheep and goats and possibly also herds of cattle, crossbreeds, and even water buffalo are taken up valley to summer pastures in the Great Himalaya above 3000 meters. The fourth pattern that traverses

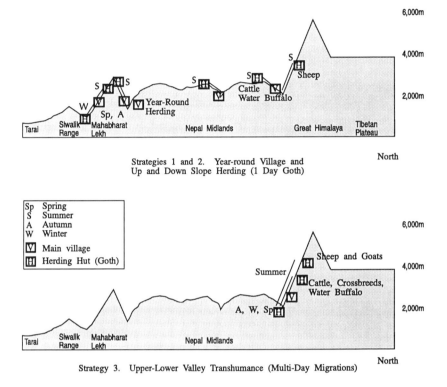

Strategies 1 and 2. Year-round Village and
Up and Down Slope Herding (1 Day Goth)

Strategy 3. Upper-Lower Valley Transhumance (Multi-Day Migrations)

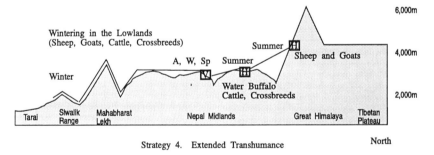

Strategy 4. Extended Transhumance

Figure 3.   Middle-Altitude Agropastoralism, Altitudinal Patterns

the ranges adds to the third pattern a longer-distance winter move as
well, one which takes herders and their sheep, goat, and cattle across the
Mahabharat Lekh and out of the midlands into the valleys north of the
Siwalik range or even beyond the Siwalik range to its southern alluvial
slopes and the Tarai.

In the Dudh Kosi valley and adjacent areas variations of middle-altitude agropastoralism are practiced by Rais, Gurungs, Magars, and several Sherpa groups including the Pharak and Shorung Sherpas. Some of these Sherpa groups have adopted land-use practices that they may have adapted from those of their non-Sherpa neighbors and which from the standpoint of Khumbu Sherpas are quite "un-Sherpa." Some Sherpa families in the Dudh Kosi region, as in the Arun watershed, for example, raise water buffalo and practice swidden cultivation. And in the Arun region some Sherpas even raise pigs and cultivate irrigated rice.

Within the range of microenvironments found in the Nepal midlands and the Great Himalaya between 1,000 and 3,000 meters a great diversity of production systems could be described through which middle-altitude agropastoralists use the resources of particular sites. In the 1,500–2,500-meter altitudinal range in which most of the main villages are located these systems include several different types of production based on rainfed permanent crops with and without small-scale keeping of cattle, water buffalo, sheep and goats such as year-round cultivation, summer-only cultivation, different types of annual and multiyear crop rotations and fallowing practices, different types of intercropping and relay cropping, food and fodder crops, subsistence and commercial crops. There are also systems based on irrigated agriculture (including rice only and rice followed by winter-irrigated or nonirrigated crops), again with and without associated livestock raising, as well as a number of different types of woodland, forest, grassland use and management. These could be further differentiated on the basis of variations in the use of agricultural inputs (seed, fertilizer, and labor especially), by the social and cultural arrangements influencing crop production, and orientation towards subsistence or commercial production. While this approach would offer valuable insights into land-use techniques and institutions, a fuller understanding of local land use would require both attention to differentiation in wealth, power and other factors that influence household and community resource-use options and decisions and the investigation of the strategies by which particular families, economic classes, communities, and ethnic groups combine sets of these microenvironmentally based production systems into household and regional economies.

Different peoples within the middle-altitude region employ slightly different production systems in very similar microenvironmental sites, and individual ethnic groups make use of more than one system even in similar sites. Crop decisions and rotations, field-fertilization practices, grazing patterns, and other aspects of production systems may vary. Such differences may reflect cultural perceptions (including crop and livestock preferences and religious prohibitions), social arrangements (especially land tenure and communal resource-management institutions), economic differentiation, and political power. There is accordingly no simple equation between given microenvironments and the production systems used in them, although environmental conditions (including altitude, slope, aspect, and the amount, type, timing, and intensity of precipitation) may broadly influence the range of what is and is not rewarding in terms of high yield and low risk.

High-Altitude Mountain Agropastoralism

Above 3,000 meters life is different for the high-altitude-dwelling peoples who make the high valleys of the Great Himalaya and trans-Himalaya their home.[16] At these altitudes crops can only be produced during the summer and generally there is not time enough in the short growing season to cultivate more than a single crop per year of a small number of plants fit for high altitude. The cool temperatures and the risks of late spring, early autumn, and even summer frosts restrict the range of staple crops to barley, buckwheat, wheat, and tubers. Rice, maize, and millet cannot be cultivated and at higher altitudes even wheat production is not possible. Pastoralism is similarly limited in options. The climate is too difficult for water buffalo and cattle may be kept year round at such heights only if much effort is made to stable them and provide them with fodder through the winter months. But yak, yak-cattle crossbreeds, sheep and goats all thrive. Figures 4 and 5 show the geographical and altitudinal land-use patterns associated with variations of the high-altitude mountain agropastoral strategy. As with the middle-altitude agropastoral strategy both extended and compressed strategies are depicted. Extended strategies make greater use of seasonal transhumance, either across the Himalaya to Tibet for winter herding of yak and sheep, down-valley winter herding of sheep, goats, and cattle in the midlands or even the Tarai, or a combination of both of these. The compressed pattern relies on year-round use of the resources of the Great Himalaya alone.

For the past century, and very likely for long before that, peoples following high-altitude agropastoral strategies have usually relied on trade to supplement their crop production. Trade for lower-altitude-grown grains has been particularly important. Until the 1960s many families devoted much of their winter to trading trips to the south in order to barter Tibetan salt and other goods for grain and take advantage of the warmer weather, the readily available food, and the good grazing. Sheep and goats were often taken on these winter journeys as pack animals, whereas families who kept yak left them with herding specialists in the high valleys or sent them across the mountains to Tibet, where the grazing is often better in winter than on the south side of the Himalayan crest

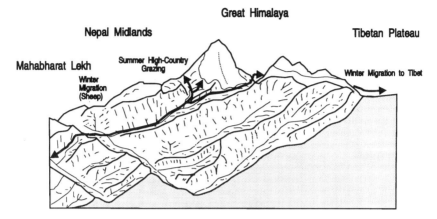

Figure 4.   High-Altitude Agropastoralism, Geographic Patterns (Base drawing adapted from Metz 1989)

since there is less snow and yak can find abundant grazing on the vast pastures. This way of life, however, is today no longer followed by many Bhotia groups. Political and economic changes in Tibet following the onset of Chinese administration after 1959 have greatly affected trade conditions for Bhotias, as did the 1962 war between China and India and its diplomatic aftermath. Some groups of Bhotias have been unable to continue trans-Himalayan trade on the old scale or to winter stock in Tibet. This has led to major changes in life for Bhotia peoples (see, for example, Fürer-Haimendorf 1975; Rauber 1982). Some have abandoned their former subsistence strategies and migration from the high country and other high-altitude peoples have developed new types of trade or become involved in the tourism industry.

Transhumance is an important feature of the high-altitude agropastoral strategy. After the crops are planted and well established in the main villages families follow the good grass into the upper valleys. Yak and sheep may be herded in mid- and late summer as high as 6,000 meters. Herders usually live in tents, often the black yak-hair tents familiar also in Tibet, moving through a series of different camps in a long-established routine modified to meet the pasture and weather conditions of the particular year. Sometimes additional fields are also cultivated in the summer herding settlements where fine crops of potatoes and barley can be grown even as high as 4,300 meters. Families who continue to trade with Tibet dovetail the demands of the agricultural cycle with one or more trips across the border during the period between spring and late autumn to obtain salt, borax, wool, and other goods for their winter trading in the south.

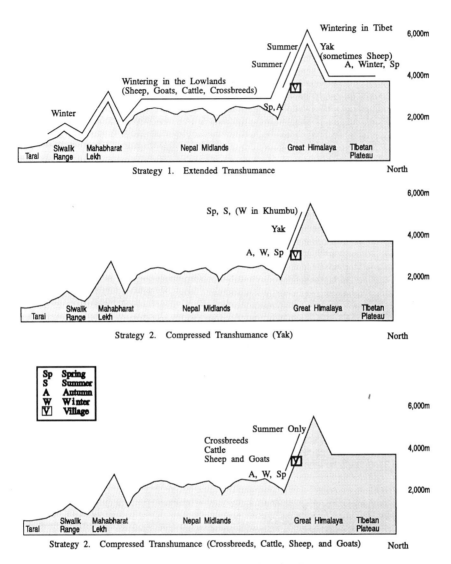

Figure 5.   High-Altitude Agropastoralism, Altitudinal Patterns

Only a few Sherpa groups follow a high-altitude mountain agro-pastoral strategy, the best known of which are the Sherpas of Khumbu and Rolwaling. The narrow valley of Rolwaling, seven kilometers in length and never more than one kilometer wide, is comparable in altitude to Khumbu. The main village, Beding, is situated at 3,600 meters. Fewer than fifty households inhabit the valley and they practice a mixed agropastoralism emphasizing transhumant yak herding and multialtitudinal crop production of potatoes and barley. Potatoes and barley are grown as high as the settlement of Na at 4,100 meters. Yak are herded in the summer to above 5,000 meters (Sacherer 1981:157–158). Khumbu Sherpa subsistence strategies are discussed later in this chapter.

Altitudinal Production Zones

Land use at the scale of an entire Himalayan valley such as that of the Dudh Kosi becomes exceedingly complex to analyze at the level of production systems. The range of altitudinal gradients and micro-environmental variation, the diversity of adaptive strategies, and the high level of cultural diversity (in the Dudh Kosi case four different Sherpa groups, at least two Rai groups, as well as Magars and Gurungs) make for a wealth of production systems. This diversity can be simplified at the regional scale for a general overview, however, by considering altitudinal production zones. These constitute broad categories of land use encompassing a number of different, discrete production systems. They narrow the focus of microenvironmental concern to a few broad altitudinal bands identified according to basic regional altitudinal variation in land-use patterns. From this perspective six altitudinal production zones can be discussed in the Dudh Kosi valley (fig. 6) ranging from irrigated and nonirrigated year-round crop production to high-pasture pastoralism. Table 2 shows the current upper altitudinal limits of the staple crops in the valley. Note that Khumbu is situated too high in altitude for the harvest of more than a single summer crop per year and is just out of the current altitudinal range of cultivation of such important Dudh Kosi valley crops as maize and wheat.[17]

Sherpas, Rais, and Gurungs each make use of several zones through middle- and high-altitude agropastoral strategies and in each practice one or more different production systems based on different institutional, cultural, and technical arrangements for raising various types of crops and livestock and exploiting altitudinally related natural resources such as forests, woodlands, and temperate, subalpine, and alpine grasslands and tundra. Within a given altitudinal production zone the choice of cultigens and domestic animals, the scale at which they are raised, and the techniques employed may vary with cultural preferences, religious beliefs, political economy, household status and wealth, and historical processes and events such as the diffusion of terracing techniques, the introduction of new crop varieties, changing trade patterns, and changes in local affluence.[18]

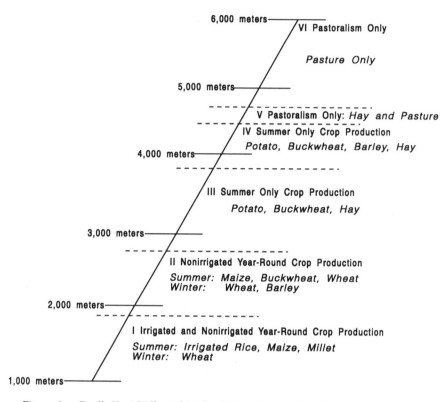

Figure 6.   Dudh Kosi Valley Altitudinal Crop-Production Systems

*Table 2.   Current Upper Altitudinal Limits of Staple Crops in
the Dudh Kosi Valley*

Potato	4,700m	(Tarnak)
Barley (*ua*)	4,300m	(Dingboche)
Buckwheat	4,000m	(Pangboche)
Barley (*jou*)	2,800m	(Jorsale)
Wheat	2,800m	(Jorsale)
Maize	2,800m	(Jorsale)
Millet	2,000–2,200m	(Kharikhola)
Rice	2,000–2,200m	(Bupsa)

Regional Linkages

Himalayan valleys are typically the homes of a number of different
peoples and groups each occupying its own territory and pursuing its
own characteristic subsistence strategy or strategies and set of land-use

production systems. These different groups are seldom independent socioeconomic islands and even where there has not been a history of political conflict (conquest, raids, tribute), migration, pilgrimage, or intermarriage there is usually a complex interaction based on trade. It is common also for the dynamics of subsistence strategies themselves to create other types of more direct resource-use interaction. Both the middle-altitude and high-altitude strategies characteristically employ tactics of multialtitudinal land use and seasonal movement between several settlement sites. There is often some overlap in the microenvironmental areas exploited, especially for pastoral patterns. Both middle- and high-altitude agropastoralists value the grazing resources found in the high summer pastures and the lower valley forests, grasslands, and fallowed fields. This sometimes leads to different groups, often of different ethnicities, seasonally making use of the same areas or of crossing other groups' territories en route to their own secondary settlement and grazing areas. To coordinate this multicultural use of particular areas arrangements are made that include joint resource management, user or transit fees, defined common property resource-use boundaries, sequential resource-use arrangements, and other examples of temporal and territorial resource partitioning. In some cases differences in resource-use goals and land-use production systems make it possible to develop complementary multiethnic resource use. A given high-altitude place may be valued as a summer herding area for the sheep and goats of a middle-altitude agropastoral group, for example, and also be an agricultural area for a high-altitude agropastoral group that may value the additional manure provided by the outsiders' sheep. Such complementary resource use is even more common in the altitudinal reaches where middle-altitude main villages are situated. Here middle-altitude households may offer cash and other incentives and compete with each other to obtain better fertilization for their winter fallow fields by having them grazed by the flocks of high-altitude herders who have come down valley for winter grazing. In other cases, however, these same resource activities are seen to compete with local herders' access to pastures, and herding families demand compliance with local grazing restrictions and the payment of grazing fees, or even ban outside herds altogether.

The subsistence strategies and regional resource-use patterns of middle- and high-altitude agropastoralists in the Dudh Kosi valley have historically resulted in a complex pattern of regional trade, seasonal movement, overlapping land use, and resource partitioning and management within a broad general altitudinal differentiation of land use and associated production systems. Rais and Gurungs from the lower reaches of the watershed have taken sheep and goats up into the high pastures, often to their own areas (some of which they retain through claims predating the in-migration of Sherpas), but in other cases to Sherpa-controlled regions where they have had to follow local grazing regulations and sometimes pay fees. Sherpas from both Khumbu and Pharak formerly took stock (at least on a small scale) into middle-altitude Rai areas, in some cases paying fees for the privilege. In other areas Sherpas have had to pay summer high-altitude grazing fees to Rais. High-altitude Khumbu Sherpas have engaged in trade with both middle-altitude Sherpas and Rais as a basic component of their subsistence economy and both middle-altitude Sherpas and Rais have also long traded to Khumbu. The scale, nature, and geographical extent of this trade, however, has varied enormously over the past century with considerable impact on local subsistence strategies.

Khumbu Sherpa Subsistence Strategies

In many respects Khumbu Sherpa subsistence is a typical high-altitude agropastoral strategy. Like other high Himalayan peoples Khumbu Sherpa have developed an integrated mix of agriculture and pastoralism based on the familiar crops and livestock varieties of the high country. And like other Himalayan peoples in these circumstances they have long supplemented the agricultural and pastoral resources of their high valley homeland by trade geared to obtaining lower-altitude-grown grains. Khumbu Sherpas traded crop surpluses and pastoral products, but they, like many other peoples who live along the Tibetan frontier, have particularly profited from playing a middleman's role transporting goods across the Himalaya between Tibet and the lower lands of Nepal and India. Since the 1960s this trade has dwindled and Sherpas have instead achieved a similar diversification of their subsistence by using cash to buy rice, maize, wheat, millet, and buckwheat grown in the lower Dudh Kosi and adjacent valleys and the Nepalese Tarai. Today, as for at least a century, Khumbu Sherpas do not need to achieve regional or household self-sufficiency in agricultural production. Instead they have the more limited goal of meeting household requirements in a small range of Khumbu staple crops.

Sherpa households' ability to obtain agricultural and pastoral products from regions below 3,000 meters through trade has enabled them to adopt a strategy of specialization in high-altitude-fit varieties of crops and livestock. At the same time, Khumbu Sherpas, like other high-altitude peoples, have made the use of a range of microenvironmental sites a prominent basis of their subsistence strategies. Within Khumbu itself Sherpas make adept use of the minor microenvironmental variations between different sites due to altitude, aspect, precipitation, and soil. Herding strategies for the various types of livestock all involve the use of seasonal transhumance between lower and higher valley common pastures. Crops are usually planted in fields at a variety of altitudes. Some direct use has also been made of altitudinal and microenvironmental regions beyond Khumbu itself. Sherpas have never developed secondary agricultural areas in lower valley areas remote from their main villages in the way that some Bhotia peoples have in the western Himalaya, but like most other high-altitude peoples they have made use of environmental opportunities in adjacent regions for herding. Earlier in the twentieth century Khumbu Sherpas, like the peoples of Dolpo and Humla in northwestern Nepal, benefited from Tibetan winter pastures and some also took a few crossbreeds and sheep south to the lower valleys to serve as pack stock on trading journeys. In the Khumbu case, however, only a few families did so and then only on a quite small scale. Before the mid-1960s many families, like other Bhotias, devoted the winter to travels which might last five months and which were spent trading and visiting sacred places and shrines.

Since at least the late nineteenth century, and probably for long before, Khumbu subsistence has revolved around the cultivation of a very small number of staple crops which can tolerate the high-altitude conditions. Buckwheat is the main grain, grown in a biannual rotation with tubers. Some barley has also been grown in those few sites where it is possible to easily irrigate the fields, and the need to irrigate barley fields rather than cultural or political economic factors accounts for the distinct lack of emphasis on this preferred food. Formerly, Tibetan varieties of turnip and radish were the staple tuber crops, but during the twentieth century potatoes have become not only the most frequently grown tuber but the dominant crop in the region. Today more than 75 percent of all crop land is in potatoes, usually monocropped and often grown year after year without rotation.[19] Regional diets are today based primarily on potatoes, which are consumed in one form or another as the foundation of virtually every meal. Adult Sherpas typically consume more than a kilogram of the tubers per day, and a family of four requires between one and two metric tons per year for self-sufficiency in this staple.[20] Grain is much less important, and a well-to-do household might live well by local standards on less than 700 kg (10 *muri*) per year.[21] Diet varies somewhat with wealth, particularly in the amount of grain consumed. Wealthier families also differ in their choice of grains, and eat rice, barley, and buckwheat rather than the millet, maize, and buckwheat of the poorer households. Rice, maize, and millet are all obtained from outside the region and much higher prices are paid for rice than for the other grains. Recent affluence has greatly increased the amount of rice being consumed today in Khumbu. During the past decade well-to-do families, especially those in Nauje, have also added more processed foods to their diet. Instant noodles, refined sugar, powdered milk, and even such things as mayonnaise have become popular.[22] Some adverse effects on health and teeth are beginning to become evident.

Over the centuries Khumbu Sherpas have developed a set of different production systems to raise potatoes, buckwheat, and barley. Each makes use of different techniques and requires different knowledge. Some are in use throughout the narrow altitudinal reach of Khumbu whereas others are employed more specifically in a particular type of microenvironmental setting within the region. All of these crop-production systems are also linked with grazing and forest-use practices that furnish crucial field nutrients. Ten different crop-production systems can be identified in Khumbu today, all but two being unirrigated. Nine are based on the cultivation of crops in main terraced fields. The two most important systems in terms of the land devoted to them and the share of local food production which they account for are both unirrigated, single summer crop production systems. In one of these potatoes are grown in a monocrop for decades with no rotation. In the other there is a two-year rotation between potatoes and buckwheat. Two other production systems that are now employed on a limited basis by a few families in some of the main villages build on these basic potato and buckwheat systems by adding intercrops. In one of these potatoes are intercropped with a Tibetan variety of radish and in the other a semivolunteer called *to* which produces an edible tuber is encouraged in either potato and buckwheat fields. There are also two production systems now practiced in Nauje which build on potato and buckwheat cultivation through following the main harvest with a second crop. Several families follow potatoes with a second crop of mustard that is grown as a green vegetable, not for its oil. A few families follow potatoes with a second crop of wheat or barley grown for fodder. In both cases the need for the second crop to mature before November limits the use of this system to the lower-altitude settlements of the region and often involves harvesting potatoes before they reach their full maturity in order to make field space available. Two further nonirrigated production systems are devoted to fodder. The most important of these is establishing and tending hay fields. In Nauje there is also some small-scale growing of nonirrigated barley and wheat as fodder crops. Finally, there are two crop-production systems in Khumbu that make use of irrigation. Both are very localized. The most important in terms of the area devoted to it is the cultivation of irrigated barley as a summer field crop. In this system barley is rotated biannually with unirrigated potatoes. The requirements of the agroecosystem for irrigation water in the late spring limit it today to a single site in eastern Khumbu. The other system is the household vegetable garden. Here water is carried by hand from the village water sources to the garden rather than being fed by a canal system to the site, and is carefully delivered to the plants through hand watering in contrast to the techniques used to irrigate barley.

Agriculture is carried out by nuclear families on privately owned crop and hay fields. Khumbu Sherpas are all smallholders. Most families own less than a quarter hectare of land and wealthy families own little more than half a hectare. The tremendous productivity of potatoes, however, makes even this small amount of land sufficient to meet household requirements in all but the worst years. Indeed, the 1,200 kilograms required by typical Nauje families can be grown on only 600 square meters of land even in a relatively bad year. Elsewhere families have no trouble cultivating enough potatoes for their needs on less than 1,000 square meters of land, and those who own a quarter hectare of land may also reap two muri or more of buckwheat (120 kg), enough for household self-sufficiency in that grain (although far from the ten or more muri of total grain stocks required by a household). In good years families who own only a quarter hectare of land may be able to exchange or sell surplus potatoes or, as is more common, use them as fodder. In some years such a household may also have a surplus of buckwheat. Surplus production is mostly exchanged in Khumbu itself, but throughout the twentieth century dried potatoes have been exported on a very small scale to Tibet and some potatoes have been traded or sold as seed potatoes each January and February to Rais and Sherpas from lower-altitude areas. A relatively small number of families, however, sell potatoes or grain on an annual basis. The great majority of Khumbu families orient their crop production entirely to family consumption. Even the few households that often raise and sell a surplus have had their crop production practices less affected by commercialization than might be expected. These farmers do not, for example, base their decisions about which crops and crop varieties to plant on the basis of market demands and commodity prices. They maintain the same monocrop potato or biannual buckwheat-potato rotations as their fellow villagers, cultivate the same varieties of these crops, and raise them with the same fertilizer and labor inputs.[23]

Pastoralism is based on the care of several species of stock that are hardy enough to winter in Khumbu conditions that are kept for their direct contributions to the family of milk, meat, blood, hair, manure, draft power, and transport and also as sources of income from the sale of calves or their use as pack stock. Yak and especially the female yak, known in Khumbu as *nak,* are the preferred stock. Cattle, yak-cattle crossbreeds, and sheep are also kept on a small scale.[24] There is a very strong cultural preference for raising yak which is deeply entwined with local conceptions of wealth and status. During the past century the major emphasis in stockkeeping has been on the raising of nak kept both as a source of milk and for crossbreeding with bulls to produce cross-breed calves that were commercially valuable for sale in Tibet and to middle-altitude Sherpas.

Herding emphasizes the use of the extensive Khumbu rangelands that are regarded as common property resources freely available to all Khumbu Sherpas. The use of some of these areas is regulated during summer by local pastoral management regulations intended to protect growing crops from livestock, reduce pressure on lower Khumbu winter pasture areas, protect areas that furnish wild grass for hay, and establish a form of rotational grazing of some high summer pasture areas. Today all herding is carried out entirely within Khumbu throughout the year. The full range of Khumbu pastoral resources from 2,800 to 6,000 meters is used. Cattle and crossbreeds are herded to high altitudes in summer and in winter are stabled in the lower floor of main village houses and fed fodder. The hardier yak and nak are taken to the very highest Khumbu summer pastures and also often spend much of the winter at high-altitude herding bases where they feed on what grass they can find in the snowy upper valleys and are fed hay that has been grown in nearby walled fields and stored in the high-altitude herding huts. This yak-herding pattern, with its distinctive reliance on year-round use of Khumbu pastures and the consequent importance of winter and spring moves to areas where hay and fodder supplies have been stored, is largely a development of the last sixty years. Previously many yak were taken to Tibet to winter and much less hay was grown in walled fields in Khumbu.[25]

Pastoralism, in contrast to crop growing, is much influenced by commercial concerns and has been for at least a century. Livestock make many contributions to household sustenance, but they are equally as important for many families as sources of income and this affects decisions about herd composition and size. The sale of crossbreed calves bred from nak was long a very important Khumbu industry, although less so in the recent past than formerly. Early in the century it was one of the most lucrative of all Khumbu entrepreneurial avenues. And at various times the keeping of yak and male crossbreeds has reached high levels when it has become especially worthwhile to own pack animals. During the past fifteen years the number of male crossbreeds kept in several Khumbu villages has increased markedly, reflecting interest in opportunities for their use as pack animals in the tourist trade.

Altitudinal Land Use

Khumbu Sherpas make use of a span of vertical environments which ranges from 2,800 to 6,000 meters (fig. 7). Microenvironments within this region vary in climate from cold temperate to arctic and in vegeta-

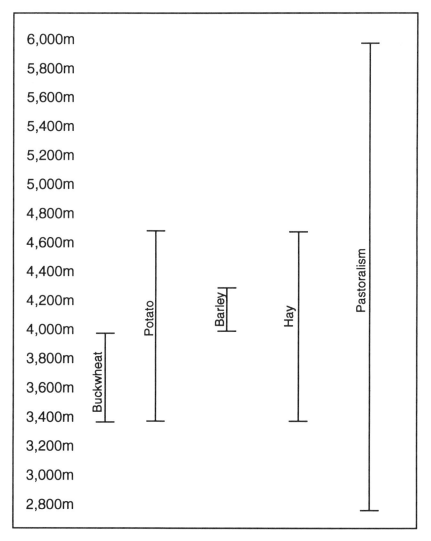

Figure 7.   Khumbu Historical Altitudinal Land Use

tion from temperate forests of pine, fir, birch, rhododendron, and juniper to subalpine forests of fir and rhododendron and high-altitude, alpine shrubland, and from temperate grassland to alpine tundra. Local knowledge of this diversity underlies agricultural and pastoral decisions in everything from terrace siting and crop selection to the timing of planting, the siting of herding huts, and the seasonal movements of herds of different livestock species. Khumbu Sherpa subsistence strategies emphasize the use of some local microenvironments more than others. The most intensive use is made of the sunny, south- and west-aspect slopes. Here substantial amounts of land even above 4,000 meters have been put into crop terraces and hayfields and considerable grazing is done on other land. Remnant woodlands are much used as sources of forest products, especially as sources of leaves for fertilizer, and they are also often valued as grazing areas. Much less use is made of north- and east-aspect slopes for farming or herding, but the extensive forests below 4,000 meters in these areas have historically been valued sources of timber, roof shakes, and fuel wood. At the higher altitudes the subalpine forest, shrub, and grassland ecosystems and alpine ecosystems of both north- and south-aspect slopes are used for grazing. Below 5,000 meters in these areas much land is also put to the cultivation of hay. Here the difference between sunny and shady aspects is less crucial. The historical emphasis on the use of south- and west-aspect slopes between 3,000 and 4,000 meters has apparently had a role in diminishing the extent of temperate and subalpine forest-woodland ecosystems and in creating the relatively large expanses of shrub and grassland in these zones.

Whereas Khumbu pastoralism, as already mentioned, is carried out through the full vertical range of altitudinal land use, agriculture is conducted in a much narrower range of altitudes. Most crop fields are situated between 3,400 and 4,000 meters with only a relatively small amount of land in higher-altitude fields. Hayfields, by contrast, have been established from 3,400 to as high as 5,000 meters. Multialtitudinal crop production is common. Individual families may own fields that span a thousand meters in elevation. There are four different altitudinal crop zones associated with different types of main and secondary settlements: the gunsa; main village; secondary, high-altitude agricultural site; and the high herding settlement or phu (map 6 and fig. 8). Multialtitude cultivation increases flexibility in labor scheduling, extends the growing season (thus increasing the amount of land that can be cultivated by individual families), expands villagers' access to land beyond the relatively scarce area in the vicinity of most main settlements, and reduces the risks of crop failure by dispersing staple crop production across several different microenvironments. The small altitudinal range between the different crop sites, however, means that no advantage is gained in terms of widening the local crop repertoire beyond buckwheat, barley, and potatoes. Families with gunsa land are able to get an early start on planting potatoes and harvest them before the later-planted, main settlement fields have matured.[26] Generally only a small amount of land is farmed at these altitudes and many families own no gunsa land whatsoever. The main emphasis is on fields in the main village and

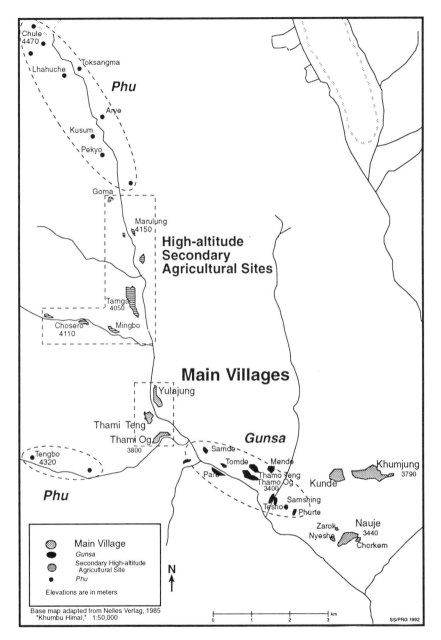

Map 6.   *Multialtitudinal Agricultural Sites, Thamicho*

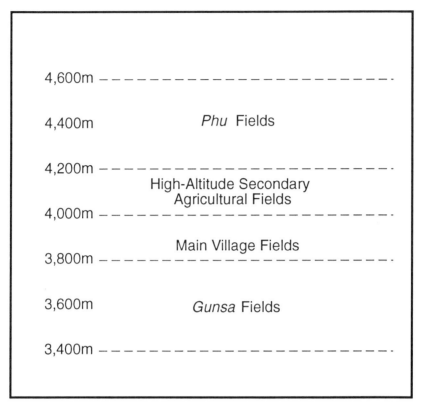

4,600m –·

4,400m   *Phu* Fields

4,200m –·
                    High-Altitude Secondary
                     Agricultural Fields
4,000m –·

                    Main Village Fields
3,800m –·

3,600m   *Gunsa* Fields

3,400m –·

Figure 8.   Thamicho Multialtitudinal Crop Production

especially those immediately adjacent to the house. Most households
have well over 50 percent of their total cropland in the main village.
High-altitude, secondary agricultural sites are the next most important
agricultural zones. Here large amounts of land are put to potatoes, and
at Dingboche a roughly equal amount of land is in barley. Formerly
buckwheat was also produced in some of these settlements. High-
altitude fields above 4,000 meters yield smaller harvests than lower-
altitude ones and usually only consist of a small patch or two of potatoes.
Harvests at these high herding settlements, however, release families
who spend a good part of the year with herds in the high pasture from
the effort of transporting as many supplies from the main village.

Distances between gunsa and high-altitude fields may be great enough
that in the often rugged terrain a day or more is required to traverse them.
Thamicho families with gunsa fields in the lower Bhote Kosi valley at
Thamo, for example, may have their highest fields eighteen kilometers up

valley at Chule or Apsona. Some Khumjung, Kunde, and even Nauje families have fields as far away as Dingboche, fifteen kilometers to the east. The distances involved often make cultivating high-altitude, secondary agricultural sites and high herding settlement fields from a single, main village base impractical, and many families shift instead back and forth between dwellings in different reaches of the valley a number of times during the growing season. These movements must also be coordinated with the requirements of herding, community regulations governing the timing of field work and grazing, and social responsibilities at two major summer festivals held in the main village and a third conducted in some high herding settlements.

Integration of Land-Use Activities

Life in Khumbu is a complex integration of the requirements of conducting crop production and pastoralism at the altitudinal limits of both (fig. 9). The use of forest resources is also interwoven with agropastoralism as well as making an essential contribution to the provision of fuel and shelter. Agropastoral practices must also be integrated with the seasonal rhythms of conducting long-distance trade and with the required commitments of time and energy for tourism work and business operations. I have not heard of any Khumbu Sherpa family during the twentieth century which specialized in trade to the total exclusion of agriculture, and no Khumbu Sherpa family today relies solely on earnings from tourism. Farming continues to be carried out by all households and ways are found to compensate for the scheduling problems and occasional labor shortages created by the demands of tourism employment. Even the families which have grown wealthiest from trade and tourism continue to place great importance on maintaining the cultivation of their land and many also keep livestock. Some are among the largest stockowners in their villages.

Khumbu crop production is intimately linked to pastoralism even for those families who own little or no livestock. Annual fertilization of potato fields is a fundamental principle of Sherpa agriculture and although everything from composted weeds and forest-floor litter to human waste is used, the most important soil additive is undoubtedly manure. This is so important that the route and timing of herd movements is decided in part on the basis of where household fields are located and when the optimal times are to supply them with manure. Families without sufficient livestock of their own either scour the slopes for dung or devote scarce cash to purchase it—in some places even doing so a year in advance in order to be sure of an adequate supply.

Crop production in turn contributes to pastoralism. Livestock are grazed in autumn on field stubble and fed fodder in winter, which includes

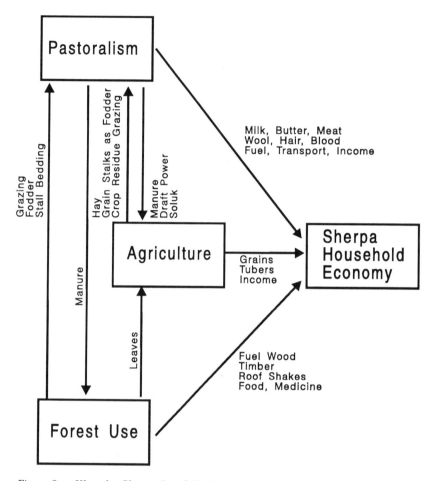

Figure 9.   Khumbu Sherpa Land-Use Integration

carefully dried and stored stalks of buckwheat, barley, and *danur,* a perennial plant believed to have medicinal and stimulative qualities for cattle. Large amounts of land and labor are devoted to the cultivation of hay that is grown on fields in the lower valleys and in some main villages as well as on a large scale above the range of productive potato cultivation.

Both crop growing and herding are further linked to forest use. Khumbu is at the upper limit of tree growth, and only its sub-4,000-meter, lower reaches support temperate and subalpine forests. Sherpas depend on these limited forest areas for timber and fuel wood and also rely on them for grazing and for *soluk,* dried leaves and needles scoured from the forest floor.[27] Open woodland in the vicinity of villages provides much-used grazing and browsing, especially in winter, so much so that its intensity may be affecting forest regeneration in some areas. Forest-floor litter is gathered twice a year and carefully used as stable bedding, after which the urine- and dung-rich material is stored until spring when it is dug into the fields. In some areas and for some families this constitutes a far higher percentage of fertilizer by volume than does manure. The intensity of foraging for leaves and needles gives the forest in the vicinity of villages a freshly swept look and in places every fir needle or birch leaf seems to have been removed. Both the gathering of dead leaves and needles and grazing were permitted even in forests that were traditionally strongly protected by Sherpas from logging or lopping.

Economic Differentiation

The social complexity of indigenous societies has often been neglected or underemphasized in cultural ecological studies. Adaptive strategies and land-use patterns have often been generalized as if regions, settlements, and even individual households maintained the same practices. Yet there may be tremendous variation in household wealth, status, and political power within settlements, as well as contrasts among villages. This may affect lifestyles, types and scale of resource use, land-use practices, access to private and communal land, and the maintenance of local resource-management systems—all of which also can have considerable ramifications for local environmental change. The analysis of local economic and political differentiation may thus be as important an element in the study of environmental degradation as is the examination of the historical impact of national government policies on land use and resource management.

There is not as much differentiation of wealth among Khumbu households as is typical in many parts of Nepal where tremendous disparities in land ownership are common. There is no class of landless Sherpa families or tenant farmers in Khumbu, although there are a few families who own very little land indeed. Nor are there any great estates here like those in some other Sherpa-settled areas such as Shorung and Chyangma. Most families own enough land to harvest a supply of potatoes sufficient for their annual requirements, although in the relatively frequent years of bad harvests they may have some shortfall. No family today owns more than thirty-five head of yak and nak and almost all families, regardless of their wealth, carry out a substantial part of all the labor on their lands.[28] All share the same types of local knowledge of environment and agronomy, the same assumptions about the culturally correct ways of working the land, the same technology and techniques, and the same basic strategies for coping with altitude and Khumbu microenvironments and their distinctive risks and opportunities. This makes it possible to discuss a "Khumbu Sherpa" pattern of agriculture and of subsistence more generally despite differentiation in wealth. Yet all Khumbu Sherpa families do not uniformly implement the same adaptive land-use strategies and practices. Differences in individual inclination and, to an even more important degree, differences in household labor, land, livestock, and fiscal resources, have led to several different ways of organizing agriculture and pastoralism. Individual families may practice more than one of these over a period of years as their resources change and make possible or impractical different subsistence practices.

Khumbu society has sometimes been regarded by casual visitors as remarkably egalitarian. Yet there are important differences of status, wealth, and power within communities and have been for many generations. There are such differences among Sherpas, between Sherpas and Khambas, and especially between both of these and Khumbu's few families of Hindu blacksmiths. And there are differences in power between ordinary families and those wealthy and powerful families Sherpas refer to as "big people." Differentiation of wealth is clear in diet, amount of fuel wood use, house size, style, and construction materials, and the amount of land and labor they control. It is also clear in subsistence strategies, not only in stockraising and in some cases in crop selection but also in such small but important things as when crops are planted and harvested and how much manure is put on them. Social differences are thus significant for describing Khumbu land use and exploring possible processes of environmental degradation both before and after the social and economic changes of the 1960s.

The wealthy lived better by Sherpa standards even during the pre-1960 era of "traditional" society. They occupied larger and more luxuriously furnished houses. They owned enough land to produce not only plenty for family consumption but also surplus for sale. If they wanted to raise yak they did, and could afford to maintain the necessary secondary herding huts and hay lands. They could hire sufficient labor to care for their fields and livestock. They had the capital to conduct the most lucrative types of trade with Tibet and to do so without the back-breaking labor of hauling their own loads over the pass. They ate better, being able to afford imported tea and sheep meat from Tibet and rice from lower-altitude Nepal. They dressed better, were better educated, and were better able to accrue merit and social status through support for religion. In many cases they were also the hereditary political leaders and generally had louder voices in village assemblies and in the smaller meetings of village "big people." Formerly there were relatively few such well-to-do families in the region, a disproportionate share of whom lived in Nauje and had acquired their wealth from trade. Tourism has led to general affluence in the region as compared with both earlier times in Khumbu and current conditions in neighboring regions. This has affected lifestyles and resource use across Khumbu. It has enabled a number of families who formerly were not among the "big people" to gain more local socioeconomic status and political power. Tourism income has not, however, ended regional economic differentiation. Some types of tourism work and business are more lucrative than others and certain families and villages have prospered more than others.

Wealth influences land use primarily through affecting crop-selection decisions. The amount of crop land a family owns is a crucial factor in its decision over whether to cultivate both grains and tubers or only tubers. Although there are some exceptions, most land-poor families tend to emphasize potato production. This tendency is spectacularly visible in the monocrop potato cultivation that has characterized Nauje where crop land has been more limited than in any other main village since the early twentieth century. Wealth may also influence decisions about the variety of potatoes planted, in that rich families can better afford to choose varieties that are highly regarded for their taste even if their yields are inferior to other varieties. Wealth may also affect the amount of manure and other fertilizers which can be worked into fields, since gathering manure tends to be more difficult for households that do not own large numbers of livestock or possess the wealth to purchase manure. Poorer households obviously cannot afford to hire agricultural wage laborers and indeed may have to postpone their own field preparation, planting, weeding, and harvesting if their poverty leads them to work as wage laborers in others' fields. It may not be merely good luck and fine agricultural knowledge that result in some of the wealthiest families in the region also having a reputation as good farmers. They often own some of the best land and can afford to take the best care of it.

Difference in wealth has also been a major factor in variation among households in pastoral practices and in changes through time in herding by individual families. During the past forty years fewer than 50 percent of Khumbu households have kept livestock and fewer still have practiced the yakherding with which Khumbu Sherpas are so identified and which is the local herding preference. There are rather at least four patterns of livestock ownership reflecting different degrees of wealth as well as types of knowledge, skill, and lifestyle preference. These pastoral patterns involve differences in herd structure and size, associated ownership of multiple-altitude herding huts and hay fields, use of hired labor or family labor specialization, patterns of movement in terms of altitude, periods and timing of absence from the home village, and impact on land in terms of areas affected and intensity. They may also lead to differing environmental impacts. Yak and especially nak herding has long been the preferred form of pastoralism, bringing with it great personal satisfaction in a culture that prizes these animals above all others as well as great social status. But yak and nak herding is a pursuit for relatively well-to-do households, requiring the ownership of high-altitude herding huts and hayfields and either the money to hire leaders or household members willing to take up a relatively difficult and lonely lifestyle for most of the year. The herding of cows and female crossbreeds (*zhum*) is primarily carried out by families who own only a few animals and make use of their milk and manure within the household. Male crossbreeds (*zopkio*) are the livestock of choice for many families in the lower parts of Khumbu, who are not heavily involved in other stockkeeping and want to invest in animals to be used to earn cash from tourism. The seasonal herding movements, grazing preferences, and fodder requirements of these different types of stock are distinct. The recent regional trend towards more families raising livestock and especially towards the keeping of small numbers of cows and crossbreeds has put greater pressures on some types of pastoral resources, especially on rangelands in lower Khumbu.

Different degrees of household involvement in trade and tourism have also made for enormous differences in cash income and wealth, which have also had repercussions for land use. Such wealth has historically enabled some families to acquire land and livestock, to hire herders, agricultural laborers, and woodcutters, and to consume more non-Khumbu-raised grains, meat, dairy products, and fruits and vegetables. Wealth from trade and tourism has allowed some families, particularly in Nauje and also now in Thamicho, to increasingly specialize their crop production by devoting their relatively small holdings exclusively to potato production. This means that all grain must be purchased.

The Demographic Cycle, Economic Differentiation, and Subsistence

Changing household demographics affect household wealth, resource use and land-use practices in Khumbu as they do in other societies based on subsistence agriculture and pastoralism.[29] Chayanov (1966), in his early work on Russian peasants, was the first to note that household financial and labor resources and economic goals were related to the ages of family members and the rearing of children. He discussed household subsistence economy in terms of a domestic life cycle in which the key factor is the number of producers relative to the number of consumers. This concept of a household demographic cycle has become basic to studies of the domestic mode of production.[30]

According to this view households are more constrained in their labor resources when children are young and then increasingly improve their capacity for agricultural production as children become older and begin to contribute more labor. Chayanov found that this extra labor capacity was not necessarily, however, fully employed. Once families achieved self-sufficiency in food they tended not to produce surpluses for sale but rather to devote more time to other (often noneconomic) activities. The demographic cycle is significant in Khumbu subsistence today. Sherpas, however, may be more inclined than Chayanov's Russian peasants to channel surplus household labor into profit-making economic activities, including trade, craft production, and tourism development.

It is not yet possible to analyze fully the role of the demographic cycle in Khumbu. At this time the necessary statistical data and the interview-based data on family decisions, the perceived economic utility of children, views on the appropriate work roles of children, inheritance practices, and other key factors are not available. Yet some pertinent observations can be made about the composition of Khumbu households and the relationship between the demographic cycle and subsistence and wealth. Theoretically each Khumbu family passes through this demographic trajectory, each with comparable experiences. In reality, of course, there are major differences due not only to family size and the chronological spacing of children but to luck, skill, inherited resources, and income-earning opportunities. Yet it is worthwhile considering the ways in which the demographic cycle can influence household fortunes and to keep in mind that temporal variation operates at the scale of individual household economies as well as in larger historical and regional settings.

Khumbu Sherpa households are mostly virilocal, nuclear families formed at the time of the completion of the final marriage ritual (*zendi*). After this ceremony a man comes into his full inheritance of land and livestock and moves with his wife into a house built for him by his parents in his home village. There are also stem families that result from the custom that the youngest son inherits his parents' house and the responsibility of caring for them in their old age. These stem families consist of the youngest son, his wife and children, and his parents. In cases where a couple has not had any sons another sort of stem household is developed. In this household one daughter remains in the home with her parents after her marriage and is joined there by her husband in an arrangement in which the son-in-law is referred to as a *maksu*. There are also a very small number of more complex households that are the result of polygynous and polyandrous marriages. In polygynous marriages each wife may be established in a separate house. Polyandrous marriages are usually entered into with two brothers, the spouses sharing a single household. In earlier eras the brothers divided family economic responsibilities, one often specializing in herding. The establishment of such a household counteracted the usual fragmentation of family land which results from the custom of dividing it equally among the sons. In the 1950s an estimated 8 percent of Khumbu Sherpa households were polyandrous (Fürer-Haimendorf 1964:68). This has greatly declined, however, and none has been established as far as I know within the past fifteen years.

Households are largely economically independent. Communities enforce some resource-use regulations, particularly for pasture and forest use, but otherwise seldom interfere in household decisions about subsistence practices and land use. Much agricultural work is carried out by reciprocal work teams formed by groups of women friends, but relatively little cooperative labor is organized by clans, lineages, or extended families. Kin are called on to assist in difficult times and for such major efforts as housebuilding, at which time a careful ledger is kept of contributions of materials, money, and labor so that this can be reciprocated when the occasion arises. Wage laborers are also employed by those households who have the means for everything from crop production to building, weaving, and household tasks.

The Khumbu demographic cycle begins at the time of the establishment of the household after the zendi ceremony. Sons, who until this time have contributed their labor to their parents' household, now inherit their share of land and livestock and establish their own households. The youngest son remains a part of his parents' household and gradually takes control of the house and land, the parents working with him as long as they are physically able to. This stem family generally stays together until it is dissolved by the death of the parents.[31] Occasionally an elderly couple decides not to remain as a part of their son's household and instead establishes itself in one of the family's gunsa or secondary, high-altitude agricultural site houses. Some old people go into religious retreat after their children's marriages and build a house for this purpose in a quiet place nearby the village or at the Tengboche monastery.

Young couples normally already have one and often two children at the time that they establish their own household. This reflects Sherpa cultural patterns of courtship and marriage, which involve a series of rituals and increasing commitments conducted over a period of years (see Fürer-Haimendorf 1979) during which time the prospective husband openly visits his fiancée at her parents' home and spends nights with her. Khumbu families thus begin their operation as a household with dependents. During the next few years additional children are usually born, and although infant and child mortality rates are high the family often increases to five or six members. During this first stage of the demographic cycle the number of consumers is high relative to the number of producers, since one pair of adults must care for several children. In this situation either agriculture production must be increased (e.g., by the expenditure of relatively high amounts of labor to increase yields by more careful and time-consuming manuring and weeding) or more wage income must be made to purchase food. In this stage of the demographic cycle there is considerable pressure on both the young parents, on men to make income for food purchases and on women to grow as much food as possible for the family as well as to care for infants. Today men work for wages in the tourist trade. Formerly they might have attempted to profit through trade or, if all else failed, may have emigrated with the hope of returning with more capital.

A second stage in the demographic cycle begins when the oldest child reaches the age of five to seven. Even at this early age children begin to contribute to the household economy and cease to be merely consumers. Young children care for still younger ones, freeing mothers for other tasks. Some play at gathering manure, forest leaves, and water with diminutive baskets and carrying jugs. By the age of ten children are contributing significantly to household labor. Boys gather water, dead wood, and manure and take animals out in the morning and bring them back in the evening following instructions given by their fathers. Girls haul water, cook, perform other housework, gather manure and forest leaves from nearby slopes, and sometimes help with the livestock. Many girls have already dropped out of school by this age and thus have considerable time to help out with household responsibilities. By the age of thirteen to fifteen both boys and girls are usually performing nearly a full adult share of household labor during school vacations (if they remain in school at all) as well as working before or after school. Even at the age of twelve some strong girls handle an adult's share of agricultural work and represent their family in reciprocal labor groups. By the time they reach fifteen years of age many boys do the full work of an adult and some even go trekking by the time they are thirteen to fifteen. By twenty all young people are performing full adult work for their families. Sons who begin wage labor in tourism donate their full salary to the household. The ratio of producers to consumers thus begins to change as the children grow older and does so much faster than in the Russian peasant families with whom Chayanov worked. Because the eldest child is usually one to three years old at the time the household is established the family begins in only a few years to escape from the early pressure on the adult couple to support and care for the entire household.[32]

Khumbu Sherpas today marry relatively late, usually not until after age twenty-five. Young men and women are expected to contribute their full labor to the household until the time that they marry and leave it and also to turn over wage earnings to the family. This means that the Sherpa household typically matures into an economic unit of four or more producers and remains together for a number of years after the children have reached adulthood. During this time there are opportunities for considerable increase in family wealth. Older males can be freed from herding responsibilities by younger ones and devote more time to wage earning in tourism, or the family may build up a larger herd now that it has the hands to care for it. The family women may be able to farm their own family land without participating in a reciprocal labor group or hiring wage laborers, and they may also engage in some wage-earning activities such as operating a lodge or shop or producing handicrafts for the tourist industry. And rather than depending on a single income from a tourism-related job the family may now profit from several, for the sons may contribute a decade or more of earnings to the family welfare. During this time the family may expand its land holdings and (more frequently) its livestock ownership, and may accumulate more capital for trade ventures or entry into the lodge or shop business. As children become engaged (the *demchang* ceremony is usually held three or four years before the final zendi marriage ceremony) the family labor situation may change slightly as young men begin to contribute some of their time to helping their fiancée's family with its herding, even to the extent of taking pack stock trekking and giving the income from this to the young woman's family. They help out in this way from six months or a year after the demchang engagement to the time of the marriage.[33] In some cases there is some reciprocity, and young women help their fiancée's families when asked to.

The family enters another stage as the children begin to marry. As sons wed and leave home they take with them an equal share of the land and livestock; daughters carry with them a dowry of cash, jewelry, and household possessions. The wealth a household has accumulated during the past twenty-five or more years begins to be fragmented, and seldom are the sons of wealthy men themselves wealthy simply through inheritance. If the sons share a wife in a polyandrous marriage this fragmentation is avoided. It may also be minimized if there is no son and a son-in-law is adopted into the family lineage and given the family house and property. In most cases, however, the nuclear family ultimately fragments.

A stem family in which a parental couple share a house with their youngest son and his daughter often has labor resources that are superior to those of the new nuclear families of the older brothers, for the parents may remain economically active for many years. The aging parental couple continues to work, and in former eras might still be major traders well into their 60s and 70s as well as sharing in the fieldwork and herding. Some men and women continue to make major income from the tourist trade into their later years, especially from operating lodges and shops. The stem family of the youngest son may ultimately be burdened more than advantaged, however, if the parents' health declines and they require much support in their old age.

3

Farming on the Roof of
the World

Khumbu crop production matches Sherpa knowledge and ingenuity
against the limited range of possibilities and the adverse conditions of
their homeland. The variety of crops that can be grown is severely
constrained by the short growing season at altitudes of more than 3,000
meters and even the most altitudinally fit crops are at great risk due to
frost, untimely rains, and crop pests and diseases. Sherpas bring to this
challenge many generations of experience with the microenvironmental
conditions of each Khumbu agricultural site and considerable knowledge of the characteristics, requirements, and capabilities of the crops
from which they live. Sherpas manage to cope successfully with the risks
and limitations of high-altitude agriculture by basing their farming on
local knowledge of both microenvironments and the performance of
cultigens in them. They consider Khumbu not so much a hard place to
survive as country where good harvests reward good farmers except in
the worst years. Yet for all their care they nevertheless experience frequent poor harvests. Entire crops of buckwheat can be lost, and as often
as one year out of three the potato harvest is likely to be at least a third
poorer than in a good year.[1]

The pattern of agriculture in Khumbu in terms of which land is cultivated and which crops are grown in it represents the decisions of both
today's farmers and the cultural and physical capital which they have
inherited from those who worked the land before them. Khumbu farming today reflects continuity of knowledge and practices across many
generations. Virtually all the terraces farmed today were created more
than a half century ago and many are much older. The long maintenance

of these terraces, the long and careful attention to maintaining and building soil fertility, and the many long-standing practices in crop selection and care all testify to continuity in farming. Agriculture has not, however, been a static art. There have been many changes in agricultural techniques and emphases even during the past century, as I will discuss in chapter 6. But the fundamental basis of Khumbu agriculture and the local knowledge that underlies it remain strongly based on cultural patterns developed earlier in the twentieth century and preceding eras, patterns of perception and land use which Sherpas have in part evolved out of the experience of life in Khumbu and which are attuned to the place, its seasonal rhythms, its climatic hazards, its topography, and its soil.

In this chapter I explore twentieth-century Khumbu Sherpa agriculture and its basis in environmental adaptation. Here I look particularly at the selection of particular crops and crop varieties for specific altitudes and sites, the use of multialtitudinal fields, and the fine-tuned agricultural calendar that is followed for different crops at particular localities. I also discuss other factors that affect agricultural practices and the annual cycle of crop growing, from the social organization of production, economic differentiation, and state policies to religious beliefs and efforts to enhance farming luck.

Altitude, Terrain, Climate, and Soil: Perception and Farming

Sherpa perception, categorization, and evaluation of microenvironments greatly influence their selection of crop sites and choice of crops and crop varieties. Especially important are the differences that are distinguished between soil types, precipitation patterns, and agricultural growing seasons. More than 97 percent of Khumbu is considered unsuitable for cultivation due to altitude, aspect, poor soil conditions, or excessive precipitation. Most of the remaining potentially arable land has never been put into cultivation and remains today in forest, woodland, and grassland. The amount of land that is actually in fields is far less than 1 percent of the total area of Khumbu, and nearly 20 percent of this small area, moreover, is in hay.

The preferred areas for crop production have an altitude of less than 4,100 meters, a south or west aspect for maximum daily sunlight during the short summer growing season, level or gently sloping sites, and loam or sand-loam soils. Although potatoes can be raised as high as 4,700 meters, almost no land is planted with them above 4,300 meters. The major centers of crop production, with the exception of Dingboche, are all located between 3,400 and 4,100 meters. Below 3,800 meters virtually every bit of relatively flat or gently-sloped land with a south aspect has long been claimed for agriculture.[2] Alluvial terraces, hanging valleys, and former lake beds at these lower altitudes are tightly patterned with small, stone-wall-girded, rectangular fields. Where adjacent slopes are only moderately steep these have often been carved into flights of terraces that may sweep across slopes in staircased tiers spanning up to 200 vertical meters of elevation. Terraced fields tend to be narrow and small; fields of 300–400 square meters are common and many are smaller still. Only on the rare level expanses are there larger fields, some of which are as much as 1,500 square meters in size. Considerable effort has gone into the creation of these terraces, for each is built with stone risers (*tsigpa*) and stone walls are also usually raised around the individual fields to protect them from livestock.[3] Most are constructed on relatively gentle slopes of less than ten degrees, although in some localities slopes of twenty degrees have been terraced.

Soil type and quality are major concerns of farmers and the focus of a great deal of labor to maintain and enhance. Regional soils vary with altitude, geomorphic and vegetative cover history, and the underlying country rock from which they have developed. The region is not blessed with exceptionally fertile soils. Most have developed relatively recently from gneissic and granitic parent material. The alpine soils of the high valleys are generally quite recent and thin, having developed in reaches of the region which were affected by glacial processes during the not-distant past. These are primarily entisols and are often less than sixty-five centimeters in depth. Below 4,000 meters spodosols have developed in extensive areas that are or were forested with a mixed fir, birch, and rhododendron subalpine and temperate forest, especially on northern-aspect slopes. The extensive grassland and shrubland areas that now cover large areas of southern-aspect slopes below 3,750 meters have inceptisol and entisol soils (Byers 1987c:210).[4] Land currently in terraces was presumably claimed for crop land from areas that were previously grassland or historically cleared forest and woodland. Some important agricultural sites, however, including Thami Og, Thami Teng, Yulajung, Tarnga, parts of Khumjung, and Dingboche, are also located on sandy, nutrient-poor, former lake sites and alluvial terraces.

Sherpas have developed a body of knowledge and techniques concerned with evaluating and conditioning soils. They discriminate between several soil types on the basis of color and texture and use this categorization in assessing land value and deciding on appropriate crop selection and manuring for fields (table 3). Across Khumbu the preferred soil is *sa nakpu*, literally "black soil." This is the black, humus-rich loam common in main fields in the villages and also in some outlying areas where care has been taken to provide plentiful manure. Black soil

Table 3.   Khumbu Soil Types and Characteristics

Name	Local Defining Characteristic	Evaluation
Sa Nakpu	black, neither sandy nor clay	excellent
Sa Seru	yellow, more sand	good
Pemi Sa	sandy	fair for potato
Dambak Seru	yellow clay	poor
Dambak Nakpu	black clay	poor

is the product of generations of effort and assiduous application of manure and other fertilizers, and indeed the soil color of village fields reflects well their manuring history and generally varies strikingly between fields adjacent to the house and those at the edges of the settlement. More common than black soil is *sa seru,* "yellow soil," less rich in humus and usually higher in sand content. Though generally considered to be less desirable, it can yield good potato harvests if adequate manure is applied. Sherpas point to the high potato yields of some fields in Thami Og, Tarnga, Phurtse, and Khumjung as an indication that yellow soil can be productive.

Several other soils are less highly regarded. These are *pemi sa,* "sandy soil," and *dambak seru,* "yellow clay soil." Sandy soil areas are not necessarily shunned for agriculture and can produce good potatoes when properly manured. Some areas of conspicuously sandy soils in the Bhote Kosi valley produce potatoes renowned for both taste and yields. Yellow clay soil, however, is considered hopelessly poor. Black clay soil is nearly as bad. A few fields in Khumjung, Thami Og, and Phurtse have black clay soil and they are considered to produce sparse, stunted crops of buckwheat and subnormal yields of potato. There are also some clay soil fields in small areas of Dingboche, Pangboche, and minor sites such as Nyeshe near Nauje. All of them are considered to be very unproductive.[5]

The general characteristics of Khumbu climate have already been introduced in chapter 1. From the standpoint of crop production, microclimate variation on a much more intimate scale becomes important. Temperature and precipitation vary within the region with altitude and topographic situation, giving Khumbu a rich variety of microclimates. Sherpas believe that climatic conditions within Khumbu vary nearly as much as soil conditions. They identify differences in precipitation among valleys, precipitation and temperature differences between lower valleys and upper valleys, and high rainfall pockets in certain areas within valleys. The Bhote Kosi valley is considered to be drier than either the Dudh Kosi or Imja Khola valleys. Areas above 4,000 meters are considered to be drier than lower-altitude areas. Some places such as Pulubuk and Nakdingog in eastern Khumbu are notoriously rainy and are considered unsuitable for cultivation. Differences are recognized in some cases between the microclimates of places that are scarcely a kilometer apart and these perceptions influence decisions about the extent and timing of crop planting. Precipitation is one factor here, but temperature and frost patterns are even more important. The role of aspect in microclimate is recognized, and shaded slopes are usually shunned as crop sites. Altitude is also considered to affect both precipitation and temperature and is taken into account in the selection of crops and crop varieties for particular sites.

Sherpas have developed a large body of knowledge about climatic threats to crops. Particular crop areas known to be susceptible to particular types of climatic problems and particular crops and crop varieties have reputations for being at different types of risk in different parts of Khumbu. This knowledge is reflected in local planting schedules and cultivation practices, as well as in religious and other protective measures.

Frost (se) is the most common climatic risk across Khumbu. It is a major spring problem throughout the region, and unusual freezing conditions in early autumn can devastate buckwheat crops.[6] The severity of risk to crops can be judged from the frequency of frost-diminished harvests during the last few years and the extent of damage. Between 1981 and 1987 frost caused major crop loss in some main villages three times.[7] One farmer from Thami Teng explained that frost affected crops there in most years, although the amount of crop damage varied considerably from year to year. A single night's frost during these critical spring weeks can lower potato harvests by a third and cause such striking damage to buckwheat fields that many farmers replant entire fields.[8]

The greatest risk from frost to both potato and buckwheat is usually in April and May. At low-altitude sites (in Nauje and in gunsa) April frosts are the main problem and are rare in the following month. For the other main villages and the phu there is greater danger in May. Here April frosts are no threat due to the later date of planting. In the phu frost danger continues even into July and begins again in late summer. Often there will be several days of frost in succession.[9] Farmers recognize that early planting increases the risk of frost damage, but planting cannot be delayed long at altitudes with such a short growing season. Planting potatoes later will yield smaller tubers at harvest time. And the later that buckwheat is planted the smaller the crop is likely to be, for with later harvests comes greater danger from early snows. A heavy September snow can wreak havoc with buckwheat in the higher settlements. Such a snow in 1968 caused great damage to Pangboche and Phurtse buckwheat despite the villagers' emergency harvest efforts.

Frost problems are considered to occur most often when a clear night follows a rainy spring afternoon. The 1986 frost, for example, was described by a Nauje man as coming on the full moon of the fourth Sherpa lunar month, *Dawa Shiwa* (generally in May), when after a rainy afternoon the night was clear. That frost affected crops in the Bhote Kosi valley from Tarnga to Phurte, and also damaged crops in Pangboche, Phurtse, Khumjung, and Kunde. Heavy frost can overnight transform fields that were green with young crops into brown, withered plots of shriveled yellow and black plants resembling, as a Khumjung woman put it, "dry tobacco leaves." Although frost damage is usually associated simply with unfortunate weather conditions many Sherpas believe that these can be triggered by human actions. A particularly bad frost in 1981 in the main Thamicho settlements, for example, is said to have been caused by an improperly performed crop-protection ceremony. On the very night that the annual rite of circumambulation of the fields was performed a severe frost caused considerable damage. The ritual circumambulation has not been performed since.[10]

Some farmers believe that certain crops in a given place are more affected by frost than others. In Phurtse one woman, for example, contended that differences in soil may influence losses from frost or the lack of them. She noted that buckwheat grown on good black soil produces a harvest even when it is hit by repeated frosts whereas crops on drier yellow soil at the edge of the village yield little under those conditions. Other farmers have suggested that different varieties of a single crop may respond differently to frost conditions. Some elderly Sherpas recall that one type of potato grown early in this century was much less vulnerable to frost than another popular variety. Some people say that there are differences in the frost hardiness of two of the common varieties grown today. They suggest that the red potato may be more vulnerable to frost than the yellow potato since it develops a stalk earlier, but note that even the red potato usually survives a single frost. It is also said, however, that the yellow potato flowers earlier and thus suffers more from frost than other potato varieties.

Untimely rains are also considered to be a major agricultural problem. Rainfall varies from year to year in timing and intensity. Drought is not considered a major hazard. A delayed monsoon can result in poor crops, especially in the earlier-planted gunsa fields, but no major crop failures due to drought are remembered. A greater regional problem is an overabundance of rain that can lead to poor crops in the main villages.[11] The greatest risks result from intense multiday rains. According to Sherpas there are two such intense rainfall periods in Khumbu. *Yerchu*, "summer rain," is a July–August (*Dawa Tukpa–Dawa Dimba*) rain that lasts five to seven days. Sherpas describe it as a period in which the cloud ceiling is unusually high and the sky atypically light, but during which very heavy rain falls. *Tenju,* "autumn rain," is a very heavy September–early October (*Dawa Gepa–Guwa*) rain that lasts up to a week (although some people say that it lasts only two to four days). Both are rare and may fall only once or twice in a decade.[12]

Yerchu rains are usually not a problem for crops, although some Sherpas feel that they can cause buckwheat to develop large stalks and leaves but produce rather poor seed. It is also possible that in unusually moist summers there is more risk of accelerated spread of fungal blight (*shimbak*). Tenju rains, however, can cause considerable damage. Terrace walls occasionally buckle following such storms, and in the spectacular 1968 case so much runoff occurred that two houses were damaged in Nauje and many villagers abandoned their homes and spent a night huddled with their most prized belongings beside the supposed protective influence of the main shrine (chorten) of the village. The same storm caused much damage to buckwheat in Pangboche and Phurtse. Tenju rains are said to have destroyed buckwheat at Pangboche and to have also damaged Dingboche barley. Autumn rains are not considered a threat to potatoes, but they can cause damage to harvests by seeping into open potato-storage pits and rotting the stored potatoes. Autumn rains of any scale, of course, can also cause havoc with hay making.

Besides their familiarity with local soil, climate, and weather Sherpas also bring to their farming a keen sense of other environmental risks to crops. Foraging mammals and birds (including Himalayan tahr (*Hemitragus jemlahicus*), Impeyan pheasants, choughs, and in lower Khumbu—although rarely today—bears and monkeys) can cause crop losses. Certain places in Khumbu are well known for the frequency of their crop losses to particular types of wildlife.[13] The destructive potential of livestock is also recognized, and hence the long-standing practice of banishing stock from villages (and also from some secondary crop-growing areas) during the height of the growing season. A yak or crossbreed can graze a buckwheat or barley field to ruin in a few hours. Livestock are said to damage potatoes even in the final weeks before they are harvested since their trampling causes the tubers to rot below ground. There are no serious insect problems, by contrast, although farmers in the lowest-altitude areas of Khumbu sometimes lose some potato plants to a worm that severs the stalks.

Plant diseases are of much more concern. Khumbu potatoes are affected both by late blight and warts, and blight is also said to infect buckwheat and barley.[14] The greatest problem is shimbak or late blight (*Phytophthora infestans*), the fungal disease that was responsible for the Irish potato famine of 1846–47.[15] Late blight is a major problem today in most of the potato-producing areas of the Himalaya.[16] Khumbu communities have taken extraordinary precautions in attempts to stave off the onset and spread of blight, believing that by observing a set of bans on activities believed to affect the well-being of plants that they could prevent the outbreak of the disease or limit its impact.

The leaves of potato plants that have been affected by blight wither and blacken as early as June. The blight is usually at its height in July and August (Dawa Tukpa to Dawa Dimba).[17] Entire Khumbu valleys can be swept by blight and in bad years crop losses can be as much as 50 percent. Throughout the twentieth century blight has been a continuing problem. Serious infestations typically occur several times per decade. In recent years there have been problems in 1983, 1987, and 1990.

Some people believe that heavy rain can encourage the outbreak of blight, particularly when it falls in late May or early June (*Dawa Nawa*). Sherpas also believe that blight can be transmitted by human or livestock contact with plants during the summer and hence communities have banned farmers from entering the fields from late June or early July until harvest and have also banished livestock from the villages during this period.[18] There are also a number of other restrictions that were once carefully observed in the villages and enforced by village officials. These included bans on bringing freshly cut timber, fuel wood, or bamboo into the settlement, construction work, drying herbs and leaves outdoors, or firing guns. These various antiblight regulations are not observed by Sherpas in Shorung and other areas and may have been unique to Khumbu.

The importance of keeping stock away from the maturing village crops in order to protect them from blight is taken very seriously. In Khumjung and Kunde, where this ban is enacted soon after the Dumje festival, some people predicted that there would be trouble with blight in 1987 because Dumje was scheduled to be held late that year due to an unusual counting of the months in the Sherpa calendar to avoid an astrologically inauspicious condition. Within those villages there was much debate over whether Dumje should be held as customary at the beginning of the fifth lunar month in order to be able to close the village to livestock as usual in June rather than in July. Konchok Chombi was concerned that "there will be blight if we wait until then to get livestock and people away from the fields." Such advice was not heeded and that summer the blight was the worst it had been for many years.

The viral disease *Synchytrium endobioticum* (Sherpa *kongsur re, ze shur re,* or *ne zakpa*), often called potato wart, is a problem in much of eastern Nepal and other areas where Darjeeling varieties of potatoes are cultivated. The disease causes crusty, scab-like areas on the potato skin, and infected potatoes can rot. Once the virus infects an area it can become established in the soil and is difficult to eradicate, and it is capable of destroying entire crops. In Khumbu the problem is especially severe at Dingboche where some people associate it with the unique emphasis on fertilizing fields with juniper needles, suggesting that this alters the soil in a way that leads to the onset of the disease. Infected potatoes also occur in other communities, however, where juniper needles are not used as fertilizer. When the problem is discovered at harvest time the usual practice is to separate infected tubers from the rest of the harvest. These tubers are then consumed first before they deteriorate further. It is considered to be very important not to mix wart-infected potatoes with good ones in the storage pits, for it is thought that the infection can spread to previously unaffected tubers.[19]

Specialization and Diversification: Crop Selection and Multialtitudinal Fields

Both specialization and diversification can be defenses against disaster in areas of high agricultural risks. Risk can be appreciably lowered by specializing in crops and crop varieties that have been found to be particularly fit for local conditions. In some parts of the world farmers even deliberately select relatively low-yielding crops and varieties to plant rather than others with which they are familiar because they are likely to give some yield even in the worst expected conditions. Diversification may also be an effective risk-minimization strategy. Diversity can be achieved in a number of ways. At the level of the individual field farmers may plant a variety of crops and varieties rather than a monocrop. Thus five types of rice, not a single one, might be planted in a field, or several different types of maize might be grown intercropped with millets. Multicropped fields provide a buffer against risks as a result of the different climate, insect, and disease tolerances and resistances of different varieties. They may also decrease the spread of disease and provide barriers to the dispersal of species-specific insects. At a higher level of the farming strategy a household might choose to plant a number of different crops in their various fields rather than to emphasize a single staple. This prevents disaster when a particular crop is laid to waste by disease, avoiding a catastrophe such as the Irish potato famine that was based on overreliance on a single crop. A household may also choose to cultivate a given staple in plots at a number of different locations with different microenvironmental conditions rather than only in the main village. By doing so they may well avoid loosing everything to drought, heavy storm, frost, hail, or disease even when these devastate some sites. The heavy rain that destroys a crop in one part of the valley may never fall in another area nearby. And a farming family can also reduce risk at a still broader level by diversifying their household economy to avoid the need to achieve self-sufficiency in food production. This can be done in many ways, from devoting some of their land and labor to producing cash crops and other agricultural products to engaging in nonfarm labor and other enterprises.

Khumbu Sherpas employ subsistence strategies that emphasize both specialization and diversification. They specialize in the most productive crops for high Himalayan conditions (potatoes, buckwheat, and barley) and have emphasized potatoes, the most productive of all, to a highly unusual degree. They have a somewhat narrow scope for diversification in the severe microclimates of their homeland, but they do practice some forms of it. Formerly families generally cultivated several staple crops, including a set of different tubers rather than simply potatoes, and despite the increasing monoculture of potatoes today many families continue to cultivate both grains and tubers. Although farmers may emphasize one or two varieties of potatoes, they commonly grow several others as well and are well aware of their altitudinal fitness, climate hardiness, and disease resistance. Some people intercrop radishes with potatoes. Typically families cultivate fields at a number of different sites around the village and many also produce crops at more distant sites at different altitudes. And Sherpas have also long integrated agriculture into a broader economic base which frees them from a need to depend solely on Khumbu harvests for sustenance.

Multialtitudinal crop production is a risk-minimization strategy shared by many mountain peoples. Poor crop-growing conditions at one site, such as drought, excess rain, crop disease, or insect pests, may not affect other sites with different microclimates in the same way, diminishing the risks of crop failure at any particular site. Good yields in some sites may compensate for poor ones in others. Multialtitudinal crop growing is a basic feature of agriculture for many (although far from all) Khumbu families. It is common to plant potatoes in fields at a number of different elevations in gunsas, main villages, secondary high-altitude, and high-herding settlements. Farmers testify that yields at different altitudinal sites often vary considerably from year to year. A late monsoon that may seriously affect the earliest-planted potato fields of the gunsa settlements may foster bumper crops in secondary high-altitude fields planted a month later. A year of higher than normal rainfall such as occurred in 1990 may support unusually fine crops in the relatively drier upper valleys while lower valley crops do poorly. The frost which affects low-altitude crops in May may not touch higher-altitude crops that have not yet germinated. Microclimatic conditions and crop performance may also vary in these cases among fields at the same altitude but on different sides of a valley. Differences in the amount of sunlight received at these sites influences soil temperature and available soil moisture.[20] The de-

gree of steepness of slopes can also be an important factor in soil moisture conditions.

Multialtitudinal crop production also has several other advantages.
Labor can be scheduled across a longer agricultural season than would
be the case if only main village fields were cultivated (Fürer-Haimendorf
1975:28).[21] Planting potatoes in the high-altitude herding settlements
saves the trouble of transporting the quantities of food to those places
from the main villages that would otherwise be necessary. And a family
that has not been able to acquire enough crop land in the main village
area where land is often scarce and too expensive to meet its needs may
find that higher valley land is their only alternative. Many Nauje families
have bought land in Bhote Kosi valley secondary high-altitude agricultural sites over the last decade for this reason.

Despite the multiple benefits of multialtitudinal crop production most
families do not farm fields in the full range of altitudinal sites. Most have
land in several sites in the main village and its immediate vicinity and
in one of the major, secondary, high-altitude agricultural sites. Most
Thamicho families also own gunsa lands, but this is not common today in
the other villages and only a small percentage of families has crop fields
in the high-herding settlements.[22] The bulk of most families' harvest
comes from main village fields. Usually well over half of a family's
harvest (often 75 percent or more) comes from main village fields. A
family that owns less than 25 percent of its cropland at different altitudes
generally cannot hope to compensate fully for a poor harvest year in the
main village. The relatively great concentration on main village farming
today may, however, have been largely a twentieth-century development. In the nineteenth century, when the Khumbu population was
much smaller and before the introduction of higher-yielding varieties of
potatoes, it may have been more typical both to have more total land per
household in cultivation and for these holdings to be more evenly distributed altitudinally.

Khumbu Crops

The range of crops that can be cultivated at Khumbu altitudes is
limited to a rather small number of Himalayan staples. As was discussed
in the previous chapter, even the lowest Khumbu villages and gunsa are
too high in altitude for the cultivation of wheat and maize and are well
over 1,000 meters above the highest-altitude Dudh Kosi valley sites
where rice and millet are cultivated.[23] Khumbu Sherpas have accordingly
long based their farming on high-altitude-fit buckwheat, barley, and tubers. No green vegetables are grown as field crops. Small amounts of
mustard (*pezu*) are grown in household gardens for their green leafs and

Table 4.   Khumbu Crops

Common Name	Sherpa Name	Nepali Name	Latin Name
Barley	*na*	*ua*	*Hordeum vulgare*
Buckwheat	*tou*	*tito phapar*	*Fagopyrum tataricum*
Potato	*riki*	*alu*	*Solanum tuberosum*
Radish	*lo*	*mula*	*Raphanus sativus*
Turnip	*tulu*	*salegam*	*Brassica rapa*
Mustard	*pezu*	*rayo saag*	*Brassica juncea*
Garlic	*gokpa*	*lasun*	*Allium sativum*
Jerusalem artichoke	*ge riki*	*gane suryamukhi*	*Helianthus tuberosus*

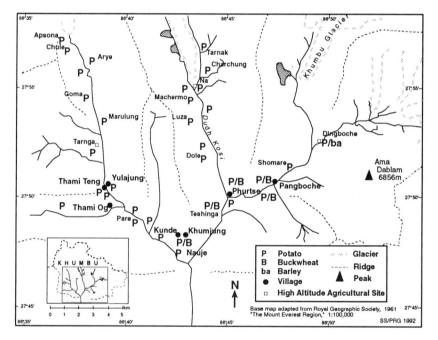

Map 7.   Crop Patterns, 1987

several types of garlic (*gokpa*) and chives are grown in window pots and gardens. Recently a few families have begun growing cabbage, carrots, and cauliflower in garden plots in some of the lower settlements.[24] Fruit growing is still less important. Although several families have attempted to raise apple trees it has not yet proved possible to produce fruit in Khumbu. The current regional distribution of crops is shown in map 7.

Grains

Buckwheat is today the most important grain grown in Khumbu in terms of the amount of land planted to it and its place in the regional diet. Sherpas, like Tibetans, esteem barley far beyond buckwheat as a food and barley is equally as altitudinally fit for Khumbu conditions. Yet barley is grown only on a very small scale in Khumbu, probably due to the need to irrigate it in May and early June and the scarcity of easily irrigatable sites. Buckwheat cultivation requires no irrigation, and formerly it was grown in all the main villages and in the high-altitude, secondary, agricultural site of Tarnga. Today it is grown in half of the main villages of the region.[25] Barley, by contrast, has since the early twentieth century been cultivated only in the high-altitude Imja Khola settlement of Dingboche (4,300m). This is one of the highest altitudes at which grain production has been reported in the Himalaya or Tibet. Buckwheat is today cultivated as high as 4,000 meters in Pangboche and was formerly grown at a similar altitude at Tarnga in the Bhote Kosi valley (table 5).

The Khumbu varieties of both buckwheat and barley are Tibetan varieties. The barley grown at Dingboche is a naked black barley known in Sherpa as *na* (Nepali *ua*). This is grown on a small scale in the neighboring Tingri region of Tibet. It is quite distinct both from the bearded white barley grown in Pharak (Nepali *jou*) and from the white barley which is the staple in Tingri.[26] Khumbu buckwheat is the Tibetan variety, Tartary buckwheat (*Fagopyrum tataricum*) that is known in Nepali as "bitter buckwheat" in contrast to the "good-tasting buckwheat" (*Fagopyrum esculentum*) of the lower altitudes. The latter variety is the buckwheat most familiar worldwide. The Tibetan variety, however, is able to withstand cooler temperatures better and Sherpas consider it the only one fit for Khumbu conditions. In Khumbu two varieties of the Tibetan type are cultivated, one white and one black. The white variety is by far the most common. In neighboring Pharak, however, the black variety predominates.

Both buckwheat and barley are rotated with tubers in a two-year sequence. A field devoted to grain one summer is planted in potatoes the next spring. Barley has the longer growing season and is planted in the first days of April, six or more weeks earlier than buckwheat. It is also harvested earlier, in late September rather than in early or mid-October as buckwheat is. The two crops also vary considerably in the care they require. Buckwheat flourishes on relatively nutrient-poor soils and without any irrigation. It alone of all the Khumbu crops is normally not fertilized. Barley, by contrast, is very carefully fertilized and is the only irrigated Khumbu crop.[27]

Table 5.   Historical Altitude Ranges of Khumbu Crops

	Lowest Altitude Grown	*Highest Altitude Grown*
Grains:		
Barley		
(white)	4,050m (Tarnga only)	
(black)	4,380m (Dingboche only)	4,380m (Dingboche only)
Buckwheat	3,400m (Nyeshe)	4,050m (Tarnga)
	3,985m (Pangboche)	
Maize	2,800m (Jangdingma)	2,800m (Jangdingma)
Wheat	3,400m (Tashilung)	3,400m (Tashilung)
Tubers:		
Potato	3,400m (Nyeshe)	4,690m (Tarnak)
		4,753m (Chukkung)
Radish	3,400m (Nyeshe)	4,380m (Dingboche)
Turnip	3,400m (Nyeshe)	4,480m (Chulungmasur)
Jerusalem artichoke	3,400m (Tashilung)	3,600m (Samde)
Other crops:		
Peas	3,440m (Nauje)	4,050m (Tarnga)
Rayo sag	3,440m (Nauje)	4,753m (Chukkung)
Garlic	3,440m (Nauje)	4,753m (Chukkung)

The special treatment devoted to barley reflects the high value that Sherpas place on it as a food. It is also considered to be suitable for use in religious ceremonies and offerings. Fields at Dingboche that yield good barley crops are considered especially valuable property and ownership of them is cause for pride. Although most of the fields of Dingboche belong to families from the nearby village of Pangboche there are also fields owned by Khumjung, Kunde, and even Nauje households. It is extremely difficult to purchase such a field for owners seldom offer them for sale. Indeed, even to be offered the chance to buy Dingboche barley from a family that may have a surplus is considered to be a mark of friendship and favor. In the early twentieth century there was so much concern over the quality of barley harvests that Sherpas hesitated to plant less spiritually pure crops such as buckwheat and potatoes in the same settlement where barley was cultivated for fear of offending the barley crop and losing the harvest. It was only in the twentieth century that this self-imposed ban was broken at Dingboche. Barley is also the only Khumbu crop that is associated directly with divinity: there is a local god of barley whose seat is the beautiful barley-grain-shaped snow peak of Cho Polu that overlooks the barley fields of Dingboche. Buckwheat has no such religious associations but is instead regarded as an inauspicious grain that can be used ritually only to hurl at ghosts in an effort to drive them off. Puffed buckwheat can also be used in a ceremony to appease feared spirits (*saptok*) who dwell in boulders. Even white buckwheat is inauspicious, for the shape more than the color of the grain is the issue. Buckwheat grains have three sides, a number which is considered very unlucky by Khumbu Sherpas.[28]

Buckwheat is ground into flour and eaten as a thick porridge (*sen*) and as an unleavened, *chapati*-like flatbread. Barley is processed into *tsampa*, the distinctive staple food prized throughout the Tibetan culture region. Tsampa is a flour produced through a multiday process of soaking, drying, and finally popping barley grains in a pan of heated sand held over an open outdoor fire and then grinding the puffed barley into a flour in a water mill. Sherpas eat tsampa in three forms. It can be mixed with salt-butter tea as a porridge (*chamdur*), kneaded with a small amount of tea into a paste (*pak*), or thrown dry by the spoonful down the throat (*chamgagyou*). Sherpas treasure the uniquely rich taste of tsampa and take great pains to produce and procure the best quality. Villagers perceive enormous differences in tsampa depending on the care with which it is made and on the type and source of the barley used. Black barley tsampa has a different color and taste from that of white barley, and in Khumbu Sherpas' judgement the best quality tsampa requires that dark Dingboche barley must be mixed with small amounts of Pharak-grown white barley. Black Dingboche barley is widely considered in Khumbu to have a more superior taste than that imported from Tibet or Pharak.[29]

Pangboche villagers note that buckwheat grown there has yields similar to Dingboche barley. Yields, of course, vary enormously from year to year due to the high susceptibility of grain crops to damage from bad weather and livestock. The return on the volume of grain planted as seed is from four to eight times.[30] A Dingboche barleyfield of 1,500 square meters planted with ten *pathi* (approximately twenty-six kilograms of seed) may yield four muri (approximately 9.6 bushels and 208 kilograms of grain at 52 kg/muri) in a superb year. Two to three muri is more typical. In the best of conditions, yields are thus only 0.14 kilograms per square meter on large barley fields at Dingboche. In Pangboche some fields produce as much as 0.7 kilograms per square meter of buckwheat in the best of years. But even this is a meagre harvest in comparison to the two to four kilograms per square meter common for potato cultivation across Khumbu. Few families own enough grain land to harvest more than a few muri per year. In 1987 five Pangboche families averaged 1.65 muri each of buckwheat production and to this could be added another two muri of barley—well under the ten to fifteen muri of grain typically consumed by Khumbu households. These figures reveal the continuing Khumbu need for obtaining substantial stocks of lower-altitude-grown grain.

Tubers

Khumbu Sherpas have emphasized the cultivation of tubers as well as of grains since at least the late nineteenth century and very possibly for considerably longer. For a hundred years or more tubers have very likely been grown on half or more of Khumbu crop land and during the twentieth century they have increasingly dominated Khumbu agriculture. Four tubers are planted: potato, radish, turnip, and Jerusalem artichoke. A fifth, which Sherpas call *to*, is harvested as a wild semicultivate.[31] The potato (*riki*) is today the most important by far of these, and is indeed the primary Khumbu crop. Although the other tubers are raised only on a quite minor scale today, they were more important in regional agriculture before the 1930s. The white, turnip-like, Tibetan variety of radish (*lo*) that is grown as high as 4,000 meters has long been a valued food and fodder crop and was once even dried for sale to nearby Rais. Today it is a minor crop grown only as an intercrop in a few potato fields. The Tibetan variety of turnip (*tulu*) was a field crop in the early twentieth century, but today is only grown in a few gardens. Jerusalem artichoke (*ge riki*) is cultivated only in a few fields in Bhote Kosi–valley gunsa by families who use it to produce a particularly potent alcohol.

The potato is by far the most important crop today in Khumbu. It is cultivated from the lowest gunsa to the highest herding settlements where crops are planted and flourishes as high as 4,700 meters. More than 75 percent of all land in food crops is in potatoes. In half the villages of Khumbu it is the sole food crop grown and in the others it is grown on 50 percent or more of the crop area. Potatoes dominate crop production at higher altitudes. Except in Dingboche they are the only food crop in the secondary, high-altitude agricultural sites and the high-herding settlements. Even in many gunsa they are today the only crop raised. Nearly all of this production is for Khumbu consumption and the potato is the central staple of Khumbu household sustenance.

No food in Khumbu is as basic as the potato. Potatoes form the basis of virtually every meal and almost every dish. Even the most common snack is a bowl of boiled potatoes. Most potatoes are eaten boiled, served in their skins (which diners then peel and discard) and dipped in salt or hot pepper, yoghurt, and garlic sauces.[32] Potato pancakes (*riki kur*) are popular, prepared by grating uncooked potatoes on a ribbed slab of stone, mixing these in a batter with buckwheat flour, and then cooking the pancakes on a flat slab of slate over a wood fire and serving them with nak butter and yoghurt sauces. A form of mashed potatoes (*rildok sen*) served with bowls of a sharp cheese soup is also popular. Potatoes usually are the main ingredient of stew (*shakpa*) and potato curry is the most common accompaniment to rice. They are also distilled into alcohol, sun dried for use in stews or for trade to Tibet, and are made into a flour (*riki karruk*) through a process of mashing, drying and stone-grinding.

The degree to which Khumbu Sherpa agriculture is today based on potatoes is probably unique in the Himalaya. Elsewhere it is more common to emphasize grains as much as or more than tubers and to grow other tubers besides potatoes. There is no simple climatic or edaphic explanation for why the peoples of high-altitude central and northwestern Nepal, including those of Mustang, Dolpo, Mugu, and Karnali base their agriculture on the cultivation of barley, wheat, and buckwheat more than on tubers and often emphasize the cultivation of Tibetan varieties of radish more than potato. Potatoes would very probably be well suited to these regions. Their lack of importance may reflect better conditions for cultivating irrigated grain crops in these regions as well as a possibly later date of potato introduction and diffusion.[33] The strong Khumbu Sherpa emphasis on potatoes, however, reflect many factors other than the region's relatively poor irrigation possibilities and longer familiarity with the crop. Local interest in agricultural intensification apparently played a key role in the historical process of focusing Khumbu agriculture around monocultured potatoes. The process took several generations to develop and is discussed in chapter 6.

Today at least nine varieties of potatoes are grown in Khumbu, and during the twentieth century Khumbu Sherpas have introduced at least fifteen varieties (table 6).[34] Sherpas have named local potato varieties primarily on the basis of their color, although other qualities such as shape and even assumed source of introduction can be used, as can be seen in table 6. Varietal names can differ among valleys, villages, and households. Black and brown potatoes, for example, are considered by some Sherpas to refer to the same variety. The name "English potato" is used by some farmers for tubers also known by other people as *kyuma* and *koru;* kyuma is also known as *hati.*

Sherpas have developed considerable familiarity with the characteristics and performance of the different potato varieties they have introduced, experimented with, and retained as part of their crop repertoire. They categorize potato varieties on the basis of tuber size, tuber skin and flesh color, flower color, leaf size, hue and growing patterns, and growing season, as well as on local evaluations of their taste, yield, altitude fitness, disease resistance, intercropping capabilities and storage qualities. Their evaluation of the performance of the various varieties includes rating them in these various characteristics at specific altitudes and in particular agricultural sites. There is a widely shared conventional knowledge about long-familiar varieties and much discussion and exchange of insights about experiences with new varieties. Women also

Table 6.   Khumbu Potato Varieties

Sherpa Name	English Name	Characteristics
*Riki moru**	red	red skin, pink flower
*Riki seru**	yellow	yellow flesh, white flower
*Riki bikasi**	development	red flesh, large tuber, purple flower
*Riki mukpu**	brown	black/red/brown skin, purple flower
*Riki nakpu**	black	black/red/brown skin, purple flower
*Riki ngamaringbu**	long tail	white and pink flower
Riki linge		watery
Riki belati	English	refers to several varieties
Riki koru	round	small, white tubers
Riki koru (2)	round	yellowish flesh, pink flower
*Riki kyuma (hati)**	elephant	long, white/yellow flesh, white flower
*Riki anka kali**	black eye	riki moru (?) with darker eyes
Riki nyungma		long, slightly redder than kyuma
Riki ngumbu		very watery, poor taste
*Riki madangshe**		similar to anka kali

*Currently grown in Khumbu

often exchange small amounts of seed potatoes so that their friends, relatives, and neighbors can test new types for themselves in their own fields. There is special concern with altitudinal suitability, climatic hardiness, productivity, taste, disease resistance, storage qualities, and fodder value. Each household reaches its own conclusions about the varieties it prefers to plant on its lands, for although there is usually a good deal of agreement about particular varieties' characteristics the way in which households weigh the relative importance of criteria varies. For some households it may matter a great deal how well a variety is rated as fodder whereas for others this may not be a factor. In recent years concern with yield and altitudinal fitness have tended to outweigh all other factors for most families. The relatively high-yielding yellow and so-called development potatoes have been widely adopted despite considerable shortcomings in some other criteria, including a poor regard for their taste.

Taste, however, is of some importance in cropping decisions, as is the production of potato varieties that are considered to be good for specific culinary purposes. A people consuming as many potatoes as Khumbu Sherpas do and for whom the tuber figures in very nearly every meal might be expected to have a well-developed appreciation of varietal variation in taste. Red potatoes are widely considered to be the finest tasting (*skakindi*) of currently grown Khumbu varieties and are used in all potato preparations. They are especially preferred for preparing boiled potatoes, the dish where the taste and texture of the tuber is most savored. Yellow potatoes are less well regarded, although their generally larger size makes them especially useful for dishes that require grinding tubers, such as potato pancakes and mashed potatoes. Some potato varieties have been briefly experimented with and rejected due to poor taste. *Riki nyumbu,* which gained a reputation for being watery and poor tasting (*shalindi*), was one of these. Many people also decided that it was bad for their health after word spread that it caused stomach pain and intestinal problems. Many families initially balked at planting the yellow potato due to reservations about its taste. As with nyumbu there were also some complaints of stomach problems associated with eating it and several Sherpas also questioned the yellow potato's nutritional qualities. One person insisted, for example, that a single load of red potatoes is worth two of yellow potatoes in terms of food value. A number of these families have since decided nonetheless to plant yellow potatoes, ultimately deciding that their high yields, better disease resistance, and good storage qualities outweighed other factors. A similar process of initial rejection on taste and health grounds and subsequent gradual reevaluation has occurred with most recently introduced high-yielding varieties of development potatoes.

Three varieties are particularly important today, the red potato (*riki moru*), the yellow potato (*riki seru*), and the development potato (*riki bikasi*). Red potatoes in Khumbu are similar in size and shape to the red potatoes familiar today in the U.S. They have red-hued skins and a generally round form, and the tubers are typically apple-sized or smaller. The red potato has been grown in Khumbu since the 1930s and was obtained from Sikkim. Yellow potato tubers are more oblong and often much larger than red potatoes, with a lighter-hued skin and slightly yellow flesh. They are neither our russet potato nor white potato although they are often similar to the former in size and shape. The yellow potato was only introduced to Khumbu in the mid-1970s from Darjeeling and from north-central Nepal, but within a few years it had become the most commonly grown variety in spite of its poorly regarded taste due to its high yields. The tubers of the development potato are large and thin-skinned, with a deeper red skin and flesh than any other Khumbu variety. They have the longest growing season of any potato grown in Khumbu, a characteristic that at first was thought by many to be a considerable drawback before interest in its high yields overcame this and other early objections to it. It is the most recently introduced of the set of Khumbu potato varieties and was brought to Khumbu by Nauje Sherpas who found it at a Shorung government agricultural station only in 1981.[35]

Historically the relative importance of Khumbu varieties has changed considerably. The red potato, which was the staple variety throughout the region from the 1930s until the late 1970s, is considered to be the best-tasting Khumbu potato and remains the primary variety grown at high altitudes. By 1987 it had been nearly totally supplanted, however, at altitudes lower than 4,000 meters by the far higher-yielding yellow potato and the development potato, and by 1990 it was losing ground to these even in the settlements at higher altitudes. During much of the 1980s the yellow potato was the mainstay of Khumbu potato production through most of the region, not only producing a much higher total yield than the red potato but also being planted on more land. The development potato has been widely experimented with and adopted by farmers in much of lower- and mid-altitude Khumbu during the past five years, however, and it now appears likely to replace the yellow potato as the region's main variety despite earlier local reservations about its hardiness and suitability for cultivation at high altitudes due to its longer growing season. The original introduction of the potato to Khumbu and the subsequent introduction, adoption, and diffusion of other varieties in the twentieth century figure prominently in the history of Khumbu agropastoralism, the subject of chapter 6.

Yield

Khumbu farmers have strong views about the productivity of potato varieties. They believe that, in general, Khumbu is a good region for potato cultivation and that yields in some parts of Khumbu, especially the Bhote Kosi valley, are excellent in comparison with those in other areas of Nepal with which they are acquainted. Contrasts are drawn between the productivity of different sites, especially on the grounds of altitude (a topic which is taken up in the following section). And farmers consider that the different varieties have very different average yields. The potatoes of the early part of the century, kyuma and koru, are widely remembered, for example, as being relatively low-yielding in comparison with those varieties that have been the staples since the 1930s. Of the early varieties kyuma was considered better yielding than koru, but many farmers recall that it yielded less than half as much as the red potato does. When the yellow potato was first being cultivated in Khumbu in the mid-1970s farmers reported yields that were quadruple or more that of red potatoes. This contrast has decreased in recent years, but double and triple yields are common at the altitude of the main villages. The development potato has been found to yield triple or quadruple the yield of the yellow potato in the main villages, making it by far the most productive variety in the region.

Estimates of potato productivity vary, of course, from farmer to farmer and place to place. At Pangboche the return on one load of red potato seed tubers is said by one farmer there to be about four loads, with a similar return at Dingboche. Two Khumjung farmers, by contrast, report returns of six to eleven loads per planted load in good years and a return of fifteen loads in the best of years. Yellow potato yields are considerably higher even at 4,000 meters when the gap between yellow and red potatoes begins to narrow. At Pangboche, for example, a load of planted yellow potatoes is considered to yield about eight loads in autumn as compared to four of red potatoes. At Nauje in good years the rate of return on yellow potatoes is about ten to one, and in such years most families could harvest enough to live on from planting only two-and-a-half loads. In Khumjung one highly successful farming family reports yields of fourteen loads of yellow potatoes per planted load in average years and a twenty-five-to-one ratio for the best years.

Variations in yields from year to year can be enormous. Many farmers report a range of up to 300 percent between extremely bad years and good ones in terms of the total number of loads harvested. Others note a similar variation in different terms, pointing out that whereas in some years a single worker is able to harvest three loads of potatoes in a day in other years three people working together cannot harvest a single load in a day. This degree of variability is described for both red and yellow potatoes, and earlier varieties are remembered as being even more prone to bad years.

There is also a widespread belief that yields of particular varieties have declined through time. Kyuma, red potatoes, and yellow potatoes are each said to have yielded larger crops during their first years of cultivation than in later years. Sherpas do not attribute this decline to loss of soil fertility or to a gradual loss of a particular variety's disease resistance or other characteristics. Instead they believe that old varieties become dispirited and lose their vitality and will to produce when farmers begin planting new potato varieties in their fields. People note that kyuma yields declined after the introduction of the red potato and that red potato yields did the same after yellow potatoes began to be cultivated. Some Sherpas are now saying that yellow potato yields are beginning to decline and they relate this to the adoption of the development potato. Beliefs about this process seldom deter people from adopting new varieties. One Dingboche-based family, however, did decide not to plant yellow potatoes for a number of years for fear that planting them would cause the disappearance of the red potato. They only began planting the new variety after being assured by the abbot of Tengboche monastery that the fate of the red potato at Dingboche would not be affected by a single family's honoring it by refusing to plant yellow potatoes. Within a few years the family converted its Dingboche cropping entirely to yellow potatoes.

Field measurements of Khumbu potato yields had not been done prior to 1987. That year I attempted to carry out measurements of crop yields at a number of different agricultural sites. This proved to be a complex endeavor. One complication was the fact that harvests occurred simultaneously in the different valleys and at many different sites and altitudes. Another, more serious factor was that many people were uncomfortable with the prospect of their harvest being measured. By enlisting the help of several of my Sherpa assistants and friends in different villages I was able to collect data on yields in eighty-two potato fields in five different agricultural sites: Nauje (seventeen fields), Thami Og (thirty-seven fields), Tarnga (six fields), Pangboche (thirteen fields), and Dingboche (nine fields). In each field three one-square-meter plots were randomly chosen and marked. All the tubers within these areas were then dug and weighed and the yields totaled and averaged. Except for a few fields at Nauje and those at Tarnga, a total of eight fields, all fields in which measurements were carried out were main fields adjacent to dwellings. Such fields are almost always well-manured and carefully cultivated and they are generally the most productive of a family's holdings. There were major contrasts in yields. Average yields in the main villages ranged from 1.5 kilograms per square meter at Pangboche to 3.8 at Thami Og. Nauje had an average yield of 2.2 kilograms per square meter. The yields at the two higher-altitude, secondary agricultural sites differed still more: Tarnga, with 4.8 kilograms per square meter, had the highest average yield of any of the sites, whereas Dingboche, with 0.9, had the lowest. These yields were much affected by the severe blight of 1987 that struck fields in some settlements more than others. Yields in Nauje were much lower than normal due to widespread blight damage. Pangboche also experienced some problem with blight, although much less than Nauje. Harvests in the other locations were minimally affected by blight and are more typical of a good year's crop.

These potato yields are quite high by national standards. They bear out not only the general perception that yields are good in the region but also that they are especially good in the Bhote Kosi valley. The average yield for main fields measured was 2.8 kilograms per square meter, or 28 metric tons per hectare. This is far above the Nepal national average of 6.25 tons per hectare (Khanal 1988:27).[36] Such extraordinarily high figures may reflect a number of factors. Potatoes are intensively grown in Khumbu, and may be better fertilized, more carefully cultivated, and more closely spaced than in many other areas. They are grown here as a summer rather than a winter crop as they are in some of the country. At Khumbu altitudes there may be less damage from some types of insects and other pests. Disease losses may also be less. Although blight and warts are certainly factors in Khumbu there may well be less loss from these and other viral diseases than in lower, warmer, and moister areas. The moderate rainfall levels and sandy loam soils of Khumbu may be particularly good for potatoes.[37] Tubers stored for spring planting material may winter more viably and disease-free in the cooler climate than in lower areas.[38] And finally, aggregate national figures may be low due to underreporting or underestimating of harvests as well as poor-yielding areas lowering the average.[39]

The 1987 field measurements also give some support to Sherpa evaluations of the varying productivity of different varieties at given altitudes. Here the data are quite limited, for of the eighty-two main fields that we were able to measure only twenty-nine were planted in a single variety of potatoes, twenty-one in yellow and eight in red potatoes. Varietal yield comparisons from the other fields are not possible since in those fields Sherpas had mixed the seed tubers of a number of varieties before planting. Twenty-five fields were planted with evenly mixed yellow and red potatoes and the remainder in various other combinations of yellow, red, development, brown, and, in a few Nauje fields, a local variety called *ngamaringbu*. The small number of fields in single variety cultivation makes comparisons of the relative productivity of different varieties difficult. While the contrasts are suggestive, the small sample size means that these findings can obviously not be given much weight. The yellow potato slightly outyielded the red potato at Nauje ($2.6kg/m^2$ to $2.0kg/m^2$) and Pangboche ($1.39kg/m^2$ to $1.33kg/m^2$). At Dingboche yellow potato yields were considerably lower ($.89kg/m^2$) and roughly equal to those of red potatoes ($.91kg/m^2$).[40] Altitudinal differences in yields are suggested by these figures, but the Bhote Kosi valley figures serve to caution against a simple equation between altitude and yield. Yields at Tarnga (4,050m), for example, were quite high. All fields measured there produced yields comparable to the best Thami Og fields and better than the highest-yielding fields in other sites. Tarnga is famous for both the yield and the taste of its potatoes and the great contrasts between yields there and at Dingboche, which has a reputation for poor yields, accorded well with Sherpa assessments of the productivity of potato cultivation at these sites.

The average yield of 2.8 kilograms per square meter from the eighty-two fields in which measurements were obtained suggests that Sherpas can achieve household potato self-sufficiency on very little crop land. This yield can be taken to represent a far lower than average year's productivity, for more than two-thirds of the fields that were sampled had been infected by blight. For Nauje households, which typically require at least twenty-five loads (1,200kg) of potatoes per year, only 600 square meters (.06ha) of crop land planted with perhaps four or five loads of mixed yellow, red, and development seed potatoes would be sufficient for family needs even in a poor year that yielded only two kilograms per square meter. This is the equivalent of only two small Nauje terraces. A household in one of the villages where potatoes are a higher percentage of the diet might require somewhat more land, but for producing forty loads (1,920kg) of potatoes at even two kilograms per square meter only 960 square meters would suffice. In average or good years even these small farms would produce a substantial surplus that most households would normally devote to fodder for their cattle.

Altitude and Varietal Performance

The performance of a given potato variety is considered to vary with altitude in a number of key criteria including yield, frost vulnerability, and taste. Some varieties perform relatively well across the full altitudinal range of Khumbu potato cropping, whereas others are not considered suitable for fields above 4,000 meters. Those with longer growing seasons are much more vulnerable to frosts at high altitudes and tend to have much lower yields even when they are not damaged by spring frosts, perhaps reflecting reduced tuber development when the harvest is carried out in late summer or early autumn. It is also believed that at high altitudes some varieties become less tasty, and farmers may forgo planting them even if their yield would be acceptable.

The red potato and the brown potato are considered to be the most altitudinally hardy of Khumbu potato varieties and are the most widely grown at altitudes of 4,200 meters and above (table 7). The brown potato is only grown in the Bhote Kosi valley on any significant scale, but it is the main potato of its upper reaches from Tarnga to Apsona.[41] Red potatoes are also grown as high as Apsona and are the potato of choice in the highest fields in the Dudh Kosi where fields are cultivated as high as Tarnak (4,690m); a small patch of these tubers was harvested for several years in the upper Imja Khola valley at Chukkung (4,753m). In the Dudh Kosi valley only the red potato is grown in the settlements above 4,000 meters. The yellow potato, which by far outproduces the red one at the altitude of the main villages, has a relatively long growing season and is considered by many farmers to therefore be unsuitable for fields above 4,000 meters.[42] At that altitude yellow potato yields are also considered to decline, and some people also consider that its taste begins to deteriorate. According to some Thamicho villagers yellow potatoes become more watery when grown at a higher altitude. This same view is expressed by many Dudh Kosi valley farmers and some Pangboche residents consider yellow potatoes too watery to plant not only at Ding-

Table 7.   Altitude and Potato Cultivation, 1987

	Bhote Kosi Valley	*Dudh Kosi Valley*	*Imja Khola Valley*
4,400m	*brown*, red	*red*	*red*, yellow
4,200m	*brown*, red	*red*	*red*, yellow
4,000m	*yellow*, red, brown	*red*	*yellow*, red
3,800m	*yellow*, red, development	*yellow*, red	*yellow*, red
3,400m	*yellow*, red, development	*yellow*, red, development	

NOTE: Italics indicate the most widely cultivated variety in a valley at a given altitudinal range.

boche but in Pangboche itself. Very few families grow the yellow potato in any of the agricultural sites above Phurtse. At Na, for example, only two of the twenty-three families with potato fields grow any yellow potatoes. No one plants yellow potatoes in Machermo and Panga, sites of comparable altitude on the west side of the Dudh Kosi.[43]

It is widely perceived that yields of both red and yellow potatoes decline above 4,000 meters, although at some sites, such as Tarnga, yields of both can be quite good even at this altitude. Although potatoes are planted in the upper Dudh Kosi valley nearly to 4,700 meters, this is very unusual in Khumbu and in general there is almost no potato cultivation above 4,300 meters. This no doubt primarily reflects an assessment of the likely diminishing returns and greater risk of such extremely high-altitude potato production. Other factors, however, may also be involved, including the lack of a need for large amounts of potatoes in the high-altitude herding settlements, sufficient agricultural opportunities at lower altitudes for meeting main village subsistence requirements, and lack of interest in producing greater potato surpluses for sale. There may also be concern that it is much more difficult to protect crops from depredations by livestock in high-altitude areas that are prime summer grazing ground.

Only a few years ago the development potato was considered to have the narrowest range of altitudinal fitness. When it was first introduced a number of people doubted that it would be an important variety at altitudes very much higher than that of Nauje due to its extremely long growing season. As recently as 1987 cultivation had only been attempted as high as Thami Teng and Yulajung (3,800m), and there was only undertaken on a very small scale and with mixed evaluations. At that time the variety was not considered suitable at altitudes higher than 3,800 meters. By 1990, however, it was being grown widely at Tarnga (4,050m) where it produced quite good yields, and it had also been introduced to Dingboche.

Other Factors in Varietal Selection

Sherpas also evaluate several other qualities in making decisions about which varieties of potatoes to plant at specific sites. The most important are their perception of disease resistance, storage quality, and fodder suitability. Potato varieties are considered to vary considerably in their degree of resistance to disease. Old potato varieties all had a bad reputation for being blight-susceptible. So did the red potato. For some years Sherpas felt that the yellow potato was relatively blight-resistant and that even if the plant was affected early in the summer its "stronger" leaves and stalk enabled it to produce larger tubers than blight-infected red potatoes would. This opinion was probably widely revised after 1987 when yellow as well as red potatoes suffered major blight damage in Nauje and the lower Bhote Kosi valley. Both yellow and red potato harvests were small. Development potatoes, on the other hand, went noticeably unscathed and fields in this variety flourished on through September surrounded by fields in yellow and red potatoes that had withered by late July.

How long and well potatoes can be stored is an extremely critical quality given their year-round role in the Khumbu diet. Potatoes are stored in outdoor, underground storage pits (*miktung*) to keep them the longest possible time, and some varieties tend to spoil in these conditions more than others. The red potato has a poor storage reputation. People point out that if a single red potato tuber goes bad in a storage pit it is likely to affect all the others stored there. It is not unheard of for pits to be opened in the spring and for no potatoes to be salvageable from them. The yellow potato, by contrast, has an excellent reputation.[44] Although I have heard of cases of yellow potato-filled storage pits going bad the general view seems to be that even if one yellow tuber totally rots in the pit the others will not also be lost. Some people believe that if red potatoes are mixed with yellow ones in a pit that rotting red potatoes may affect the yellow tubers. This was considered to be the case in 1987 by some Tarnga people who suffered major losses of their stored potatoes. Opinion has not yet solidified regionally on development potatoes, but some disquieting stories are being told of storage rot. Some Bhote Kosi and Phurtse families have had entire storage pits of development potatoes rot. In one case this involved the loss of sixteen loads of potatoes, enough to feed a family for half a year.

Few families feed large amounts of potatoes to livestock other than in years of unusual surplus harvests. The production of potatoes for fodder, therefore, influences cropping decisions for only a very small percentage of Khumbu farmers. Those concerned about fodder production, however, consider that red potatoes are far superior as fodder than other current varieties, especially yellow and development potatoes. This has to do with the supposed nutritional value of the red potato, which is considered to be much greater than the other two varieties, and also with it suitability for intercropping with radish. Radish is also highly valued as fodder and is grown intercropped with potatoes by those families that cultivate it. Many farmers are convinced that radish cannot be successfully intercropped with yellow potatoes, for it is felt that the large leaves and long growing season of this potato variety lower radish yields by shading out the intercropped radish late in the summer and early autumn. Development potatoes would have the same shortcoming. Red potatoes, however, complete their growth cycle much earlier and die off, allowing the radish crop more light.

The Social Organization of Agriculture

Agriculture is a social enterprise as well as a cultural and economic one. Khumbu crop production is greatly shaped by customs concerning land tenure and inheritance, assumptions about the proper division of labor and appropriate forms of individual, household, and communal work, and traditions about the correct boundaries between individual rights and community responsibility in deciding how land is to be used. Some of these social values have varied historically and regionally within Khumbu, and households within a village may vary in the degree of their conformity to social ideals. Yet there are so many levels of shared belief and practice that one can identify long-standing Khumbu characteristics of the social organization of agriculture.

Khumbu agriculture is based on private land ownership and subsistence farming by nuclear households.[45] Land can be freely bought and sold both to fellow villagers or to Sherpas from other settlements. It is uncommon, however, for land to change hands other than through inheritance.[46] According to Khumbu custom crop land is divided equally among sons, each coming into his share at the time he establishes his own household.[47]

All Sherpa families in Khumbu own at least some crop land. Tenant farming (*pijin*), with the harvest shared fifty-fifty, is unheard of today. A small amount of land is rented (*torin*), much of it owned by the Tengboche monastery, with payment due in cash or the equivalent amount of grain.[48] Rented land very seldom, however, constitutes the major component of a family's land. Rent is usually paid in cash on an annual basis. This is generally the equivalent of a quarter to a half the market value of a good year's harvest from the field.[49]

The State and Farming

As remote as Khumbu has been from Kathmandu government concern and supervision for much of the past 200 years, land use has nevertheless been influenced in some ways by central government policies. This has been especially true in the twentieth century when Kathmandu edicts have had an impact on both agriculture and forest use. Government tax policies, land-registration regulations, and development planning have all affected Khumbu crop production.

Since the early nineteenth century, and possibly for some time before that, Sherpas have paid tax in cash to Kathmandu on both land and houses. Khumbu families also had to contribute unpaid agricultural labor (wulok) to the local pembu[50] which amounted to three to five days' work per year, usually met by women working in the pembu's fields. In some societies similar tax policies have been employed by governments to pressure farmers to cultivate cash crops and they sometimes have led to indebtedness and loss of land. In Khumbu tax collection has not been used as a tool to influence crop selection. But in the nineteenth and early twentieth centuries tax burdens may have accented the poverty of some households and contributed to the emigration of many families to Darjeeling, Rolwaling, and other regions. Although it is likely that land fragmentation in Khumbu and the lure of the chance for wealth and fame in Darjeeling were greater factors, taxes were certainly an additional burden.

In the late nineteenth and early twentieth centuries land taxes were much heavier than they are today. The hereditary Rana prime ministers who ruled Nepal from 1847 through 1950 had a reputation for exploiting the country's resources to amass family wealth and one of the avenues they used was a tax on land holdings. Tax roles by household were compiled for Khumbu based on an estimate of the amount of maize seed required to plant fields. In 1939 the regional land revenue was 4,000 rupees (Fürer-Haimendorf 1964:119), twice the current level, at a time when the rupee was worth a great deal more than today and income was much lower. In that era day labor paid less than half a rupee per day and rice was less than two rupees a pathi (compared with 110 rupees per pathi today). Land taxes that averaged perhaps eight rupees per family thus were substantial. Most Sherpas would have met them with the profits from trans-Himalayan trade, by working as porters for wealthy traders, or by agricultural day labor.

In some parts of Nepal the men who collected taxes on the Ranas' behalf grew wealthy and established considerable estates or were given these estates and the right to collect taxes to support them by the Ranas. In Shorung and in the Chyangma area several Sherpa pembu ultimately amassed large estates that were worked in part with the corvée labor due to them by their tax clients. Nothing of this sort, however, occurred in Khumbu.[51] That tenant farming did not become widespread in Khumbu as it may have in some other parts of Solu-Khumbu is probably due to several factors. Wealthy Khumbu pembu may have preferred to put their cash into trade ventures rather than into land. Land in Khumbu may have long, as it is today, been offered for sale relatively seldom. The lack of any major commercial crop possibilities in Khumbu also may have discouraged the accumulation of land. And the absence of tenant farming may have reflected the means Khumbu Sherpas had for raising cash for paying taxes through trade and wage labor (*lamay*) as well as the ability of poorer families to emigrate rather than be forced into servitude by increasing debt.

The land tax in Khumbu today is paid only on land owned in the main village.[52] The tax now never amounts to a great deal because government policies have rolled back regional tax rates to half the level of the 1940s in order to compensate for the difficulties of agriculture in the remote, high-altitude area. Inflation since the 1940s has rendered the value of the resulting taxes small indeed. Even very large (by Khumbu standards) landowners are not assessed more than about sixteen rupees per year, less than a day's wage at the poorest day-labor rates in the region. Corvée labor taxes have also been halted as one result of the land-reform measures that were implemented in Khumbu in 1965.

Khumbu agricultural land use has also been influenced by government land regulations. At one time it was apparently relatively easy for Sherpas to establish new fields on uncultivated village lands. Recent immigrants from Tibet may have first had to gain the permission of a local Sherpa pembu before establishing new fields (Fürer-Haimendorf 1979:124), or at least had to find a pembu who would be willing to place the new fields on his tax rolls. In the early 1940s, however, the Kathmandu government began to implement a national land-registration system and a set of accompanying policies that had the effect of curtailing any further expansion of Khumbu crop areas in subsequent decades. Land that had not been registered could henceforth not be claimed and cultivated without making the proper arrangements with the government office at the district center, and for many years there was a moratorium on new land claims. This prevented a number of immigrant families from obtaining land by carving new terraces near the villages or even from claiming any of the long-abandoned terraces that are plentiful in some parts of Khumbu. In 1965–1966 there was a further major change in national land-tenure and registration regulations (*bhumi sudar*) as part of a government land-reform program. At this time Sherpas had an opportunity to claim and register abandoned and tax-delinquent fields. But claims on previously uncultivated land were not allowed and even the opportunity to register and resume cultivation on abandoned land was only offered for a few years. Farmers who are short of land have sometimes attempted to get around these regulations by surreptitiously enlarging their fields or restoring the protective walls around abandoned terraces and resuming cultivation on them. Yet even these tactics have sometimes been unsuccessful, especially in the late 1980s. Officials of the government land-tax office in the district center rarely if ever come to Khumbu and given the very rough descriptions of field locations and sizes in the tax documents would probably be unable to detect these minor, local adjustments of field boundaries and areas under cultivation. But from about 1984 until 1989 Sagarmatha National Park administrators zealously enforced the regulations. Some park administrators viewed even the resumption of cultivation on abandoned terraces as violations of park control of all noncultivated lands. Here they may have been on tenuous legal grounds, for technically all the villages and settlements of Khumbu are outside the park's jurisdiction having been deliberately left as islands of private and community property within the national park in order to allow villagers to continue their customary forms of land use and to enable them to make their own choices about future development. But the several local residents who had their new fields destroyed by park staff had no immediate recourse to contest this possible abuse of power, although one Thamicho villager spoke of bringing a court case against the national park. By 1991, however, park administrators were no longer blocking agricultural reclamation, and several Sherpas had found that they could indeed register large numbers of abandoned terraces by paying some of the back taxes on them.

Khumbu crop production has been only indirectly affected by government agricultural development efforts. Agricultural extension services have been established in the Solu-Khumbu district, but the nearest office is at Phaphlu, close to the district center and a three- or four-day walk from Nauje. No programs from this office have ever been extended to Khumbu. Government agricultural development efforts have only had an impact on Khumbu crop production indirectly through Khumbu Sherpa interest in new potato varieties. A few Sherpas have visited the Phaphlu agricultural development office and it was there that the seed potatoes of the new variety that Khumbu Sherpas call development potato were obtained.

The government establishment of the weekly market in Nauje in 1965 has also affected Khumbu agriculture. So far this has had relatively little fundamental impact on crop production in Khumbu. There has been no major shift to greater commercial production across the region, and regional exchange has not focused solely on the market, for much direct barter and cash sales of agricultural surpluses among families continues. But the weekly market has become an important forum for the sale of surplus potatoes to lodges, government officials, and families with insufficient harvests from their own lands. The continuing high demand for potatoes may be a factor in some farmers' decision to abandon the rotation of buckwheat with potatoes in order to specialize in potatoes, although this is never the reason farmers give for this change. It is also true that the market is now the major source of grain for Khumbu. In this dimension to some degree it now merely fulfills the function that earlier barter trade did. But much more grain is delivered directly to Khumbu than before when many Sherpas instead themselves hauled grain home from down valley. The convenience of this new system may have been a factor in some families' decision to give up cultivating grain and devote all their land to producing their supplies of potatoes.

Agricultural Labor

Khumbu farming revolves around family and reciprocal labor. It is extremely rare even for the wealthiest traders, lodgekeepers, or landowners to depend entirely on hired labor. This remains true today even in Nauje where the most use has been made of immigrant non-Khumbu laborers since the mid-1970s. The myriad tasks of subsistence life—hauling water from the spring and fetching firewood, gathering dung and forest leaves for fertilizer and fuel, tending crops and herds—are carried out by the entire family, with tasks for everyone from children under ten years of age to grandparents in their eighties. Some old people retire from the world, retreating to the family shrine room or a hermitage during their last days to devote themselves to religion. But everyone else works at subsistence as an integral part of daily life.

According to Khumbu social customs household agricultural tasks are strongly differentiated by sex. Men herd and perform certain other tasks such as plowing, hauling fertilizer to the fields, helping with grain and hay harvesting, and carrying the harvest to storage places. Women perform most of the agricultural work and assist with some aspects of pastoralism, especially milking and processing milk as well as handling most domestic chores and childcare. These roles are not totally rigid, and men may occasionally be found digging potato fields and women may herd.[53] Flexibility thus remains an important facet of household-labor allocation. People do work that needs to be done, even though this may require cutting across usual gender roles. Table 8 illustrates the general sexual division of agricultural, pastoral, and forest labor.

The sexual division of labor described above means that cultivation is for the most part carried out by the women of the household with minor and largely specialized assistance from the men. Usually women join together with female relatives and friends in reciprocal labor arrange-

Table 8.   *Gender Division of Labor*

Task	Labor Contribution
Field Preparation	
Gathering manure	women, occasionally men
Gathering soluk	women, occasionally men
Rebuilding field walls	men
Transporting manure/soluk	men and women
Digging potato fields	women
Plowing	men
Planting	
Broadcasting seed	women
Planting potatoes	women
Crop Care	
Irrigation	men and women
Weeding	women, occasionally men
Harvest	
Potato harvest	women
Transporting potatoes	men, women
Harvest grain	women, occasionally men
Forest Work	
Fuel-wood gathering	men, occasionally women
Lumber cutting	men
Pastoralism	
Herding nak/yak/zopkio	men, very occasionally women
Herding zhum/cows	men and women
Milking	women
Herding sheep/goats	men, occasionally women
Hay harvest and storage	men and women
Stall feeding and labor	men and women

ments to work the fields. These reciprocal work groups are called *ngalok* and are organized for a variety of tasks including field preparation, planting, weeding and harvesting.[54] Typically women from six or more households (sometimes from as many as fifteen) form a group. Each family contributes one laborer from among its members or household servants (*lawa*). The group alternates days of work in the fields of its members. The order in which fields are planted and harvested depends in large part on the horoscopes and lucky and unlucky days of their owners. The owner of the fields that are to be worked on a particular day has to provide food and drink for the work group for the day. Assembling in groups makes the work go quickly, for a work group can complete the planting or harvesting of a field in a few hours which could take the women of an individual family days to complete. It also makes the work a social occasion, and the work is punctuated with conversation, singing, and joking.

Not all Khumbu field labor is performed by reciprocal labor groups. Hired labor has also long been a feature of Khumbu agriculture. Families may hire agricultural day laborers to take their places in reciprocal work groups, but more often hired labor frees families from joining such groups and enables them to carry out the farming of their lands entirely on their own schedule. Cash wages from agricultural work have been an important source of income for some of the poorest Khumbu families.[55] Women from these households continue to work for more well-off families at planting, weeding, and harvest times. Men work less often as agricultural day laborers. Those who do such work usually cut hay and wild grass. During the 1940s and 50s many new immigrant Khamba families relied on such agricultural day labor to make a living. Since the mid-1970s the prospect of relatively good wages as agricultural day laborers and year-round household help has lured many young Sherpas, Tamangs, and Rais to Khumbu from lower-altitude areas of up to a week's journey by foot.[56] Some of these young people are employed as year-round household servants, but field work is one of their main responsibilities.[57] Others come to Khumbu only for a few weeks' work at planting or harvest time. These migrant workers now compose a significant part of the total agricultural work force in Nauje. Many also work in Khumjung and Kunde and in the past few years some have begun working for a few weeks in spring and autumn in Pangboche and Dingboche. They do not as of yet, however, work in Phurtse or Thamicho. In all Khumbu villages, however, some Khumbu Sherpas continue to work in the fields for day wages. Some Phurtse people take agricultural day labor also in Khumjung and Kunde. Sherpas who today work as agricultural day laborers tend to be people who are unable or unwilling to work for the far better wages available in the tourist trade. While agricultural wages have increased in recent years, they have not kept pace with the pay offered for even the lowest-paid tourist jobs.[58] In Khumbu trekking and mountaineering porters were generally paid at least a hundred rupees per day in 1990. Agricultural day laborers, by contrast, received thirty to forty rupees per day.[59]

The Agricultural Cycle

The sequence of operations and techniques and the timing required to prepare fields, plant and tend crops, and bring in and store the harvest represents a considerable intellectual achievement. The annual agricultural cycle involves scores of decisions, each of which can greatly affect field yields and the ultimate sustenance of the farming family. Accurate evaluation of environmental conditions, intimate familiarity with the capabilities of particular crop varieties, and command of appropriate technology and techniques are all required, and a string of activities must be carefully orchestrated with exact timing. The necessary local environmental and agronomic knowledge and repertoire of crop varieties and techniques, moreover, also has to be integrated within a broader socioeconomic context that includes religious beliefs and practices, ethics, the policies of community land-use regulations, lifestyle preferences, family demographic and economic situations, and customs concerning the social organization of agriculture. Khumbu Sherpas, like other agricultural peoples, have developed through time a complex body of knowledge and practices that shapes the form and rhythm of their agricultural cycle and makes it distinctively their own. This is in large part shared across Khumbu, although details in timing vary with microclimatic and other site-specific conditions. This group of shared practices, values, and goals constitutes a culturally transmitted set of instructions, one that is constantly being reshaped by local experience with new crop varieties and techniques, the acquisition of new knowledge, and the development of new values and customs. Significant differences in agricultural practices can be discerned historically and even within relatively recent history. There have also long been some regional differences in agricultural practices. Today these regional variations reflect differences in community customs and institutions and household decisions based on different circumstances of wealth, labor, opportunity, and individual tastes, goals, perception, and knowledge.

Field Preparation and Planting

In Nauje and the gunsa settlements the agricultural year begins each spring in late February and early March, so early that the waterfalls frozen against the northern rock faces of the gorges have not yet thawed and another major snowfall or two may still be ahead. At this time, a full month before potatoes are planted, long lines of women begin redigging the terraced fields and the work of preparing and fertilizing the fields for another summer crop season gets under way. By mid-March potato fields are being prepared all over lower Khumbu (fig. 10). Men and sometimes also women now carry basketloads of manure (Sherpa *cha*) to the potato terraces, depositing conical loads in long lines at intervals of one to three meters. Meanwhile women wielding Nepali-style hoes (*tokzi,* Nepali *kodalo*) double dig each field.[60] The first digging is a simple pass to loosen winter-hardened topsoil and to get moist earth to the surface. During the second digging, which takes place several days to a week later, women mix fertilizer into the topsoil and simultaneously plant potatoes.

Potatoes and barley are always manured and special care is taken to heavily manure barley. Manure put on barley fields, moreover, is carefully pulverized first by beating it with a pole.[61] Several sources of fertilizer are employed and much effort, thought, and sometimes cash is put into procuring them. Manure, forest leaves and needles, and composted human waste are all used for fertilization. Composted toilet wastes are considered the richest of these, followed by sheep and goat manure and cattle dung. The amount of manure put on a given field varies depending on how much manure a family has available, the composition of the fertilizer, the amount of labor that can be devoted to the task, and beliefs about the manure requirements of particular crops. Generally fields closest to the house are more heavily manured, and the richer, dark soils of these fields reflect generations of careful attention.

Nak manure is the most extensively used fertilizer across Khumbu due to its great availability. Manure from dung that is collected in late summer or autumn is considered to be the best, for at this time the stock are well fed and it is felt that their waste now has more energy than at other times of the year. This dung also has a chance to age before use in the spring, which also increases its value as fertilizer. Sherpas refer to this richer aged dung as *temcha.*

Fields are sometimes directly fertilized in the autumn and spring by corralling livestock in them at night or penning them there for some days to feed off crop residues. Much manure is also gathered from surrounding slopes, and in spring and autumn men, women, and children can often be seen moving about on the slopes with baskets gathering dung for use as fertilizer and fuel. In the Nauje area in the spring a diligent worker can gather about sixty kilograms of dung per day from slopes within a half hour's walk.[62]

There is a market for manure in a number of communities where farmers have cash but lack the livestock or labor resources to obtain sufficient supplies of field fertilizers. In most main villages a load of manure (thirty to forty kilograms) could be obtained in 1987 for five to seven rupees. In some places such as Thami Og, Thami Teng, and Tarnga demand has pushed prices up to nine and even ten rupees and it can be extremely difficult to find any for sale. In the Tarnga-Chosero area the degree of demand for purchased manure makes it necessary to place orders a year in advance. This allows Thamicho stockowners time to build up greater reserves of fertilizer by placing more forest leaves and needles under their stock and composting the resulting dung-rich mixture.

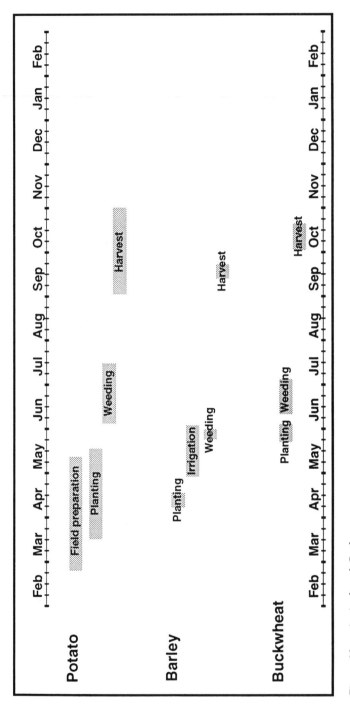

Figure 10. Agricultural Cycle

Forest leaves and needles are used in several ways to enrich fields. Leaves can be put directly onto fields, in which case they are usually first burned and then the ash is dug into the soil. This is probably done mostly by people who have no livestock and lack the opportunities to follow the preferred approach of enriching forest-floor products with dung and urine. This composting practice is very popular among Khumbu farmers, as it has been among many people who practice mixed agropastoralism in other parts of the world. It is a common technique in Nepal among middle-altitude agropastoralists. Khumbu Sherpa place great value on the use of birch and rhododendron leaves and conifer needles for fertilizer, and have a wealth of knowledge and beliefs about this facet of farming. They consider rhododendron and birch leaves to be the best resources. Both of these, they observe, rot more rapidly than conifer needles. The best of all for fertilizer are said to be the small leaves that have already decomposed on the forest floor for several years. Such leaves collected from under rhododendron bushes are preferred since they compost especially quickly. Sometimes leaves and needles are put directly onto fields, in which case the mounds are usually burned and then the ash is mixed into the soil. The usual practice, though, is to first use them as livestock bedding and compost them. The resulting deep rich fertilizer is called *mandur*.

Little use is made of green manures, although some weeds and crop residues such as barley stubble may be plowed into the fields in the autumn. Winter cover crops are not planted, nor are legumes. No chemical fertilizers are used. Ash, however, is added to fields. Some families set aside ash from the family hearth in a special place for use in the spring. Some, as already mentioned, also burn vegetation gathered from nearby woodlands directly on the field.

By the middle of March potato planting is underway in Nauje and the gunsa.[63] Teams of women work the terraces, most armed with hoes and others carrying small baskets of small whole tubers and sometimes also baskets of cut pieces of potatoes.[64] Groups of women form double, facing lines, each woman working in team with a partner. The women in one line each hold a small basket with seed potatoes in it, the women in the other line each churn the soil with their hoes, simultaneously mixing fertilizer with soil, breaking up clods, and covering the potato seeds that their partners lob into their work area. As the women wielding hoes work back and forth over patches of earth their partners casually but deftly toss potatoes into the pockets momentarily opened up in the soil. From time to time partners change tasks. The line of workers sweeps across terraces, covering in a day ground that could take a week or two for an individual family. The work as a whole is easy, reckoned by some to be less arduous than spending the day laboring in a tourist lodge.

Days tend to be short, beginning late and ending early, a contrast to the more intense pace of harvest work. Breaks are frequent for tea, snacks, and conversation with one another, other groups, and bystanders.[65]

It is considered best to plant as soon as possible after the soil has been initially broken in the spring in order to provide the potatoes with optimal moisture conditions for sprouting. Seed potatoes, small whole tubers specially set aside from the last harvest, are the main planting material.[66] These are selected for their small size and average only 2.5 to three centimeters in diameter.[67] If there are insufficient seed potatoes farmers resort to cutting and planting pieces of larger tubers, which are usually planted within a few hours of being cut. Usually these tubers are simply cut in half, each half having one or more eyes. Seed potatoes and tuber pieces are planted across the entire field. No rows or beds are made. Planting is usually very dense with only ten to fifteen centimeters between seed potatoes.

It is common to plant several varieties in a field, mixing the seed pieces up before planting to create a homogenous planting pattern, then separating the types again at harvest for storage and consumption. A single family may plant as many as five or six varieties in a field. Typically, though, such fields are predominately in one or two varieties planted with seed potatoes stored from the previous autumn's harvest, whereas the rest of the planting material consists of a very small number of seed potatoes obtained from exchanges with friends and grown as experiments or for fun.[68] Other crops are usually not intercropped with the potatoes, although occasionally families interplant radish into the fields ten days to four weeks later.[69] Radish planting is not carried out by work groups, for it is a simple matter for a woman to plant a field herself, pausing every meter or two to dig in a heel or a trowel-sized weeding tool (*koma*) and drop in a radish seed.

Potato planting usually proceeds from lower-altitude settlements towards the high-altitude sites. Nauje and gunsa planting take place first, followed about two weeks later by planting in the other main villages. As families finish fields in the main village they begin to move higher into the secondary agricultural sites and phu. In the Bhoti Kosi valley, for example, by the end of April planting is finished at Thami Og and Teng, and by the first days of May is underway at Chosero, Mingbo, Tarnga, and Marulung farther up the valley.[70] Planting at the still-higher-altitude sites of Goma, Arye, and Chule does not take place until late in the month. The full potato-planting period may last up to eight weeks for families with both gunsa and high-altitude fields.[71]

In a region with such a short growing season and where agriculture is carried out in so many different altitudinal and microclimatic sites, the timing of planting at particular sites becomes a critical decision. Sherpas have accordingly refined site-specific planting calendars that integrate local knowledge of microclimates with knowledge of local crop varieties. Their intimate familiarity with the climatic conditions at specific places, the growth characteristics and fitness of different crop varieties, and the risks to plants at different stages of their growth enable them to choose planting times that minimize the chances of major damage from particular Khumbu climatic stresses. Farmers are especially concerned with avoiding the dangers of late and early frosts and untimely heavy rains. Crops planted too soon run the risk of succumbing to spring frosts. Buckwheat planted too late may be more vulnerable to midsummer rains and autumn frost and snow. Late-planted tubers do not achieve full size, and late-planted buckwheat may not develop mature grain. But other factors besides weather and the growing season are also important in decisions about the timing of planting, including household evaluation of labor availability, the scheduling of other household economic and social responsibilities, and proper astrological conditions. All decisions must be workable from each of these standpoints. Compromises are sometimes necessary, and if need be families divide forces to work in several different parts of a valley simultaneously.

Individual families make their own decisions about when to plant and they develop their own means by which to decide on the correct date for a particular site and crop. Environmental markers are often used to judge when the best planting window has arrived for a particular place and crop. Spring phenomena such as the blooming of certain flowers, bird migration, and the thaw of winter snow and ice are used in some cases. In Nauje and Khumjung the blooming of iris (*themi mendok*) is taken as one indicator. Some Khumjung families note the timing of the passage of ducks en route to Tibet. Some Nauje families formerly took the spring breakup of the frozen waterfalls that drape Kwangde as a sign to plant potatoes. Most common of all is the use of sun and shadow markers. In Phurtse, for example, people keep track of the shadow of Ama Dablam, a peak east of the village, on the slopes of Khumbu Yul Lha. There are three different shadow points. The first two mark the time for potato planting at the main village and the high-altitude settlements respectively, and the third, reached ten days after the second, indicates that it is time for buckwheat planting in Phurtse. Dingboche planting time is chosen by consulting the shadow of Ama Dablam as it falls on a small shrine near Orsho, between Dingboche and Pangboche. Once this event has occurred families choose a specific day to begin barley planting based on their personal horoscope.[72] At Marulung potato-planting time is read from the sunrise light coming over a particular boulder. At Phurte, in the lower Bhote Kosi valley, people note the sun's position, watching where the morning sun breaks over the ridge.

Some families observe the sunlight within their own houses; it is time to plant potatoes when a sunbeam passes through a particular window or when light filtering through a roof opening strikes a certain standing beam. At Samde people take their guidance from the location of the setting sun.

Not all families perform their own calculations to know when a planting period has arrived in an area. Some are content to follow the lead of others, and in a particular community the judgments of certain individuals may be especially respected. Planting knowledge such as the reading of environmental indicators is passed on between generations and shared openly with others. It may change through time, and indeed must change in order to adjust to the different growing season requirements of new crop varieties.

Environmental markers do not determine the exact day on which a family plants. They simply provide an alert that it is no longer too early to plant a particular crop in a certain place. The choice of the exact date is based on household labor commitments, the decision of the women's work groups as to the order of the fields they will plant, and the horoscope of the male head of the household. Luck (*yang*) is considered to play a major role in agriculture, and learned men and Tibetan almanacs may be consulted to arrive at the best possible day to plant. People also may trust to their "lucky days" or avoid their "unlucky days." Every Sherpa has both lucky and unlucky days of the week, the particular days involved varying with the day of the week on which he or she was born.

Most families have to dovetail the scheduling of potato planting with the requirements of preparing and planting grain. This is relatively simple for buckwheat which, due to its susceptibility to late frosts, is not planted until late May. Barley planting requires more schedule adjustment. In order to be mature for harvest in late September the crop must be planted at the beginning of April, a time when families with land at Dingboche would otherwise be preoccupied with potato planting in their main village fields. Pangboche families cope with this conflict by interrupting preparation of the potato fields in the main village in order to prepare and plant Dingboche barley fields. Most of them also plant potatoes at Dingboche once they have the barley in, risking frost danger for the convenience of thereafter being able to focus entirely on Pangboche potato and buckwheat planting for the rest of the spring.[73] Some families divide their efforts once the barley is planted, with some family women remaining at Dingboche to plant potatoes while other female family members return to resume Pangboche operations.

Khumbu techniques of preparing and sowing grain fields require much less time and effort than does planting potatoes. Buckwheat requires especially little work since it is not manured and the task is usually completed in a day or two. Both barley and buckwheat fields are plowed twice with a zopkio- or yak-drawn scratch plow. The first pass with the plow breaks the ground. During the second seed is broadcast.[74] According to local custom men guide the plow while women broadcast the seed. The sower walks behind the plow carrying a basket or tin can full of grain, hurling seed down at the furrows with a forceful overhand delivery intended, perhaps, to bury the seed deep in the fresh earth. As often as not, however, the seeds simply ricochet back into the air. Barley fields are then smoothed by using a zopkio or yak to drag a small log across the field or else by hauling a juniper bough across it by hand.

Barley is the only irrigated Khumbu crop.[75] Between late April and the arrival of the summer monsoon rains in mid-June it is necessary to supply barley fields with water on three or four occasions, and if the monsoon rains arrive late irrigation may be continued for several more weeks. Water is taken from the Imja Khola and led in a small, unlined ditch more than a kilometer to the head of the large, alluvial terrace on which the settlement is situated. Here the flow is directed into two channels, one running down the center of the settlement and the other along the foot of the slope on the northern side. Small intake ditches plugged with rocks, rags, and mud lead from these channels into individual fields. Here the water is led into a set of furrows that dissect each field into a number of three-meter-wide beds. It is then splashed onto the crop with a specialized long-handled wooden tool (*ongbu*). There is no community management of water use. Families draw what water they need whenever they desire, even though this sometimes creates shortages of irrigation water for families whose fields are farthest from the head of the system. There is communal organization, however, of the maintenance of the main irrigation channels and their opening each spring. Each April the community celebrates *kachang* ("ditch beer"), a day on which all the families of Dingboche gather to worship the gods, prepare the irrigation system, and witness the transfer of office between two community officials (*nawa*) charged with enforcing a set of local regulations affecting pastoralism and agricultural practices (local resource management institutions are discussed further in chapters 4 and 7).[76] On this day each household must contribute at least one laborer to the crews readying the irrigation system. Male volunteers move boulders that may have choked the outtake from the river, while women clean the ditch that leads to the village. The new nawa choose the day on which irrigation will begin.

Planting season culminates with a ritual protection of the crops. In Khumjung-Kunde, Phurtse, and Pangboche (and formerly also in Thamicho) a circumambulation of all the fields in the settlement is performed each May after planting is completed in order to ensure the safety of the crops.[77] Groups of villagers, monks, and village lamas carrying prayer flags, sacred books, temple statues, and ritual implements circle the outer edge of the settlement fields, accompanying themselves with cymbols, horns, and drums and stopping from time to time to plant prayer flags and recite prayers to the local gods. The rite is known variously as *Tengur, Chokor* (from *kor,* referring to circling) and *Orsho* (from the practice of collecting grain from each household to finance the ceremony). In Thamicho, where it is called *Chokur,* the circumambulation was a multivillage enterprise and was carried out differently in alternate years. One year the monks from Thami monastery circumambulated the fields of the three settlements of Thami Og, Thami Teng, and Yulajung and their outlying hamlets. The next year the same route was followed by the head lama of the Kerok temple. Khumjung and Kunde also hold a joint protective ceremony (here called *Orsho*) in which all the fields of both villages are circled each year, but the saying of the associated *lapsong* prayers rotates annually between the two. Phurtse and Pangboche villages each carry out their own circumambulation, which is known in this part of Khumbu as *Tengur.* In Pangboche the fields of both the upper and lower Pangboche village are circled.

The ritual is taken very seriously. Some of the most sacred things enshrined in the village temples are taken out at this time and paraded. Care is taken to collect grain from each family of the community and to extend protection to each—farmers whose fields were inadvertently omitted from the blessed boundaries would be very angry indeed. The costs of an improperly performed ritual can be high, for it is believed they can directly endanger the crops of all villagers.

Despite the careful timing of planting and the conducting of protective rituals it is not uncommon for May frosts to damage both potatoes and buckwheat. Different families respond to this situation in different ways. One common reaction is to do nothing and make the best of what crop survives. With buckwheat (but not potatoes), however, some families replant the crop. Good results, however, are not guaranteed by this effort, for the late-planted crop may be endangered by heavy, early-summer rains, and at best yields will be low given the short length of the Khumbu growing season. In 1986 about half the families of Phurtse chose to replant and half chose not to. That year those who did not replant had better harvests.[78]

Crop Care

Women, again usually working in reciprocal work groups, carry out the weeding of potato, barley, and buckwheat fields. In the main villages this is mainly a June activity, although it begins in late May in Nauje and Dingboche barley is also weeded in late May. In 1987 weeding was underway in Nauje on May 26 and by May 29 many families were at work in the fields. Weeding was completed here on June 10. In Phurtse, by contrast, weeding was primarily a late-June activity. The precise timing depends on families' assessments of the size of the weeds and the weather as well as their other work priorities.

Usually fields are only weeded once, although potato fields may, if time allows, be weeded a second time. Families who conducted a second weeding in Nauje in 1987 did so two weeks after the first. Many families also weed again in autumn before later weeds have the opportunity to seed. These weeds can then be dug into the fields as green manure. Weeds can also be useful in other ways. Some are valued as human food (*lu cherma*), although one of these must be double cooked first to make it palatable. Weeds may also be used as fodder.

Weeding is considered to be one of the more laborious field tasks. This is due in part to the Sherpa custom of employing a short-handled hoe or a weeding tool for the task, both of which require constant stooping or squatting. Weeding a potato field can occupy twice the time as planting it, and weeding buckwheat is considered far more laborious than weeding potatoes. A 1,000-square-meter field planted to potatoes can be easily weeded in a day by three people, but the same area in buckwheat is likely to require four days. The time required for weeding potatoes is slowed down, however, if women also take the opportunity at this time to mound earth (*sa kongduk*) around potato plants. This is considered to benefit both plant and tuber growth, but many people forgo it if their free time is limited.

All weeding is completed by the Dumje festival, a seven-day celebration which culminates on the full-moon night of June–July. After this festival all further field work was once banned in all the Khumbu villages until harvest. The Dumje rites thus come at an important time in the agricultural cycle, the point where the crops and the well-being of the villages are thereafter entrusted to luck and the will of the gods. The protective rites of the festival are indeed the great religious event of the year in Khumbu. Dumje is the one festival considered by Khumbu Sherpas to be distinctly Sherpa and it is a powerful expression and reinforcement of village solidarity.[79] Its celebrations include a number of masked dance performances as well as daily communal feasts and nightly parties. But the heart of the festival is a set of exorcism rites that protect the village and its inhabitants from evil including, presumably, such calamities as crop failure.[80] It is held today at Thami Teng, Khumjung, Pangboche, and Nauje, and a similar celebration (known as *Chojen*) is conducted by a separate group of Thamicho families at Kerok.[81] Since the timing of the festival is fixed by a lunar-based calendar, it varies from year to year by up to several weeks.[82]

After weeding is completed another set of Khumbu community measures aimed at safeguarding crops also goes into effect in some parts of Khumbu. These are the regulations aimed at protecting crops from blight. It is believed that these must be activated each summer by late June or early July (early Dawa Tupka), and some communities put them into effect immediately after Dumje. They are considered important enough that enforcing the regulations was the primary responsibility of community nawa, some of whom also were in charge of enforcing seasonal pasture exclusions. While today the full range of these regulations is only maintained in Phurtse, Pangboche, and Dingboche, they were also implemented in living memory in Khumjung, Kunde, Nauje, and Thamicho. These Khumbu measures were apparently unique among Sherpas. They represent an example of close, community cooperation and the overruling of individual family freedom in economic decision making on behalf of the welfare of the community itself.

There were several different types of blight-protection measures. One set of regulations attempted to restrict contact with crops to prevent the outbreak of the fungus and slow its diffusion. To accomplish this communities banned people from entering fields until harvest and excluded livestock from the settlement area until after harvest in order to keep them out of the fields. Keeping people away from crops also helped keep fields free of bad smells that were believed to offend plants and lead to blight. Stories are told of how blight has been brought on by people passing near or through fields, who smelled strongly of soap, garlic, or milk. Many people note that this accounts for field edges near trails often becoming infected with blight, and villagers can tell tales of specific incidents such as the Pangboche case where a line of blight-killed plants traced the path taken by a villager who had crossed a field after bathing.

There were also rules concerned with preventing the outbreak of blight through halting sympathetic responses by crops to events involving the death or drying of plant material in the village area. Among these were prohibitions on cloth dyeing (for which vegetable dyes were common), drying edible, wild plants, and bringing freshly made bamboo mats and freshly cut fuel wood into the village. In some places these rules were still more stringent. In Phurtse and Pangboche no fuel wood, fresh or dried, could be brought into the village once the danger season for blight had arrived in July–August (during the sixth month, Dawa Tukpa) for fear that green wood might be mixed inadvertently with dead wood in fuel-wood loads.[83] In several settlements concern over the importation of freshly cut wood apparently underlay customary summer bans on roof repair (and in some areas on house building). Phurtse today continues to ban outdoor fires, including burning juniper in religious rites, after the fourth day of the sixth lunar month (a date known as *Dawa Tukpa Seshi*). The ban remains in place until the beginning of the potato harvest. At Dingboche this is taken another step further, and on the same day of the sixth month all fires are banned in the settlement until harvest time, including hearth cookfires. Habitation is thus impossible for more than two months. This custom is still deemed to be important to enforce. Many people with Dingboche fields were unhappy and some were outraged when the nawa failed to prevent a 1987 U.S. Everest expedition from camping in the community and allowing their porters to light cookfires at a time when this should have been forbidden.

The rationale underlying these preventive measures seems to be twofold. Some bans are clearly related to fear of contagion and keeping humans and livestock from accelerating its spread. Others are concerned with keeping manifestations of death and decay away from fields. These bans seem to reflect a concern about offending the spirit of plants by polluting fields with the presence of plant death and bad smells. They parallel Sherpa beliefs that crop plants can take offense to the introduction of new varieties or offensive species; for Sherpas plants are sensitive and show their unhappiness by withering and losing their productivity.[84]

Harvest

Harvest season in Khumbu is the busiest time of the year, the only season when work is carried out throughout the daylight hours regardless of the weather. Women move to and from the different settlements in which the family owns crop and hay fields while men add hay (and sometimes also barley and buckwheat harvesting and threshing) to their herding and tourism work. This requires a complex scheduling of family labor across a two-month-long period and often across a variety of different locations. Khumbu families move up and down the valleys from mid-August potato harvesting in the gunsa settlements to September harvesting in the high-herding settlements and the final harvest in late October and early November of a few last fields in the main village. Formerly a number of villages as well as Dingboche enforced community restrictions on the day harvest could begin. Families who began earlier were subject to fines. This practice is still followed in Phurtse and Dingboche. Otherwise the decision is entirely up to individual preference. Household decisions about when to harvest particular agricultural sites during this time are linked to evaluations of the maturity of both potato and grain crops, communal restrictions on resuming field work, increasing risk of crop damage from bad weather, the need to harvest crops ahead of livestock herds descending from the high summer pastures, household labor resources, and astrological concerns.

The first potatoes are harvested as early as the first of August in the gunsa settlements and in Nauje, as families eager for new potatoes dig a few loads for their immediate needs.[85] Harvest is underway in earnest in the gunsa by mid-August, by the end of the month in Nauje, and a few weeks later in the other villages. Most families have completed harvesting most of their main village fields two weeks to a month later.[86] It is not uncommon, however, for families to leave some of the main village potato harvest to complete later, after the barley, buckwheat, and higher-altitude–settlement potatoes are safely in. By early October many families are at work in the higher-altitude potato fields of the valleys and by the middle of the month this is usually completed. For most families that marks the end of potato harvest. The few families who never found time to complete the digging of all their main village potato fields, however, work on. They finish up the harvest only at the end of October or during the early days of November, sometimes working in the snow.

Fields are mostly harvested by reciprocal labor groups of eight to fifteen women. Harvesters move in a line across a field, digging up tubers with their hoes and then deftly flicking or tossing them into lines of baskets set up ahead of the advancing team. Different varieties and sizes of potatoes are tossed into different baskets so that they can be stored separately. Tubers of the small size preferred for seed potatoes are carefully set aside. So too are potatoes that are infected with wart or scab virus; these are eaten immediately before they rot. Men carry the filled baskets in from the fields to the house or hut where the potatoes are air-dried for a week or so before they are stored.

Several different storage techniques are employed. Small quantities of potatoes are cut and sun-dried (*riki shakpa*), and small amounts of potato flour may also be set aside. At low altitudes, including in Nauje, it is possible to store potatoes indoors through the winter. The usual technique is to store tubers indoors in the lower floor of the house either in open, wooden bins (*riki dom*) or inside large, cylindrical containers made from bamboo mats. Care must be taken in cold weather to prevent potatoes from freezing and during bad winters hay may be packed around the wooden bins as insulation. It is too cold in the other main villages and the higher-altitude settlements for this type of storage and in most of Khumbu potatoes are put into underground storage pits (*mik-tung*). Storage pits are approximately 1 to 1.25 meters deep and are lined with straw or juniper boughs and made waterproof with a top layer of straw capped by a firmly packed-down layer of mud. If moisture is kept out underground storage can preserve potatoes for ten or eleven months.[87]

The season for harvesting grain overlaps with potato harvest, and when barley and buckwheat reach the correct degree of maturity their harvest becomes of the utmost urgency. Buckwheat is very susceptible to major damage by late autumn snow or heavy rain, and there is also considerable danger that livestock, wildlife, or birds may destroy the crop. When snowfalls occur while the buckwheat crop is still in the fields some villagers attempt an emergency harvest. Rather than first cut the stalks, carry them to a threshing place, and thresh the grain from the stalks, they instead speed up the process and simply beat down the buckwheat stalks with poles right in the field. If the snowfall proves to be less heavy than feared this tactic can misfire and result in less grain being harvested than would have been otherwise. This was the experience of some Phurtse families two decades ago. In nearby Pangboche, however, heavier snows devastated the harvest of those families who did not rush to harvest what they could as quickly as possible.

When early snows do not rush the process buckwheat harvest begins in late September and lasts for several weeks.[88] Barley harvest also takes place in mid to late September. The two grains are not equally mature at this time. Buckwheat is only harvested when fully ripe. Khumbu Sherpas prefer to harvest barley, however, a week or more before it fully matures. The origins of this custom are not clear. It may be that farmers are reluctant to risk the crop longer to the danger of heavy rains or early snows or that they do not want to delay any longer opening the Dingboche area to grazing. In either case farmers may feel that by harvesting slightly before maturity there is less risk of grain being lost during the process.

Both men and women work together at harvesting grain, although here again most of the labor is contributed by women. Reciprocal work groups are common, but some families prefer to harvest grain as a smaller, family operation. The two grains are harvested using different techniques. Buckwheat is cut with the same type of sickle that is used to cut hay. Workers grasp the stalks with one hand and with the other wield the sickle and cut the stalks close to ground level. Barley, by contrast, can be harvested in several ways. Whereas some people cut the grain from the stalk in the same manner as buckwheat, most simply pull the entire plant out by the roots. This is an easy operation. At Dingboche a group of harvesters can very quickly harvest a large field, a crew of four uprooting barley while two other workers bundle it into stacks. Men and women work their way in a line across a field, uprooting barley stalks and placing handfuls of them behind them for a second crew of women who shake the dirt off, line up the grain heads, and tie together clumps of barley with straw. These are then stacked in small pyramids in the field and later consolidated in large, outdoor stacks.

Buckwheat threshing takes place immediately after harvest. Here again several techniques are employed. Some people prefer to beat or rub the grain free on rocks. Others beat the stalks with a short, forked stick. Often a bamboo mat enclosure is set up to keep out wind and to capture flying grain. Buckwheat stalks are valued as fodder and after threshing they are piled, dried, and stored. Barley, by contrast, is not threshed until a month or more after it has been harvested and is beaten with a jointed flail (*geli*). In late October and early November groups of women carry out the threshing with poles on large community threshing grounds in Dingboche.

The agricultural year draws to a close in mid to late November. The few families who have neglected to complete the digging of their potato fields due to more pressing business now dig in the frozen soil for the final tubers, sometimes working while the snow falls. Some families also redig their potato fields now or plow the fields that had been in grain to loosen the soil in the belief that this will improve the coming year's crop. Some families in Phurtse note that this process also makes the spring's work easier, for by deep digging the soil of next year's potato fields in the late autumn, when there are no other agricultural tasks, they save themselves effort during the busier days of spring. In the spring it is then sufficient to simply dig the fields once rather than the usual double digging. Planting can be done simultaneously with loosening the spring soil and digging in manure.

Labor Commitments

The amount of labor required to produce Khumbu crops varies considerably among crops. Potatoes require by far the greatest labor, demanding a double hand-digging of fields for preparation and planting and a third digging at harvest as well as one or more weedings. Buckwheat is much less trouble in terms of field preparation and planting, for it is not manured and a few hours of plowing and broadcasting serves to plant the crop. Weeding the buckwheat, however, is extremely time consuming as is harvesting it with sickles. Barley requires more work than buckwheat, for it must be manured and provided with irrigation. It is, however, much less trouble to harvest.

The amount of time devoted to producing a crop also varies among different fields. Considerably more effort is usually put into the care of fields nearby the house than those located elsewhere in a settlement, much less those a half an hour, a half a day, or several days away. Fields adjacent to houses are usually the best-manured and the most carefully weeded. Special attention may also be given to fields in the high-altitude, secondary agricultural sites. The crops in the high-herding settlements, by contrast, are sometimes given less care. Here weeding especially may be relatively neglected. The amount of time devoted to potato production in both main and secondary fields also varies considerably among households. Many families only conduct a single weeding whereas others perform this task twice. Some require more time and work to prepare fields because of their greater supplies of manure. A few families in Nauje and Khumjung have their potato fields plowed rather than digging them with hoes, thereby saving a great deal of labor. And the amount of work a particular family must invest in a given field also varies from year to year. Here the size of the harvest makes a considerable difference. A bumper crop may require much more effort to harvest and store. But a poor potato crop can be highly time consuming to harvest in terms of the commitment of time relative to the returns in stored food. In a good field in a good year a single worker can readily harvest three or four forty-kilogram loads of potatoes in a single day. On poorly producing land a day's labor may yield only a single load. One Nauje family farming relatively marginal land outside of the main village found in 1987, for example, that twenty-five person-days of labor were required to harvest 26.5 loads of potatoes from seven small terraces.

The relative labor requirements of different agricultural tasks and the degree of variation in the labor invested in them can be seen in a comparison of the farming practices of two Nauje families in 1987. Family *A* farmed seven small terraces in the vicinity of the village (Mishilung and Nyeshe). Figures are given for the five fields at Mishilung only. Family B

Table 9. Nauje Potato-Cultivation Labor Inputs

Task	Person-days of Labor	% of Total Labor
Family *A*		
Field Preparation	13	35
Planting	4	11
Weeding	8	22
Harvest	12	32
Total	37	100
Family *B*		
Field Preparation	6* (22)	9 (26)
Planting	16	23 (19)
Weeding	12	17 (14)
Harvest	35	51 (41)
Total	69 (85)	100

*These fields were plowed rather than being hand dug. Digging them would have required another sixteen person-days of labor. Figures in parentheses refer to estimated adjustments for hand-dug fields.

farmed eleven small fields in Nauje itself. Family *A* was part of a recipro-cal labor group whereas family *B* relied mainly on wage labor. Family *B* also plowed some of its larger fields which accounts for the discrepancy in terms of labor used. For family *B* I have shown a second set of figures which is adjusted to show labor investments if the fields had not been plowed.

Nauje family *A* harvested only eighteen loads of potatoes in 1987 from its five fields at Mishilung with an investment of thirty-seven days of labor. Terraces of Nauje family *B,* by contrast, yielded 64.5 loads in 1987 (down considerably from nearly ninety in 1986 as a result of blight, but still far more than the family itself required) with an investment of sixty-nine days of labor. There are thus considerable differences in labor efficiency relative to yields as well as to the relative amounts of time the two families devoted to different agricultural tasks. Nauje family *A* de-voted considerably more time than family *B* in preparing its fields and weeding them, but harvested only eighteen loads of potatoes with thirty-seven person-days of labor investment (2.1 person-days of labor per load). Nauje family *B* harvested nearly twice as many potatoes per day of work invested, harvesting 64.5 loads with sixty-nine days of labor (1.1 person-days of labor per load). At this rate of return on labor input, a family could attain self-sufficiency in potatoes with less than a month of labor devoted to the effort. Less than half that much labor would need to be spent on grain crops.

4

Good Country for Yak

Pastoralism in Khumbu must cope with the challenges of severe winter temperatures and meagre winter and spring range resources. Stockkeeping at these altitudes can be a highly risky endeavor and in bad years some stockkeepers can loose much of their herd to starvation, cold, and disease. Their choice of which types of livestock to herd, the means by which they can best exploit different altitudinal and seasonal range resources, and the ways to best minimize stock loss in the difficult winter-spring period represent adaptive responses to the high-altitude conditions of Khumbu. Yak can thrive at these altitudes, and Khumbu is good country for yak, but even yak require special attention if they are to survive all year in Khumbu conditions. Sherpas have had to develop an ingenious set of pastoral practices and seasonal herding patterns and make a major commitment to cultivating hay in order to safely support their yak herds all year on Khumbu resources.

Environmental adaptation thus certainly underlies and to some degree broadly shapes characteristic Khumbu patterns of herd composition and herding movements. Khumbu pastoralism, however, is a complex activity that also reflects a number of other cultural, social and economic factors. Cultural values and lifestyle preferences influence decisions about herd composition and size, the degree of multifamily cooperation in herding, and the role of community regulation of livestock management. Differences in household wealth, subsistence orientation, and involvement in commercial production create different herding strategies, scales, and goals. And cultural knowledge, assumptions, and beliefs about livestock, predators, pastures, climate, disease, luck, and spirits also influence herding decisions through the role that they play in herders' interpretations of environment, resources, and risks.

This chapter surveys contemporary Khumbu pastoralism. It begins with an introduction to the types of stock herded, household and village patterns of herd composition and size, herding strategies and goals, and regional range resources. The second half of the chapter then focuses on the diverse factors that influence family and village seasonal herding patterns. It explores the ways in which these annual pastoral cycles of transhumance, like the agricultural cycle, are complex cultural achievements in which Sherpas have successfully integrated knowledge of local environments and the characteristics of different types of livestock in a set of pastoral strategies. These strategies, in turn, will be seen to reflect sociocultural values, communal resource-management efforts, and individual households' conditions of wealth, labor availability, and lifestyle preferences.

Yak, Nak, Crossbreeds, and Cattle

Khumbu Sherpa pastoralism is primarily based on the herding of a set of different varieties of cattle (*chungma*), as shown in table 10. For many generations it has been especially characterized by the keeping of yak, particularly the female nak. In recent years households in some villages (especially in Nauje, Khumjung, Kunde, and Thamicho) have begun keeping increasing numbers of yak-cattle crossbreeds and cows. Since 1957 the number of nak has declined by nearly half (from 2,061 in 1957 to 1,121 in 1984), and today yak and nak constitute 58 percent of Khumbu cattle rather than the 79 percent share they comprised in 1957. Yet they remain the most numerous and the most valued large livestock in the region. Yak and nak have long been an important symbol of wealth and the keeping of a large herd confers considerable status on its owner. They are prized not only for their economic and social utility but also out of appreciation for their beauty.[1]

Yak herding is a central facet of subsistence in much of highland Asia. The great, shaggy, black, brown, multihued, and occasionally white bovines are one of the most distinctive and widespread residents of the high, alpine lands of the Great Himalaya and the vast expanses of the Qinghai-Tibetan plateau. Yak herding is important virtually wherever there are Tibetans or their close cultural cousins, from the higher reaches of the Himalaya across Tibet to the high country of Gansu, Sichuan, and Yunnan, and even beyond in an arc encompassing Mongolia, the Mongol-inhabited Lake Baikal country of Siberia, some Kirghiz (Kyrgyz)-inhabited areas of Xinjiang and Kyrgyzstan in Central Asia, and portions of the Pakistan-administered Karakorum range. Most of

Table 10. *Khumbu Cattle Numbers and Emphases, 1984**

	Kunde	Pangboche	Khumjung	Phurtse	Thamicho	Nauje	Khumbu
Yak	62(19%)	69(22%)	62(17%)	67(17%)	109(12%)	44(15%)	413(16%)
Nak	85(26%)	132(42%)	88(24%)	261(68%)	525(56%)	30(10%)	1,121(43%)
Zopkio	56(17%)	21(7%)	82(22%)	7(2%)	178(19%)	138(48%)	482(18%)
Zhum	80(24%)	32(10%)	60(16%)	27(7%)	41(4%)	18(6%)	258(10%)
Pamu	42(13%)	57(18%)	74(20%)	16(4%)	57(6%)	53(19%)	299(11%)
Lang	5(2%)	7(2%)	7(2%)	5(1%)	35(4%)	3(1%)	62(2%)
Total	330	318	373	383	945	286	2,635

*Numbers of stock and percentage of cattle by type.
SOURCE: Data derived from Brower 1987:189

the world's yak graze the high country of the Qinghai-Tibetan plateau. In Qinghai Province alone there are 4,920,000 yak, by one estimate a full third of the world yak population (Zhang et al. 1987:88), and in the Tibetan Autonomous Region there are more than an additional four million head of stock (Yan 1986:241). Another 700,000 are herded in one Tibetan-inhabited county in southwest Gansu, Gannan Tibetan Autonomous Prefecture (Li 1986:133). The Himalaya supports far fewer head. The entire range may be the home of fewer than 100,000 yak, and estimates of the number of yak in Nepal range from fewer than 9,000 to somewhat more than twice that (Brower 1987:202 n. 1).[2]

Yak are not likely to be confused with anything else in Asia.[3] Large-bodied, long-haired, long-horned and humped, yak are one of the world's most physically distinctive cattle types. Domestic male Nepalese yak attain a size of 230 to 360 kilograms and females from 180 to 320 kilograms (Brower 1987:166), rather smaller than the 1,000 kilograms that wild yak are said to sometimes reach but still a large animal in comparison to other varieties of Himalayan and Tibetan cattle. They are also notable for the deep, grunting, staccato bellow that gave them their Latin appellation *Bos grunniens* ("grunting cattle") as well as for their heavy coats. This coat ends in a distinctive fringe of long hair on the legs and flank and is complemented by an extraordinarily bushy tail—so striking that yak tails were accepted as tribute gifts by the Chinese imperial court and were in demand as a trade good in India for use in Hindu temples.[4]

Yak are superbly suited to year-round life at high altitudes. They can endure altitudes and harsh weather in which cattle or crossbreeds could not survive and are even able to thrive in the cold of a Khumbu winter. They do not require stabling even at 5,000 meters in January. In mid-winter yak are able to forage in snowy country by seeking out patches of ground that the wind has swept free of snow or by pawing through the snow to feed off quiescent grass. Sherpas believe that yak are so much creatures of the heights that they will sicken and die if taken to altitudes of less than 3,000 meters during the warm-weather months. Below 3,000 meters they are thought to succumb rapidly to "low-altitude sickness," a malady Sherpas associate with both altitude and unhealthy water sources. Research on yak physiology has found that they are physically adapted to cope with high-altitude conditions. Besides their heavy coats they have larger lungs and thoracic cavities than other cattle, and their unusual blood composition may also enable them to adapt more effectively to high-altitude conditions (Brower 1987).

Yak and nak provide a number of valuable products and services.[5] Both males and females provide meat that Sherpas prize above all other. Like other Tibetan people Sherpas do not subscribe to Hindu beliefs about the sanctity of cattle and have no qualms about eating beef.[6] Before the 1960s old animals were culled each autumn and winter, slaughtered, and butchered by itinerant specialists of the lowest social status who came for this purpose from Tibet. After socioeconomic change in Tibet during the 1960s these butchers ceased coming to Khumbu and culling was virtually abandoned.[7] One Tibetan butcher, however, is now living in Khumbu and occasionally practices his craft. Some Sherpas suggest that the problem with culling cattle today is not so much the absence of butchers as it is the need to avoid offending government officials who reside in Nauje.

More important to subsistence than yak meat is nak milk. Of its many uses, the most important is the production of butter (*mar*). Some milk is also processed into cheeses, including both a soft cheese (*shomar*) that is consumed fresh and spread on potato pancakes and a dried, hard cheese (*churpi*) that is a valued trail food. Butter is perhaps most often consumed in tea, for it is an essential ingredient of the basic beverage of the region, the *solcha* or salt tea familiar across the Tibetan culture region. The making of butter is also an important strategy for storing milk, for little fresh milk is available in the region other than in summer. Butter is much preferred over cheese in this respect and it can be kept nearly year round in Khumbu's cool climate, sometimes wrapped in skins for insulation. Yak and nak manure is also regarded as an important resource, both for fertilizer and fuel, and for many families fertilizer is the most crucial contribution of yak to the domestic economy. Both male and female yak provide hair that is shorn, spun with drop-spindles by men and women, and woven into *chara,* a heavy material that has many uses. Hides can be used as boot soles. And both males and females are used as pack animals and some have been trained as draft animals.[8]

Yak and nak also can provide income. Surplus meat, dairy products, and manure are bartered and sold locally. Butter is an acceptable currency with which to hire agricultural day laborers and formerly, according to oral traditions, was used to pay taxes to Tibet. Yak-hair chara are sold in Khumbu and in other nearby Sherpa regions. Considerable income can be earned from hiring out pack stock or from training stock to pull a plow. Trained teams are rare and many farmers pay a good deal of money to hire plow-teams and their drivers. And the breeding and sale of nak-cattle crossbreed calves has historically been so lucrative that it was probably the most important factor in the regional prominence of nak.

Khumbu Sherpas keep a small number of common cattle of a Tibetan variety known locally as *kirkhong* (*Bos taurus*). These today number fewer than five hundred (Brower 1987:169,171). Despite their relatively thick coats Tibetan cattle are not hardy in Khumbu winters and must be stabled in the lower stories of village houses and fed considerably more winter fodder than yak or nak require. Most of the stock kept in Khumbu are cows (*kirkhong pamu*) that are valued for their milk and manure. During the past twenty years, as the standard of living has increased with tourism development, it has become increasingly popular for families to keep a cow or two to provide milk for their household. The small number of bulls (*kirkhong lang*) play a vital role in Khumbu pastoralism, for they are bred to nak to produce the crossbreeds for sale to Tibet and to Shorung Sherpas.

The crossbreeds that are bred in Khumbu from nak mothers and Tibetan bulls display a combination of traits reflecting their parentage. In general they differ from yak in size, coat, and their lesser fitness in extreme high-altitude conditions (particularly in winter). The males, known as *dimzo* (or *dim zopkio*), are highly valued as draft and pack animals whereas the females, known as *dim zhum* (or *dzum*), are considered excellent milch stock.[9] Fürer-Haimendorf reported that zhum yield more milk per lactation than nak, although it is not as rich. According to his information Khumbu crossbreeds produce about ten kilograms of butter per year per head compared to the nak's seven kilograms per year per head.[10] Crossbreeds herded in Shorung were said to yield twice as much butter as those pastured in Khumbu (Fürer-Haimendorf 1975:50). It should be noted, however, that higher milk yields for Khumbu zhum compared to Khumbu nak may reflect differences in fodder feeding and the relatively greater parts of the year that the zhum are herded in the lower-altitude areas of Khumbu. Zhum could not survive, much less yield large amounts of milk, in the conditions in which nak live.

Tibetans have long practiced yak-cow crossbreeding. Tibetans along the eastern fringe of the Tibetan culture region, for example, gave such crossbreeds in tribute to the Tang emperors of China (Schaffer 1963:74). In some parts of Tibet, however, crossbreeding is not conducted due to a belief that the forced cross-species mating is offensive to local gods such as yul lha and lu.[11] The Tingri region just to the north of Khumbu is one of the areas where crossbreeds are valued and used, but are not bred. This has provided Khumbu herders with a major market for crossbreed dimzo calves and a complex trade developed in them still continues today on a limited scale. Dim zhum, considered finer milk producers than nak, although less hardy, have long been sought after by Shorung Sherpas and in the past twenty years have begun to be kept by Khumbu Sherpas in large numbers to provide family milk supplies.

In recent years a third form of crossbreed has been important in Khumbu. This is the *urang zopkio,* the male offspring of lower-altitude *pamu* or *palang pamu* (*Bos indicus*) cows and yak. Urang zopkio are not bred in Khumbu, but are purchased at considerable expense from lower-

Female Parent

		nak(f)	pamu(f)		zhum(f)
			kirkhong	palang*	
		(m)yak	(m)dimzo**	(m)urang zopkio	(m)koko yak
	yak(m)	(f)nak	(f)dim zhum	(m)urang zhum	(f)koko nak
Male Parent	lang(m)	(m)dimzo	(m)lang	(m)lang	(m)zhumi tolu
		(f)dim zhum	(f)pamu	(f)pamu	(f)tolmu***

*   only found in Pharak and other low-altitude areas
**   also known as dim zopkio
***also known in Pharak as pakhim

Figure 11.   Major Khumbu Cattle Breeds

altitude areas including Shorung, Kulung, and Pharak. They have become important in Khumbu as plow animals and especially as a source of pack-stock income from tourism.[12] Their value in trekking work reflects the importance of the airstrip at Lukla as a tourist entrance to the area. Sherpas do not like to take yak down into the low-altitude reaches of the Dudh Kosi gorge in order to reach the Pharak airstrip. But they have no hesitation in taking the low-altitude-fit urang zopkio to Lukla, and the crossbreeds are also capable of making the haul through Khumbu to the foot of Mount Everest. The increase in the number of zopkio since the mid-1970s has been the most spectacular recent change in Khumbu pastoralism. By 1984 18 percent of all Khumbu cattle were zopkio and the great majority of these were urang.[13] They now far outnumber yak.

Sheep and Goats

Sheep and goats have also long been a component of regional livestock keeping. Both are valued for their hair and wool, their meat, and their manure.[14] According to oral traditions sheep were an element of early Sherpa pastoralism in Khumbu, and goats may also have been raised here for centuries. But for at least the past hundred years both have been raised on a far smaller scale than nak and during the past forty years regional numbers have not far exceeded 1,500.[15] Today there are perhaps 500 in all (Brower 1987:171) and a flock of 30 head is considered large. Only a Tibetan variety of sheep is kept. The wool and meat of the Tibetan sheep are far preferred to those of the less hardy Himalayan varieties that Gurungs once herded each summer in Khumbu. These sheep must be fed fodder in the winter, but they do not need to be stabled at the altitudes of the main villages. The Khumbu lack of emphasis on sheep keeping contrasts strikingly with high-altitude pastoralism in northwestern Nepal (Goldstein 1974, 1981; Goldstein and Messer-schmidt 1980; Rauber 1982), Tibet (Ekvall 1968; Goldstein and Beall 1990), and Mongolia, where sheep are the most numerically important form of livestock.[16] Sheep also outnumber yak in Nepalese high-altitude pasture areas to the immediate south and east of Khumbu. Most families who keep yak or other cattle do not also herd sheep, and sheepherding is primarily a pursuit of relatively poor households that cannot afford other stock. The lack of interest in the region in sheep keeping seems to reflect a cultural attitude rather than either a perception that the region is poor sheep country, as Fürer-Haimendorf (1979:13) and Brower (1987:171) have suggested, or a lack of interest in sheep products. Wool, dried sheep meat, and dried fat are imported from Tibet and bring high prices.

As far back as Sherpas can remember goatherding has been con-ducted on an even smaller scale than raising sheep. In 1983, when goat keeping was banned in the area as a conservation measure, there were fewer than 300 head, most of which were kept by non-Sherpa, black-smith families.[17]

Household Pastoral Emphases and Herd Structure

There are currently several patterns of livestock ownership in Khumbu that differ not only in scale but also in the type of livestock herded, the degree to which pastoralism is a subsistence or a commercial activity, the altitudinal span of Khumbu rangelands used, the seasonal migrations followed, and the amount of wealth and labor required. There are four main patterns of herding (fig. 12). These emphasize the keeping of nak, cows or crossbreed zhum, yak or zopkio, and sheep. There are also sev-eral major patterns of keeping mixed stock herds.

Nak herders primarily keep herds of a size large enough not only to meet household requirements for milk and other products but also to produce enough crossbreed calves to make the enterprise worth the effort. Herds generally are almost entirely composed of nak, with one or two Tibetan bulls kept for breeding with them. Usually more than ten head of nak are kept, for smaller herds yield a very small return of calves. Earlier in the twentieth century there were herds of as many as eighty head, which were tended by hired herders, but today herds of more than thirty head are rare. These animals are kept much of the year in the high reaches of the valleys, often only spending a few weeks or months in the autumn and early winter below 4,000 meters. Usually the stock require large amounts of fodder as supplementary feed in winter and especially in early spring before the new grass greens. Nak herders as a result usually grow hay as well as collecting and drying wild grass for

Livestock	Herd Size	Economic Goals	Requirements
Yak	2-6	Income from pack stock	High-altitude herding huts Hayfields
Nak	15-35	Income from sale of crossbreeds Manure, dairy products	High-altitude herding huts Hayfields
Zopkio	2-8	Income from pack stock	Winter stabling
Cows/Zhum	1-6	Household dairy supplies Manure	Midaltitude herding huts Winter stabling in village
Sheep	10-40	Wool, food, manure	None

Figure 12.   Khumbu Pastoral Strategies, 1990

hay. The need to invest in stock, hayfields, and high-altitude herding huts to act as winter and summer bases and barns limits involvement in this kind of herding to relatively well-to-do families. It also requires a major lifestyle commitment on the part of herders, who must be away from the main village for eight months or more of the year. Usually this task is taken on by one member of a family, who is known as the family *nakpa* ("nak person"). In summer these herders are often joined by the entire family and these months in the high pastures are considered by many people to be the best time of the year. But few envy the nak herders the many winter months they must spend at high altitudes in the bitter cold in the remote corners of the region. Some families who once were large-scale nak owners have had to break up their herds as the nak herders aged and were unable to interest sons in carrying on the work. In a few cases daughters have taken over the task. Today this job is rarely entrusted to hired herders, and Sherpas especially feel that low-landers are not capable of caring for nak properly. Nak herding is felt to require considerable skill, for to produce a high number of crossbreed calves a herder must know his animals well.

Households that keep cows and female crossbreeds keep them in small numbers as a source of dairy products for household consumption, and this style of pastoralism requires relatively little wealth and relatively minor lifestyle commitments. Usually only two or three head of such stock are owned and the largest herds in memory numbered fewer than twenty animals. Larger herds are kept by Sherpas in Shorung and some other areas, but there the production of milk, butter, and cheese is a commercial enterprise. In Khumbu virtually no dairy products are sold and the region has long imported butter from Shorung and Kulung. Keeping cows and female crossbreeds does not require ownership of hayfields, long periods of residence in the high pasture, or winter migra-tion to the high herding huts. During most of the year it is simply necessary to send cows out to the village's adjacent slopes in the morning and to round them up again in the evening. During summer the stock must be taken outside of those villages (Khumjung, Kunde, Pangboche, and Phurtse) where the settlement area is closed to grazing to protect the growing crops from livestock. But it is not necessary to make long migrations or to maintain hay fields or huts in high-altitude settlements. Typically cows are taken to areas just outside the restricted zone and herded there the entire summer from a single hut or temporary shelter located just an hour or two away from the main village. Khumjung and Kunde villagers who keep cows may move their cows as little as five kilometers to a summer base near Mong. Many owners of cows and female crossbreeds do not grow their own hay, but instead rely on harvesting wild grass, storing crop residues, and purchasing additional hay if necessary. The need to stable cows in winter and to feed them considerable fodder in the winter and spring does, however, demand considerable labor and expense.

Earlier in the twentieth century a few families kept small herds of yak to use as pack animals. These yak herds usually consisted of fewer than ten animals and the largest was probably that of a famous Nauje trader who kept as many as thirty yak to use in his caravans. Today families that emphasize herding yak and their use as pack stock in the tourist trade commonly own only two to four head of stock. Yak herding is in some ways much simpler than herding nak, for in summer they do not need to be corralled at night or milked daily and can instead be allowed to wander in the highest reaches of Khumbu. Herders simply seek them out every week or ten days to feed them salt and check on their condition. If relatively small numbers of stock are owned these can be wintered in lower Khumbu without the necessity of maintaining extensive hayfields and high-altitude herding huts.

Far more common today than yak herding, however, is owning a small number of urang zopkio to use as pack stock on mountaineering and trekking journeys. Here again the herd size is small, typically with two head of stock per household. Zopkio can also be kept without any need to own a high-altitude herding hut since they can be allowed to range freely for the entire summer with occasional day-trips up from the village to feed them salt. In the autumn, whenever they are not needed as pack animals, zopkio can be left to graze freely in the lower forests until the snowy season begins. Zopkio must be stabled and fed fodder through the winter, however, and this requires either an investment of time in collecting wild fodder or maintaining hayfields or the expenditure of a proportion of the income made from the pack stock on purchasing fodder. Fodder must also be supplied to stock when they are on upper-valley trekking trips during the late autumn, winter, and spring.

Sheep (and formerly also goats) are raised in small numbers, primarily by poor families who are unable to afford cattle but who want manure for their fields, meat, wool, and the possibility of a small income. Manure is often reckoned to be the most important benefit. Sheep and goat keeping requires less capital and lifestyle investment than any other form of Khumbu pastoralism, although in villages where summer exclusion of livestock is practiced it is still necessary to practice transhumance and herders must either spend several months living in temporary shelters and caves or maintain herding huts.

There are also several patterns of keeping mixed herds. These include household herds that emphasize nak but also include yak or zopkio, herds in which both nak and zhum and/or cows are kept, and herds that have a combination of zopkio or yak with zhum or cows. Families who keep nak may also keep a small number of zhum or cows that provide milk to the household during seasons when the nak may be herded in distant parts of Khumbu. Nak-herding families may also keep a few yak, that are useful for breeding nak to keep up herd numbers and as pack stock for income from tourism. A few urang zopkio may also be kept as income-generating pack stock. And families who keep a few zopkio or yak may also keep one or two zhum or cows to provide milk.

Herding Strategies: Village Contrasts

Besides contrasts among households in herding practices there are also some broad contrasts among villages. Nak herding predominates over the other two patterns of cattle herding in Phurtse, Pangboche, and the Thamicho villages. In Kunde, Khumjung, and Nauje, however, keeping cows and zhum as well as yak and urang zopkio herding are both more important than nak herding (table 11).[18] Phurtse families especially emphasize nak over all other types of cattle keeping whereas Nauje families emphasize it the least. In Kunde and Khumjung keeping cows and zhum is virtually balanced by herding yak, urang, and dimzo, and in these two villages very few families today herd nak.[19] Nauje families display by far the greatest emphasis on the herding of male pack stock, and especially on the herding of urang zopkio.[20]

These different village herd-ownership patterns are also reflected in different patterns of seasonal use of Khumbu rangelands. Nak and yak are kept for much longer periods of time at high altitudes than other cattle. The Thamicho villages, Pangboche, and Phurtse all accordingly keep 64 percent or more of all village large livestock in the high country for most of the year. In the case of Phurtse this reaches 85 percent. The other villages, by contrast, have more than 50 percent of their large livestock in crossbreeds or common cattle, all of which spend all but the

Table 11.   Village Cattle-Herding Styles by Percentage of Cattle, 1984

	Nak/Lang	*Zhum/Pamu*	*Yak/Zopkio*
Kunde	28	37	36
Pangboche	44	28	29
Khumjung	26	36	39
Phurtse	69	11	19
Thamicho	60	10	31
Nauje	11	25	63
Khumbu	45	21	34

SOURCE: Data derived from Brower 1987:189.

summer months in the immediate vicinity of the main villages. This tendency is most pronounced in Nauje where male and female cross-breeds and cows have mostly grazed close to the village year round since the practice of banishing stock from the settlement in summer was abandoned in 1979.

Village Patterns of Regional Pasture Use

Village herding patterns are also shaped by land ownership and customs of pasture use. Since the nineteenth century all Khumbu has been open range for Khumbu Sherpa livestock except where seasonal prohibitions on grazing are in effect, and theoretically any family can herd in any part of Khumbu it wishes.[21] But there are old traditions of village territories (*saja*) which continue to guide herders today, and some influence is also exerted by inheritance. Households tend to herd in areas where they have herding huts, hayfields, crop fields that require manuring, and local knowledge of terrain, grass, predators, and microclimates.

The upper Imja Khola valley and the Lobuche Khola area are regarded by Pangboche villagers as their village land even though Kunde families also herd in the area and own many herding huts. The eastern side of the upper Dudh Kosi valley is considered to be Phurtse land and the western side (except for one small Phurtse area opposite that settlement) to be Khumjung land. The Bhote Kosi valley is considered to be Thamicho land except for the lower eastern valley area that belongs to Nauje.[22]

Herders from particular villages tend to herd their stock in the same areas making use of the same valleys or the same sides of valleys.[23] These village herding areas have remained roughly similar during the twentieth century except for the area utilized by Nauje herders, who previously made more use of high-altitude pastures in the Imja Khola and Bhote Kosi valleys (see chap. 6).[24] Recent herding patterns are shown on map 8.

Thamicho families from all the Bhote Kosi valley settlements herd throughout that valley. Here herding settlements are usually shared by families from several different villages. In the past a few Nauje families also used the upper Bhote Kosi valley pastures for nak and since the early 1970s Nauje families have made increasing use of Langmoche and Chosero as summer herding bases for urang zopkio. The lower Bhote Kosi valley, including areas considered Thamicho land, has long been grazing for Nauje livestock of all types except during the three to four months when it was closed to grazing by now-abandoned Thamicho communal-management regulations. Now Nauje stock graze there year round. The most important current pasture areas for Nauje stock are the slopes immediately around the village and extending for short distances up the Bhote Kosi and Dudh Kosi valleys. Most of this area is considered Nauje village land. The Gyajo valley has also long been used by Nauje families for summer dimzo herding. There are no herding huts in the valley, and urang zopkio are turned loose there to graze on their own for weeks on end. Kunde and Khumjung herders do not make much use of Gyajo, and Thamicho herders who take herds there only do so in the early spring of years when grass and hay are very scarce.

The upper Dudh Kosi valley is the grazing ground of Phurtse and Khumjung herders. With only very few exceptions herders from each village keep to their own side of the river.[25] Phurtse families make use of the large rangelands around the main settlement and extending along the Imja Khola as far as the border between Phurtse and Pangboche land. Here a line of stone slabs was set up along a spur to mark the boundary between Phurtse and Pangboche territory. Just to the north of Phurtse in the Dudh Kosi valley is the major hay-production site of Konar where many families have herding huts. During the post-Dumji period the lower valley is closed to grazing up to a point slightly north of Konar. All of the upper eastern Dudh Kosi valley, however, is open to grazing throughout the summer. Many Phurtse families have one or several herding huts in the numerous high-altitude herding settlements and secondary high-altitude crop-production sites in the upper valley.

Khumjung and Kunde livestock are both herded part of the year in the pasture and forest area adjacent to the villages. Khumjung herds also graze the western Dudh Kosi valley up to remote pastures among the moraines and glaciers of the head of the valley. The highest herding settlement, Gokyo, is situated on a lakeshore at 4,750 meters, but yak are allowed to wander for weeks still higher up valley at places such as

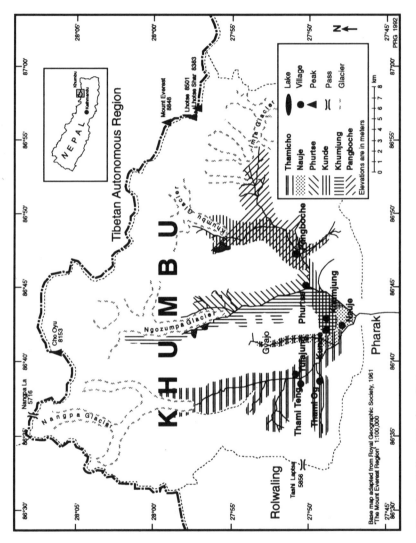

Map 8.   *Village Herding Patterns*

Gyazumpa and there are some resa at altitudes as high as 6,000 meters. Kunde yak- and nak-herding families, by contrast, rely on the high-altitude pastures of easternmost Khumbu.[26] Most concentrate on the valley of the Lobuche Khola where they share most herding areas with Pangboche families. Kunde families alone, however, use the Melinang area east of Tugla, the Lobuche area, and the higher ground towards the foot of Mount Everest. Some of the differences of emphases in herding areas between Pangboche and Kunde as suggested by patterns of herding-hut ownership can be seen in map 11. Kunde families especially base in the Pheriche-Pulungkarpo-Dusa area in the early summer before this area is closed to livestock and then use the Tugla area for a base. Pangboche villagers make more use of the western side of the Lobuche Khola, the upper Imja Khola valley, and the environs of Pangboche itself. Pangboche herders also use the Yarin area whereas some families from both villages use the adjacent Pulubuk region, as did some Nauje herders until the 1960s.

Communal Regulation of Pastoralism

Since at least the middle of the nineteenth century Khumbu Sherpas have maintained village- and valley-based agropastoral management systems aimed at protecting crops and pastoral resources through controlling grazing and the cutting of wild grass for hay.[27] We have already encountered these systems when reviewing Khumbu agriculture, for the same community nawa officials who implement the regulations designed to protect crops from blight also enforce a form of rotational grazing that protects growing crops from livestock depredation, limits the use of some high-altitude pastures, and protects crucial winter grazing and fodder resources.[28] The system operates on a zone basis. In each valley a set of zones is sequentially opened and closed to specific activities. Opening or closing a zone to an activity such as grazing is known as opening or closing the *di* for livestock, *di* being a ban on a certain type of land use. Zonal systems such as this are not common in Nepal and are very rare even among other Sherpa groups.

During the past century five different regional nawa systems have been in operation (map 9). In eastern Khumbu a system of four zones is administered by nawa chosen in Pangboche and Dingboche. In the Dudh Kosi valley there are two different administrative systems, a Phurtse-based one that supervises two zones on the eastern side of the valley and one area immediately across the river from Phurtse and a Khumjung- and Kunde-based system that regulates the use of two zones that extend from these villages as far east as the Mong ridge overlooking the confluence of the Dudh Kosi and Imja Khola rivers. Nauje had its own set of nawa until 1979, who supervised community agropastoral regulations in a single zone in the immediate neighborhood of the settlement. The rest of the Bhote Kosi valley has been the responsibility of nawa chosen in Thamo, Pare, Thami Og, Thami Teng, Yulajung, and Tarnga. These nawa administer six zones in the lower and middle altitudes of the valley. It should be noted that not all of these systems are "village" pastoral management institutions, for most are actually multi-village institutions. The operation of the livestock bans in the Bhote Kosi valley requires coordination between nawa elected in several different villages, as does the Khumjung and Kunde system. The Imja Khola valley system relies on both Pangboche and Dingboche nawa, and at Dingboche all residents of the community including families from Pangboche, Khumjung, Kunde, and Nauje, hold office in a rotation of responsibilities.

As indicated in map 9 much of lower Khumbu is included in these five different networks. In the Imja Khola and Lobuche Khola valleys a substantial amount of subalpine and alpine pasture area is managed. Elsewhere in Khumbu little high-altitude land is subject to communal constraints on pastoralism. Herding restrictions in most regulated areas are only in operation from late June through mid to late September.

Management zones are opened and closed to livestock, hay and wild grass harvest, and a variety of activities associated with protecting crops from blight. A particular zone can be closed or opened to different activities at different times. It is common, for example, to close the village area to livestock before closing it to all work in the fields and to open it in the early autumn for crop harvesting and hay cutting several days or weeks before allowing the return of the herds. In most cases nawa have little latitude in deciding when to open and close zones. These decisions are generally linked to the Sherpa lunar calendar and the regional festival cycle. The regulations that nawa enforce are also determined not personally but by custom and community decisions. The zones themselves have been defined since before 1900 and are clearly demarcated in local consciousness in terms of particular landscape features.

In most cases nawa are selected by village assemblies, although in some settlements and at some times the real selection has been made by gembu, pembu, or groups of powerful and wealthy families (table 12). In a number of cases there is a rotation system with the office annually reassigned by village election to a different household and no villager allowed to hold office more than one year at a time. Rhoades and Thompson (1975:541–542) have seen in this a parallel with Swiss institutions, but it should be noted that not all Khumbu villages rotated nawa and in some settlements where rotation of office was practiced not all

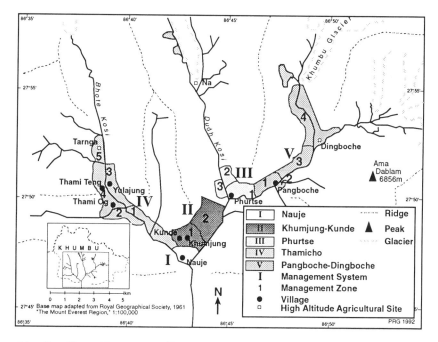

Map 9.   Nawa Management Systems

resident families were necessarily considered eligible.[29] Some Sherpas view the rotation system as an attempt to diffuse the burden of an onerous office. The nawa's work is sometimes jokingly referred to as "crazy job," a thankless obligation that involves immersing oneself in what is bound to be a great deal of trouble policing and fining one's neighbors and relatives. Other Sherpas see the rotation of the job not just as a way to share a distasteful task equitably but as a way in which abuses of power can be minimized.

In most cases nawa are chosen or announced each spring in the fourth month, usually at a village gathering at the end of planting. This commonly is held on the day when the protective ritual circumambulation of the fields is performed. In Pangboche and Phurtse this day also marks the closure of the village to livestock.[30] Elsewhere in Khumbu no zones are closed until during or after the Dumje celebration in the following month. In Nauje, for example, before the system there was abandoned in 1979, the village was always closed to stock during the week after Dumje. It is up to the nawa to meet and choose the exact day, which is usually announced several days before the deadline. Any stock still in the area after the deadline are subject to fines. After closing the area to grazing the nawa become responsible for responding to reports of viola-

Table 12.   Khumbu Communal Pastoral Management

Place	Total Zones	Total Nawa	Altitudinal Coverage
I Nauje	1	3	Low altitude
II Khumjung-Kunde	2	4	Low altitude
Khumjung	2*	2	Low altitude
Kunde	2	2	Low altitude
III Phurtse	3	2	Low to middle altitude
IV Thamicho	6	12	Low to middle altitude
Pare	1	2	Low altitude
Thamo	1	2	Low altitude
Thami Og	1	2	Low altitude
Thami Teng	1	2	Low altitude
Yulajung	1	2	Low altitude
Tarnga	1	2	Middle altitude
V Imja Khola	4	4	Low to high altitude
Pangboche	3	2	Low to middle altitude
Dingboche	1	2	Middle to high altitude

Low altitude defined as below 4,000m
Middle altitude defined as 4,000–4,500m
High altitude defined as above 5,000m

*Kunde and Khumjung nawa share responsibility for these zones.

tions and policing the zone if they consider it necessary. Nawa seek out offending Sherpas or the owners of ban-violating livestock and demand compliance with the regulation and the payment of a fine. Livestock owners are required to offer the officials *chang* (the local beer), and sometimes also pay a small fine that is usually put toward village projects such as maintaining temples, bridges, and trails. Fines escalate in size with continued refusal to comply. Nawa can do no more, however, than issue further fines. They are not authorized to confiscate livestock. If someone refuses to comply with repeated warnings a nawa can inform other community members of this. Community social pressure generally ensures compliance with the regulations.

All Sherpas are required to obey the nawa's injunctions regardless of whether they are residents of a particular valley or not. This responsibility balances the right given to all Khumbu Sherpas to make use of all pasture lands in the region regardless of their village affiliation and the boundaries of particular villages. In order to make allowances for herders' occasional needs to cross through zones that have been closed to grazing all herders are permitted to remain with their stock a single night within the boundaries of a closed zone. This privilege is sometimes abused, however, when stockkeepers judge that they can escape the notice of nawa for a few additional days. All owners of livestock who are found violating a local zone closure are subject to the same fine regardless of whether or not they are "outsiders." The same regulations have also been applied to non-Sherpas making use of Khumbu for summer transhumance. In former times when Gurungs herded in Khumbu they were also expected to adhere to the nawa's regulations in the areas in which they were allowed to herd.

Regional Grazing and Fodder Resources

The summer richness of the Khumbu high-country pastures has historically attracted transhumant shepherds from lower-altitude regions as well as fattened Khumbu herds. But winter and spring are a far more challenging time, and for at least a century the numbers of livestock kept by Sherpas have exceeded regional range carrying capacity during these times of the year. Half a century ago Khumbu herders, like those in much of the rest of the Himalaya, responded to this seasonal scarcity of grass by migrating out of Khumbu to areas that offered better grazing. Many of Khumbu's nak and yak were taken across the Nangpa La to Tibet where the winter grazing is far superior and Gurung shepherds returned with their flocks to their homes in the lower Dudh Kosi valley. The current Sherpa pattern of year-round herding in Khumbu is a twentieth-century accommodation to social and political factors rather than to environmental conditions. During the past sixty years herders have developed new fodder resources rather than scale back stock-keeping. Yet despite their efforts to put in large stockpiles of wild grasses and herbs and cultivated hay many herders risk serious losses each winter and early spring when heavy snowfalls can lead to a shortfall of fodder and the starvation of even such hardy stock as yak.

Khumbu range resources vary with altitude, season, aspect, micro-climatic conditions, and grazing pressure. Much of the skill of herding lies in evaluating local range resources and guiding stock through a number of finely timed moves among grazing areas. Through many generations of experience in particular valleys herders have developed an intimate knowledge of microregional and temporal variations in pasture quality. This knowledge is but one of a number of different elements that influence decisions about the annual round of herd movements. But, like the more formalized operation of the community pastoral management institutions, such local knowledge represents an adaptive link between environmental perception and pastoral practice.

Khumbu valleys provide pasture at several distinct altitudinal levels. The south-facing slopes of the gorges of the lower valleys below 3,800 meters offer rangeland grazing and open woodland foraging that is especially useful to crossbreeds, cattle, sheep, and goats during winter and spring. While northern-aspect slopes in this altitude range are often densely forested and relatively little grazed, the southern-aspect slopes offer vast expanses of grass and shrubs that are a major pastoral resource. Above 3,800 meters there is good summer grazing in the large expanses of subalpine and alpine grassland on the floors and slopes of the wide, glacier-carved upper valleys. There is also good grazing in this season on the slopes of long-stable lateral and terminal moraines. The quality of grass and other palatable forage in a particular pasture or woodland area varies from year to year, but some places have long-established reputations for offering fine grazing at particular times during the year. The quality of the grazing is considered to reflect not only seasonality, altitude, local rainfall, aspect, and level of grazing intensity, which affect the grass itself, but also local shade and ground moisture conditions that are believed to affect the health of the livestock. Wet and muddy sites are believed to lead to hoof infections and stomach problems whereas shady sites are thought to cause stock to lose their conditioning as a result of their spending more time resting and less time eating.

Khumbu cattle, including yak and nak, must be fed fodder in order to survive the winter and early spring. Like other Himalayan peoples Khumbu Sherpas make use of field residues. In the autumn livestock are turned into the barley fields to graze on the field stubble. Buckwheat and barley stalks are carefully dried and stored to feed to stock in winter and spring, and surplus tubers are also given to them. A small amount of fodder, especially bamboo, is also obtained from local woodlands and forest.[31] These resources, however, are inadequate for the levels of stock kept in the region and herders instead rely primarily on hay. This is obtained in two ways. Wild grass (*ri tsa,* mountain grass) is cut and dried from meadows that have been protected from grazing in the late summer. Hay (*tsi tsa*) is cultivated in stone-walled fields. The production of cultivated hay on the scale harvested today in Khumbu is rare elsewhere in the Himalaya, and hay is not cultivated in the adjacent part of Tibet.[32]

In Khumbu hay is grown from 3,400 meters to as high as 4,700 meters (map 10), with most of the area in hay situated above 4,000 meters. In the high-herding settlements of the upper valley hay is the primary crop and in many of them it is the only thing grown. The highest settlements with walled hay fields are Bibre (4,600m), Dusa (4,503m), Gokyo (4,750m) and Apsona (4,600m). There would undoubtedly be considerably more land in hay today, given the scarcity of fodder regionally,

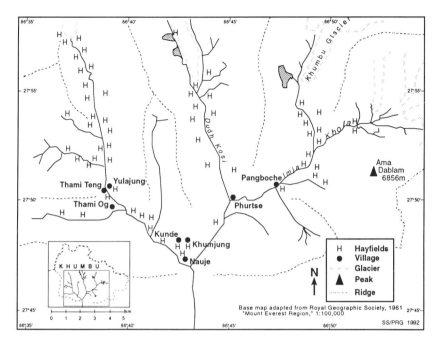

Map 10.   Hay Cultivation

except for the government policies that prohibit agricultural expansion. Some land was surreptitiously put to hay in the 1960s and 1970s and in some Thamicho villages land is now being taken out of food crops and put to hay.

Khumbu hayfields are normally planted only once. The autumn before a hayfield is to be established herders collect the seed of wild grasses that are preferred for hay. The following spring a field is plowed from pastureland and then sown with the wild grass seed. The final step is to erect a stone wall for protection against livestock.[33] Once they are established hayfields need only be manured annually. Each spring nak and yak are corralled at night on the hayfields to collect their manure and dung is also gathered from nearby slopes. Some families, however, begin collecting manure as early as the previous autumn. This dung is first pulverized and then spread thinly over the field. Little else is done to encourage a productive yield. A few families weed hayfields of plants that are unpalatable to stock, but most herders do not bother. Until recent Sherpa experiments in Nauje and the upper Lobuche Khola valley there was no irrigation of hayfields.

The main effort in haymaking comes at harvest. Sherpas consider the cutting, drying, and storing of hay to be some of the hardest agricultural work of the year. Both men and women cut hay, but most of the work is done by men. This may in part reflect the fact that women are usually busy with potato and grain harvesting at this time of the year. It requires a full day and a group of five workers to cut the grass on a single 500-square-meter field. The grass is laboriously cut by the handful with a sickle. This can be dangerous work, especially when the grass is short and the sickle must be wielded close to the hand grasping the grass. When the grass is high the work goes much more quickly, for less care has to be exercised to avoid accidents. Households normally cut hay-fields near their homes themselves, but it is very common to hire help for more distant fields. Arrangements for laborers must be made several weeks ahead and relatively high wages must be paid. Wages are paid by the field cut rather than on a daily basis and the price agreed on usually does not include labor for drying or storing. Payment is often in butter rather than in cash.[34] Usually the field owner is also obliged to provide food and drink for laborers. Once cut the grass is spread on the field for several days to dry and then stored indoors. Sunny weather is critical to this operation.

Hay made from wild grass is also stored in large quantities. Vast areas of lower Khumbu pasture are protected from grazing for much of the summer and some high-altitude pasture areas in the upper Imja Khola and Lobuche Khola regions are also protected for a few weeks in late summer. The grass in these areas grows long, and a week or so before the nawa announce the reopening of these zones for grazing they are first opened for grass cutting. Men and sometimes also women move about the countryside with huge baskets and long knives cutting the green grass and then hauling it back in baskets to the herding huts and main houses to dry and store. Throughout the region there is real compe-tition to gather as much as possible from the most convenient areas, and in recent years there have been many cases of people defying the nawa in order to start harvesting grass a few days earlier than the rest of the community. Harvesting begins in the Nauje area as early as late August (August 18 in 1987) now that there are no nawa there to forestall grass cutting until the pasture grass is mature and seeding.[35] In most of the rest of Khumbu there are still grass-cutting regulations. Here the high-altitude grass is cut first, ahead of the herds that descend the valleys in sequence with the opening of the various zones to grazing. In the Dudh Kosi valley, for example, grass cutting in hayfields and pastures begins in unregulated areas such as Gokyo and Na in the high reaches of the valley in mid-August (August 15 in 1987). In late August (August 24 in 1987) the grass- and hay-cutting restriction at Konar is lifted and shortly thereafter hay making commences around Phurtse itself.

Families who herd cows and crossbreeds often do not cultivate hay.

Their stock requires a good deal of fodder in winter and in early spring, but households with just one or two animals can usually meet most of this requirement by collecting crop residue, grass from around the edge of fields, and wild grass and other fodder.[36] Households with more cross-breeds and cows, or that have less time and energy to collect wild fodder, may have to use some hay. A few Nauje families in this situation now cultivate hay in the Bhote Kosi valley at Samde, Tashilung, and Tarnga.[37] They and other families also rely on purchasing some hay, primarily from Thamicho villagers. Although Brower (1987:262) has suggested that zopkio are quite cheap to keep and require little fodder feeding, the experience of some Nauje herders is different.[38] Here it is considered ideal to feed urang zopkio four to five kilograms of wild-grass hay in the morning and again in the evening (as well as providing *bantsa,* household scraps, and *kole,* a grain porridge) when the pasture is poorest in spring, and the survival of stock depends on fodder supplies.[39] But many owners are unable to collect sufficient wild grass for hay even by augmenting their family's efforts with hired labor. They find themselves obliged to purchase hay from those with a surplus to sell. Nauje families especially look to Thamicho families for hay. When possible Nauje families make agreements during the summer for hay that will be cut that autumn, for this can be obtained at lower prices. By the end of the 1980s, however, fewer Thamicho families were willing to make such deals, preferring to sell in the autumn at higher prices. There is also increasing importation of hay from Pharak, although it is widely agreed that Pharak-grown fodder is lower in nutritional value than that produced in Khumbu.[40]

Field-grown hay is more expensive than hay made from wild grass, for it is considered to have more energy as livestock fodder. The most costly of all is hay from high-altitude fields (*phu tsa*). This is believed to have more energy than hay made from the same grass species but grown lower in Khumbu (*rong tsa*). In 1987 the going rate was 100 rupees per forty-kilogram load for fodder of mixed hay and wild-grass hay, a price then equal to two or three days' wages for portering. This, however, was the price for hay that was paid for in advance during the summer. The price had risen to 150 rupees per load by cutting time. By spring two hundred rupees was thought to be a very reasonable price and could be obtained only if one picked the hay up at the site rather than had it delivered. Prices also vary regionally depending on the quality of the harvest in a particular year and the dynamics and anticipated dynamics of supply and demand in a given place. In spring 1986 hay was 250 rupees per load in Kunde and at Dingboche it reached 300 rupees. But even at these prices it was nearly impossible to obtain in any quantity.[41] By autumn 1990 hay prices had increased yet further. In Thamicho advance-contract hay was 150 rupees per load and was being sold in October-November for 300 rupees per load on the open market. Pangboche hay was 400 rupees per load in November 1990, 150 rupees more than it had been at that time the previous year, and was expected to increase to 500 rupees per load the following month. In Nauje in 1991 hay was selling for 500–600 rupees per load, and small amounts were available from people with hayfields at Pare, Tesho, and Samde.

Yak and nak herding are considered to require hay growing. This is entwined with herding patterns and the ownership of herding huts in pasture areas far from villages. Hay is rarely transported in large quantities to main villages and instead is stored in the high settlements where secondary houses reflect the need to have both a secure storage place to keep hay dry and a place for herders to base during the bitter months of winter. The winter migration of nak and yak into the high country results from a simple shortage of grazing and fodder in the main villages and the fact that it is more practicable to bring the livestock to the hay than to haul it back down to the main village. Hay is occasionally brought to the village when a family considers it important enough to go to the trouble in order to be able to stay on in their main house for a few days or weeks longer. The time and energy required to transport large amounts of hay, however, limits this practice.[42] In spring the location of hay stores and hayfields patterns herd movements. This results both from the need to take the stock to the places where fodder has been stored and keep them there while supplies last and the need to fertilize the hayfields each spring.

Herding Patterns

Seasonal pasture use and patterns of annual pastoral movement among pastures reflect herding households' resources of land, labor, and capital, herd size and composition, the location of pastures traditionally used by one's fellow villagers (and in some cases also by nonvillage kin), the operation of the nawa system, lifestyle preferences, family customs, and a number of other factors that vary in importance from family to family and herder to herder. Pastoral movement patterns also reflect processes of past and present environmental adaptation at both the level of individual decisions and the development of collective resource-management institutions. The definition of community management zones and the timing of their opening and closing to livestock grazing and hay cutting reflects an intimate awareness of local range resources. Broad seasonal herd movements and day-to-day decisions about grazing represent evaluations of opportunities and risks that are also based on profound local knowledge of pasture conditions, climate, predators, and the requirements and capabilities of stock.

Pastoral movement patterns are discussed in depth in the remainder of this chapter. First the annual round of pastoral movement is explored as it relates to local knowledge about range conditions and livestock, the operation of the nawa system, and other factors. Then I discuss in detail the contrasting herding movements of two families, one from Kunde and one from Pangboche, who both herd nak in the upper Imja Khola and Lobuche Khola. This highlights some of the factors herders consider in their movement decisions and the ways in which different household priorities and resources can result in different patterns of pasture use that nonetheless also embody generally similar seasonal and altitudinal strategies.

The Annual Round of Transhumance

Seasonal movements of livestock to different Khumbu altitudes and microenvironmental regions are based above all else on evaluations of the fitness of different types, genders, and age groups of stock for different climatic and pasture conditions. Different general patterns of herd movements can be identified for several basic types of local pastoral styles and herd structures. For several types of stock raising, including the keeping of crossbreeds and common cattle, the annual round of transhumance is very similar to the summer movement into the high pastures and the winter retreat into lower-altitude regions that is so common in mountain regions throughout the world. But this is not the case for yak and nak. Yak and nak are well enough acclimated to the high Himalaya that they can cope with even the higher-altitude regions in the height of winter.[43] This offers possibilities for a rather different type of pastoralism in which seasonal altitudinal shifts are less pivotal to transhumance than movement based on other factors.

The seasonal shifts of herd movements for these different styles of transhumance can be traced from any point in the annual round. Two points in the cycle, late summer and the weeks of late autumn and early winter, are perhaps a more convenient starting point than others. Only at these two times are all Khumbu stock located in roughly similar regions, in the middle- and upper-altitude pastures during late summer and in the main villages in late autumn and early winter. I will trace the round beginning in late autumn (fig. 13).

Virtually all stockowners bring their herds down to the main villages during the clear, cool, late-October and November weeks following the end of the potato and buckwheat harvest and hay making. The zone around the main villages is then open to grazing for the first time since it was closed to protect the growing crops from stock depredations back in

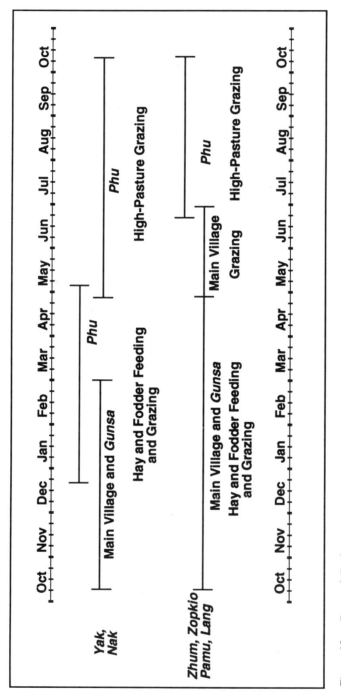

Figure 13. Pastoral Cycle

late spring or early summer, and there is good grazing on grass that has been protected for several months. Families which have cultivated grain may also allow their herd to graze on crop stubble, which contributes to field fertility as well as nourishing the livestock. Families which live in two-story houses begin to stable stock overnight in the lower story of the house, putting forest leaves and needles underneath them to be enriched by their urine and dung and composting this outdoors for use in the spring.[44] As the weather grows colder and the first snows fall in late September or October, indoor stabling becomes necessary for cross-breeds and cattle, although yak can remain outdoors.[45] During the days the herds are led out to graze adjacent rangeland and woodland.

As winter settles the grazing becomes less rich in the lower valleys. Stock are still led out to graze the lower Khumbu grass and shrublands and to forage in the forests, but it is taken for granted that both cross-breeds and ordinary cattle will begin to require supplementary fodder by November and that they will have to be fed large amounts of fodder until June. What fodder is fed them depends not only on availability but also on many beliefs about the appropriate forms and amounts of fodder to be given to particular types and ages of stock. Depending on the condition of the grazing near the villages, yak and nak may be shifted to new herding bases as early as late November or as late as January. They are moved up into areas of the upper valleys at this time to which cross-breeds and cattle will not be taken until June or later. In the era when Khumbu stock was herded in Tibet in the winter it was not necessary to feed yak or nak supplementary fodder during the winter months, for they were fully capable of subsisting on senescent grass. This is not possible today. Herders must feed fodder to both yak and nak to supplement what grass the stock can find uncovered by snow or reach by pawing through a thin snow cover. Herders move their nak and yak during these months among the places where they have stored hay during the autumn, staying in each as long as fodder supplies hold out. Even for yak, however, some areas are considered too cold in mid-winter. It is not usual, for example, for Thamicho herders to take yak to Tengbo in the upper Thami Chu valley in mid-winter. This area is considered to be too cold for stock since it is situated in a very narrow valley that gets little sun during mid-winter months. Yak and nak are only driven up to Tengbo in the spring.

Fodder becomes important for yak and nak in December. During the remainder of the winter they are given hay on a daily basis, and the amount of fodder given them increases as winter continues. At first only pregnant stock are given supplementary fodder, but soon all are given a ration. In early winter nak and yak are given hay once each day in the evening. By January, when the coldest days of the year come on, two feedings of hay per day are necessary. Pregnant nak now are given fodder three times per day. In February and March nak and yak are still fed hay twice per day, but as supplies run short the amount given during each feeding is decreased. In many years hay feeding continues to be necessary into April or May depending on weather, snow cover, grass, and hay stores. All other stock are also fed fodder in the winter and often receive not only hay but also bantsa, a mix of household refuse and grain.

The arrival of spring finds the same contrast between yak and nak at the higher altitudes and other stock stabled in the villages. Yak and nak movements during March and April remain strongly related to the availability of hay stores in different areas. An additional factor now, however, is the need to manure hay and high-altitude secondary crop areas. Both these priorities keep the movement of the herds geared to the location of household hay stores until supplies are exhausted. For yak and nak these are the difficult months when hay begins to run short and the new grass has not yet arrived, the time which Sonam Hishi calls the "months when rich men cry." In bad years stock may starve. In the worst recent years some herders have lost half their herd.

By May the new grass is returning. The key to the quality of the grass (and to the coming autumn's hay crop) is considered to be the amount of spring rain in April and May. Some people consider the rainfall in May to be the crucial factor. If there is no rain during those months both grazing and hay will be bad. Some places such as Pulubuk and Dzongla are noted for having better grass even in dry springs due to particularly rainy microconditions. Gyajo, a narrow valley just north of Kunde and west of the sacred mountain Khumbila, is also sometimes used as an emergency grazing area. The U-shaped floor of the valley offers limited pasture at an altitude of 4,000 meters, which some Thamicho herders resort to in springs short of fodder.[46]

In late spring major differences can be seen in the way in which different types of stock are moved to exploit new grass. Yak and nak are taken high to places such as Chukkung, Mingbo, Dzongla, Gokyo, Charchung, Arye, and Chule, where they can use their unique ability to lap up extraordinarily short grass that is beyond the capability of crossbreeds or cattle to graze. Other stock are kept in the main village, but the need to feed them fodder is past.

At the outset of Dawa Nawa, the fifth month, in June or early July, herd movements may reflect a herder's interest in being close to the main villages for the Dumje celebration. Many Pangboche herds, for example, are brought back down to lower altitudes for this occasion. Yak and nak may be taken to Yarin, Ralha, and Orsho in the Pangboche area. Herds remain in Yarin and other nearby areas until after Dumje when the area around Yarin is closed to livestock.

After Dumje there is much new grass in the high country. During the rest of the summer herd movements hinge on perceptions of the altitude fitness of the herds, the closure of zones to grazing, and household priorities as to where to be based at the time of the Yerchang festival.[47] During the summer yak and urang zopkio are considered to be capable of subsisting on their own in the highest, remotest pastures in Khumbu. Yak especially are taken up into the high corners of Khumbu and turned loose for weeks, and yak and nonmilking nak can be found during mid-summer in the uppermost Imja Khola valley far beyond Chukkung and above Tarnak and Gokyo in the Dudh Kosi valley. Less remote and slightly lower areas such as Gyajo, Tengbo, and Langmoche are very popular for turning zopkio free to graze on their own.

The timing of herding in different parts of the upper valleys may vary, reflecting perceptions of the value of grass in different places as well as household circumstances such as ownership of huts and resa. Consider-ations about the state of the grass must be tempered also by other evaluations of grazing conditions. Some places such as Dzongla, Tugla, and Pulubuk have a reputation for being especially rainy areas and some herders do not consider them healthy places for livestock.

Cows and zhum may also be taken high in summer. Zhum, for exam-ple, are taken as high as Chukkung. Some people note that in the summer zhum can be taken anywhere where nak are herded. Cows are also herded at relatively high altitudes, although I have not seen them as high as 5,000 meters. Zhum and cows, however, are not always herded at high altitudes in summer. Many families, on the contrary, take them to places such as Mong, the Omoka-Pulubuk area, and Ralha which are just outside of the closed zones. Kunde (and formerly Nauje) zhum are often taken to Pulubuk in the summer, and more than fifteen Khumjung and Kunde families with zhum and cows base themselves at Mong and nearby places.

In autumn the lower-valley zones are reopened to grazing and all stock are brought back down valley. They now graze areas adjacent to the main settlements which have been closed since at least July. Cows and crossbreeds stay based in the main villages until summer. Yak and nak, by contrast, shift grazing locations considerably during the follow-ing months. They are usually herded at the altitudes of the main villages in late October for a month or two. In late autumn they may be taken briefly to the high country to graze on the grass that grows up following the cutting of the hayfields. They are then returned to the main village for a few more weeks or months until the move is made to return to the high herding huts for the remainder of the winter. The timing of the winter move varies. Pangboche herders may move yak and nak up valley in December. Some Thamicho herders may begin even earlier, going up valley for the grass that grows up after the hayfields are cut, then grazing the high valleys wherever they judge the grass to be good until they are forced to begin feeding the stock fodder and must base in herding settlements where they have hay stores. Other stock owners may delay the up-valley move and keep their herds in the lower Bhote Kosi valley into December. Phurtse herders move their stock up valley rather late, usually in mid-February. Once moved up to high altitude in winter, nak and yak remain until autumn unless they are brought down briefly in the spring to graze new grass or at Dumje time.

The herding pattern outlined above is probably an old one for crossbreed and cow owners regionwide, and has also been long used by Phurtse and Pangboche herders who have relied on the cultivation of hay and on year-round herding within Khumbu itself. Many large nak herders of Khumjung, Kunde, Nauje, and the Thamicho settlements, however, historically relied much less on hay and sent their stock to Tibet for the winter rather than to high-altitude herding settlements within Khumbu.

Communal Regulations and the Timing of Herding Patterns

The opening and closing of livestock-management zones in parts of the Khumbu valleys during the summer and early autumn establishes a limited form of rotational grazing that gives a basic shape and timing to herd movements in all the valleys for three to five months of the year. The operation of the nawa system does not prescribe a fixed pattern for regional herding. Families still exercise considerable choice over their herd movements. Many families decide to move their stock into other areas before a particular zone is officially closed by the nawa, and in some years a few families choose to move their stock into zones before they are officially opened. But the sequential closing and opening of zones to grazing certainly is a major factor in herding decisions and transhumant patterns during the months when the system is in effect.

The first closures of zones to livestock take place in eastern Khumbu in May when all stock is excluded from the immediate vicinity of Pangboche and Phurtse following the end of planting and the ritual Tengur circumambulation of the fields (map 11).[48] Elsewhere no areas are closed until June and the time of the Dumje festival. The zones of lower Thamicho are the next to be closed, an event that takes place each year on the tenth day of the fifth month in the midst of the Dumje festivities.[49] Nauje and Khumjung-Kunde close the area in and around their villages to grazing only after the end of Dumje.[50] Since the date when villages are closed to grazing is tied to the lunar calendar the date

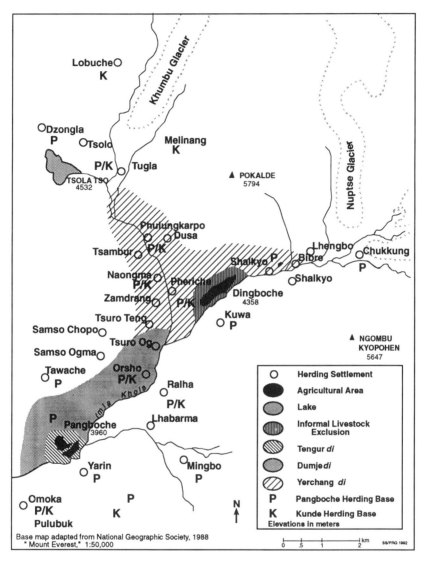

Map 11.   Imja Khola Pastoral Management

of closure can vary from year to year by several weeks. In some years Dumje and the closure of the village to stock comes in June, in other years well into July.

There are usually enough families in the lower-altitude Khumbu villages who keep their cows and crossbreeds in the villages until the last possible minute that there is a general upward movement of people and stock to the high country just before the lower valleys are closed to grazing. The stock spend the rest of the summer up valley until after the main village crops have been harvested.

Within a few days after Dumje the nawa in Khumjung and Kunde announce that the area around the villages will be closed to livestock five days hence. All stock must be gone from the zone near the villages and the gunsa of Teshinga by the first day of the sixth month, two weeks after Dumje. Nauje nawa also formerly announced the imminent closure of the village area to stock just after Dumje and also gave a five-day notice that they shouted out from the top of a big rock in the center of the settlement. In Thamicho, before nawa regulation was abandoned in the late 1980s, the lower Bhote Kosi valley was closed before Dumje as far north as the bridge just south of Thami Og, and after Dumje the ban on livestock was extended as far as the Mususamba bridge on the Langmoche Chu. Phurtse stock at this point in the year have already been excluded from the settlement, but now the herds must also be cleared from the lower valley to a point beyond Konar. The summer livestock ban at Pangboche has similarly closed the area immediately adjacent to the village for several weeks before Dumje. Now Yarin and other areas of the lower Imja Khola are closed.

Families choose their own moving day from within the period allotted before the zone is closed to grazing. The day chosen by each family usually is a day of the week or a particular date that is considered particularly auspicious. This may be one of the days considered lucky each month, such as the fifteenth, the day of the week that is lucky because it was the day on which the head of the household was born, or a day that has been determined to be lucky by consulting horoscopes or a Tibetan farmer's almanac. On the morning of the move juniper boughs are burned outside the house sending fragrant clouds of smoke skyward. To this offering the family head adds a few prayers, and then the windows are latched and the door locked and the family sets out on foot, driving their stock ahead of them and leading a few yak or zopkio laden with the household goods and supplies they will require for the summer. The lead member of the party carries a prayer flag on a length of bamboo. This will be set up at wherever the family settles and is moved with them during the summer from herding hut to herding hut.

For the next three or four months families camp out in the tiny herding huts and resa, moving from pasture to pasture according to their own whims and perceptions of pasture conditions. These high country months are a time of the year that many look forward to eagerly, a time of green slopes sparkling with wildflowers, crystal mornings beneath the great peaks, plentiful milk and yoghurt, and, best of all for many people, freedom from the gossip, factionalism, and social demands of village life. Up in the high-herding settlements there is time for life to revolve around one's own family and those of a very few neighbors.

Once beyond the border of the closed zone a family may set up its base wherever it chooses and some families go little further the entire summer.[51] Other families move between a number of high-herding huts and temporary resa. Only in the Imja Khola area does anything other than family decision making dictate any particular structure to the rest of the summer's movements. Here the operation of another livestock-exclosure zone closes a large area of good grazing ground after the Yerchang festival one month after Dumje. Most herders tend to herd in this zone until it is closed and usually base at this time in the settlement where they will celebrate the festival. Family custom and preferences for whom they want as neighbors at the festival have a good deal to do with the choice of site.

In most of Khumbu Yerchang is a major event combining communal rituals with an intense period of socializing.[52] The festival consists of a day of rites, several of which involve prayers believed to ensure the health and prosperity of the herds. All the members of a herding settlement, regardless of home village affiliation or clan, gather together to offer prayers to the great god Khumbu Yul Lla and the other deities of Khumbu. These rites include the dedication of *torma* (consecrated flour and butter images) representing the livestock associated with Khumbu Yul Lha: yak, sheep, and goat. At this time some families may make some of their livestock *chetar,* dedicated to the gods, and these animals are hence forever free of the threat of being killed by humans.[53] The crowning event is a rite in which each household plants a pole bedecked with prayer flags at a common outdoor shrine.[54] A number of days of celebration follow. Each family in the settlement hosts a party at its hut. One party takes place each day until the round is completed. The celebration of Yerchang thus marks one of the most intensive communal periods of the year, a time of major interaction with a few families in shared ritual, feasting, drinking and dancing. Yerchang is only carried out at certain high-herding settlements, and families choose where they want to celebrate it largely on the basis of the people with whom they want to share the festival. Most families return to the same place each year making this one of the few fixed points in the pastoral calendar.[55]

Herders in most of Khumbu may simply remain where they are after Yerchang since no further livestock exclosures are implemented. Many families in these areas, however, shift base, generally moving farther up valley if they have places there. In the Imja Khola and Lobuche Khola areas herding families must disperse after Yerchang due to the subsequent closing of most of the Yerchang sites to grazing. For the next six weeks or more most herders in the upper Imja Khola and Lobuche Khola valleys move up into the highest-altitude areas of eastern Khumbu.[56] Within this remaining open area families are free to move as they will, and the places most popular for grazing vary slightly from year to year. Kunde families emphasize the Tugla area in some years, for example, and in other years move to Melinang when the grass is best there.

The opening of areas closed to livestock in September and October influences herd movements in all the valleys. Although no herder is obliged to move his herd into an area simply because it has been opened, in practice most herders are eager to let their stock graze areas that have been protected from grazing for as many as three months. Some are indeed all too eager, and nawa are kept busy issuing reprimands to those whose stock violate the final days of the grazing ban. In each valley several zones are sequentially opened and livestock accordingly move down valley in several stages, all of which takes place within a two- or three-week period between late September and mid-October. Usually each zone is opened first for grass cutting and a flurry of activity then takes place as hay is harvested and wild grass cut. Zones where there are also agricultural sites are next opened for harvesting, and only then are livestock permitted to enter them.

In the Bhote Kosi and Dudh Kosi valleys there are, or recently were in the case of Thamicho, fewer stages in the down-valley pastoral migration than in the Imja Khola valley. In the Dudh Kosi valley Tarnga was opened first, then all the area down to the bridge below Thami Og, and finally the lower valley. Hay and crop harvests preceded herding in each of these areas. In the eastern Dudh Kosi area the timing of the opening of the zones to various activities is based on the ripening of the first buckwheat in Phurtse in certain fields in the center of the village. In 1987 the first white buckwheat grains were seen here on August 19. Five days later the cutting of hayfields and of wild grass was allowed in the Konar area. Normally this might have been allowed somewhat sooner, for in this year the opening was delayed to coincide with an auspicious day. On September 10 that area was opened to livestock and Phurtse was opened to potato harvesting. Buckwheat harvest began on September 23 in Phurtse and the area was then also opened to cutting wild grass. On October 1 the area between Phurtse and Pangboche villages along the Imja Khola valley (but not including Phurtse itself) was opened to livestock. The settlement itself was finally opened to stock on October 10 following the completion of the buckwheat harvest. Khumjung stock move first into the area just north of Mong once Phurtse has lifted the seasonal ban it enforces there on grass cutting and grazing, then into the Teshinga area once that area has been opened to grass cutting and harvest at the gunsa has been completed. The ban on livestock in Khumjung and Kunde itself is the last to be relaxed, an event that takes place in mid-October after the completion of the buckwheat harvest and most of the potato harvest. The management of livestock exclusions in Nauje, before this was discontinued in 1979, was the simplest system of all. Here there was only a single zone to be concerned with. Grass cutting was allowed in September, after the fifth day of the eighth month, Dawa Gepa. The exact date was chosen by the nawa with regard to the maturity of wild grass for hay making. The area was reopened to livestock in October, Dawa Guwa, following the harvest.

The process in the Imja Khola valley is somewhat more complex. Here there are more zones involved and the harvest at Dingboche also complicates the sequence. The high-altitude zones in the upper Imja Khola and Lobuche Khola valleys are opened first, in the reverse sequence of their closure. Each is opened first for wild grass and hay cutting and only then for herding. The order of these openings is fixed and the timing between the opening of different zones for different activities throughout the valleys is scheduled according to a standardized, lunar-calendar-based sequence. The sequence is initiated by the arrival of mid-summer which, according to the Sherpa reckoning of the seasons and their relationship to the lunar calendar, can come rather late in the year. In 1987 the mid-summer point in the Sherpa calendar came on September 1. When mid-summer arrives the first zones in eastern Khumbu begin to be opened for grass cutting. The Ralha area comes first in the sequence.[57] A week later, on September 7 in 1987, grass cutting is allowed in the zones that were closed after Yerchang. A week after that (September 14, 1987) these zones are opened to grazing. On this day the nawa also allow the barley harvest at Dingboche to commence.[58] Precisely one week later Dingboche is opened to grazing. In 1987 all families had completed their harvest here by that date except for one, which finished the work by mid-afternoon. Dispatch is important in order to avoid crop loss, for large numbers of livestock are immediately moved into the area as soon as it is opened for grazing. Despite the relatively late start of this sequence of zone openings in 1987 barley was still harvested somewhat earlier than Sherpa farmers would have preferred. Khumbu barley is always harvested several weeks before it fully ripens, perhaps as a safety precaution against damaging weather in October or simply as a compromise in order to allow the grass cutting, crop harvesting, and transhumant cycles in the valleys to proceed more smoothly. But the 1987 harvest had to be carried out when the grain was even less mature than usual. Some farmers noted that harvest took place at the right time on the calendar but at the wrong time according to the plants.[59]

After stock return to Dingboche there remains only the lifting of the remaining grass- and hay-cutting, crop-harvesting, and grazing restrictions in that reach of the Imja Khola valley between the confluence of the river with the Lobuche Khola at the Dolimsampa bridge and the Pangboche area. A week after stock moved into Dingboche, on September 28 in 1987, many herders moved their stock down valley beyond the Dolimsampa bridge into the remaining two zones outside of Pangboche.[60] Pangboche itself was opened to grazing on October 9, after all the buckwheat and most of the potatoes had been harvested.

Family Herding Patterns

The patterns of herd movements already discussed are broadly characteristic of most herding families in Khumbu. This level of generalization illustrates a number of the important factors involved in household decisions about herding. But a closer inspection of the decisions made by individual families is necessary to appreciate the complexity of factors that herders must weigh. All the herders from a given village do not simply all take their stock to the same areas where the grass is good and the areas open for grazing. Individual families may even follow different herding patterns from year to year. Yet for all the complexity of household pastoral movements key factors in their herding strategies can be identified and generalizations at this microlevel can be made about the seasonal movements up and down the valleys. As an example of this the following section compares and contrasts the herding decisions and pastoral movement patterns of two families who herd nak in the upper Imja Khola region. One family is from Kunde, the other from Pangboche.

Both of these households are well-to-do by Khumbu standards. Both are only minimally involved in tourism and concentrate instead on crop growing and herding. They both also predominately herd nak and sell crossbreed calves. Each has a herd that is large by Khumbu standards. The Kunde family, one of the six families in that village who herd nak, owns four yak, sixteen nak and ten other head of cattle (urang zopkio, zhum, cows). It has herding huts at Lhabarma, Dusa, Orsho, Dingboche, Pheriche, Phulungkarpo, and Teshinga as well as resa at Dzongla, Tsola, and Melinang. As one of the largest stock-owning families, the Pangboche household has about twenty-five head of cattle, primarily nak, herding huts at Dingboche, Bibre, Selum Che, Yarin, and Chukkung, and resa at Mingbo and Pulubuk. The following account focuses on the movement of the nak.

By early spring the stock of the Kunde family, like that of the other three Kunde families who herd nak in eastern Khumbu, is already in the high country, where for some weeks it has been driven between the places where the family owns hayfields and keeps hay stored in huts. The amount of time spent at each of these varies from year to year with

Table 13. Household Herding Pattern (Kunde)

Midwinter/Spring
  December/January through May

Dingboche	20–30 days
Orsho	30 days
Pheriche	15–30 days
Lhabarma	30 days
Phulungkarpo	20–30 days
Dzongla	15–30 days

Summer
  June through August

Dzongla	30 days
Dusa	30 days
Tugla/Melinang	30 days

Autumn
  September through November

Tugla	7 days
Pheriche	7 days
Dingboche	14 days
Orsho	14 days
Kunde	45 days

Early/mid-Winter
  December through January

Kunde	45 days
Dingboche	20–30 days

---

that year's hay harvests and is just long enough to finish the fodder and
contribute as much manure as possible to the hayfields to promote the
coming year's growth (table 13). Each family has its own route and all do
not share the same herding settlement bases. In most years the first stop
of the family we are following is Dingboche. Here the herd is based for
twenty to thirty days in late autumn and early winter in *Dawa Chuchikpa*
and *Chuniwa,* the eleventh and twelfth months of the Sherpa calendar.
The length of time spent herding here depends both on the snow cover
and the amount of stored fodder that is available. From there the herd is
moved on to Orsho (one month), Pheriche (two weeks to a month), and
Lhabarma (one month). The final move of this sequence is to Phulung-

karpo where the nak are pastured for twenty to thirty days. Here they finish the last of the hay in May, Dawa Shiwa. This movement pattern also enables the family to manure their hayfields at Orsho, Pheriche, Lhabarma, and Phulungkarpo and their crop fields at Dingboche.

All four Kunde nak owners complete their spring herd movements at Phulungkarpo. There they finish the last of their hay supplies before dispersing to begin seeking out the good, late-spring grass. One family then bases at Pheriche, one at Phulungkarpo, and two shift to the higher area of Dzongla (4,843m). In bad years when hay supplies are exhausted early all four families rely on Dzongla grazing in the early spring. The family with whom we are concerned herds either at Dzongla or in Chukkung in the upper Imja Khola valley at this time. The choice between the two depends on grass conditions and on predators. For many years their custom was to begin herding near Dzongla at Tsolo, where the early grass is said to be especially good, and then move to Dzongla. They abandoned this pattern during the 1980s, however, because of problems at Tsolo with loss of stock to wolves and their perception that grazing at Dzongla by more than fifty nak from the government yak farm at Shyangboche had depleted the grass. They shifted for a few years to grazing at Chukkung instead. In 1987, however, they returned again to the old pattern of late-spring grazing at Tsolo and Dzongla. During the intervening years wolves had ceased to be a concern in that area, and the family had decided that even with the government stock at Dzongla the grazing was still very fine there. The good grass at Dzongla is said to reflect the abundant rainfall there, which is considered to be greater than that in the adjacent Lobuche Khola valley.

May and June, when the herd is grazed at Dzongla, is the longest period that the family bases in one place during the summer. It is also their highest-altitude grazing of the year. During mid-summer nak are actually taken to Dusa, which is situated at a slightly lower altitude. Here the household bases for a month in late July and early August and celebrates Yerchang.[61] This allows the nak to graze in an area that will thereafter be closed to grazing and enables the family to share the Yerchang rites and parties with the other Kunde herding families who reunite to spend these weeks at Dusa. The grass at Dusa at this time of the summer is considered to be quite good, suitable even for nak who need plenty of grass to produce good milk. They also consider Dusa to be a healthier place for stock at this time of the year than Dzongla where wet conditions are believed to cause hoof and stomach problems. Some Pangboche families, however, keep their herds based at Dzongla longer into the summer.

After Yerchang, the Dusa area is closed to grazing. The family now moves to nearby areas in the Lobuche Khola valley which are just outside the boundaries of the closed zone. In some years they base at Tugla, in others at Melinang or Linjen. The site chosen in a given year has to do with grass conditions. In all three places they live in resa, putting a tarp over a simple circle of unmortared stone walls. Pangboche villagers also use Tugla, but Melinang and Linjen are only used by Kunde herders.

In late August or early September, September 7 in 1987, the zones in the upper Lobuche Khola and Imja Khola that were closed after Yerchang begin to open, first to grass cutting and then to livestock. Our family, along with the others, quickly moves its stock into the newly opened zones. In 1987 the stock was herded on September 14 into the Pheriche area and the following week to Dingboche. In both cases the move was made as soon as each place was opened to grazing.[62] In the first days of October the area of the Imja Khola valley immediately below the confluence of the Imja Khola and the Lobuche Khola is opened to grazing and the family moves its herd briefly to Orsho, and then to Kunde when the main village opens to livestock.[63]

Several months can be spent based at Kunde before the winter move back up valley. The timing here depends on the amount of snow at Kunde in the early winter. If there is a good deal of early snow the herd is driven up as early as December, *Dawa Chuwa,* to Dingboche. Here nak are fed barley-straw fodder. If there is no snow in Kunde they may stay on in the main village until January. Large amounts of fodder are not stored at Kunde, so the decision to move up depends on the grazing conditions near the village. With the move to Dingboche the winter-spring round of high-altitude herding begins, and the location of hay stores becomes the key factor governing decisions about herd movements until the new grass greens.

The herding pattern of the Pangboche nak-owning family also begins with a move from the main village to the high-herding huts. For them, however, this move comes somewhat later than for the Kunde family. Only in March or April do they move the herds up valley (table 14). During the next month the stock is shifted between several bases, from Pulubuk to Bibre, and then to Mingbo. At each place they base for five to twelve days, depending on the amount of hay harvested and stored at each place during the previous autumn and the amount of manure required there for fertilizing the hayfields for the coming summer's grass. By May this household, like the Kunde one, has exhausted its hay supplies and is ready to follow the new grass. Again, as for the Kunde herders in the Lobuche Khola area, this means a move in most years up to the highest part of their herding range. Normally the herd is now taken up to Chukkung, although in times past the family has kept the yak at Mingbo. Some other Pangboche nak herders prefer Dzongla. In

Table 14.   Household Herding Pattern (Pangboche)

Late Winter/Spring
  February/March through May

Pulubuk	5–12 days
Bibre	5–12 days
Mingbo	5–12 days
Chukkung	60 days

Summer
  June through August

Yarin	14–21 days
Bibre	28 days
Chukkung	28 days

Autumn
  September through November

Dingboche	7–10 days
Tsolungche	7–10 days
Pangboche	30 days
Dingboche	7 days ?
Chukkung	7 days ?

Winter
  December through February

Pangboche	60–90 days
Pulubuk	5–12 days

---

years when the grass is poor in these high areas the nak are taken down to Pulubuk instead. At Pulubuk there is usually abundant rainfall and good early grass.[64]

In the early summer the Pangboche family moves its herds down to Yarin, just across the Imja Khola from the main village. This move breaks with the Kunde nak-herding pattern that keeps the herds in the high pastures. The Pangboche moves revolve not around grass but around Dumje. It is only a short move for herders to be closer to the main village for the festival preparations and events and many take advantage of this opportunity. Other Pangboche families also base at this time at Yarin as well as at other nearby places such as Orsho, Lhabarma, Ralha, and Tawache.

After Dumje any stock in the Yarin area must be moved up valley, for that zone is soon closed to grazing. Most herders go directly to the area where they will celebrate Yerchang a month later.[65] The Pangboche family we have been following now moves to Bibre. There it remains with its nak through Yerchang, celebrating the festival there with relatives who also base at Bibre at this time. After the festival this area is closed to livestock and the herd is taken up to Chukkung.[66] They base here with their nak for the rest of the summer and send yak up higher in the Imja Khola valley on their own.

In the autumn, when the Imja Khola nawa-regulated zones begin to open to grazing, the herds are taken down valley in sequence with the zone openings. The herds graze first at Dingboche in late September, then at Tsolungche, across the river from Pangboche. They are taken into Pangboche itself in early October. The nak remain in the main village, however, only briefly. During November, when Kunde herds are still based in their village, Pangboche stock is taken up valley. Usually Dingboche is the first stop, and then Chukkung. The time spent at each varies with the condition of the grazing. In December the stock is again taken back down to Pangboche. There the herd remains for a month or two, at least until Losar, the lunar new year. If hay and buckwheat fodder supplies are sufficient the stock are kept in Pangboche for the rest of the winter and are only taken up valley three months later. Then in April it is time again to fertilize the higher-altitude hayfields and make use of hay stores in the high-altitude herding huts.

While the Kunde and Pangboche families described here obviously differ in the fine detail of their patterns of seasonal pasture use in the Imja Khola valley there are also a number of shared characteristics in their altitudinal herd movements. These include:

1. Late winter-early spring circuit of family hay-growing areas for fodder feeding and hayfield manuring
2. Early summer move to high-altitude pastures
3. Yerchang site grazing
4. High-altitude summer herding
5. Return down valley following the sequential opening of nawa-regulated zones to livestock
6. Winter in the main village.

The two most important differences in the pattern result from the Pangboche family's additional shifts of altitudinal base. These are the shift from high-valley grazing to the vicinity of the main village at Dumje time and the late autumn–early winter, upper-valley grazing trip. In the Bhote Kosi valley where main villages are also relatively close to the high-country herding settlements a similar late-autumn move is also made. The contrasts between the Pangboche and Kunde patterns may reflect above all other factors the simple fact that Pangboche village is much closer to the high pastures of the upper Imja Khola and Lobuche Khola than is Kunde and it is thus easier to maximize opportunities for enjoying social activities and for making optimal use of altitudinal variations in seasonal pastoral resources. Whereas for Pangboche villagers to move nak to and from the high pastures is simply a matter of a few hours, from Kunde two days are required.

5

Sacred Forests and
Fuel Wood

Khumbu is borderline country for forests, and much of the area is above the tree line. Only about 2 percent of Khumbu, approximately 2,200 hectares, is forested, but for Sherpas this small area is a critical component of their homeland and way of life. The temperate and subalpine conifer, birch, and rhododendron forests are a fundamental subsistence resource. Beyond that they are also a significant component of the spirit-filled cosmos within which Sherpas build their houses and gather their fuel wood. The sacred trees and forests that are such striking elements of Sherpa village landscapes represent substantial gestures of faith in a land where trees are so useful and so scarce.

The economic and religious importance of forests in Khumbu's tree-line terrain has led Sherpas to devise a range of local institutions to protect forests. These indigenous systems are the most celebrated in the Himalayan literature. First reported by Fürer-Haimendorf (1964: 110–113), the Sherpa *shinggi nawa* forest guardians have been widely cited as an outstanding example of an effective traditional forest-management system (Byers 1987b; McNeeley 1985; Rhoades and Thompson 1975; Schweinfurth 1983; N. Sherpa 1979; Thompson and Warburton 1985).[1] The reported demise of this traditional system after Nepal's 1957 nationalization of forests (Fürer-Haimendorf 1975) has been seen by some as a classic case of an unfortunate undermining of effective local resource management through efforts to establish central-ized, standardized systems of resource control (Thompson and Warburton 1985). The perception that local management had been abandoned also influenced national park planners to conclude that it was urgent for them to implement new regional resource-management regulations (Garratt 1981).

These depictions of the characteristics and effectiveness of traditional Sherpa forest management and the reasons for its supposed abandonment, however, have been based on only a few casual observations. The diversity of Khumbu Sherpa local management systems has not been appreciated and their resiliency has been underestimated. The reports of the collapse of traditional forest regulation have been premature and exaggerated. And the early evaluations of the goals and historical effectiveness of local forest management have also been made prematurely, without sufficient understanding of the Sherpas' diverse objectives in using and protecting forests or recognition of the environmental impact of historical and geographical patterns of forest use relative to forest protection.

In this chapter I reexamine traditional Khumbu forest use and management. I briefly survey regional forest resources and the range of different uses Sherpas make of them and then focus on the historical development and organization of what was not a single traditional institution but rather various types of traditional Khumbu forest management. This will provide the basis for the reevaluations of past and present Khumbu forest management and the role of subsistence forest use in environmental change which are important themes of the last three chapters of the book. In chapter 7 I will discuss environmental change during the period before the nationalization of Khumbu forests and show how patterns of environmental impact have been related to the specific goals of Sherpa resource use and management and the particular approaches they took to regulating land use. Chapter 8 looks closely at Khumbu forest nationalization and Sherpa responses and comes to some very different conclusions about the strength and resiliency of local institutions and values. Chapter 10 takes up recent changes in forest change and management, the question of whether or not trekking and mountaineering tourism have precipitated a deforestation crisis, and Sherpa response to new demands on local resources.

The Forests of Khumbu

Much of Khumbu is high alpine country, a realm of alpine tundra and lichen. Forests are found only in the low-altitude, southern rim of the region (map 12). Tongues of forest follow the valleys north but die out even in sheltered sites by 4,000 meters, far from the termini of the great valley glaciers. The upper limit of forest is readily visible also on the steep slopes of the valleys where forest forms a low skirt on the mountains below the high alpine expanses, cliffs and ice walls. Most of the region's

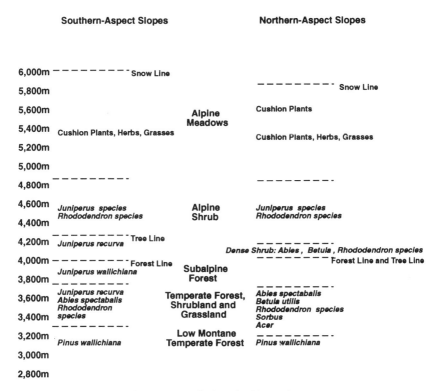

Figure 14.   Altitudinal Zonation of Khumbu Vegetation (Adapted from Byers 1987b)

forest and woodland consists of temperate forest and woodland located between 3,200 and 3,800 meters and subalpine forest and shrubland between 3,800 and 4,200 meters. There is only a small area of lower montane temperate forest in an altitudinal reach between 2,800 and 3,200 meters in lowermost Khumbu (fig. 14).[2] Forest composition varies with altitude and aspect. The lower valleys are conifer country. On slopes up to 3,400 meters, and especially on south-facing slopes, blue pine (*Pinus wallichiana*, Sherpa *metong*), is common either in pure stands or mixed with silver fur (*Abies spectabilis*, Sherpa *tashing*). Juniper (*Juniperus recurva*, Sherpa *shukpa*) forms isolated stands on south-facing slopes as high as 4,000 meters in places where it has been protected as a sacred tree or associated with a temple or hermitage. Fir is more widely distributed, and is a main species from 3,000 meters to the forest line at 3,800–4000 meters. Mixed forests of birch (*Betula utilis*, Sherpa *takpa*) and rhododendron (*Rhododendron campylocarpum, campanulatum, arboreum,* and others, Sherpa *tongmar* and *kalma*) cover extensive northern-slope areas

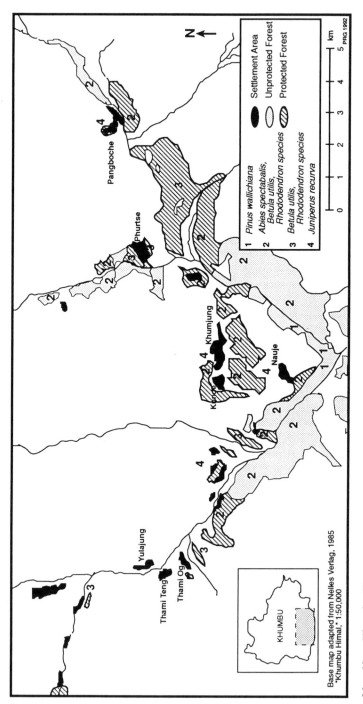

Map 12. Khumbu Forests

Table 15.   Major Khumbu Tree Species

Species	English Name	Sherpa Name
Abies spectabalis	silver fir	*tashing*
Pinus wallichiana	blue pine	*metong*
Betula utilis	birch	*takpa*
Rhododendron campanulatum	rhododendron	*kalma*
Rhododendron arboreum	rhododendron	*tongmar*
Juniperus recurva	juniper	*shukpa*
Juniperus wallichiana	juniper	*pom*
Salix sikkimensis	willow	*changma*

between 3,000 and 4,200 meters. Fir and juniper may also be found in these temperate and subalpine forests. Just beyond the forest line, which here occurs at 3,800 to 4,000 meters, there is a thin zone, 50 meters in altitudinal range, in which shrub-high birch, rhododendron, and willow predominate (Byers 1987*b*).[3] Beyond this are scattered low-growing, alpine shrubs including juniper (*Juniperus wallichiana,* or *Juniperus indica* and *Juniperus recurva,* Sherpa *pom*) and other alpine vegetation including dwarf rhododendron (*Rhododendron setosum, nivale, lepidotum,* and *anthopogon*) which thrive to above 4,900 meters on north-facing slopes.

Forest distribution has been strongly affected by historical human use as well as by patterns of topography, precipitation, and altitude. The south- and west-facing slopes of the valleys, where the villages are, are mostly in open woodland, shrubland, and grassland rather than forests, and where forests are found they usually represent traditionally protected areas. Forests are most extensive where they are hardest to reach, separated from settlements by major rivers or situated on particularly steep slopes. The large forest areas in southern Khumbu on the west side of the Bhote Kosi river and on the south and east side of the Dudh Kosi and Imja Khola rivers are clearly shown in map 12. It appears likely that at one time, perhaps as recently as the early days of Sherpa settlement, forest cover was much more extensive in the region, and Sherpas report some changes in the extent, composition, and condition of forest during the past century. These changes are discussed in chapters 7 and 8.

The Khumbu dialect of Sherpa is not rich in forest terminology, but Khumbu Sherpas do discriminate between dense and open forest as well as differentiating forests on the basis of dominant species and describing the trees in them as notably large, small, old, or young. Forests are often referred to in terms of their dominant species (e.g. *takpi nating* and *tashing nating* for birch and fir forests). The Sherpa for dense forest is *nating tukpu,* "thick forest," while open woodland is *nating shreme,* "thin forest." Nating shreme refers to areas with quite open canopy and even to areas with very sparse tree cover. Nating tukpu generally refers to forest with a more closed canopy. Sherpas draw a distinction between the high juniper scrub (*pom*) and forest (*nating*). Local, common, and scientific names for major Khumbu tree species are given in table 15.

Forest Use

Forests are sources of fuel wood, food, timber, fodder, grazing, fertilizer, and material for many of the articles of daily life: wood for the local manufacture of furniture, tools, plows, saddles, kitchen utensils, containers, masts (*gotar*) placed in front of each house to honor the birth of a son, house poles (*chotar*) decorated with a prayer flag as an indication of religious faith, incense, and bamboo wands for mounting prayer flags on house roofs and high places.[4] Forests also provide a number of valued food and medicinal resources. These were formerly especially important to poorer families and during times of food shortage. Many families recall, for example, gathering the leaves of certain forest shrubs to use to make tea before the recent rise to affluence spurred by tourism, when they could not afford to buy the Chinese brick tea that was imported from Tibet for the well-to-do. Mushrooms are avidly collected in the forests today, as are wild herbs. The collection of forest leaves for fertilizer and the use of forests for fodder and grazing have already been discussed in earlier chapters. The use of local forests for sources of fuel and construction timber is treated in detail below.

It is worthy of emphasis that no timber has been exported from Khumbu within the remembered past, as it has in some other northern areas of Nepal that market lumber to Tibet. Some medicinal herbs were formerly collected and taken south even as far as India for sale. This no longer continues today. Formerly there was also some business locally in the making and selling of charcoal, which was manufactured by Khumbu non-Sherpa blacksmiths. Since the establishment of the national park charcoal making has almost entirely shifted to Pharak, where it is responsible for considerable use of oak.

Fuel Wood

Firewood (*me shing*) is the main fuel used in Khumbu wherever it is easily obtained. Wood supplies energy for both cooking and heating, mostly for the former since most Sherpas make no attempt to heat their homes and rely for warmth even in mid-winter solely on small cooking fires, multiple layers of clothes, and drinking a great deal of hot tea.[5] Formerly the hearth consisted simply of an open fire over which was set a stone tripod or an iron grate for cooking, but today this is seen only in herding huts. Main houses are virtually universally equipped with simple, stone woodstoves that have been adopted mostly during the past thirty years. Many families burn some dried dung in these stoves as well as using fuel wood. In Nauje, for example, it is common for a family to burn a third to half a load of dung per day (10–15kg), most of it dried yak or cattle dung (*che*). Dung becomes still more important in the high country beyond the forests. Even in the highest huts, however, firewood is still used to some degree although Sherpas must travel several kilometers to gather shrub juniper.[6]

Birch is the preferred fuel wood, prized for its slow-burning qualities as compared with pine and fir, but what is burned in any particular hearth usually has more to do with what grows nearby than with preferences and birch makes up rather a small part of the woodpiles of most of the families of Khumbu.[7] Across Khumbu fir is probably the most commonly used wood, and much rhododendron, juniper, and even pine is also used in some areas. The shrub *Cotoneaster microphyllus* (Sherpa *pemba koptok*) is considered a possible fuel and its roots are said to burn with a hot flame. The amount of labor required to dig it, however, makes it a fuel of last resort.[8] *Masur,* a shrub rhododendron, is burned as a fuel at high-altitude sites and, like juniper, is burned to release its fragrant smoke as an offering to the gods. It is not, however, as significant a source of fuel as juniper.

Household firewood use varies seasonally and also with wealth, household size and labor resources, and the accessibility of forests.[9] No studies have yet been done to measure household fuel-wood use and its regional and seasonal variation. A survey of Sherpa estimates of their wood use found that 77 percent of the Khumbu households surveyed reported their wood use to be about half a load of fuel wood per day.[10] The remaining 23 percent reported heavier use of one load per day. Village differences in reported fuel-wood use were significant, and higher use was reported in Kunde, Khumjung, and the Thamicho settlements than elsewhere (Sherpa 1979:19).[11] Nima Wangchu Sherpa and others have taken a half a load of wood a day per household as a reasonable regionwide estimate, which suggests wood use on a scale of 5,000 kilograms of fuel wood per household per year and a Khumbu-wide requirement of more than 3,000 metric tons per year.

It has been reported that Sherpas traditionally only burned dead wood (Bjønness 1983:270). This ascribes to them, however, a conservation practice they do not follow: As far back as villagers remember they have gathered dead wood when and where it was readily available and have obtained whatever else they required by lopping branches or felling trees.[12] Before the ban on felling trees for fuel wood that was imposed in the late 1970s by Sagarmatha National Park administrators it was common to fell trees for fuel wood. Some Nauje Sherpas recall cutting ten to fifteen fir trees per year as recently as the mid-1970s. Usually trees were split into smaller pieces at the site and then allowed to dry before being hauled in baskets back to the settlement. Tree felling for fuel wood was traditionally permissible anywhere except within the boundaries of locally protected forests (*kyak shing* or "closed [for] wood"), where tree felling was allowed only for special purposes or banned altogether.

The distances villagers had to travel to reach unregulated forest varied. Residents of Nauje, Pangboche, and Phurtse could cut trees for fuel wood within a few minutes' walk of their houses. Nauje residents nostalgically recall being able to obtain three or more loads of fuel wood in a day since so little time needed to be spent in travel time to the cutting sites. Elsewhere, where forest was rarer or protected areas more extensive, the journey to and from cutting areas could require as much as several hours. Fuel wood was obtained from different areas during different eras, reflecting changes in the protected areas, the availability of dead wood, and the condition of forests outside the protected areas. The general pattern throughout the century has been toward longer journeys for fuel wood. Some areas reliable for fuel wood half-a-century ago 'or even thirty years ago ceased to be important sources before the 1970s. These included, for example, the slope south of Phurtse and above the Imja Khola, areas along the Imja Khola west of Pangboche, and pockets of forest along the slopes north of the Dudh Kosi river between Nauje and Kenzuma. Since the establishment of national park forest regulation the distance traveled for fuel wood has increased still further for some communities such as Nauje, Khumjung, and Kunde. Officially only dead wood can be gathered, and villagers are forced to seek out secluded areas far from national park patrols in order to obtain freshly cut wood to supplement the less-than-adequate supplies of dead wood available in the village vicinity.[13] Today residents of these three communities descend 500–700 meters to the Dudh Kosi river and climb a similar distance up the far slopes in order to reach areas where fuel wood can be safely obtained. Khumjung and Kunde villagers cross the Dudh Kosi and go high above Phunkitenga for fuel wood whereas the Nauje woodcutters cross the Dudh Kosi and climb up onto the slopes of Tamerku or cross the Bhote Kosi and climb an hour above the Satarma area. The trip to and from these sites requires a full day, and the arduous task of hauling a year's supply of fuel wood home in thirty to forty kilogram loads can require as much annual labor as farming potatoes does.[14] Across Khumbu men probably carry out most fuel-wood collection, but women do a great deal of this work as well and it is also a responsibility given to many teenagers.

Construction Timber

Sherpa houses require lumber for beams, floors, rafters, window framing, doors, and furniture. Fir shake roofs, moreover, are the most common regional roofing, even though slate is preferred whenever there is a quarry site within reasonable hauling distance and corrugated iron is increasingly being used in Nauje, Khumjung, and Kunde. Typical two-story houses today require eighteen cubic meters of timber, not including wood for furniture (Hardie et al. 1987:38). A relatively modest house requires fourteen trees for beams and joists, twenty to twenty-five trees for rafters depending on the type of roofing (larger, more closely spaced rafters being required to support slate roofs), plus several dozen more trees for boards for floors, window framing, doors, shelves and cabinets, and furniture. Recently new fashions of domestic architecture have begun to become popular which require even more timber for ceilings, paneling, and interior walls. Hardie et al. (ibid.7:39) estimate that this can increase the amount of timber needed for house construction by five to six cubic meters, or a full third.

All timber, including shakes for roofs, customarily was cut inside Khumbu. In some forests there were restrictions about what kinds of timber could be cut, but in others there were not. Certain areas were traditional sources of boards, rafters, shakes, and beams for particular communities. During the twentieth century Nauje families, for example, were accustomed to procuring beams from an otherwise-protected forest adjacent to the village, rafters and other timber from the forest immediately outside the restricted area, and shakes from forest half-a-day's walk away on the far slopes of the Dudh Kosi. The Nakdingog, Chuar, and Zamnangma Parken areas south of the Dudh Kosi (shown on map 18) have been renowned sources of timber and shakes for families from Khumjung, Kunde, Pangboche, and Phurtse since early in the century whereas beams were procured from closer at hand. Both Pangboche and Phurtse villagers were also accustomed to obtaining rafters and boards from the forest at the Tengboche monastery both before and after its establishment. Relatively little restriction was placed on this until the early 1960s when the head lama began to direct villagers to fell trees only in the Nakdingog part of monastery land.[15] It was usual to cut and saw timber at the site using simple hand tools, and there was some business in cutting and delivering lumber.

Demand for construction timber increased considerably in the twentieth century, placing new levels of demand on forests near villages. The major population growth since 1900 has meant that more structures have had to be built and rebuilt than in former eras. Khumjung, Phurtse, and Nauje have double or more the number of houses today than elderly residents remember from their youths. Today's houses, moreover, bear little resemblance to those of sixty or seventy years ago which were nearly all one story. By the end of the 1980s increasing affluence had made single-story houses extremely rare throughout the region. Houses are increasing in size as well. There is a saying now in Khumbu that whereas once, when a Sherpa became rich, he spent his money on religion (thus accruing merit for his rebirth), he now builds a new house.

Historically forest use has been greatly affected by the availability of certain types of trees and by changes in local and national government regulations concerning tree felling. Changes in the types of trees used for beams, for example, suggest that the depletion of sizeable trees of certain types near some villages in the nineteenth and early twentieth century led to a switch from juniper to fir. In the late nineteenth and early twentieth centuries juniper for beams was available in the immediate vicinities of Phurtse, Nauje, Thami Og, Thami Teng, Khumjung, and Kunde. Now fir alone is used at all of these places due to the absence of large juniper. Indeed, today at Thami Og and Thami Teng there are no trees of any type that are suitable for beams, and villagers from these two communities now cut and haul fir from sites lower in the Bhote Kosi valley such as Samshing and Phurte as far as five kilometers away from their homes. Government regulation rather than local depletion, by contrast, has affected the use of fir for building material. Fir was long the preferred building material for most uses, but in the 1970s national park policies banned almost all tree felling for building and forced Khumbu Sherpas to import pine from Pharak. In 1990 a major policy change at Sagarmatha National Park returned some authority in forest management to Sherpa communities and relaxed some of its forest-use restrictions. Village committees and the national park are now allowing slightly higher levels of tree felling for construction purposes. Although more Khumbu fir is being felled today than in recent years most timber continues to be imported from Pharak.

Sherpa Protected Forests

No one knows exactly how local Khumbu traditions of forest management began. It is not clear whether or not Sherpas brought the concept of community forest management with them four centuries ago from Kham, borrowed the notion of protected forests from sacred groves they may have heard about or seen during their migration across Tibet to Nepal or from their new neighbors in the Dudh Kosi region, or invented the idea themselves in Khumbu. There are oral traditions, however, that some forests were regarded as sacred and put off limits from tree felling very early in the Sherpa settlement of Khumbu. Other forests were protected for religious reasons during succeeding centuries. Only in the relatively recent past, in the early twentieth century, however, did Sherpas begin to establish other kinds of forest preserves as well. It was the management of one of these recently protected secular forests, the forest adjacent to the villages of Khumjung and Kunde, that Fürer-Haimendorf (1964:110–113, 1975:97–98) described as an example of the traditional Khumbu Sherpa system of village forest regulation. This forest, however, had not been protected since time immemorial but rather for less than fifty years, and represented not a typical village-managed forest but only one of several types of Khumbu protected forest. Khumbu sacred groves and other types of protected forests went unreported in the anthropological literature.

Through time a rich diversity of forest-management systems was developed in Khumbu. Particular forest areas belonged to or were administered by villages, monasteries, temples, and pembu. There eventually came to be at least seven types of communally protected forests, encompassing nearly half of Khumbu's twentieth-century forest cover (map 13).[16] More than twenty separate forest areas were regulated. These forests were set aside at different times for different purposes and administered according to different rules. The Sherpa term *kyak shing* encompassed not only forests protected so carefully that no sharp instrument could be taken into the area but also forests in which trees could be felled for house beams. Some protected forests were administered so as to keep villagers safe from divine wrath or evil fortune, others simply ensured that families would continue to have a source of certain forest resources conveniently close by. They had in common the fact that in all of them there were at least some restrictions on the purposes for which trees could be felled. In all other Khumbu forests no such rules applied.

Sacred Groves

Sacred trees and forests are an integral part of the landscape of Khumbu, an expression of the historical depth of Sherpa Buddhist faith.[17] Villages and fields are dotted with trees believed to be the homes of lu, spirits worshiped by particular families who pass down the caretaking of the tree, spirit, and shrine through generations. Temples are surrounded by sacred groves. Other forests, called *lami nating* (lama's forest) are set apart because certain lamas sanctified them as places where no tree must be cut and into which no cutting implement might be taken. These sacred forests and groves were the earliest, most strictly regulated, and most enduring of the protected forests. They encompassed substantial areas of trees in what were often prime sites for harvesting timber and fuel wood adjacent to villages.

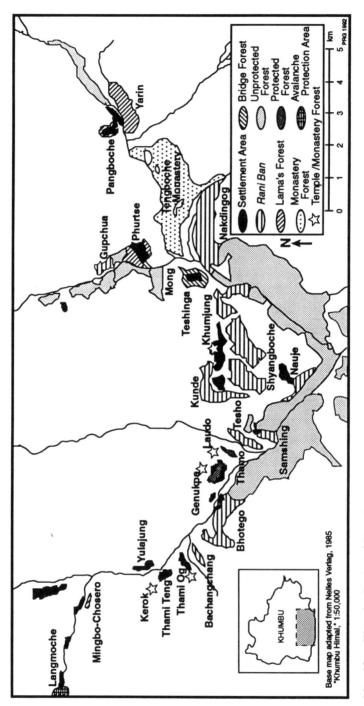

Map 13. *Khumbu Protected Forests*

Settlement Area
Rani Ban
Lama's Forest
Monastery Forest
Temple /Monastery Forest
Bridge Forest
Unprotected Forest
Protected Forest
Avalanche Protection Area

N

0 1 2 3 4 5 km

PRG 1992

KHUMBU

Base map adapted from Nelles Verlag, 1985
"Khumbu Himal," 1:50,000

Langmoche
Mingbo-Chosero
Yulajung
Kerok
Thami Teng
Thami Og
Bachangchang
Genukpa
Laudo
Bhotego
Thaqno
Tesho
Samshing
Nauje
Shyangboche
Kunde
Khumjung
Teshinga
Mong
Phurtse
Gupchua
Tengboche Monastery
Wakdingog
Pangboche
Yarin

Sherpa sacred forests may have evolved from Sherpa and Tibetan beliefs about the spirits known as lu. Sherpas believe that several types of these half-human, half-serpent, female spirits live in Khumbu, inhabiting springs, boulders, trees, shrines, and houses.[18] The lu of springs control the flow of water and can, if offended, withhold it.[19] Boulders can be inhabited by a male *saptok,* similar to a lu. These beings have an evil reputation for causing harm to people who pass near them and are much feared by travelers caught between villages by nightfall. Most houses have a lu and a special shrine (*lu khang* or "lu's house") is built inside the home for it. These are usually small, stone shrines tucked in unfrequented corners of the house, usually in a lower-story corner. These lu can influence family health and luck for good or ill and hence must be very carefully respected and given regular offerings. Tree lu sometimes live in trees near springs and sometimes in the forests. It is said that these forest spirits sometimes follow people home and take up residence in trees near their house. Lu trees within villages are a distinctive phenomenon. These may be few or many and can be of any species. Each belongs to one of the families whose house is nearby and the women of that family have responsibility for carrying out rites at it.[20] Often a small shrine is built at the foot of the tree.

Beliefs in lu are widely shared by many different Sherpa groups in northeast Nepal. Individual lu-inhabited trees and small groves can be seen in Pharak, Shorung (where there is an especially fine grove at Junbesi), and in the upper Arun valley.[21] Khumbu, however, seems to be particularly densely populated by lu, for nowhere else in the Sherpa world are lu-inhabited trees so common. These are the trees, usually old juniper (*Juniperus recurva*), but occasionally rhododendron, willow, or fir which so distinctively dot Khumbu villages and fields. Often they are virtually the only trees left standing in the settlement. Many have shrines at their base, but even when lu trees are not marked by shrines they are still trees that are well-known to all villagers and which they treat carefully. To offend a lu is considered dangerous, for serious illness and other family misfortune can follow.[22]

The beliefs and bans surrounding what is reputedly the earliest sacred forest, the grove at Pangboche, greatly resemble those associated with lu-inhabited trees. It is believed that to cut trees in the grove brings bad luck and illness. These juniper trees near the Pangboche temple are said, in fact, to be inhabited by lu as well as to be sacred because they were created four centuries ago by the great early Sherpa religious hero, Lama Sanga Dorje, who here scattered a handful of hair to the wind which, on falling back to earth, took root as forest. To cut these trees even with the best of intentions is dangerous. Pangboche villagers still talk, for instance, about a famous carpenter called in from Nauje to direct a temple-restoration project half a century ago, who requested that junipers near the temple be cut to provide building material. This was done, although very reluctantly due to fear of offending the lu. According to villagers the spirits were indeed angered and this resulted in the carpenter's death only three years later.

Lami Nating: "Lama's Forests"

The largest sacred forests in Khumbu are those that were established by the personal intervention of revered local religious leaders. There are several examples of such lama's forests including those in the vicinity of Yarin, Phurtse, Mong, Mende, Thami Teng and Mingbo-Chosero (map 13). The Yarin forest, said to have been declared kyak shing by Lama Sanga Dorje, is the largest of these lama's forests, extending more than a kilometer along the upper Imja Khola. The smallest, a tiny grove near Mingbo, was protected only in the late nineteenth century by Lama Zamde Kusho, who then had a hermitage in the area.

The most famous of the lama's forests among the Sherpas is at Phurtse where all remaining forest within an area bound on the south by Thakso ridge, the east by the trail still called "the lama's trail," the north by Luriphungga creek, and the west by the Dudh Kosi was declared protected forest by a lama well over a century ago. According to the legend a lama who lived in a hermitage just above the village warned villagers during an especially powerful ceremony that grave misfortune would attend the family of anyone who took any sharp instrument into the forest, be it an axe, a *kukri* (the Nepali machete), or a knife. This prohibition apparently also originally included the forest on the other side of the Dudh Kosi below Mong. Much of the Phurtse lama's forest has been very stringently protected, and here at 3,800 meters there are more than ten hectares of birch, some probably a century-and-a-half old, whose gnarled and mossy limbs show no sign of ever having been lopped. The island of protected trees south of the village, however, has not fared so well (see table 16) and neither has Mong. Control of the Mong area was disputed between Phurtse and Khumjung. After a series of rulings early in this century by local pembu and distant Nepali courts the villagers of Khumjung gained rights to the area. Phurtse residents claim that Khumjung people began cutting trees in the Mong forest despite the beliefs attached to it by people from Phurtse. Illness, they also say, and years of very heavy snowfall followed. In 1990 Phurtse villagers, led by young people, agreed to allow dead-wood collecting in the Phurtse lama's forest. This is the first time that this has been permitted in living memory. But the community continues to ban tree felling in that forest. Today, as for a century or more, two villagers each year

Table 16.   Khumbu Sacred Forests

Lama's Forests

Name	Location	Period Protected
Pangboche	Pangboche	400 years
Yarin	near Pangboche	400 years
Phurtse	Phurtse	more than 100 years
Mong	west of Dudh Kosi	more than 100 years
Mingbo-Chosero	Langmoche Chu	100 years

Khumbu Temple Groves

Name	Location	Approximate Age of Temple
Khumjung	Khumjung	150 years
Thami	Thami Og	400 years
Kerok	Kerok	400 years
Genukpa (tsamkhang)	Above Thami Og	100 years

continue to act as shinggi nawa, ensuring that the local regulations are followed.

Temple and Monastery Groves

Trees growing around temples are respected, and small groves are associated with the temples of Khumjung and Kerok and several small monasteries and nunneries, most notably at Thami Og and at Tengboche.[23] Temple and monastery groves (with the exception of Tengboche, discussed below) are much smaller than lama's forests. The largest, Kerok, encompasses little more than a hectare of old birch and rhododendron. There is a conspicuous lack of oral traditions regarding these groves, suggesting that these trees acquired sanctity from the founding of the temple rather than the existence of already-sacred groves having been the reason for the building of religious structures at those sites. Temple groves in Khumbu, unlike those in some other Sherpa areas, have not been planted within memory. It thus seems possible that some temple groves are relics of presumably once-more-extensive forests or woodlands. Felling trees in temple groves is regarded as an inauspicious act, and no special forest guards are considered necessary. Few trees in such groves show signs of having had limbs lopped.

Tengboche monastery, founded in 1919, controls an extensive forest area that is often assumed by Western visitors to be sacred forest. This forest, however, is not venerated as a result of its association with legendary or historical acts of gods, spirits, or lamas. It is rather the largest privately owned forest in Khumbu and has been ever since the area was granted to Lama Gulu by the eight Khumbu pembu in 1919.[24] The monastery allowed tree felling in its forests before Lama Gulu's death in 1934, and during the administration of those who managed the monastery during the youth of his incarnate successor (1934–1957). Phurtse families recount that there were no objections to their felling trees on its land for house rafters and boards. The present, stricter regulations on logging date to the period after the current head lama returned from training in Tibet in 1957. Tree felling on the Nakdingog side of monastery lands, however, continued to be authorized until quite recently.

Secular Preserves

According to oral traditions the only protected forests in Khumbu before the early twentieth century were sacred forests (with the possible exception of one area of Phurtse which is discussed below). During the past eighty years, however, a number of protected forests were established which had no religious basis and were administered by local officials or community assemblies. Unlike the strictly protected sacred groves and forests these secularly managed preserves allowed limited logging for special purposes. In some of these forests trees could only be cut for beams for housebuilding. Other protected forests supplied timber for important bridges. In one case the cutting of shrub juniper was forbidden in order to guard against avalanche danger.

Rani Bans

The most significant of these new, secular preserves in terms of both numbers and area were the *rani ban*. Between 1912 and 1915 eight of these protected forests were designated in Khumbu protecting substantial areas in the vicinities of the villages of Nauje, Khumjung, Kunde, and Thami Og, the Bhote Kosi valley gunsa of Thamo, Phurte, Samshing, and Pare, and the upper Dudh Kosi herding settlement of Konar. Management aims, regulations, and enforcement varied considerably among these forests, but in all of them unauthorized logging was prohibited by the eight pembu under authority vested in them by the Rana government in Kathmandu.

Until the early twentieth century the pembu had nothing to do with forests. This changed when prime minister Chandra Shamshere Janga Rana's Kathmandu government ordered village headmen and other local leaders including the pembu of Khumbu and Shorung to each take

Table 17.   Khumbu Rani Ban

Name	Responsible Pembu	Location of Forest
Nauje	Murmi Yulha	Nauje
Gupchua	Mun Puri	upper Dudh Kosi valley
Nakdingog*	Gyaliwa	valley of Phunki Chu
Samshing	Gardza	Bhote Kosi valley
Tesho	Karta	valley of Tesho Chu
Bachangchang	Yulha Tarkia	Thami Og
Khumjung-Kunde	Yuli Putar	Khumjung-Kunde
Bhotego	Thamo Mum Dorje	Pare

*Part of this area was transferred to the control of the Tengboche monastery at the time of its establishment.

responsibility for the administration and protection of a forest in which all hunting would be prohibited.[25] Each leader was instructed to select an area, state its boundaries, and send the information to Kathmandu. About two years later, in Solu-Khumbu beginning as early as 1911–1912, each received a document that set down his name, the boundaries of the forest over which he had jurisdiction, and the rules to be enforced there.[26] These regulations strictly limited logging. Villagers were required to get permission from the responsible pembu or other local government official for felling in these forests. Tree felling for special uses, such as building a bridge, could be agreed on by the pembu and his villagers. Other special requests, including house timber, were supposed to be passed on to a district office (*gosara*) in Olkadunga, an administrative center several days' walk south of Shorung and a week from Khumbu. Offenders were also to be reported to officials in Olkadunga, who would rule on their cases and administer fines. These forests became widely called *rani ban* in Solu-Khumbu and other regions.

Pembu throughout what is now the Solu-Khumbu district chose forests and implemented regulations for their use. In Shorung some pembu selected vast areas of forest, far larger than any of the regulated Khumbu forests. These, however, were situated in areas remote from villages and were not important sources of local fuel wood and timber. In Khumbu something else altogether took place. Khumbu Sherpa pembu creatively interpreted the Rana rules to create institutions that were different from those envisioned by Kathmandu officials, but were more useful in the environmental and social context of Khumbu. They developed their own forest regulations and their own ways of enforcing them. Unlike their counterparts in Shorung they for the most part chose responsibility for forests that were close to settlements and were then important sources of fuel wood and timber.

In all Khumbu rani ban some subsistence activities were allowed. There were no restrictions on gathering dead wood, collecting soluk for fertilizer, lopping for fodder or fuel, or grazing. Logging was totally banned in only one rani ban, Bachangchang, a forest that already had long been protected as a lama's forest. Elsewhere trees in rani ban could be felled, but only for specified purposes. In most rani ban tree felling was allowed only for obtaining beams for house building and then only with the authorization of the responsible pembu—or in the case of the Khumjung and Kunde rani ban, with the approval of the community-designated forest guardians.[27] The rani ban were thus the first Khumbu forests in which tree felling was regulated in any other way than by simply being banned, and the first forests in which felling was restricted to particular sizes of trees for specific purposes.

Enforcement of these regulations was handled differently than the Rana officials had envisioned. Offenders were not sent to Olkadunga. Appropriate fines were determined locally by pembu or village meetings and were locally levied. For the Khumjung-Kunde and Bhotego rani ban community management institutions were established which had precedents in the guardians traditionally chosen for the Phurtse lama's forests. These were the shinggi nawa made famous by Fürer-Haimendorf's depiction (1964:110) of them as forest guards chosen by every Khumbu village. Only three of the eight villages actually selected shinggi nawa and such officials administered only a quarter of Khumbu's rani ban and only a seventh of all the region's protected forests.

Some Sherpas consider that the establishment and form of the Khumbu rani ban were a response to forest change and particularly to the increasing scarcity of certain types of forest resources in particular areas. In some parts of the region there was a growing concern in the early twentieth century with the increasing scarcity of large trees in locations convenient for the provision of beams. By 1900 the preferred junipers used for house beams had been depleted in the vicinity of all the main villages and within a few more decades none could be found anywhere in Khumbu—the only remaining trees of sufficient size having long since been recognized as sacred, lu-inhabited trees. Fir became the main timber for beams, but even this was scarce in some areas. Some Sherpas interpret the declaration of rani ban adjacent to Khumjung, Kunde, and Nauje and in the lower Bhote Kosi valley as a measure to protect the supply of beams close to those communities, and the protection of the rani ban at Bachangchang as an attempt to preserve critical supplies of soluk. Doing so required creative, local interpretation of the forest-protection rules issued in Kathmandu, which made no provision for such a degree of household forest use or for administering forests with locally elected officials.

Khumbu Sherpas demonstrated considerable ingenuity in tailoring a national institution to local environmental and social conditions, and the Khumbu rani ban clearly illustrate a greater concern with protection of scarce resources than do the rani ban declared at the same time in Shorung.[28] It is worth noting, however, that only half of the eight Khumbu villages had a village system of forest management that regulated subsistence use of adjacent forest. Thami Teng and Yulajung had no protected forest of any type. Pangboche and Phurtse had only adjacent sacred forests, and Thami Og a former sacred forest that was similarly protected after being designated a rani ban. The range of forest uses that were regulated by the rani ban was also somewhat narrow, a characteristic that has had important ramifications for twentieth-century forest-use patterns and regional forest change.

Since the early twentieth century local management of most of the rani ban has eroded. The forests at Nakdingog, Samshing, Tesho, and Gupchua ceased to be well protected before the 1950s and guardians for them may not have been chosen or appointed after the deaths of the pembu who took original responsibility for them, if indeed their protection lasted that long. Tesho may have been protected somewhat longer by local customs even if these may have ceased to have institutional enforcement. The Nauje and Khumjung-Kunde rani ban ceased to be protected in the late 1960s after forest nationalization and the shinggi nawa administration of the Bhotego rani ban was abandoned in the late 1970s following establishment of Sagarmatha National Park. Local management, however, began to revive in the late 1970s in the Khumjung-Kunde rani ban.

Bridge Forests

At least two forests were protected in order to ensure that a continuing supply of bridge timber would be available. Until the 1960s when cable suspension bridges were first built in the region, Khumbu bridges often had short life spans. Most were cantilever, timber spans that extended only a few feet above the water. Rivers in monsoon spate or swollen by floods caused by the sudden emptying of glacial lakes by avalanches or the collapse of retaining moraines claimed many bridges, and some bridges that obviously had little chance of surviving the flood season used to be taken down each spring and reconstructed in the autumn. Phurtse was probably the first place where an area of forest was protected specifically to provide bridge beams. Here logging was allowed in one corner of the lama's forest for timber to maintain the bridge across the Dudh Kosi. This custom may predate the beginning of the century, and if so would be the first example of forest management in Khumbu for non-religious reasons. The bridge forest at Teshinga was protected after the establishment of the Tengboche monastery at the request of Lama Gulu who wanted to ensure an adequate supply of beams to keep the bridge over the Dudh Kosi near Phunkitenga in repair. This bridge was the main link between the monastery and western Khumbu. Later the area became administered as part of the Khumjung-Kunde rani ban.

Avalanche Protection

Protected forests in Khumbu were normally not set aside with the aim of guarding villages against avalanches as has sometimes been assumed (McNeely 1985). Forest above villages was instead generally the first cut, since avalanches have apparently not been considered to be a risk and carrying beams and timber downhill to house sites was enticingly easy. Protected forests tended to be located on the downhill or lateral margins of settlements. In one case, however, an area of high-altitude scrub juniper was deliberately protected as a defense against avalanches.[29] During the 1940s the seasonal herding settlement of Langmoche (on a tributary of the Bhote Kosi in northwestern Khumbu) was twice struck by avalanches. The second time two houses were destroyed. After that disaster residents decided that from then on no juniper (*Juniperus wallichiana*) shrub should be cut anywhere on the slopes overlooking the village on either side of the stream. A shinggi nawa system was set up in the late 1940s to enforce the regulation and was still operating in 1985. This was the only area of high, alpine shrub juniper in Khumbu which was administered as kyak shing prior to the recent protection of juniper around the monastery of Laudo (near Mende above the lower Bhote Kosi).

Administering and Enforcing Forest Regulations

Fürer-Haimendorf described one system of administering and policing protected forests, the shinggi nawa forest-guard system. According to his account this was characterized by annual selection of several villagers to act as forest guardians. Officials were chosen by a small group of powerful villagers and could be reelected for an indefinite number of years in succession. Shinggi nawa were granted the power to levy fines on violators. The amount of these fines was socially approved and the penalties were administered publicly at a once-a-year gathering during which the fines (paid in beer) were consumed amidst an assembly of villagers who listened to the confessions of the accused (Fürer-Haimendorf 1979:110–113).[30]

Elsewhere in Khumbu the shinggi nawa institution, however, was not identical to that described by Fürer-Haimendorf. In Phurtse shinggi nawa were selected for overseeing the lama's forest by a village assembly rather than in a small group meeting. This was also done in Pare for the Bhotego rani ban. At Pare, and formerly also in Phurtse, the office rotated among households and was not customarily held for more than a year.[31] The rules that shinggi nawa enforced were formulated for some forests by pembu or lamas rather than by villagers. The number of guardians was not standardized and varied from two to four. Still more surprisingly, the responsibilities and powers of the shinggi nawa, even in Khumjung and Kunde, seem to have been exaggerated in the literature. They did not patrol forests, select trees to be cut, or ordinarily inspect wood stocks for overcutting. Instead they were solely concerned with fining those people who were discovered felling trees in protected areas. These fines were slight, often paid only in beer, and the enforcement of the regulations depended largely on social pressure. In most cases, other than in Khumjung and Kunde however, there was no ceremony at which villagers were publicly humiliated and fined before the assembled community. Instead nawa paid a visit to offending families at their homes to collect cash fines and to consume fines paid in beer in lieu of or in addition to cash fines. Community meetings, or at least meetings of a large number of villagers, were held only to select nawa for the coming year (Nauje, Pangboche) and sometimes to review the regulations they would enforce.

The shinggi nawa system was not the only way in which protected forests or even rani ban were managed in Khumbu. Nawa rather than shinggi nawa were in charge of administering forest regulations in the Pangboche and Nauje areas. In both cases these officials were given forest-watching duties in addition to their usual responsibilities of enforcing seasonal grazing closures and blight defense regulations.[32] At Nauje, however, permission to cut trees in the rani ban adjacent to the village had to be obtained from the pembu responsible for that forest rather than from the nawa. For the Samshing forest the responsible pembu gave the task of enforcing the forest rules to one of his household servants. Some pembu kept full responsibility themselves for the protection of the rani ban. Table 18 identifies the management systems formerly in use in the various Khumbu protected forests. Table 19 categorizes these local administrative arrangements into seven systems.

Table 18.   Khumbu Protected Forests (Pre-1965 Administration)

Forest	Administration
Lama's forests:	
Phurtse	Shinggi nawa
Yarin	Nawa
Mingbo-Chosero	None prior to 1984
Pangboche	Temple watcher
Temple forests:	
Kerok	Lama
Khumjung	Lama
Laudo	Lama
Private forests:	
Tengboche	Lama and council of monks
Thami	Private households
Rani ban:	
Khumjung-Kunde	Shinggi nawa
Nauje	Pembu and nawa
Bachangchang	Gembu/pembu
Tesho	Pembu (lapsed)
Samshing	Pembu (lapsed)
Nakdingog	Pembu (lapsed)
Gupchua	Pembu (lapsed)
Bhotego	Shinggi nawa
Bridge forests:	
Phurtse	Shinggi nawa
Teshinga	Shinggi nawa (Khumjung)
Avalanche protection:	
Langmoche	Shinggi nawa
Uncertain status:	
Thamo	Shinggi nawa (lapsed)

Traditional Forest Management Systems Re-examined

Instead of the single "village forest" management system so often described for the Khumbu region there were thus a number of diverse forms of forest management. Sherpas developed an extraordinary number of these systems and protected an unusually large percentage of local forest and woodland through them, including a significant area of forest

Table 19.   Forest Administration Systems

Systems:

    1. Village management with shinggi nawa
       (elected, multiple terms possible)
    2. Village management with shinggi nawa (rotated)
    3. Village management with nawa (rotated)
    4. Pembu appointment of shinggi nawa, nawa, or
       guards
    5. Pembu direct regulation of forest use
    6. Temple management
    7. Private forest (group management)

System 1.

Lama's forests:	Phurtse
	Mingbo-Chosero
Rani ban:	Khumjung-Kunde

System 2.

Lama's forests:	Phurtse (formerly)
Rani ban:	Bhotego
Other:	Langmoche

System 3.

Lama's forests:	Yarin
Rani ban:	Nauje*

System 4.

Rani ban:	Samshing
	Kunde?

System 5.

Rani ban:	Bachangchang
	Gupchua?
	Tesho?
	Nauje (with nawa enforcement)

System 6.

Lama's forests:	Pangboche juniper
	Yarin
Temple forests:	Kerok
	Thami Og
	Mende
	Khumjung

System 7.

Private forest	Tengboche (council of monks)

*Formerly pembu granted permission to fell trees in the Nauje rani ban. Nawa, however, were solely responsible for enforcing the summer bans on the import of freshly cut wood, which protected the village from the risk of blight.

that was a continuing source of subsistence forest products. This is un-usual in my experience of other areas of northeastern Nepal. Although sacred forests are common in much of Nepal and rani ban were estab-lished in many areas, it seems rare to protect such a relatively high proportion of local forestland. None of the other Sherpa regions with which I am familiar have anything resembling the diversity or extent of Khumbu forest protection. In some of these there are protected forests, including temple forests in the Salpa pass area, Katanga, and Helambu, lu forests in Shorung and the Arun region, and rani ban in Shorung, Gora, Golila-Gepchua, and Olkadunga. But these are, with the excep-tion of a few of the rani ban, all quite small scale, and the rani ban other than those in Khumbu were often established in remote areas where the regulation of subsistence use was not a critical social or environmental concern. Khumbu Sherpas have thus achieved something rather special for Sherpas and very probably extraordinary in the context of Nepalese forest management more generally.

Each of the seven Khumbu systems of forest management has its own goals, rules, and institutional arrangements. None corresponds directly with the "village forest" previously presumed to have been the character-istic form of local forest management in the region. Contrary to what has been assumed most Khumbu protected forests have not been protected since time immemorial. Of the seven types of protected forest only two, the lama's forests and temple forests, were definitely protected before 1900, and only one bridge forest may date to that period. The gradual development of different types of protected forests at different times resulted in a regionally uneven pattern of management. Some villages had adjacent regulated forests, others did not. The state of forest cover that existed in the early twentieth century influenced which regions could be designated rani ban, and past patterns of forest use and degra-dation meant that not all villages had adjacent forests left to protect. And no Khumbu forest-management system administered any "village forest" in the sense of a commons in which residents obtained all of the forest products necessary to their subsistence in a regulated way. Sherpa protected forests were not intended to regulate the entire spectrum of any community's forest use, nor to be the sole or even primary source of any settlement's forest resources. Throughout the period before the establishment of Sagarmatha National Park there was other forest close at hand to exploit which was not regulated at all, and this was where use was concentrated. The significance of this selectively protective, tradi-tional resource management for historical environmental change will be taken up in chapter 7.

Part Two

Economic and Environmental Change

6

Four Centuries of Agropastoral Change

It has widely been assumed that Khumbu agriculture and pastoralism were relatively static, traditional practices for many generations before they were transformed by a variety of changes after 1960. The crop- and stock-raising practices that Western visitors first observed in the 1950s were believed to be a traditional set of practices that had been practiced since well before 1900. As far as was known or suspected the history of Khumbu agriculture consisted of three major phases. The first of these was an early period about which little was known, but during which Sherpas presumably practiced a form of mixed agropastoralism that they had brought with them from Tibet. Following this was a period that began in the middle of the nineteenth century and lasted until the 1960s and which was characterized by the distinctive reliance on the potato as the main staple crop of the region and the nak as the mainstay of pastoralism. This period, finally, was superseded during the past thirty years by an era in which traditional crop-growing and stock-raising practices have been undermined and dramatically transformed by the far-reaching effects of tourism on regional economy and land use (Bjønness 1983; J. Fisher 1990; Fürer-Haimendorf 1984).

Sherpas do not support this view of their past. Khumbu oral traditions and oral history contradict the idea that the pre-1960, potato-based agriculture and nak-based pastoralism were a relatively static way of life for many generations. They tell a story instead of a more dynamic history of innovation and adaptation. Sherpas suggest that since the late nineteenth century few aspects of crop production or pastoralism have remained static. The crop repertoire, cropping patterns, total amount of land under cultivation, technology, agricultural knowledge and belief, and community regulation of agricultural practices have all changed significantly.[1] Khumbu pastoralism historically has been even more dynamic than crop production, with major changes during the past century and a half in the types of stock raised, herd size, the goals and operation of local pastoral management systems, and seasonal herd-movement patterns. Yet at the same time Sherpas also testify to considerable continuity underlying these economic transformations. They dispute the contention that the changes of the past thirty years constitute a fundamental break from their long-characteristic subsistence strategies and practices. Khumbu agriculture and pastoralism are changing, but Sherpas are changing them within a context of continuing affirmation of local knowledge, cultural traditions, and valued land-use practices and lifestyles.

This chapter surveys Khumbu agricultural and pastoral history from the Sherpa perspective from the early days of settlement until the recent past. The oral traditions and history relating to changes in the subsistence economy, like those concerning the settlement history which I discussed in chapter 1, become richer in detail in the nineteenth and twentieth centuries. Although there are suggestions of what life was like and how it changed in earlier periods, it is only possible to analyze economic change in depth over the past hundred years. This, though, is still plenty of ground to work with—sufficient to establish both the dynamism of change in the "traditional" subsistence economy before 1960 and the processes involved. This in turn also establishes a better baseline for evaluating the changes of the recent past, some of which reflect continuing processes such as agricultural intensification, crop diffusion, and deemphasis of commercial stockbreeding which have been at work in Khumbu for a long time. I leave for chapter 10 the reassessment of what has and has not been transformed during the past thirty years by the impact of tourism.

Early Khumbu Agriculture

Khumbu Sherpas have probably practiced settled, mixed agropastoralism since the early days of their settlement of the region. Oral traditions about the first settlers and early life in the Bhote Kosi valley mention the growing of crops, the keeping of yak and sheep, and the use of manure for fertilizer. Not much more can be said of what was grown or how in those early years, or of when the pattern familiar for the past century and more of movement among main villages and herding settlements was established. In the early days it seems very likely that herders lived in black yak-hair tents in the summer rather than in the now-characteristic stone herding huts. Such tents figure in early stories from

Dzongnangpa's era and according to oral history were used in both the upper Dudh Kosi and the upper Bhote Kosi valley as recently as the late nineteenth century by some families. The idea that Sherpas were nomadic prior to the introduction of the potato in the mid-nineteenth century (Bjønness 1980a:61; Hardie 1957) may have come from stories of black-tent pastoralists. But abundant oral traditions testify that Sherpas have been a village-based people since the early days of their settlement of the region, and all evidence indicates that the main settlements familiar today had all been developed by 1830.

Oral traditions and oral history begin to give a firmer portrait of regional crop-growing practices in the late nineteenth century. By then the pattern of multialtitudinal crop production in the gunsas, main villages, secondary high-altitude crop-production sites, and high-altitude herding areas had already been established. All of the current main and secondary herding settlements had been settled and the crops familiar since then as the staples of Khumbu agriculture—potatoes, barley, and buckwheat—were all being grown.[2] But crop-production emphases were rather different. Potatoes were just one of several tuber crops rather than the dominant crop of the region and were not being grown at all in some of the high-altitude herding settlements or at Dingboche. Radish and turnip were being grown as major field crops, suggesting that they may have been the staple pre-potato tuber crops.[3] Buckwheat then as today was the most widely grown grain, but it was grown on a much larger scale than in the twentieth century including in places such as Tarnga, Thamicho, and Nauje where it is not grown today. Barley too was more widely cultivated and was being grown at several places in the Bhote Kosi valley.

Barley was apparently grown more widely in the nineteenth century than it has been in this century, although the need to irrigate it must have always made it of less regional significance than rain-watered buckwheat. In the late nineteenth century barley was cultivated not only at Dingboche but also at Tarnga in the Bhote Kosi valley and perhaps other places as well.[4] An elderly Thami Teng couple recalled seeing barley at Tarnga before 1920 while herding yak there. Other Sherpas recall that gembu Tsepal of Nauje was supposed to have owned barley land at Tarnga as well as at Dingboche in the early years of the century.[5]

Tarnga may have been the major agricultural center of Khumbu at one time and in the nineteenth century may have equaled or exceeded Dingboche as a barley center. There are large numbers of abandoned terraces above the settlement which were probably irrigated in an earlier time, and much of the land that is currently in potato terraces may have once been irrigated with a system that was far more complex than that employed at Dingboche. Elderly Tarnga residents have heard that formerly there was an irrigation ditch leading from the Chu Nasa creek north of the settlement.[6] Higher up the creek there is evidence of an old diversion and ditch that lead more than a kilometer from the creek to the head of the slope above the settlement. From here the irrigation network appears to have branched. One channel contoured along the slope to the west whereas the other descended through a flight of now-abandoned, upper-slope terraces. This system could conceivably have watered much of the field area of Tarnga.

The decline and disappearance of barley cultivation in the Bhote Kosi valley remains a mystery. There are no oral traditions describing the decline of barley growing in Tarnga. People have heard that formerly there was a prohibition on growing buckwheat there for fear that it would offend the barley. Barley, they note, has indeed disappeared from the area, and suggest that it might have declined on account of the introduction of buckwheat. Buckwheat is remembered to have flourished at Tarnga early in the century when it was a noted center of buckwheat production.[7] Beyond this they have no explanation.

Several other factors could conceivably have caused a decline of barley production at Tarnga and the abandonment there of many terraces. Crop disease and declining soil fertility might have figured in farmers' decisions; so too might the increasing scarcity of available land as the population rose, making higher-yielding buckwheat a more attractive option. Buckwheat could also have produced better crops on less fertile soils. Yet it is difficult to envision the total abandonment of barley for these reasons. Barley is still cultivated on a large scale at Dingboche despite the apparently poorer soils there and greater land scarcity than there would presumably have been in the nineteenth century at Tarnga. The cultural value and social status involved with growing barley remain strong. Yet even wealthy Sherpas with land at Tarnga have not grown even small patches of barley for many decades.

The most likely cause for the decline of barley cultivation at Tarnga may have been a failure of the irrigation network. The creek flow during May when the irrigation of barley is most critical is minor, barely enough to send a trickle of water through the ditch that supplies drinking water to the settlement. At one time more water was presumably available to feed the far more extensive earlier irrigation network. A declining flow of irrigation water could well have led to disputes over access to water and to the gradual abandonment of barley cultivation on lower-slope terraces. Once the available water declined significantly it is possible that there might not have been enough community enthusiasm to keep the out-take channel and ditches clear, and barley cultivation even on a limited scale would then not have been possible. Terraces would have been converted to a rotation of rain-watered buckwheat and tubers, and the upper-slope terraces that had formerly been valued for their ready access to irrigation may have been considered to be too marginal in terms of soil fertility to be worth planting to unirrigated crops.

The Introduction of the Potato

It has been widely assumed that the potato reached Khumbu about 1850, but nothing is known of the original variety introduced, its source, the exact date and place of its introduction, or the pace of its diffusion within Khumbu.[8] Fürer-Haimendorf believed that Khumbu Sherpas began planting potatoes "about the middle of the nineteenth century". According to his interpretation of two oral-history accounts the new crop had diffused even to such a "conservative" village as Phurtse "soon after 1860" (Fürer-Haimendorf 1964:9). Although his analysis of these oral-history accounts may have been incorrect, as I discuss below, an introduction before the last quarter of the nineteenth century is still likely. Presumably potatoes were well established in Khumbu before 1866–1876 when Rolwaling oral traditions reported by Sacherer testify that they were brought to Rolwaling by settlers from the Bhote Kosi valley (J. Sacherer, "The Sherpas of Rolwaling: A Hundred Years of Economic Change," 1977, in Seddon 1987).

Potatoes could have reached Khumbu in the nineteenth century from several sources including Kathmandu, Darjeeling, and Tibet. William Kirkpatrick (1975:180), an early English envoy to Kathmandu, reported potatoes there in 1793 and noted that the seed potatoes had to be brought each year from the Patna region of India. They were probably introduced to Darjeeling sometime soon after the British established a hill-station settlement there in 1835 and were certainly being cultivated there by 1848 when they were noticed by Sir Joseph Dalton Hooker (Hooker 1969:230). By the mid-nineteenth century potatoes could also have been grown in southern Tibet. George Bogle, who journeyed to Shigatse in 1774, had been instructed by the governor general of India, Warren Hastings, to introduce potatoes to regions through which he passed (Sandburg 1987:103). It is not known whether Bogle did establish potato growing in Tibet, but he is thought to have successfully introduced the crop to Bhutan (ibid.:103) and it might well have reached Tibet from there. Grenard (1974:247) reported potatoes in Lhasa in 1893.

The case for a Kathmandu origin for potatoes in eastern Nepal dates back to Hooker's explorations of northeasternmost Nepal in 1848. He noted that potatoes were being grown in the high-altitude fields at Yangma and at Kambachen in the Gunsa area northeast of Taplejung where he was given "some red potatos, about as big as walnuts" (1969:247). At Yangma, a place very near the Tibet border and inhabited by "Tibetans" (possibly Sherpas or Bhotias), he remarked that

> there was no food to be procured except a little thin milk, and a few watery potatos [*sic*]. The latter have only recently been introduced amongst the Tibetans, from the English garden at the Nepalese capital, I believe; and their culture has not spread in these regions further east than Kinchinjunga [Kachenjunga], but they will very soon penetrate into Tibet from Dorjiling [Darjeeling], or eastward from Nepal. (ibid.:230)

Hooker apparently rejected the idea that the Yangma potatoes had come from Darjeeling, despite the proximity of the two regions, because he had not found them cultivated in the intervening area.

Fürer-Haimendorf (1979:8–9) noted that Kathmandu and Darjeeling were the two most likely sources for the diffusion of potatoes into eastern Nepal. He raised the possibility that Hooker was wrong and that Darjeeling was the origin for both Khumbu and Yangma potatoes. Oral traditions he heard in Khumbu in 1953 seemed to indicate an introduction there after 1850, a date that would correlate well with earlier cultivation at Yangma if potatoes had reached eastern Nepal from Darjeeling (ibid.:9).

Both Fürer-Haimendorf's and Hooker's speculations about the origin of potatoes in eastern Nepal are based on assumptions that diffusion takes place in an orderly progression and constant pace across space. Such assumptions do not seem to be very well supported by what is known of the history of the diffusion of crop varieties in the region in the twentieth century. Diffusion is a complex process and is often difficult to predict. Other factors besides location and distance are clearly important. Peoples, villages, families, and individuals differ in their degree of receptivity to new crops. Regional patterns of trade and travel may bring one people into contact with a new crop rather than another people who may inhabit a region closer to the source. It only takes a single accidental encounter for someone to recognize a new crop and introduce it to his or her home region. Twentieth-century introductions of other potato varieties into Khumbu illustrate how important a factor chance can be. No analysis of typical Sherpa trade patterns would have concluded, for instance, that during the 1970s a potato found in a teashop on a trail near Darjeeling would come to dominate Bhote Kosi valley agriculture or that an apparently identical variety would be found just a few years later in a monastery kitchen north of Kathmandu and would rapidly diffuse from Khumjung to become the major variety grown in eastern Khumbu. Nor would it have seemed likely that in at least three twentieth-century cases Khumbu Sherpas would have acquired new potato varieties from Sikkim, Darjeeling, and central Nepal before other Sherpa groups and Rais who lived considerably closer to those source areas, or that potatoes brought from these remote areas to Khumbu would have subsequently diffused *from* Khumbu *into* areas that might have been expected to have had them long before.

In the mid-nineteenth century Khumbu Sherpas did not frequent either Kathmandu or Darjeeling. Khumbu was politically and culturally more linked to the Tibetan world than to the Kathmandu valley and very few Khumbu Sherpas went to Kathmandu for trade or pilgrimage. Before 1850 they also probably had little to do with Darjeeling, although later in the century Sherpas went to the British hill station to trade and to work as porters and at other occupations and in the twentieth century many Khumbu Sherpas went to Darjeeling to work for mountaineering expeditions. The lack of major contact with Kathmandu or Darjeeling does not mean, however, that a single Sherpa trader or a pilgrim visiting either area could not have brought back a few potatoes to Khumbu, or that Sherpas might not have encountered the crop in Shorung or other areas where Khumbu Sherpas traded and traveled frequently and which were in closer contact with Kathmandu.[9]

As important as the introduction of the potato is to Khumbu economic history there are unfortunately no surviving oral traditions today in Khumbu which describe the event. Fürer-Haimendorf (1979:9), however, had the opportunity to hear two accounts of the introduction of the potato when he began fieldwork in Khumbu in 1953. One of these testimonies was given by an eighty-three-year-old woman of Thami who is now deceased. The other was offered by a Phurtse resident, Sun Tenzing. From these two testimonies Fürer-Haimendorf (ibid.:9) arrived at a hypothesis that the original introduction of the potato probably took place not long before the 1860s. It is worth looking more closely at both the accounts and their interpretation.

It is immediately evident that the two testimonies differ considerably in detail and that in both cases it would be useful to know more about the varieties involved and the time referred to. Whereas the Phurtse account specified the location of the first fields planted and gave a fairly specific time frame, the Thami account was extremely general. For Thami we learn only that an eighty-three-year-old woman of that village (Thami Og?) "told me in 1953 that potatoes were brought to her village by people of her father's generation" (ibid.). By itself this account does not establish very much, for the language used makes it difficult to narrow the suggested time frame. A considerable span of years could fall within the era of "her father's generation." The introduction might have occurred before her birth, during her childhood, or conceivably even during the time before she married and left her parent's household. In the latter case, the time being discussed could be the very end of the nineteenth century.

The Phurtse account was rather different. Fürer-Haimendorf records that:

> In 1953 Sun Tensing of Phortse, then a man in his middle forties, told me that as a young boy he knew an old man of over ninety of whom it was said that he had first planted potatoes on Phortse land and I was shown the plots of land on the right bank of the Imja Khola, roughly opposite Milingo, where these first potato fields were supposed to have been. (ibid.:90)

Here again, however, crucial facts are missing. The account does not, for instance, indicate how old the man referred to was at the time he planted these potatoes, which variety of potatoes he planted, where he obtained them, or how long it was until potatoes were planted in the village of Phurtse itself.

In 1985 I had the opportunity to seek these details from Sun Tenzing, with whom I had a number of discussions. According to Sun Tenzing a man named Zoa Dolma did indeed introduce a new variety of potatoes at Tsadorji, a gunsa settlement east of Phurtse on the northern, right bank of the Imja Khola and across the river from Milingo. Zoa Dolma had brought nine small, long, white potatoes to Khumbu which he said that he had obtained from Darjeeling and which he believed had come from Belait (England). This variety, which is today most often referred to as a type of kyuma, was therefore also called *riki belati* (English potato) by many people. He brought this potato to Khumbu, however, half a century later than Fürer-Haimendorf had supposed. The potato variety had not been introduced long before Sun Tenzing's birth but when he was a boy of about five years of age, around 1914. At that time Zoa Dolma had been in his fifties, for in his earlier account Sun Tenzing had meant to convey that Zoa Dolma would have been in his nineties had he been alive in 1953 when he had told his story to Fürer-Haimendorf. Equally as startling, in the context of further questioning he also pointed out that while Zoa Dolma had brought the first kyuma or Belati to the Phurtse area he had *not* introduced the first potatoes there. Before the long white potato was introduced by Zoa Dolma, he noted, Phurtse villagers were already planting a small, round potato known as *riki koru* (white potato).[10] The introduction of this potato took place before his birth and he knew nothing of the circumstances or date. I was not able to uncover any of these details from other Khumbu Sherpas either.[11]

Table 20.   Khumbu Potato Introductions

Variety	Period	Source	Khumbu Site
?	mid-19th century?	?	?
Koru	late-19th century?	?	?
Kyuma	late-19th century?	?	Thamicho?
	ca. 1915	Darjeeling	Phurtse
Koru 2	1920s?	?	Phurtse
Kyuma 2 or 3	1930s?	Lazhen, Sikkim	Thami Teng
Moru	1930s	Lazhen via Pharak	Thami Og
	1930s	Lazhen via Pharak	Khumjung
			Kunde
Seru	1973	Darjeeling	Yulajung
			Pare
	1976	Singh Gompa	Khumjung
Bikasi	1981	Phaphlu	Nauje

There were thus apparently several early introductions of potato varieties to Khumbu (table 20), at least one of which came from Darjeeling. The earliest variety was introduced before 1900 and was being cultivated at least several decades earlier, if Sacherer is correct in her hypothesis that potatoes reached Rolwaling from Khumbu by 1876 ("The Sherpas of Rolwaling: A Hundred Years of Economic Change," 1977, in Seddon 1957). But the original date and source of introduction is more uncertain today than ever.

Early Twentieth-Century Agriculture

By the beginning of the twentieth century the potato was apparently being grown in all Khumbu villages and in most of the secondary sites where it is grown today.[12] By the 1920s, and perhaps for quite a while earlier, kyuma was planted in the Bhote Kosi valley throughout the altitudinal range from the low gunsa to high phu. In the Dudh Kosi valley it was grown not only in the main village of Phurtse but also in Na and other high-altitude areas east of the river (although apparently it was not cultivated at west-bank sites such as Luza and Dole where today it is grown on a small scale). Other potatoes were also being grown widely. But although widespread and important to regional agriculture and household subsistence, the potato had not yet come to dominate Khumbu crop production in the way that it would a few years later and does today. At this time potatoes were being grown on less than 50 percent of the cultivated area and indeed were only one of several tuber crops rather than the sole tuber grown. Elderly Sherpas maintain that during their youths (ca. 1910–1925) four crops were important in Khumbu: barley, buckwheat, potato, and radish. A fifth crop, turnip, was then being grown on a small scale. In most of Khumbu potatoes were less important a crop in the first part of the century than they have been in recent decades. In the Bhote Kosi valley there are reports that radish and turnip were grown on as large a scale as potatoes were in Thami Og and Thami Teng and that unlike today large areas of land were also in buckwheat. Estimates of the amount of buckwheat cultivation vary from half of all land in Thami Teng and Tarnga to perhaps only a quarter of the cultivated area of Thami Og. Radish was also being grown at Tarnga as a major crop, and the area was then considered to be excellent for radish growing just as it is for potato cultivation today. From Phurtse there are also reports that in the early decades of the century the potato was just one of several tubers, not the most important one. In Khumjung, by contrast, several elderly Sherpas remember that even during the 1920s the potato was the main tuber, and by then it was the major crop of Nauje and was being monocropped by many families.

Not one but rather three potato varieties were important in this era: the long kyuma whose introduction to the Phurtse area was discussed by Sun Tenzing, the early, round, white potato grown there (henceforth referred to as *koru*), and a second variety of white, round potato (*koru 2*) which was apparently introduced after kyuma in the 1920s.[13] These varieties seem to have diffused at different paces in different valleys. Kyuma, for example, was evidently grown decades earlier in the Bhote Kosi valley than at Phurtse, for Thamicho Sherpas older than Sun Tenzing do not remember the introduction of kyuma during their lifetimes and have not even heard stories about it from their parents. The same is true in Khumjung and in Nauje.

Many elderly people remember both kyuma and koru 2 as the potatoes of their youths.[14] Kyuma was higher yielding and more reliable, although it was also notorious for producing extremely poor crops in bad years and for being highly susceptible to devastating blight infections. Many farmers recall times when a full day's work at harvesting kyuma would not yield enough potatoes to serve the work crew for lunch. Accounts about the productivity of kyuma, however, are inconsistent, for the variety did supplant the earlier koru and was still grown on at least a small scale in much of Khumbu until about twenty-five years ago. Good harvests are also reported, and clearly some families produced surpluses of kyuma. Some dried kyuma tubers were even exported to Tibet on a small scale.[15] That farmers remember the productivity of kyuma differently may reflect differences in experience as well as in memory. Harvests could have been better in some areas and fields than in others due to differences in local soil and microclimatic conditions, altitude, disease problems, seed stock, or planting and manuring practices. Some Sherpas may tend to best remember the disastrous years and others the good ones. Some may best remember the variety from more recent years when it may have produced more poorly than it formerly did, possibly reflecting a loss of disease resistance or a regional decline in soil fertility. And others may base their evaluation not on productivity per field but on recollections of harvest size relative to family requirements, an equation that would have been different in earlier periods when so many families spent the winter outside of Khumbu on trade and pilgrimage journeys to lower regions of Nepal. Certainly on a Khumbu-wide level a surplus would have been achieved earlier in the century at a much lower level of regional production than at present, for the population of Khumbu was much smaller then and mostly only resident in the region for part of the year, and locally grown potatoes were not also consumed by thousands of tourists as they are today.

The Potato Revolution Reexamined

It has been commonly assumed that the potato was widely and quickly adopted throughout the region and that this radically transformed local agriculture in the nineteenth century in ways that also had major implications for regional demographic and sociocultural change. As already mentioned Hardie has suggested that before the potato was introduced Sherpas were nomadic pastoralists, a view that has been recently echoed by Bjønness (Bjønness 1980a; Hardie 1957) whereas Fürer-Haimendorf remarked that "it is difficult to imagine conditions before potatoes found their way into Khumbu" (1979:8). Fürer-Haimendorf ascribed to the adoption of potato cultivation both a population boom and a new level of prosperity that made possible a flowering of Sherpa culture in the late nineteenth and early twentieth centuries which included the building of an unprecedented number of temples, shrines, and religious monuments and the founding of the region's first monasteries (ibid.:10–11).

Links between the Asian and European adoption of New World food crops and major population gain have been noted in a number of parts of the world, including the adoption of potatoes in Switzerland (Netting 1981), and maize in Nepal (MacFarlane 1976). Fürer-Haimendorf attributed a similar growth of population in Khumbu to potatoes, noting that

> the population of Khumbu was a fraction of its present size until the
> middle of the nineteenth century and there can be no doubt that the great

increase of the last hundred years coincided with the introduction and spread of the potato. . . . No great imagination is required to realize that the introduction of a new crop and the spectacular increase in population must have been connected. (1979:10)

Even leaving aside for the moment the basic question of the degree to which potato cultivation transformed Khumbu agriculture in the nineteenth century, it seems that this issue deserves further exploration for the data on which to base such conclusions are rather slender. Indeed, the only available data on regional population size during the entire period from 1800 to 1957 are Fürer-Haimendorf's count of Khumbu households in 1957 and a count that he derived from 1836 tax documents (1979:5, 11, 118). The 1836 document lists only 169 households in Khumbu whereas in 1957 the count was 597. The regional difference of 428 households established over this 121-year period represents the supposed major population growth that has been presumed to have been precipitated by the introduction of the potato in the mid-nineteenth century. This tripling of the population (assuming that the average regional household size was similar throughout this period and that the number of households was tallied equally well on both occasions) is roughly comparable to some estimates of the population growth during the same period in Nepal as a whole, where demographic change is also believed to have been related to agricultural innovation, the introduction of maize, and the diffusion of the practice of growing irrigated rice in permanent terraced fields (MacFarlane 1976).[16]

Although the potato (along with increased trade for grain) undoubtedly provided the means to support the higher regional population density that developed after the early nineteenth century it remains premature to assert that the adoption of the crop caused this population growth. It is not clear how much of the increase in regional population can be attributed to natural population growth and how much is due to immigration. Natural rates of population increase may indeed have been raised by higher food availability and better nutrition, which in turn may have lowered mortality rates, an argument that Netting (1981) has made for the introduction of the potato in Switzerland. But it is also certain that a great deal of the regional population growth in the late nineteenth and twentieth centuries has reflected migration from Tibet, as is indicated by the high percentage of Khamba households in the region in 1957. According to Fürer-Haimendorf's 1957 data (1979:102–104) (and assuming that at least half of Nauje's households then were Khamba) at least 45 percent of the households in five of the eight main Khumbu villages were Khamba immigrants from Tibet. This means that the number of Sherpa households in Khumbu less than doubled over the 121 years from 1836 to 1957. This does not seem like a major population boom. It may, however, still represent a significant increase in population growth rates over earlier centuries. And it may also be that greater potato cultivation was a factor in the increased immigration during this period. Surplus Khumbu food production and the ability to pay wage laborers in food, for example, would have helped create a regional economic climate attractive to immigrants. But a great number of other factors were also very likely involved and the role of the potato in fostering immigration should probably not be overly stressed.

Beyond the issue of the degree of natural population increase there is the more fundamental question of whether the potato was at the center of a revolution in agricultural production in nineteenth-century Khumbu. I have suggested that the adoption and diffusion of potatoes was not as rapid as had been assumed. The potato-based agriculture familiar to foreign visitors since the 1950s was not universal in Khumbu even earlier in this century, much less in the nineteenth. The potato's rise to preeminence in Khumbu agriculture was a much more complex and lengthy process than has been thought. Not only was the area in potato cultivation during the late nineteenth century and early twentieth centuries apparently considerably less than that planted to the tuber during the past fifty or sixty years, but yields were also apparently lower and poor harvests due to disease more common.

The initial diffusion of the potato was still in progress as late as the 1920s and in at least one important crop-producing settlement there remained powerful resistance to the adoption of the new crop. At Dingboche at that time the planting of potatoes was forbidden out of fear that the crop might offend the barley.[17] The ban was upheld into the twentieth century by community understanding (a community which then comprised individuals from at least four Khumbu villages). Gembu Tsepal of Nauje, Khumbu's major political leader of the time and a large landowner at Dingboche, and his son and successor gembu Pasang are both said to have been particularly concerned with upholding this ban.[18] This attitude toward potatoes was not universally shared by Dingboche residents, and the lama Kuyung Rinpoche is said to have moved from the area after a conflict with gembu Tsepal over potato planting. Eventually the degree of community agreement on the ban deteriorated to the point that it was no longer enforceable. But it was only in about 1925 that Dilu, the mother of the powerful Kunde pembu Ang Chumbi, precipitated the abandonment of the ban by advising other women at Dingbochc to plant whatever they wanted to, insisting that gembu Pasang had no authority to tell them what they could or could not plant in their own fields.[19]

The process of the adoption and diffusion of the potato thus did not take place overnight in Khumbu, but involved incorporating the new tuber in an earlier cropping complex where it supplemented but did not immediately replace tubers cultivated earlier. Adoption of the new crop involved the development of new knowledge and new beliefs, including the abandonment of a belief that potatoes threatened barley yields and the invention of community measures to guard against devastating outbreaks of late blight. Not one but several different introductions of different varieties of potatoes were involved, and although the process of adapting the new cultigen to local agricultural patterns had advanced considerably even as early as the beginning of the twentieth century, the potato did not become the mainstay of Khumbu agriculture until the widespread cultivation of a newer, higher-yielding variety, the red potato, in the 1930s.

This reevaluation of the historical process of the adoption of the potatoes has significance for the second part of Fürer-Haimendorf's potato-revolution thesis as well, the idea that there was a connection between the adoption of the potato and the flourishing of religion in Khumbu. According to him:

> the foundation of monasteries and nunneries as well as the construction of new village temples and many religious monuments have taken place within the last fifty to eighty years. This points to economic events which favoured a sudden spurt of non-productive activities and in my opinion there can be little doubt that these events were brought about by the introduction of the potato and the resulting increase in agricultural production. (1979:10–11)

Here too, however, a more complex process appears to have been involved, and the role of higher-agricultural yields may have been relatively minor. Closer examination is necessary of the circumstances of the construction of religious buildings and monuments, and especially of the sources of capital and labor which made this possible. Labor for these projects seems to have been, then as now, primarily volunteer labor from nearby communities.[20] The ability of communities to mobilize this labor is based on whether local leaders can inspire participation or demand it by virtue of their offices. There is no direct link beteen local food supplies and the building of religious monuments. Even the availability of leisure time is not really an issue, for in the Khumbu agropastoral system there has long been a substantial part of the year in which there are no pressing demands. Khumbu life is built on the basis of a considerable amount of leisure, and the development of the local festival cycle can only be understood in this light. The establishment and maintenance of monasteries, temples, and monuments, however, also requires substantial capital, for institutions must be endowed, specialists employed in their construction and sometimes in their maintenance, and many precious materials obtained for building and blessing the shrines. Yet here a larger role was often played by a few wealthy individuals rather than by the community as a whole. These sponsors were not men who had become wealthy as a result of potato growing, but for the most part were large-scale traders in other merchandise. In some cases they were not even Khumbu Sherpas: Shorung Sherpas were especially important in the original establishment of the Tengboche monastery (see Ortner 1989).[21]

There is a problem too with the chronology of the development of monumental religious architecture in the region. It is not entirely accurate to emphasize the post-1880 period only, for most temples, chortens, and smaller religious monuments were actually established earlier. All of the village temples other than that at Nauje date to 1830 or earlier. Almost all the large chortens were built before 1880 (the main exceptions are two smaller chortens at Khumjung, one of which dates to the 1920s and the other to 1984, and three chortens in the Bhote Kosi valley). Most major *mani* walls (religious monuments composed of slabs of stone into which have been chiseled prayers and sacred texts) also predate 1880, although parts of the Khumjung mani wall and the mani wall at Tengboche were built after 1915. Most of the prominent expressions of religion faith in the Khumbu landscape thus seem to have been there well before the potato came to be a central focus of local agriculture in the early twentieth century. It may be that both before 1880 and since then the galvanizing force for the construction of religious monuments has not been the potato but rather the leadership of charismatic individuals who were able to inspire the financial support of well-to-do Sherpas and mobilize the volunteer labor of entire communities. And rather than a single period of major elaboration of monumental religious architecture there appear to have been a number of such periods stretching far back into Khumbu history.

Although the introduction of the potato may not have immediately been the seminal event in Sherpa history as has been assumed, it was nevertheless an important agricultural innovation that ultimately had far-reaching impacts on land use. As potatoes began to be more widely cultivated in the late nineteenth century they must have begun to increase regional agricultural production while at the same time lowering the amount of land required to meet household subsistence requirements. They provided a means of agricultural intensification, and the Sherpa response to this possibility may account not only for historical shifts in cropping emphases but also for changes in the area in crops and the abandonment of some fields in many parts of Khumbu.

Abandoned Terraces

By the late nineteenth century a great deal of land was under cultivation in Khumbu despite the relatively light (even by Khumbu standards) population density. In the Bhote Kosi valley more land seems to have been cultivated during the nineteenth century than has been in production since, a phenomenon reflected in the considerable amount of terraced cropland there that has been abandoned for many decades. Although some of these old fields were returned to cultivation later, a great many have never been reclaimed.[22] Abandoned terraces are also found today in widely scattered sites throughout much of Khumbu, as indicated in map 14. Table 21 identifies sites with abandoned terraces and attempts a rough estimate of the periods when different sites were taken out of cultivation.[23] The most extensive abandoned areas are located in the Bhote Kosi valley. It is clear from this map that there is an extraordinary number of sites with abandoned land in this part of the region and that these areas are situated through much of the altitudinal spectrum of Khumbu farming. More than forty terraces were abandoned at Tarnga. At Samde much terraced land has been neglected and there are the ruins

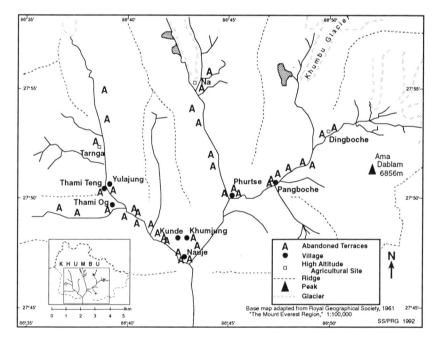

Map 14.   Abandoned Terraces

Table 21. Khumbu Abandoned Terraces

Site	Probable Period Abandoned			
	Pre-1900	1900–1930	1930–1960	Post–1960
Bhote Kosi Valley				
Thengbo			x	
Kure		?		
Marulung	?	x		
Lungden		x		
Tarnga	x	x		
Yulajung	x	x		
Thami Teng	x	x		
Thami Og	x	x		
Leve	x	x		
Pare	x	x		
Samde	x	x	x	
Thomde	x	x		
Tesho		?		
Samshing	?	x	x	
Tashilung		?	x	x
Nyeshe		?	x	x
Phurte		?	x	
Jangdingma		?	x	
Mishilung		?	x	
Imja Khola Valley				
Pangboche (upper west side)		?		
Pangboche (near Milingo bridge)		?	?	
Pangboche (west of Shomare)		?		
East of Shomare		?		
Orsho		x		
Across the river from Orsho		?		
West of Dingboche		?		
Dingboche		x		
Dudh Kosi Valley				
Phurtse Tenga			x	
Kele			x	
Dole			x	
Machermo			x	
Shomare			x	
Charchung			x	
Na			x	
Phurtse			x	
Dawa Futi Chu (east of Phurtse)		x		
Khumjung (east of village)		?		
Kenzuma (west of settlement)			x	

of eleven abandoned houses as well. Five houses and associated crop-land have been abandoned at Samshing where only two houses remain. And all of the more than twenty houses at Leve were abandoned before 1930 and not a single one of the scores of terraces there is now farmed. There is also a considerable number of abandoned fields at Thami Og as well as some in the other Thamicho main settlements

The immediate cause of abandonment can be determined in some cases, especially for those areas that were cultivated as recently as 1930. Diverse factors were involved. At Samshing and Samde emigration and the death of childless couples were factors. In Phurtse some terraces were also abandoned by families who emigrated to Darjeeling. Some fields abandoned near Thami Teng in the early years of the century are said to have been neglected because they had been producing poor crops. Terraces near Kenzuma in the Dudh Kosi valley were abandoned by Nauje settlers, it is said, because of repeated crop losses due to the depredations of langur monkeys more than half a century ago. Problems with Himalayan tahr were a factor in the gradual abandonment of crop production at Tashilung in the lower Bhote Kosi valley near Nauje over the past fifty years. Tahr were also a factor in some families' decisions to give up farming at nearby Nyeshe.[24]

The reasons for the abandonment of terraces before 1930 are harder to evaluate. The abandonment of the settlement of Leve, situated only a few minutes' walk from the major villages of the Bhote Kosi valley, for example, remains somewhat mysterious. There are legends describing bad luck at the place, but few details of how the houses and terraces came to be neglected. Some traditions regarding a few of the families who aban-doned the settlement, however, suggest that emigration was a factor in some cases. Two families left Leve for Rolwaling and other families sim-ply gave up growing crops at the site and concentrated instead on their nearby fields at Thami Og and other valley locations.[25] It is conceivable that the underlying factor in all of these cases was a perception that Leve had become marginal for crop production. This would also explain why no one has taken up growing crops there since. But there is no insight into this in the oral traditions. It is possible that disease may also have played some role. When traveling through Khumbu in 1885 Hari Ram reported that there had been an outbreak of smallpox in the 1850s (Ortner 1989:208, n. 4). Again, however, there is no hint in surviving oral tradi-tions about an epidemic in Khumbu, much less a link between this and the subsequent abandonment of any settlements.

Agricultural intensification may also have been a factor. Sacherer cites this as the explanation for old abandoned terraces in Rolwaling, which she attributes to the adoption of potato cultivation. The new high yields possible with the new crop, she suggests, rendered cultivation of some marginal land superfluous.

> There can be no doubt at all that the introduction of the potato brought about an economic revolution in Rolwaling where today one can easily observe the presence of abandoned fields in the more rocky and inaccessible high altitude areas despite the fact that the valley population has increased since the conversion from barley to potatoes. (J. Sacherer, "The Sherpas of Rolwaling: A Hundred Years of Economic Change," 1977, in Seddon 1987)

In Khumbu a similar phenomenon could well have occurred in the late nineteenth and early twentieth century. Even if the adoption of the potato in the region was a more gradual process than has previously been assumed and if harvests were relatively small and variable by the standards of recent decades, the new crop may still have been more productive than the other available tuber crops of that era.

The location of abandoned terraces in the Bhote Kosi valley may lend some support to an intensification hypothesis. Two kinds of sites are most common: small numbers of old, unfarmed terraces located either at the edges of current settlements or at a slight distance beyond them and more extensive, abandoned terraces in some gunsa settlements. Abandoned terraces at the edges of main villages and secondary high-altitude sites may well have been considered marginal land. The abandonment of so much gunsa land requires more detailed consideration. The degree to which fields have been carved out in the steep slopes of the lower Bhote Kosi valley for gunsa fields is unparalleled elsewhere in Khumbu. The greater historical intensity of land use in the Bhote Kosi valley may well reflect greater nineteenth- and early-twentieth-century population pressure there than in Khumbu's other valleys, which led people to cultivate even quite minor patches of relatively easily terraced land. The demand for land also would have been greater in this period because more land would have been required per household for subsistence in the time before the introduction of higher-yielding varieties of potatoes. In this era it was also more common to cultivate grain as well as potatoes and to try to harvest enough to fulfill most family grain requirements.[26] Families may thus have needed to farm up to twice the amount of land that they do today. Gunsa land would have been highly appealing in this situation due to the advantages that it offers for labor scheduling. The attempt to cultivate substantially more land in the main village could have strained family labor resources to carry out planting during the relatively brief spring planting period even if the field area had been available. Making use of gunsa lands would have expanded the agricultural season by nearly two months. Families from Thami Og, Thami Teng, and Yulajung could have planted large amounts of land as early as the beginning of March rather than waiting until late April.

The later adoption of higher-yielding potato varieties and the ability to grow more food on less land might well have led to a consolidation process as families devoted less time and energy to farming and relied on the larger harvests now possible on their best fields in main villages and high-altitude secondary agricultural sites. Presumably the further introduction of even higher-yielding potato varieties meant that many terraces were never returned to cultivation, although some were certainly sold to new immigrants.

Agricultural Change 1930–1973

Agricultural change can be reconstructed in much more detail for eras within living memory, and far more careful cross-checking of sources is possible for the period after 1930 than for earlier eras. This oral history testimony suggests that the middle of the twentieth century was characterized by an increasing emphasis on potato cultivation. This was associated with the adoption of another variety of potato, *riki moru* (red potato). This round, red-skinned, pink-flowered potato was both better yielding and more consistent in yield than earlier Khumbu varieties had been. It was adopted throughout the region and supplanted almost all other tuber cultivation. Turnip and radish ceased to be grown as field crops, kyuma was grown on a smaller scale than it had been, and other potato varieties began to disappear from the region (maps 15a and 15b). With the adoption of the red potato thus began the era of potato-dominated Khumbu agriculture so familiar to foreign visitors since 1950. There were also other less spectacular changes. The area in crops expanded on a small scale in some areas such as Phurtse, Nauje and lower Pangboche where terraces were constructed on previously unfarmed slopes.[27] Draft plowing was also more adopted for grain farming during this period, especially after the 1950s.

The Introduction of the Red Potato
(*Riki Moru*)

There are several accounts of the introduction of the red potato. These may reflect several different introductions to different parts of Khumbu over the course of two decades. According to one Khumjung account the red potato was originally brought to Pharak from Lazhen (a valley in Sikkim) by Karke Tikpe and Ang Pasang, who planted it in Tsermading (half a day's walk south of Nauje) more than sixty years ago. A second Khumjung resident recalls that Ou Sungnyu of Kunde first brought the red potato to Khumjung and Kunde close to fifty-five years ago. One Thami Teng resident who was seventy-seven years old in 1985 recalled getting his first red potato seed directly from Pharak when he was about twenty-four years old (ca. 1932). He also remembers hearing that Pharak Sherpas had gotten the variety from Lazhen. This Thamicho man recalls that from his first small pot of seed potatoes he harvested a single basket load and from this small beginning (and a few potatoes brought from Pharak by other Thamicho Sherpas) the new variety spread throughout the Bhote Kosi valley. The red potato may have been introduced to eastern Khumbu slightly later.

When moru first appeared it had a very high yield, comparable in some people's memories to that of the yellow potato when it was first grown in the 1970s. In good years some large landholding families harvested so many red potatoes that they were unable to sell their surplus even at a quarter-rupee per eleven kilograms (versus a 1991 price of fifty rupees/11kg) and had to throw many tubers out even after they had used all they could as livestock fodder. The new crop was susceptible to frost and to blight, but represented a much more dependable food source than had earlier potato varieties.

The red potato did not totally supplant kyuma in the region, which was still cultivated on a small scale in main villages until the 1960s and even more recently in some high-altitude settlements. Many Sherpas maintain that kyuma was abandoned not simply because a higher-yielding variety became available, but also because its yields began to decline. This is widely believed to have been related to the introduction of the red potato. Many people believe that the introduction of a new potato variety into a settlement adversely affects the characteristics of varieties already being grown there. Yields decline because plants feel offended or neglected by the attention being transferred to the new variety. The resulting poor performance may hasten the old variety's abandonment. The same explanation is also given to account for the later decline of red potato yields and the recent perceived decline of yellow potato yields.

The red potato was not the only potato introduced between 1930 and 1975, but it was the only variety that was widely adopted. Some families experimented briefly with *linke,* but this white, watery variety developed a reputation for causing stomach problems which led to its total rejection in Khumbu. The brown potato (*mukpu*), which is regarded so highly in the Bhote Kosi valley, may also have been introduced during this period. According to some poeple it was originally obtained from Pharak Sherpas.

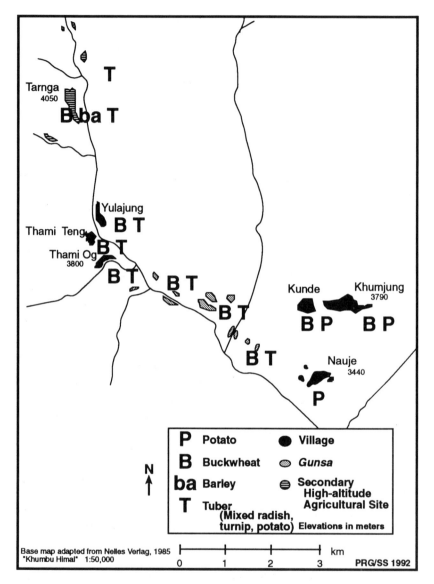

Map 15a.   Agricultural Change: Bhote Kosi Valley Agriculture, circa 1920

Technological Change

Prior to the 1950s plowing in Khumbu was usually done by teams of men rather than by draft animals. It is unclear why this practice persisted as long as it did throughout the region. Draft plowing was the usual prac-

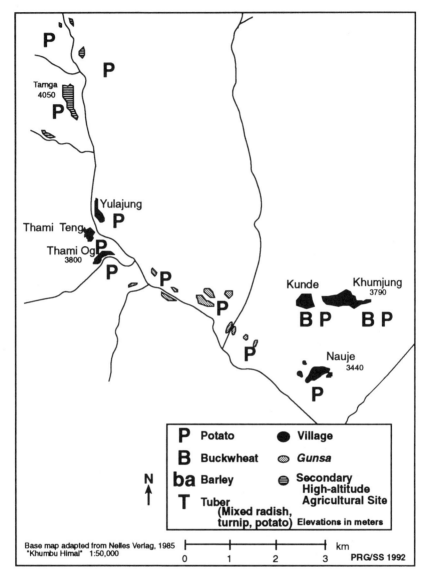

Map 15b.   Agricultural Change: Bhote Kosi Valley Agriculture, 1987

tice in Tibet, and indeed the dimzo raised in Khumbu for sale there were primarily in demand as plow and pack animals. Yak are also used as plow animals in Tibet and today some Sherpas plow behind yak. Yet earlier in the century neither yak nor zopkio seem to have been widely used in this way in Khumbu. It was only during the 1950s that draft plowing began to supplant plowing by human teams throughout the region. Men still pulled plows in Phurtse until only a few years ago. In Phurtse the increasing adoption of draft traction was the cause for some tension within the community. Men who had made good wages by pulling plows resented the loss of this opportunity. This had been the best paying day labor in the area, for each of the three or four men pulling a plow was paid double to triple the normal wage for agricultural day labor and was given better quality food and beer than normal laborers. Some families who began hiring yak- and zopkio-pulled plow teams to prepare their buckwheat fields are said to have been told by men who formerly had plowed their fields that they could have the livestock do their weeding and harvesting as well, for their families would not work for them again.

The adoption of draft plowing may still be underway in Khumbu today. During the 1980s a few families in Nauje and in Khumjung began having large potato terraces plowed, breaking with the custom of only plowing grain fields. This was judged to be cheaper than hiring agricultural laborers to perform the same work. It is not yet, however, a very widespread practice. Virtually all families continue to rely on either reciprocal labor arrangements or hired labor to dig and prepare their potato fields with hoes. A cultural factor may also be involved, for some people feel that digging potato fields gives more energy to those fields than plowing does, producing better crops. This impression could conceivably be related to a difference in the depth to which the soil is worked. The Khumbu scratch plow probably does not work the soil as deeply as hoe digging does.[28]

Post-1960 Agricultural Changes

In the past thirty years there have been a number of changes in Khumbu agriculture. These include changes in values that have affected community agricultural management practices, the introduction of additional new potato varieties, the decline of buckwheat production, an increase in the attention given to growing fodder crops, and continuing experimentation with new techniques and crops.

Blight-prevention Practices

During the 1960s the enforcement of regulations aimed at preventing blight declined across much of Khumbu except Phurtse, Pangboche, and Dingboche. In Nauje few of the prohibitions had been enforced with any stringency for some years and even in the 1960s the only bans carefully enforced were the exclusion of livestock from the village after Dumje and the prohibition on bringing freshly cut wood into the village after the fourth day of the sixth month, Dawa Tukpa, two weeks later. The ban on bringing freshly cut wood into the village during the time of maximum blight danger ceased to be enforced around 1965 and after that the Nauje nawa only enforced the livestock-control measures. During the late 1960s and early 1970s many blight-protection provisions were also abandoned in Thamicho, Khumjung, and Kunde. By the mid-1970s only Phurtse, Pangboche, and Dingboche enforced prohibitions against bringing in freshly cut wood after the fourth of Dawa Tukpa. All settlements, however, continued to enforce the mid-summer bans on grazing near the villages.

The decline in the enforcement of some blight-related rules in several villages appears to go back to well before 1960 and reflects changing beliefs. But the striking decrease in enforcement of restrictions on freshly cut wood in the late 1960s and early 1970s in Nauje, Khumjung, and Kunde may be related to changes in local resource management after 1965. The implementation of new forest regulations by a government office in Nauje led to the abandonment of local enforcement of tree-felling rules in those three villages. Perhaps it was decided that all other forest-related controls had also been made unenforceable by the government's announcement that it would be responsible for forest management.

The Introduction and Diffusion of the Yellow Potato (*Riki Seru*)

In the middle of the 1970s yet another new potato variety was separately introduced to Khumbu by two men from different parts of the region who encountered it in very different parts of the Himalaya. Both had recognized the variety as something new, a potato of a different color and texture which had a reasonably good taste and produced a high yield of large tubers. Both brought back a few seed potatoes to grow at home, one planting these in Thamicho and one in Khumjung. Following these introductions the tuber became the most important variety throughout the main villages of Khumbu within ten years and diffused from Khumbu to Pharak, Shorung, Salpa, and even the Rai regions of the Hinku and Hongu Khola and the lower Dudh Kosi.

The yellow potato was brought to Khumjung by Pemba Tenzing about 1976.[29] He encountered the new variety at Singh Gompa north of Kathmandu on the trail to Langtang National Park. While employed by a tourist trekking group he made an overnight stop at the monastery and met an old Tamang friend who was then working in the monastery kitchen. His friend served him some boiled potatoes which Pemba

Tenzing noticed were unusually large in size. He asked if he could have half a load (15–20kg) of them to take back to plant in Khumbu.[30] The Tamang was unwilling to part with that many, but did give him five potatoes. From the five potatoes that Pemba Tenzing and his wife planted they harvested five tins (ca. 50kg) of the yellow-tinted, slightly watery, large, oval tubers. From these few potatoes, he notes, the yellow potato spread all over Khumbu.

That they did spread throughout Khumbu and far beyond in remarkably little time had a good deal to do with Pemba Tenzing's neighbors, Konchok Chombi and his wife Ang Puli. Konchok Chombi and Ang Puli obtained two yellow potatoes from Pemba Tenzing's wife. She cut eighteen eyes (*mik*) from these and planted them, harvesting nearly a tin (11kg) of potatoes. These were planted the next spring and yielded four loads. During the next few years they obtained successively greater amounts of seed potato and before long were able to convert almost their entire Khumjung production to the new variety. Soon they were harvesting more than 200 loads per year.

Konchok Chombi realized the importance of the new variety and spread the word widely about the discovery. Equally as important, he was able to provide seed potatoes for others to buy and introduce into their own fields and villages. By Khumbu standards Konchok Chombi is a large landowner with four large fields at Khumjung which can produce harvests of up to 400 loads of yellow potatoes in the best years, far above his household requirements. Not only could he provide seed potatoes, but in order to meet the widest possible demand and diffuse the new variety as rapidly as possible he limited purchases to a small amount of seed potatoes per customer. He would sell no more than three tins (33kg) to any one family, enough for them to harvest sufficient seed potatoes the following year to plant much of their fields in the variety. Beyond this he even promoted the new potato in other regions, sending gifts of seed potatoes to the Sherpa villages of Golila, Gepchua, Mure, and Tingla west of Shorung, to Chaunrikharka in Pharak, and to Sherpas in the Katanga and Kulung regions. Villagers from Pharak began to seek him out at home for the new tuber. Within a few years the potato became known in much of the region as "Au (uncle, a respectful term of address) Chombi's potato." Pemba Tenzing's discovery of the variety has been nearly totally forgotten.[31]

Also largely unknown is the fact that there were not one but two introductions of the yellow potato to Khumbu. Outside of Thamicho most Sherpas are totally unaware of the separate introduction of the variety to Pare and Yulajung. About 1973 Dorje Tingda, a Thamicho man who has houses in Pare and Yulajung, discovered a potato in Darjeeling which is now considered to be identical to the Khumjung tuber.

He brought back a small quantity of potatoes that he had first noticed while having lunch in a tea shop near Darjeeling while guiding tourists on a local pony trail. It is said that for three or four years he kept knowledge of the new variety to himself because he was worried that if other families also began producing the new, yellow variety it might affect the business he was then doing selling his crop to the Japanese-operated Everest View Hotel near Nauje. According to some Thamicho residents he only began selling villagers potatoes after three local Sherpa panchayat officials asked him to sell them a few. One of these officials, a Yulajung Sherpa, noted that from the single tin of potatoes he obtained he harvested four large loads (more than 160kg) the first year. He had thought that this was an unparalleled harvest until he learned that one of the other officials had harvested five loads from his single tin. After that, he remarked, the new potato spread rapidly through Thamicho.

The pace of the diffusion of the yellow potato through Khumbu can be roughly charted. Thamicho acquired the variety mainly through its introduction to Pare, but it was not until the early 1980s that there was very wide diffusion of the tuber because of the difficulties in acquiring seed potatoes.[32] By 1981 the potato was becoming well established in Nauje as well as Khumjung and Kunde. In these places it was grown on an approximately equal basis with the red potato by 1985, and by 1987 dominated crop growing. In eastern Khumbu the yellow potato became established slightly later. It was first planted in Phurtse in 1981, but many families there did not begin planting it until 1983 or 1984. It has, however, become the major variety grown there during the past few years. The yellow potato was first planted at Dingboche in 1982 and may have been tried at Pangboche a few years earlier.

The rapidity with which the yellow potato was adopted and at which it supplanted the red potato in the main villages reflects Sherpas' interest in experimentation with new varieties and with adopting high-yielding varieties. In this case the degree of interest in greater harvests out-weighed a number of early reservations that many people had about the variety. A considerable number of people initially disliked its taste and many farmers initially refused to grow it. It was also noticed very early on that the variety had a different growing pace and different leafing characteristics than the red potato, and some farmers were concerned that this would diminish the yield of intercropped radish. Some families for whom radish production for use as fodder was important were reluc-tant to cultivate the new variety as a result. There were also questions about its performance at higher altitudes. It was felt that the yellow potato's longer growing season put it at greater risk than the red potato in the short summer season of the high-altitude settlements. Some peo-ple who experimented with it in the Dudh Kosi and Imja Khola valleys at altitudes above 4,000 meters also found that the tuber seemed to become still more watery when grown at that altitude. This reputation for poor taste at high altitude led a number of families in eastern Khumbu to decide not to grow the variety at sites above the main villages.

Many of these early evaluations, however, were modified within a few years as the process of testing and evaluating the yellow potato continued. Families experimented with the variety at different sites and people noted with interest the crop experiments of the pioneer growers in each locality. Word of good harvests spread quickly. Farmers who said in 1984 that they would never plant the yellow potato were planting it by 1987. The reservations about its taste became less pronounced, although the red potato continued to be proclaimed the finest-tasting potato in Khumbu.[33] It even began to be grown more widely in high-altitude settlements. By 1987 the variety was being planted by many families at Dingboche and also by several families at Na despite earlier reports that yellow potatoes that had been planted there had had poor taste. By 1987 it had also become the major variety planted at Tarnga. The yellow potato, however, has still not been universally accepted as fit for high-altitude planting. Some people refuse to plant it at Dingboche and in the upper Dudh Kosi and others are reluctant to plant it at Bhote Kosi valley sites higher than Tarnga. The increased wateriness of the tuber at these altitudes is still the factor most often noted as the main factor in this decision.

The Introduction of Other New Potato Varieties

Following the introduction of the yellow potato at least two and possibly three more varieties have been introduced to Khumbu by Sherpas who encountered them elsewhere. The most important across Khumbu is development potato (*riki bikasi*), which is said to have been introduced about 1981 from Phaphlu in Shorung by a Nauje man who brought back some from the agricultural extension office there.[34] Thus far the development potato has been grown primarily in lower Khumbu, especially in Nauje, Khumjung, Kunde, Thami Og and Thami Teng. In the Bhote Kosi valley it has been experimented with as high as Tarnga. Very little was being grown in 1987 in Yulajung (although families there have been experimenting with it for at least two years), but it is being grown fairly widely in Thami Og and Thami Teng. The adoption of the variety was slowed down in Yulajung by concerns over its storage qualities and in eastern Khumbu both by doubts about its hardiness at higher altitudes given its long growing season and by a lack of seed potatoes.[35] There have been reservations also about its taste and its high degree of wateriness. Some people suspect it causes stomach problems and others are convinced that it is lower in food energy than other potato varieties. But the yields of development potatoes are usually so outstanding even in comparison to the yellow potato that the new variety continues to be experimented with by many farmers and to be adopted increasingly widely. By the late 1980s there were fewer reservations about both its altitudinal fitness and taste, although many people still consider it far inferior in taste to the red potato, the yellow potato, and kyuma.

It is possible that the brown potato (*riki mukpu*) was also introduced in this period, although some people put the date of its introduction earlier. The brown potato is a round, dark-skinned variety of uncertain origins, which some Sherpas maintain was introduced in the late 1970s from Pharak.[36] It has developed a reputation for yielding well at high altitudes and is much sought after by people who want to experiment with a few plants in their fields. The secondary high-altitude agricultural site of Goma in the upper Bhote Kosi valley is very well known as a center of brown-potato cultivation, but thus far the variety is little grown outside of the upper Bhote Kosi valley.

The latest addition to the Khumbu potato repertoire was introduced by Nauje families (including Sonam Hishi's family) from Pharak in 1987. It has no commonly accepted name. Sonam Hishi's wife Chin Dikki is calling it, with much amusement, long tail (*ngamaringbu*) after a hairlike protrusion from the base of the tuber.

The Decline of Buckwheat Cultivation

Until the 1970s substantial amounts of buckwheat were being grown in all Khumbu villages other than Nauje.[37] In Khumjung, Kunde, Pangboche, and Phurtse buckwheat continues to be grown on a great deal of land and in Phurtse and Pangboche it is planted on nearly 50 percent of the cropland. Formerly considerable buckwheat was grown in the villages of the Bhote Kosi valley as well as in some of the gunsa and at Tarnga. But buckwheat has long been less emphasized there than in the other buckwheat-growing areas. According to elderly residents even in the early decades of the century buckwheat was grown on as little as a quarter of the land in Thami Og. During the past ten years interest in planting buckwheat in the Thamicho villages has plummeted. In 1986 it was being grown in Thami Teng in only three fields and the following year was only cultivated in three small patches. In 1987 only a single, small field was planted in Yulajung and no buckwheat whatsoever was grown in Thami Og. The Bhote Kosi valley has become monocropped with potatoes from the gunsa to the high-altitude herding settlements.

There may be several factors in this increased emphasis on potato cultivation in the Bhote Kosi valley. One may be population pressure and a response to it by intensifying production. As food demand in the Bhote Kosi valley increased during the twentieth century it might be expected that potato production would be further emphasized, for it produces much more food per hectare than any other Khumbu crop and can form the bulk of a household's diet if necessary. For at least half a century families with little land have tended to put their limited land to potatoes and to grow less buckwheat. One of Khumbu's oldest residents, a man of Khumjung, noted that when he was young if a family had a good deal of land it would plant half in kyuma and half in buckwheat and rotate them annually, but that if it was poor it would primarily plant potatoes. A concern with intensification probably also led families in Nauje, where land is in very short supply relative to the size of the population, to emphasize potato monoculture. There grain cultivation was abandoned very early in the century.

Yet there remain some questions about the role of response to population pressure in the conversion of land from buckwheat to potatoes. It is unclear how great recent population growth has been in the Bhote Kosi valley and to what degree fragmentation has affected land ownings per household. It does not seem likely that land shortages there are significantly greater than in eastern Khumbu where buckwheat has not been abandoned. Even if the average size of household land holding has declined valleywide there are still problems with a simple intensification explanation, for not all families are equally land-poor, and there is no doubt that many Thamicho families do own amounts of land which are substantial by Khumbu standards. That even these households have chosen to specialize in potato production suggests that factors other than population growth are involved in the current monoculture of potatoes in the valley.

Commercialization might be a factor, although this is difficult to substantiate given the lack of pertinent household and land data. Thamicho has long exported small amounts of dried potatoes to Tibet and since the early 1970s it has played an increasingly important role in the small-scale regional exchange of potatoes within Khumbu itself. Many Nauje families have depended for generations on purchasing some potatoes to augment their own production. Before the mid-1970s much of the Nauje demand was met by Khumjung production. During the mid-1970s more attention was turned towards the Bhote Kosi valley when Khumjung suffered a series of disastrous harvests. During the last years before the widespread adoption of the yellow potato Bhote Kosi valley surplus potato production became important not only for Nauje families but also for many Khumjung and Kunde households. Demand was so high that tension broke out between Nauje villagers and those of Khumjung because some Khumjung residents were intercepting Thamicho farmers on their way to the Saturday market at Nauje and buying out their entire supply of potatoes.[38]

The adoption of the yellow potato eased the shortage of potatoes in Khumbu, but demand for tubers continued to increase as a result of growth in the numbers of Nepali residents in Nauje and dramatic increases in the scale of tourism. Tourists consume large quantities of locally grown potatoes that are an important component of both the food offered by local lodges and that cooked by commercial camping tours. Potatoes are sold to families, lodgekeepers, and trekking groups at the Saturday market at Nauje as well as continuing to be bartered in direct family-to-family exchanges. Some Nauje households, for example, make deals with Thamicho farmers for potatoes months before the harvest, offering cash, tea, kerosene, and other commodities bought at bulk prices at the Nauje market or in Kathmandu.

The possible relationship between greater opportunities to sell potatoes locally and increased Thamicho emphasis on their production requires further study. As of now I cannot evaluate how common it is for Thamicho farmers to sell surplus potatoes or the role that interest in producing a surplus plays in crop decisions. When I attempted to pursue this line of investigation in 1987 I found that Thamicho farmers unanimously denied that an interest in selling potatoes had anything to do with their decisions to discontinue the cultivation of buckwheat. They also did not cite land shortage as a factor. This does not mean that commercial motives and interest in intensification are not factors in land-use decisions, for certainly an interest in higher yields has driven the recent widespread adoption in Thamicho of the yellow potato and the development potato, and some families must make substantial income from the sale of potatoes. But it does suggest that these were not the most immediate reasons for the relatively rapid and extensive recent decline in buckwheat growing.

According to Thamicho farmers the decline of buckwheat in their region was caused by the increase in the number of crossbreeds herded in the valley and the breakdown of local pastoral managment regulation. This has made it increasingly risky to cultivate buckwheat. Yak and zopkio have always been considered threats to buckwheat crops and urang zopkio especially are considered very apt at slipping down valley and getting into fields. The great increase in the 1980s in the number of crossbreeds kept by Thamicho villagers magnified this risk. Thamicho people complain that whereas yak and nak seem content to graze in the high pastures of the upper valleys in late summer that some urang zopkio move down valley during the night and break into buckwheat fields. By the time the damage is discovered in the morning entire fields can be ruined. This risk is further increased in late August and early September when many Thamicho herders take pack stock to Nauje to meet mountaineering expeditions. According to local regulations they are only allowed to spend a single night in the zones closed to grazing around the main villages as they move their stock through to Nauje. But many herders abuse this custom and keep their stock based in the villages for days. These violations and the nawa's inability to control them have not only seriously threatened buckwheat crops but have also been an important factor in undermining the effectiveness of the entire system of herding regulation in the Bhote Kosi valley during the mid-1980s.

Fodder Crop Production

Another trend during the past two decades has been an increase in fodder-crop and hay production. This has been especially marked in the Bhote Kosi valley and in Nauje. In Nauje more terraces are being planted to fodder crops and a few fields in Thami Og and Khumjung have also been planted in fodder crops recently rather than in food crops. Many new hayfields have also been established, especially in the Bhote Kosi valley. Some were created from pastureland in the upper valley, whereas others in the main villages were converted from cropland. Large numbers of such converted fields can be seen at Yulajung where they represent a major hay-growing resource for settlement families. There are also examples in Thami Teng and Thami Og as well as in Tarnga, Marulung, and some lesser settlements.[39] The interest in producing more hay and fodder in Thamicho probably reflects both the end of winter herding in Tibet and the recent decline of the nawa-enforced regulations that formerly protected some winter pasture and areas where wild grass was collected for hay from year-round grazing.

In Nauje some potato fields have been converted to production of barley and wheat grown for fodder. These fields are planted late, some as late as the first of August when grain is put into fields from which the earliest potatoes in Khumbu have already been harvested. When planted this late there is no opportunity for the grain to ripen, and the early harvest of potato also diminishes the yield of the potato crop. Barley and wheat stalks, however, can be dried and make fine hay. The second cropping of grain in potato fields is a new phenomenon and is mainly restricted to Nauje where the first barley fields were planted in 1984. It is not widespread even there, and less than 5 percent of the village field area is involved. There are only two examples elsewhere in Khumbu of growing grain for fodder, a single wheat field in Thami Og and a barley field at Khumjung. But in Nauje there is increasing interest in growing barley. In

1984 only four fields were planted in barley, all of them on terraces that were used in the spring and early summer as trekking campsites and hence could not be planted in potatoes. In 1987 twelve fields were planted to barley, six of them as second crops following potatoes. One field was planted to wheat. Since then the trend has continued and both barley and wheat are being grown on a small scale by still more Nauje families. Interest in Nauje in fodder cultivation reflects the increasing scale of livestockkeeping there, the greater fodder requirements of urang zopkio as compared with yak, the impact on local pastoral resources of the post-1979 pattern of year-round grazing in the Nauje area, increasing shortages of wild grass to dry as fodder, and escalating hay prices. Villagers' lack of hay land in the higher valleys is also undoubtedly a factor. Few Nauje families own hayfields, although one family has recently begun to grow hay at Tarnga and others at Samde and Tashilung.

Agricultural Experimentation

Sherpa experimentation with new crops, crop varieties, and agricultural techniques is a continuing process. Women constantly trade seed potatoes and take keen interest in the productivity and other characteristics of these potatoes when they are grown in the microenvironmental conditions of their own fields. A few women and men also take an interest in other forms of agricultural experimentation and try out new cultigens, varieties, and techniques, the results of which their neighbors observe with great interest.

Although potatoes have been the major focus of Sherpa agricultural experimentation in the twentieth century, some Sherpas have also experimented with maize at Jangdingma (a gunsa site below Nauje on the Bhote Kosi), with wheat in Nauje, Thami Og, and Dingboche, with barley at Nauje and Khumjung, and with wheat, buckwheat, peas, and white barley at Dingboche. Families in Nauje have recently adopted a variety of new household garden plants, including cabbage, cauliflower, carrots, and spinach. They obtain seeds for these plants from Kathmandu and from foreign friends. A few families in Thami Og and Pangboche are experimenting with these and other vegetables and one family has unsuccessfully attempted to raise cabbage and cauliflower at Dingboche. There has also been much interest in fruit. Apple trees have been planted Nauje and in the lower Bhote Kosi valley for several decades, although thus far without any success.

There seems to be more interest in experimenting with crops and crop varieties than with new agricultural techniques. Two exceptions are the current efforts at Samde and at Phulungkarpo to irrigate hayfields. The Samde experiment is the first Khumbu use of polyvinylchloride pipe for delivering irrigation water. There have been no experiments thus far with mechanized tools or pumps, nor with insecticides, herbicides, or fungicides. A couple of farmers have brought small amounts of chemical fertilizer from the agricultural development office in Shorung and tried them out, but the transport involved and the expense of the fertilizer have limited the use of such fertilizers to a few experiments.

Pastoral Change in Traditional Times

Khumbu pastoralism has commonly been portrayed as having recently undergone a major transformation from the traditional yak-herding practices of the 1950s. There has indeed been very significant change in pastoralism since 1960, and the economic, social, and environmental implications of this are only beginning to be understood. Many of these recent changes are related to tourism development and will be discussed in the final chapter of this book. But before taking up the processes and implications of recent change it is necessary to examine more closely the notion that as recently as 1960 Sherpas had long-standing herding practices that were then suddenly transformed by national and international political and economic developments. In the remainder of this chapter I reexamine historical change in Khumbu pastoralism from the early days of Sherpa settlement until the 1960s.

Early Herding in Khumbu

It is very likely that Sherpas arrived in Khumbu already familiar with the herding of yak, sheep, and goats. These animals are raised throughout the Tibetan world including the Kham region. Oral traditions about the Sherpa migration from Kham to Khumbu speak of the immigrants bringing their yak with them to Nepal. There are accounts, for example, of conflicts that these yak precipitated with both the inhabitants of Tibet and the Rais who then lived in the country around the Pi Ke peak in the Golila region. Another oral tradition relates how one of the first Sherpa settlers used to bring sheep manure each spring from his winter gunsa in a cave near Phurte to his Tarnga fields.[40] The tending of goats is not mentioned in oral traditions about the early decades of Sherpa life in Khumbu, but it may nevertheless also be a very old Khumbu practice. Goats, along with yak and sheep, are considered to be under the special protection of Khumbu Yul Lha. Paintings of the god in temples and the one on the huge boulder above Nauje which figures centrally in Dumje ceremonies also depict a goat, a yak, and a sheep. And at the summer Yerchang ceremonies barley flour representations of all three animals are ritually offered to Khumbu Yul Lha in asking him to guarantee the welfare of the herds.[41]

During the early centuries of Khumbu pastoralism the present system of using stone-walled huts as gunsa and high-altitude herding bases had apparently not yet evolved. Several legends mention the use of black tents (*ri bu*), a practice associated today with Tibetan pastoralists and some Tibetan-culture Himalayan groups (e.g., the people of Dolpo), but not with Sherpas.[42] These were still in use in parts of Khumbu until the turn of this century, and in some high summer herding settlements such as Dole, Marulung, and Tarnak huts may have first been built only in the late nineteenth century. Some families in the late nineteenth and early twentieth centuries set up elaborate camps in the highest pastures and rich families erected as many as ten or twelve tents, each of them requiring a yak to transport.[43]

It is not clear when the first herding huts were built in the high country and at what pace they replaced the use of tents in the different valleys. It could well be that tents were in common use for many generations in the highest pastures and were replaced by huts only where hay growing developed and a need arose for some permanent, roofed structure to serve as a barn. Today herders who want to base in an area where they have no hayfields are often satisfied with using resa for a few weeks and are content with a simple bamboo mat or a tarpaulin thrown over a low structure of rough rock walls. But at places where they grow hay the same herders have more permanent, roofed huts where they can store hay through the winter and which they can use as a base in the winter and early spring when they take their stock up to the often still-snowy high country to feed them hay and fertilize the hayfields. The building of herding huts in the phu may thus very well be closely connected with the development of hay cultivation in various high-altitude herding areas.

The origins of the Khumbu practice of growing hay in stone-walled fields, however, is also unknown. Oral traditions are silent about either early hay cultivation in Khumbu or its later introduction. The growing of hay could be a tradition dating to Kham or it could have been developed as recently as the nineteenth century. Without archaeological evidence there is no way to know. Neither Tibetans from the Tingri region nor the Rais and other peoples of the Dudh Kosi region grow hay in walled fields, and this practice also seems to be rare elsewhere in Nepal.

Some Sherpas speculate that hay growing was probably much less significant to Khumbu pastoralism before the late nineteenth century. Before then, they suggest, the total number of livestock would have been much less in the region and there would have been abundant grazing in both summer and winter without the need to story hay. This is only speculation, for there is no way to be certain that early Khumbu household herds were not larger than herds have been during the past few generations. But elderly people believe on the basis of what they have heard from parents and grandparents that the number of stock in the second half of the nineteenth century was less than that in the twentieth century, and also note that the practice of taking yak and nak to Tibet for winter grazing meant less stress on Khumbu pastoral resources.

In the high-altitude settlements where hay is grown today there were hayfields as early as the late nineteenth century. Sherpas believe that their development was related to two factors. One was a harsher winter and spring climate with heavier snowfalls. In the late nineteenth and early twentieth centuries there were also very heavy snowfalls in some years, events that Sherpas call *kaumuche*. In such times so much snow fell that people climbed out of their houses through upstairs windows, and there were avalanches that destroyed houses and took human and livestock lives.[44] These snows made grazing impossible for weeks at a time even in lowermost Khumbu and Pharak and made it vital to have hay supplies. The second factor that people believe led to increasing cultivation of hay was a regional increase in the number of livestock. This increased the competition for winter and spring grazing and made it advisable to grow hay as well as to dry and store other types of fodder such as wild grasses.

The Origins of the Nawa System

The nawa system has generally been assumed to be an ancient, traditional Sherpa system. It, like so many other facets of Khumbu land use, could date to premigration practices in Kham. But a Kham origin may be unlikely, for if the practice was that old it might be expected to be typically Sherpa rather than an institution that is unique to Khumbu. Sherpas in other areas have some seasonal regulations regarding the movement of livestock, but with the possible exception of the Rolwaling region none of them has a system of opening and closing a series of zones to different land-use activities or of coordinating these in valleys that are the homes and grazing grounds of people from several different main villages. In Khumbu multivillage zonal systems may be no older than the middle of the nineteenth century. Before then grazing regulation appears to have been the province of individual villages that zealously restricted the use of their village pastures to stock belonging to village residents and which may have had regulations regarding times of the year when grazing was allowed or forbidden in the vicinity of the village and surrounding winter pastures.

Konchok Chombi believes that the distinctive nawa system familiar in twentieth-century Khumbu must have been developed no earlier than about the middle of the nineteenth century. He thinks that it originated after an intervillage grazing dispute when Kunga Hishi of Thami Og was the gembu of Khumbu. According to an oral tradition Kunga Hishi married a Phurtse woman of the Sherwa clan. She was given a nak by her parents at the time of the wedding and took this nak with her to her new home in Thami Og. The nak, however, soon tried to return to its familiar Dudh Kosi valley grazing grounds. On its way to Phurtse it passed through Kunde-Khumjung village lands and at Zarulungbuk, the ridge between the Bhote Kosi valley and Kunde, it was killed by Khumjung villagers. A major court case ensued. The infuriated gembu demanded compensation for the nak. Khumjung villagers maintained that they had been within their rights, for at that time grazing areas were controlled on the basis of village boundaries rather than by the open range and zone system familiar today.

The dispute worked its way up through a series of Nepali courts. In that era the government court at Olkadunga had not yet been established and the case went first to Those and then to Charikot before reaching Kathmandu itself. Here a decision is said to have been passed down by the king himself. This edict informed the people of Khumbu that the decision should be made by the region's leading local political authority—gembu Kunga Hishi! Gembu Hishi thereupon declared a verdict in his own favor and further ruled that villages no longer had the right to limit the use of their village lands to village residents. Henceforth all of Khumbu was to be open range and the rangelands of Khumbu were to belong to all Sherpas and their stock with the proviso only that all herders were responsible for keeping their charges out of fields and hayfields.

This verdict threatened the very basis of the communal regulation of resources of that era. Villages did not give up the old system quickly, as several subsequent court cases attest. Later in the nineteenth century, for example, Pangboche villagers made a formal appeal to the government to halt Kunde-Khumjung livestock from being herded in their area. This appeal, it is said, was presented to a visiting Nepali official for judgment. He is said to have appeared to favor the Pangboche cause, for he told Pangboche villagers that he had decided in their favor and would present them with documents to verify their rights to restrict grazing on their village lands. He did give them a document, written apparently in a bureaucratic Nepali that no Pangboche villager was able to read. The villagers presented the official with a number of expensive, woven, yak-hair and wool tarpaulins. Indeed, according to one tradition they gave him these before he had ruled in their favor and at his request or demand. They thought the matter over until the dispute flared up again a few years later and they took the case to court. Here, when they confidently presented the document that established their case, they found that it did not win the day, for Kunde-Khumjung villagers had been given an identical document. Neither document upheld the right of villages to exclusive grazing lands. Pangboche people had to acquiesce to herders from other parts of Khumbu making use of their rich pastures and ever since Kunde herders in particular have relied on "Pangboche" land.

There have also been some cases in the twentieth century of attempts to limit village lands to grazing by residents' stock. In the 1940s, for example, Phurtse villagers attempted to keep Nauje stock from being herded on their lands. One wealthy Nauje trader had sent his large herd of zhum to the upper Dudh Kosi Phurtse pastures in the care of a hired Phurtse herder. Phurtse people technically could not ban Nauje stock from their area, but they could decide that henceforth no villager could act as a herder for stock owned by non-Phurtse people. This put an end to the Nauje zhum grazing in that area. But in this century villagers have otherwise had to accept the open range policies established more than a century ago. In recent years many Thamicho, Khumjung, and Kunde residents have been unhappy with increased grazing by Nauje stock on their lands, but they have not had any legal ground to ban Nauje urang zopkio from the Mende, Phurte, Khumjung-Kunde, Langmoche, Gokyo, and Tengbo areas. Bans on outside stock were imposed successfully, however, against non-Sherpa herders. During the early twentieth century a number of Sherpas strongly resented Gurung grazing in Khumbu and there were efforts to stop it and to put pressure on the Khumbu pembu who had authorized it. In one case a Khumjung pembu was pelted with stones by fellow villagers for having allowed Gurungs to bring their flocks into the region. Yet, while for much of the early twentieth century Gurungs as well as Sherpas benefitted from the open range tradition, their grazing access was ultimately ended. Thirty years ago Nauje residents banned Gurung sheep from entering Nauje village lands. Villagers posted a sign on the bridge below the village which forbade the Gurungs to continue with their flocks, and some residents stole into the Gurung camp at night to show them that they were serious about closing Nauje to Gurung grazing by carrying off some of their sheep.

The development of the nawa system may represent an ingenious response to the regional undermining of village control of village lands. Konchok Chombi speculates that the development of the system of zonal regulations was a reply to the impact of Kunga Hishi's ruling. Something had to be done to decrease the risks of crop losses from livestock and the need to preserve winter fodder and grazing areas, and villages were interested in setting their own regulations for their own areas. The establishment of the nawa system would have enabled communities to establish strict controls on herding and the cutting of wild grass without contradicting the new requirement that access to an area could not be limited solely to local residents. The nawa system affirms that access is open equally to all, but it also makes everyone subject to the same regulations that may restrict certain activities in certain areas for specified periods of time. Villages could thus largely continue to decide which activities they would tolerate on the lands immediately around them. The opening and closing of village areas to grazing had to be coordinated with the operation of zonal restrictions by other communities in the valley, and where several villages shared grazing grounds it might be necessary to jointly administer zones. Crops, haylands, and winter pastures, however, could be protected from grazing. If such a sequence of institutional change did take place in the nineteenth century it represented a superb example of an adaptive response to both environmental and social conditions. This is true even if the concept of village officials charged with enforcing community relations such as seasonal exclusions of stock from villages predated Kunga Hishi's decision. The killing of the gembu's wife's nak may have resulted precisely because it had violated such a village summer ban on stock, perhaps damaging crops in the process. In this case the development of the zonal systems now in use would have been a creative elaboration of new institutions from an older institutional base in response to changing social conditions.

The establishment of the current form of the nawa system certainly predated 1900 and very probably took place at least a generation before that. This did not, however, end Khumbu concerns with the boundaries of village lands and the control of pasture and forest areas within them. Since the late nineteenth century there have been a number of major disputes over boundaries, for control of areas determined which villages decided on which resource-use rules would be enforced there. The shift of a boundary line could in effect change a forest from being a carefully protected, sacred forest to one in which there were no prohibitions of any kind on tree felling, or a pasture area from one closed to grazing all summer to one open year round. A late-nineteenth-century dispute between Khumjung and Phurtse over the Mong area resulted in the end of strict protection of the nearby sacred forest. And during the twentieth century Nauje villagers have been involved in several long-standing disputes with Khumjung-Kunde and Thamicho villagers over village land boundaries that would make an enormous difference in how some pasturelands are regulated. Nauje villagers have been trying to gain control over the Shyangboche area from Khumjung and Kunde for over half a century so that they could remove the summer grazing restrictions there.

They have similarly tried to extend their control north in the Bhote Kosi valley from the present Nauje village boundary south of the gunsa of Phurte. In the Shyangboche case Nauje villagers have been arguing for at least two generations that the traditional boundary between their village and Khumjung-Kunde is unfairly located just above Nauje. This, they say, reflects a decision reached a century-and-a-half ago when Nauje was a very small place and not the second largest village in Khumbu. They would like to move the boundary about a kilometer from the outskirts of the settlement to the watershed between their village and Khumjung and Kunde. Both pasture access and forest control were issues here, for the disputed Shyangboche area was within the boundaries of the Khumjung-Kunde-administered rani ban. This dispute has on several occasions very nearly sparked violent conflict and once, a few decades ago, almost led to the collapse of both the Nauje and the Khumjung-Kunde nawa systems before villagers were persuaded by a Nepali official to make peace with one another and resume nawa regulation. But though the boundary remains, the Khumjung-Kunde ban on summer grazing in this area is often ignored by Nauje herders and their stock. This continues to be a cause of great tension between the villages and threatens to ultimately undermine the continued operation of the Khumjung-Kunde nawa system.

Yak and Nak Herding in the Early Twentieth Century

In the early decades of this century Sherpas kept herds of a size that has never been equaled since. A number of families had herds of more than sixty head of yak and nak and a few had herds of eighty head— more than twice the largest Khumbu herds of the past forty years. These large, early-century herds were owned by men who were interested in breeding and selling crossbreeds and thus were primarily composed of nak. But some families also kept relatively large numbers of yak to use as pack stock on trading expeditions to Tibet. Many of these large herds of nak and yak were tended by hired professional herders. Usually these men were recent Tibetan immigrants, some of whom also kept sizeable herds of their own. There were so many head of large stock in Khumbu in that era that the winter range and wild fodder resources of the region were inadequate to support them. Rather than go to the trouble to produce vast amounts of hay to winter their herds, many big stock-owners sent their herds to Tibet each winter to graze on the vast grasslands beyond the Himalaya.

The great herds are well remembered today in Khumbu by people old enough to have seen them or who heard about them from their parents and grandparents. One of the greatest herds was kept by the Mendoa family of Khumjung, who at one time owned more than eighty nak and yak. People say that in the autumn when Mendoa sent livestock north to Tibet he usually sent more than sixty nak and yak, and that he had five yakherders (*yakpa*) to drive them. His fellow villager, U Kunggu, some-times sent sixty-one yak and nak north to Tibet and he liked to boast that he had one more head of stock there than Mendoa. Sundokpa, Pemba Kitar, Ang Chumbi, Thaktoa, and Yulha Tarkia all also had large numbers of yak early in the century. Some Thamicho families also kept big herds. Men in their seventies remember that early in their herding careers and in their fathers' time there were so many nak and yak in the Bhote Kosi valley that it was difficult to feed all the livestock. Here too many people sent their stock to Tibet for the winter with professional herders. Samshing Kitar and Gardza were especially famous for their large herds. People today still talk of the long line of stock that Samshing Kitar sent north each year to Tibet.

Nauje families also owned large nak herds in those days. Today the village has a reputation as the Khumbu village least involved in nak herding, but early in the century some of the greatest herds of all were kept by Nauje families. The greatest of the Nauje herds belonged to gembu Tsepal. He owned more than eighty head of stock, including yak, nak, dimzo, and zhum, and kept two lang (Tibetan bulls) for breeding purposes. These animals were cared for by several hired specialists. A Pinjo herded the nak and Yakpa Tundu the yak (as well as herding fifteen or twenty head of his own yak). A woman named A Droma was responsible for the milking. At that time Nauje families made much use of the rich grazing on Pangboche lands in the upper Imja Khola valley. Tsepal had a house and four fields at Dingboche and herding huts at Pheriche, Bibre and Chukkung. His herd was so large that when it was driven from Pheriche to Dingboche the line of stock stretched over a kilometer from the moraine above Pheriche all the way to the eastern edge of Dingboche. People used to come out of their houses to see the spectacle and excited children would shout "the gembu is coming, the gembu is coming." Other Nauje families also kept large herds. Ang Dorje kept thirty to forty nak. He herded them in the summer at Tsolo (near Dzongla) and at Lobuche. Other village families took their nak and yak to the upper Imja Khola valley. Oungu (Ang Dawa's father) had ten to fifteen nak and herding huts at Chukkung, Dzongla, and Bibre as well as fields at Dingboche. Nauje Urken's parents and Ang Gelgen's grandfather also had nak-herding bases in Chukkung. Another Nauje resident, Guru Nima, who immigrated to Khumbu from the famous yak-raising Chang region of Tibet, also had a large number of yak and nak. Unlike other Nauje villagers he kept his stock most of the year in Tibet in the Rongshar valley. The nak that comprised the greater part of the herd were based there year round and cared for by hired herders who in exchange kept a half-share of the milk and calves. Even the yak, which numbered more than thirty, were mainly kept in Tibet and were only brought to Khumbu when Guru Nima needed them as pack animals on his trading trips.

Until the 1930s it was typical for the really large herds of yak and nak to be taken to Tibet each winter. The yak and nak of smaller-scale herders also often went with them, for the professional herders who worked for the big stockowners would take other stock for a fee. Yakpa Kasare, who worked for the grandfather of the present head of the Khumjung Mendoa family, for example, not only took the large Mendoa herd north but also the stock of other herders. He charged half a rupee per head for herding yak and nak throughout the winter, a fee then equivalent to about a day's pay for unskilled labor, and his service was popular with families who wanted to devote these months to trading trips into southern Nepal and southern India rather than herding on the Tibetan plateau. On his annual herding trip north Kasare took hundreds of head of nak and yak belonging to families from Thamicho, Kunde, Khumjung, and Nauje. Kasare set out for Tibet each year in late October or November (Dawa Chuwa, the tenth month) and returned five or six months later in March (Dawa Sumba, the third month). In those days there were no grazing fees in Tibet, since the Tibetans did not resent sharing the grasslands as long as the Sherpa herds were not kept longer than a week in one place. Different Khumbu herders used different pasture regions. Kasare preferred to use Chakpakok, Surcho, and sometimes Lungar, whereas other Khumbu herds focused on Shalung and other nearby regions. For three months Kasare kept the stock on the move through the vast Tibetan grasslands north of the Nangpa La, shifting his base every two or three days. Then in mid-winter he turned and retraced his route back towards the Nangpa La. In early spring he would send a message back to Khumbu alerting the stockowners that it was time for them to reclaim their animals, and they would meet him in Tibet. Many of them then put their yak and nak to good use, buying Tibetan salt in Ganggar and Kaprak and hauling this with the pack stock back to Khumbu.

The big herds had mostly been developed as commercial enterprises, especially for breeding of crossbreed calves. The male crossbreeds, dimzo, were in great demand in Tibet as pack stock and there was a market for the female crossbreeds, dim zhum, in Shorung where they were esteemed as milch stock. Khumbu breeders had no trouble selling their calves to the Shorung livestock dealers who annually came to Khumbu on buying trips. Some Sherpas had long-term business arrangements with these dealers and made advance contracts for the delivery of a particular number of calves. The Shorung dealers then took both the dim zhum and dimzo south to Shorung where they were grazed for one or several years on the good grass of that region before being sold again. The dim zhum remained in Shorung and nearby areas, but the dimzo were mostly sold back to Khumbu men who then traded them in Tibet. Business was good enough that Khumbu herders could focus exclusively on breeding crossbreeds rather than also trying to raise nak to keep up their herd size. Nak were readily and cheaply available from Tibet and could be obtained in Khumbu from both Sherpa and Tibetan traders if one was not interested in taking the trouble to scour Tibet for good stock.

The Decline of Large-scale Nak and Yak Herding

During the first decades of the twentieth century some Khumbu stock-owners thus had herds of as many as eighty yak and nak, while several others had herds of more than fifty head and a number had herds of more than thirty. The big herds were broken up, however, by the 1930s and since then the ownership of even thirty head of stock has been very rare. A number of different factors could have precipitated this change. The trade in crossbreeds, for example, could have become less lucrative. There was excellent money in raising crossbreeds when the complex trading circuit that it was a part of was operating smoothly and the ultimate prices paid for dim zhum in Shorung and dimzo in Tibet were good enough. But a change in any one of a number of elements could have made it much less profitable. A rise in the price or a decline in the availability of Tibetan nak would have a long-term impact on Khumbu nak herding. A more immediate impact on the profitability of keeping large nak herds could have come from such diverse changes as a decline in the price offered for crossbreed calves, a decline in access to Tibetan pastures or the imposition there of grazing fees, difficulty in recruiting hired herders, herd taxes in Nepal, trade fees, new Nepalese regulations against exporting crossbreeds or importing nak, Tibetan regulations against nak sales or crossbreed imports, or closure of the border to trade. A number of these factors were involved in a further decline of nak herding in the 1960s, but none of them seem to have been the cause of the breakup of the big herds of the early century. Sherpas remained free to use Tibetan pastures for herding until the 1970s. Tibetan immigrants who were interested in herding for low wages continued to arrive in Khumbu during subsequent decades. Nak continued to be imported from Tibet on a substantial scale until new Chinese regulations were implemented in the 1960s and trade in crossbreeds in Tibet remained lucrative until then as well. Nak herding remained important in Khumbu and owning them still conveyed great status. But the great herds none-theless disappeared.

It may be that the business of breeding crossbreeds gradually became less attractive than other forms of commercial enterprises. By the 1930s and 1940s wealthy Sherpas were heavily involved in the trade of butter, paper, and other goods to Tibet and the return trade of salt, wool, and Tibetan luxury goods to Nepal and were moving beyond trade merely with Ganggar to dealing also in Shigatse, Lhasa, and as far afield as Calcutta. The owners of the great herds may have gradually been at-tracted by the opportunity to make greater profits by shifting their capi-tal away from pastoralism and into trade. There may have been a genera-tional factor at work as well, for it may be that the sons of these larger herders were less inclined to perservere in large-scale pastoralism and more inclined to trade. When sons inherited their share of the family herd some may have been content with herding this smaller number of stock on their own without hired herders. Others may have sold off much of this inheritance and kept only a few pack or milk animals. By the 1940s those lineages that continued to raise livestock raised much smaller herds and the wealthiest households among them were not those that had the largest herds but those that most successfully devoted their energies and capital to trade. Nak herding was left to families who were interested in it as a lifestyle and as a source of minor income from calf sales. Families who kept herds of more than twenty or twenty-five head of nak were still regarded with great respect and some envy by many Sherpas, but some of the truly wealthy families now kept no nak at all.

There was also a decline before 1950 in the keeping of yak as pack stock. A trend away from yak keeping can be traced back quite far in parts of Khumbu. In the Bhote Kosi valley elderly herders recall that they heard from their parents and grandparents that there had been a shift in their lifetimes away from the keeping of yak and towards a greater emphasis on nak. Au Puta of Thami Og, for example, a man in his seventies, commented that in his grandfather's time many more yak were kept and that these were useful as pack animals in the Tibet trade. In his father's time they cut back on yak to emphasize nak for crossbreed-ing. This same process appears to have taken place in Khumjung and in Nauje as well, although in Nauje some traders such as Guru Nima kept large numbers of yak for pack stock well into the 1930s. A decline in the scale of the bulk trade in iron and salt during this period may have been one factor in the declining emphasis on yak.

The Nepal government also played a role in the decline of Khumbu yak numbers early in the century. In July 1903 a British military force under the command of Francis Younghusband set out from Sikkim for Tibet. He was charged with discussing trade terms and international political issues with the Dalai Lama and ultimately, after several military encounters with Tibetan troops, reached Lhasa. The government of Nepal had pledged to support the Younghusband expedition and one of the ways in which it did so was by providing large numbers of pack stock. Khumbu was ordered to contribute yak. Large numbers of yak, for which the Nepal government is said to have only paid a pittance in compensation, were dispatched for Kalimpong. They were sent under the care of professional Khumbu yak herders, but the low-altitude route through the Nepal midlands proved to be fatal to most of the stock. Almost none reached Kalimpong alive.

Another important factor in the declining interest in keeping large numbers of yak during the early twentieth century was the availability of Tibetan pack stock. Until 1959 Tibetan traders came to Nauje each year with large numbers of pack yak to haul salt. On the return trip to Tibet there were insufficient loads for this many animals and the Tibetans accordingly offered to transport loads for Khumbu traders. This worked out very well for many Khumbu traders whose trade revolved around taking bulk goods to Tibet and importing higher-value commodities back to Nepal. Khumbu traders could also hire Thamicho yak for transport and there were also always Khumbu Sherpas willing to work as porters on the trade route for low wages, at least until the 1950s when higher-paying work for mountaineering expeditions became available in Khumbu. These opportunities for arranging hired transport for goods made it unnecessary to bother with the labor and expense of maintaining one's own pack stock. During the 1940s and 1950s Nauje traders relied entirely on hiring either pack stock or porters.

The breaking up of the large nak and yak herds may not have meant that the total number of head of large stock in Khumbu declined, for the stock may simply have been divided among heirs or purchased by other smaller-scale Khumbu herders. Regional grazing pressure nonetheless increased, for with the end of the large herds the practice of winter herding in Tibet declined dramatically. It was no longer easy for any herder to send even a few head of stock north by making arrangements with the hired herders of the rich stockowners, for there were no more big herds and no more professional hired nak and yak herders. Only Thamicho herders still took their stock to Tibet to winter. Herders throughout the rest of Khumbu began keeping their livestock year round in Khumbu instead. This meant that each head of stock now required more grazing than before. The number of stock that could be supported by regional range resources would have been lowered, and there would especially have been increased risk that the carrying capacity of the winter and spring pastures of lower Khumbu would be exceeded unless much greater amounts of hay and other fodder were stored for this time of the year. In the absence of a regional decline in stock numbers herders had no option other than to devote more effort to hay cultivation and run greater risks that the survival of their stock would be endangered by overgrazing, a poor haying season, or a snowy winter in which their fodder supplies might become exhausted too early.

The Declining Importance of Sheep in Khumbu

Although the keeping of sheep and goats goes far back in Khumbu Sherpa traditions, in the twentieth century these animals have primarily been kept by poorer Sherpa households. For such families both sheep and goats offered a critical source of manure and small amounts of wool or hair and meat. They were also sources of income. During the past twenty-five years the number of sheep in the region has probably never exceeded 1,000 and today is not much over 500. There have been even fewer goats, and at the time they were banned in 1983 there were only about 300 of them in all of Khumbu, many of them kept by non-Sherpa, blacksmith-caste families. Yet historically sheep were much more important in regional pastoralism. Some Sherpas kept sheep in larger numbers than today and large numbers of sheep were brought into Khumbu each summer from lower-altitude regions.

Today a flock of twenty sheep is large. In Pangboche in the 1930s and 1940s, however, there were three or four families who kept quite large flocks, one of which had more than eighty sheep and goats. In the era before 1950 there were also several Nauje families that kept large flocks, some of them as large as forty head. A few Nauje families used to use sheep as pack animals in the way that is also common in Tibet and in many Bhotia-inhabited, high Himalayan areas. No sheep are used for this purpose today, but as recently as the early 1960s a Nauje family used sheep to carry rice back to Khumbu from places as far afield as Namdu and Kabre, villages more than a week's walk toward Kathmandu.

The decline of Sherpa sheep and goat raising appears to have been directly associated with increasing affluence. Some Nauje families who kept sheep and goats during the 1950s and 1960s abandoned the practice in favor of cattle keeping once they could afford dairy animals. A similar process may well have taken place in other settlements. Sheep are currently kept only by very poor families. Some of the largest flocks, which at present number only as many as thirty head, are kept by non-Sherpa, Nauje blacksmith families who took up shepherding after the recent local government ban on goat keeping.

The relative lack of recent Sherpa emphasis on sheepherding might be thought to reflect a local perception that Khumbu is not well suited for sheep. Several observers have indeed concluded that Khumbu is poor sheep country (Brower 1987:171; Fürer-Haimendorf 1979:13). Brower (ibid.) has suggested that this is related to precipitation, that Khumbu pastures are less well suited for sheep than the rain-shadow Tibetan pastures. Yet Sherpas, Gurungs, and Rais all raise sheep in wetter areas and at equally high altitudes in both the Annapurna range and the upper Arun region. Along most of the span of the Himalaya pastures such as Khumbu's are highly valued by middle-altitude shepherds who see in them a promise of rich, high-country summer grazing and escape from the heavy rains, treacherous ground, disease, and leeches of the monsoon season in the lower mountains. Areas as high in altitude are regular summer herding destinations for Gurung and Rai shepherds in both the Annapurna-Lamjung ranges and in eastern Nepal. In the Modi Khola valley Gurung shepherds have made a two-week spring migration for generations up through a heavily forested gorge in order to reach 4,000–6,000-meter pastures far less extensive than Khumbu's. Arun region Rai shepherds similarly use the high Kempalung pastures, and those in the Hongu Khola, a Dudh Kosi tributary southeast of Khumbu, take flocks of more than 300 sheep each as far as Pharak as well as to the Naulekh area near Mera. Gurungs from the lower Dudh Kosi valley and adjacent areas today send sheep to Pharak, the Mera region, and northern Shorung, all of which are wetter in the summer than Khumbu. Their sheep in recent years have grazed on the pastures on the shoulder of Kwangde and even today graze high on Tamserku in view of Khumbu.

Oral traditions suggest that Khumbu may indeed have been formerly used as summer pasture by Rais and was certainly used by Gurung shepherds from at least the late nineteenth century until less than thirty years ago. The oral traditions about early Rai herding are few and sparse in detail. There is little to go on other than the idea that the Rais once came to Khumbu in the summer and that certain apparently long-abandoned ruins in the upper Dudh Kosi valley are said to be early Rai herding huts. Some Sherpas speculate that Rai summer herding could have been established before the Sherpas arrived in Khumbu and that it might have continued during the following centuries. In some other nearby areas such as the upper Hinku Khola valley and the Arun-Barun river areas Sherpas today share summer herding grounds with non-Sherpas, and in the Pharak, Kulung, Salpa, and Arun regions they have made arrangements to herd on Rai lands.[45]

Although the legends of early Rai pastoralism in Khumbu may be few and vague, the same cannot be said for nineteenth- and twentieth-century Gurung transhumance to Khumbu. Sherpas alive today remem-

ber the Gurung herds in Khumbu and have heard that in their parents'
and grandparents' time still-larger Gurung herds grazed even more extensive areas of Khumbu. Some point to the ruins of old high-altitude
herding huts or corrals in the upper Dudh Kosi valley, sites high on the
slopes above the highest yak-herding huts as possibly old Gurung or Rai
sheepherding bases. Others note that many of the placenames in the
upper Imja Khola valley and the valley of the Lobuche Khola seem to be
associated with Gurung sheepherding in that region, Tugla, Dusa, and
Ralha among them. Dusa, for example, is said to be from *dumsa* and to
refer to a place where rams and ewes are allowed to graze together to
breed. A number of elderly Sherpas who have herding bases in the
nearby region remember hearing stories of previous Gurung use of that
area for just that purpose, the main herds of ewes being kept between
Pheriche and Dingboche in the Chajung area and the rams being taken
up to Tugla-Palung until late summer when the herds would meet at
Dusa at mating season. This seems to have been a nineteenth- and
perhaps very early twentieth-century practice, for during the lifetimes of
eighty-year-olds the Gurungs have not summered at Tugla or Dusa, but
instead have been restricted to the area south of the Imja Khola.

In the twentieth century Gurungs primarily herded sheep in the Imja
Khola valley, although there was some use also of pasture in the Dudh
Kosi valley and in the Bhote Kosi valley. Phurtse people recall Gurung
sheep moving from Pangboche above Phurtse and across to Dole on
several occasions forty or fifty years ago, and have heard that more than
a hundred years ago sheep used to be taken up the east side of the Dudh
Kosi into the high country.[46] In the Bhote Kosi valley Gurungs herded
sheep for several years in the Bhotego area on the west side of the river
in the 1940s and long after that took sheep up onto the southern shoulder of Kwangde.

The main center of Gurung grazing during this century, however, was
the south side of the Imja Khola valley, particularly the Mingbo and
Ralha areas (map 16). Each spring several Gurung herders came up the
Dudh Kosi valley and into Khumbu, each with as many as 200 or 300
sheep. In the late 1940s and early 1950s as many as 1,000 to 1,500
Gurung sheep reportly grazed in the Mingbo and Ralha areas each
summer. Their migration path did not take them through the main
Sherpa villages. Instead they traversed high above the Dudh Kosi to
near the confluence of the Dudh Kosi and the Imja Khola and then
continued on to the southern Imja Khola valley pastures.[47] Sherpas
gained little from the Gurungs. They did not demand any grazing tax,
although it is possible that Gurungs may have made gifts to Sherpa
pembu in order to obtain their blessings on grazing in the region. Local
villagers were not interested in purchasing wool from the Gurungs, for

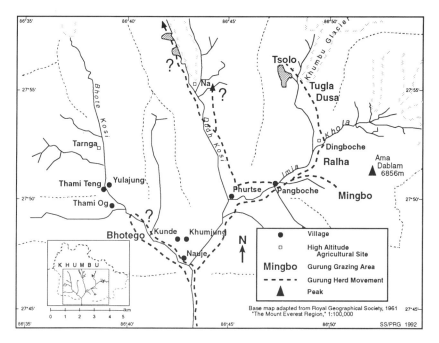

Map 16.   Gurung Twentieth-Century Herding in Khumbu

Sherpas consider the wool from the Gurung sheep to be far inferior to that of the Tibetan breeds of sheep that Sherpas raise. Sherpas imported wool from Tibet rather than from the Gurungs. The only real gain most Sherpas realized from sharing their pasture areas with the Gurungs was manure. In the spring the sheep dung was valued for manuring hayfields in the high-herding settlements and Sherpas often asked Gurung shepherds to pen their sheep at night on their hayfields. In return they fed both the shepherds and their dogs.

Sherpas resisted several Gurung attempts to enlarge their grazing areas. It is remembered that Dimal Gurung, a powerful and rich Gurung sheep owner, brought sheep up across from Orsho in the upper Imja Khola for three years to the great resentment of local Sherpas and that once he tried to bring his sheep into Dingboche. Some Sherpas had apparently invited him to bring his flock to Dingboche so that they could benefit from the manure, but others were strongly opposed to the idea. A community meeting was held and as the level of hostility rose Dimal is said to have fled to avoid a beating.[48] Dimal subsequently attempted to purchase grazing rights in the upper Imja Khola valley from the government. He is said to have offered a fortune—five loads of coins, a thousand coins per load—for these rights, but was unsuccessful. In the Bhote Kosi valley

Gurungs attempted to move up into the Konyak area, but were turned back by Sherpa villagers. In about 1960 Nauje Sherpas also closed their region to Gurung grazing and posted a notice on the bridge into Khumbu below their village advising that Gurung sheep were banned.

Gurung sheepherding in Khumbu ended about 1960.[49] A number of Nauje Sherpas raided the Gurung camp near the confluence of the Dudh Kosi and Bhote Kosi rivers and carried off their sheep. The Gurungs called in the police from the post at Nauje, who searched village houses, arrested four Sherpas, and returned the sheep. The Gurung shepherds then turned back to Pharak and no Gurung flocks have come to Khumbu since.

Although Gurung sheep are no longer taken to Khumbu they do still make a summer migration as far as Pharak and the Naulekh and Mera areas on the nearby upper Hinku Khola (map 17). Sheep from Jubu and Duwe (south of Aislalukarka) are brought up to Naulekh and Mera each summer in large numbers. Each shepherd is in charge of several hundred sheep and the total number involved exceeds 1,000, more than double the number of Rai sheep in the area. These pastures are also used by Sherpa herds. Both Gurungs and Sherpas must pay grazing taxes to the Rais. Gurungs are charged sixty rupees for the right to establish a herding base in one of the high pastures and an additional seven rupees per sheep for three months of summer grazing. Gurung sheep from the Rumjatar area west of the Dudh Kosi are taken up into Sherpa-inhabited northern Shorung to areas north and east of Junbesi.

Pastoral Change in the 1960s

There have been several major changes in Khumbu pastoralism during the past thirty years. One of these was a further regional decline in the importance of nak herding that began in the early 1960s and still continues today. A second major trend has been an unprecedented increase since the mid-1970s in the keeping of crossbreeds, and particularly in the ownership of male urang zopkio. This is directly linked to the increasing involvement of Sherpas in the tourism industry which will be discussed further in chapters 9 and 10. A third change that has already been mentioned was the end of goat herding in the region in 1983.

The decline in nak and the increased emphasis on crossbreeds show up very clearly in counts of regional livestock taken in 1957, 1971, 1978 (table 24), and 1984 (table 10). It is clear from these statistics that there was a pronounced decline in the number of nak herded in the region between 1957 and 1971 and that this was particularly important outside of the Bhote Kosi valley. In Thamicho there was also a decline in nak keeping, but this was much less sizeable and occurred primarily after

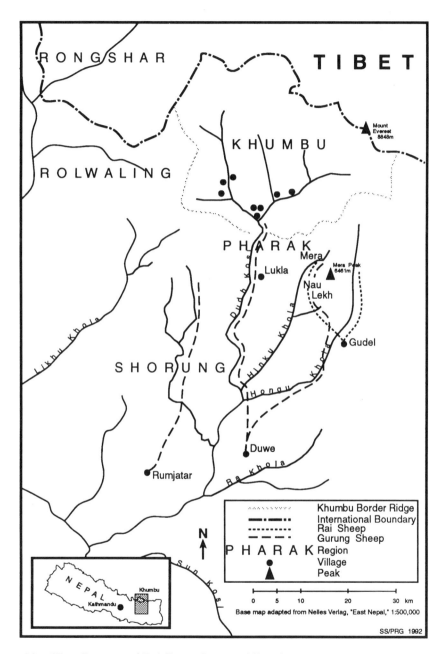

Map 17.   Gurung and Rai Sheep, Summer Migration

Table 22.　Village Cattle-Keeping Emphases, 1957

Percentage of Village Cattle by Type						
	Nak	*Zhum*	*Pamu*	*Yak*	*Zopkio*	*Lang*
Kunde	81	1	4	1	9	4
Pangboche*	90	1	5	1	3	1
Khumjung	73	3	12	3	5	3
Phurtse	92	—	3	3	—	2
Thamicho	76	3	5	0	9	3
Nauje**	29	36	35	—	—	1

*Pangboche 1957 stock totals provided by Fürer-Haimendorf seem low in comparison to other villages and recent Pangboche herding patterns. I use these figures and those of the zopkio totals in the other villages with reservations.

**I have subtracted the 238 head of zopkio included by Fürer-Haimendorf in the Nauje figures since these were undoubtedly only in transit to Tibet. Those zopkio tallied for other villages are also probably suspect, but I have included them here. I have also derived these percentages from a total of 2,678 (2,916 minus 238 Nauje zopkio head of stock) since the village and regional totals in Fürer-Haimendorf (1975:40) are incorrect. In Fürer-Haimendorf (ibid.) the number of cattle owners was inadvertently added to the number of cattle to make village totals.

SOURCE: Data derived from Fürer-Haimendorf 1975:44.

Table 23.　Village Cattle-Herding Styles by Percentage of Cattle, 1957

	Nak/Lang	*Zhum/Pamu*	*Yak/Zopkio*
Kunde	76	16	8
Pangboche	91	6	4
Khumjung	76	16	8
Phurtse	93	3	3
Thamicho	79	8	9
Nauje	29	71	—

SOURCE: Data derived from Fürer-Haimendorf 1975:44.

1971. Despite this regional variation, the Khumbu-wide trend was a massive loss of nak with the total number of stock in 1984 (1,121) barely half that of 1957 (2,061). The second clear pattern is that after 1971 the increase in crossbreeds virtually balances the continuing decline in nak.

The dramatic decline in nak herding in the 1960s was the result of Sherpa responses to changing conditions in Tibet and in particular to changes in Tibetan economy and trade policies after Chinese administration of the area north of Khumbu was established in 1959. During the next few years the Chinese introduced new regulations that seriously undermined the international trade in dimzo on which Khumbu nak

Table 24. Cattle Ownership, 1957–1978

		Kunde	Pangboche	Khumjung	Phurtse	Thamicho	Nauje
Yak	1957	—	—	14	39*	2	—
	1971	—	—	32	109	46	—
	1978	40	75	62	95	134	51
Nak	1957	—	—	333	454	647	—
	1971	—	—	194	240	618	—
	1978	109	175	121	260	496	31
Zopkio	1957	—	—	24	0	79	—
	1971	—	—	39	5	140	—
	1978	61	25	60	7	134	80
Zhum	1957	—	—	15	0	27	—
	1971	—	—	98	7	32	—
	1978	56	24	98	12	103	49
Pamu	1957	—	—	56	17	72	—
	1971	—	—	40	—	123	—
	1978	37	39	80	31	126	82
Lang	1957	—	—	13	8	27	—
	1971	—	—	13	—	30	—
	1978	5	7	8	3	31	8

*Fürer-Haimendorf gives a tally for Phurtse yak of 39 in one place (1975:58) and 16 in another (ibid.:44).

SOURCE: Derived from Fürer-Haimendorf 1975:44–58 and Bjønness 1980a:66.

herding was largely based. On the one hand the Chinese hampered the export of Tibetan nak to Nepal and this affected Sherpa herders' ability to restock and build their herds, but the most serious impact on Khumbu herding was a decline in the profitability of dimzo trading. Khumbu traders continued to take dimzo to Tibet for many years, but it became much less lucrative a venture than it had been. By the 1970s the Chinese had developed new policies that required that dimzo could be sold only to a government office established in Ganggar, and at controlled prices lower than what Sherpas had been accustomed to receiving. They had also begun to encourage Tibetan herders to breed dimzo themselves despite their religious scruples against such breeding practices. With the decline in the demand and price for dimzo in Tibet the Shorung demand for Khumbu crossbreed dimzo calves evaporated, although there continued to be interest in zhum calves. But an important source of income from keeping large nak herds had been lost and it was probably this above all other factors that led many herding families to abandon nak herding altogether or scale down the size of their herds.

While nak keeping was in general decline during the 1960s there was also a minor counter trend for a few years in some places, especially Nauje. Again the economic and political changes in Tibet after 1959 were responsible. With the arrival of large numbers of refugees in Khumbu in 1959–1960 it suddenly became possible to build up a herd of yak and nak very inexpensively. Many refugees had brought their herds with them and selling stock was one of the few sources of support they had in Nepal. Many were also eager to sell because the great increase in livestock quickly exhausted grass in many areas of Khumbu and herders tried to sell stock before the animals starved. Sherpas talk of how in those days the forest was thickly littered with rotting sheep carcasses. One Sherpa witness thought that 99 percent of the Tibetan's stock ultimately starved to death, with the sheep dying first, then the goats, and finally the nak and yak.

Yak and nak were cheaper than they had ever been.[50] A number of families built up herds at this time, especially by purchasing yak. Yak remained useful as pack animals as well as conferring prestige. Some Nauje families that had never been able to afford yak before now built up small herds, turning them loose to graze in summer in the nearby Gyajo valley and keeping them in the winter in the Nauje area. During the middle 1960s these families kept a total of more than 100 head of yak. Within a few years, however, most of these families decided that yak herding was more trouble than it was worth and sold off their stock.

While grazing became difficult in Khumbu due to the influx of Tibetan stock it became better in the border areas of Tibet. There were several Thamicho families who had continued to go north each winter with their herds and who were able to benefit from the now richer Tibetan pastures. These families did not rely on professional herders, but instead either cared for their own stock or worked out a cooperative arrangement with other local families. During the 1960s several Tibetan areas including Shalung, Jalung, and Melungjang continued to be popular Thamicho winter herding grounds. It now became necessary, however, to pay a small fee to Tibetan villagers for grazing privileges. Eight or nine families continued to go annually to Tibet despite the new fees. These families also took the livestock of their relatives and even that of a few non-Thamicho people. Maila, a Nauje man, for example, used to send his many yak to Tibet with Thamicho herders. In the late 1970s, after many Tibetans had returned to Tibet and livestock numbers there increased again, the grass was no longer so exceptional north of the border. Grazing fees climbed and with administrative changes it became difficult to determine the right people to whom to pay the fees. In some areas where good grazing was in short supply Tibetans were reluctant to allow the Sherpas to herd. Thamicho herders grew discouraged by these conditions and by 1980 the old Khumbu tradition of herding on Tibetan winter pastures was a thing of the past. As herders turned to grazing year round in Khumbu they found that they had to put in much larger stocks of fodder to winter their yak and nak. Some of them had not bothered with cultivating much hay before and had put so little care into their hayfields that they had not even bothered to manure them. In Thamicho today herders pay as much attention to such details as they do everywhere else in Khumbu.

There have been other changes in the past twenty years in Khumbu agriculture and pastoralism. Tourism especially has been a new catalyst of change. I discuss tourism and its impact on land use in chapters 9 and 10. Before turning to tourism, however, there is another dimension of historical Khumbu land use in which the relative roles of tradition and change deserve more careful attention. The next two chapters look in detail at historical change in Khumbu resource management and the role of Sherpa subsistence practices in environmental change.

The village of Nauje (3400m) in the Bhote Kosi valley, and Kwangde peak.

Sonam Hishi of Nauje.

Konchok Chombi of Khumjung.

Khumjung village (3790m), with Mt. Everest, Lhotse, and Ama Dablam.

Pangboche village (3985m) and Tawache peak. Sacred juniper trees flank the temple in the center of the settlement.

Phurtse village (3840m), high above the confluence of the Dudh Kosi and Imja Khola. The forest which extends from the left side of the settlement to the river is a sacred lama's forest.

Phurtse in mid-summer. The dark-hued fields are potatoes and the lighter-hued ones are buckwheat.

Potato fields in the village of Kunde (3840m).

Potato varieties. Upper row, from left to right: brown (mukpu), red (moru), and yellow (seru) potatoes. Lower row, left to right: kyuma (from the Salpa region, southeast of Khumbu), "development" (bikasi), and "English" (Belati) potatoes.

Buckwheat.

Dingboche (4358m), in the upper Imja Khola valley. Lhotse is at the left, Imja Tse or Island Peak in the valley center, and Cho Polu to the right.

Yak in the upper Imja Khola valley (4700m) in spring.

Pheriche (4272m), a high-altitude herding settlement in the upper Lobuche Khola valley. On the far side of the river are the herding settlements of Tsamdrang (left) and Naongma.

The forest at Yarin across the Imja Khola from Pangboche. No tree could be felled which could be seen from the Pangboche temple at the lower right.

The lama's forest at Phurtse. In the foreground is the bridge forest where tree felling has increased since the early 1960s.

Looking west across southern Khumbu from near the Tengboche monastery. The lower Dudh Kosi valley is in the foreground, and the Bhote Kosi valley lies at the foot of the skyline ridge. Khumjung village is in the center of the photograph, and the gunsa settlement of Tashinga is at the lower right. Note the contrast in forest cover along the ridge crest above the Dudh Kosi valley which forms the southern border of the Khumjung-Kunde rani ban.

Abandoned terraces near Nauje at Nyershe (3400m) in the lower Bhote Kosi valley.

Fir forest (Abies spectabalis) at Namdakserwa (3400m) near Nauje. This area was formerly part of the Nauje rani ban. A major woodcutters' trail can be seen descending the slope, which before the mid-1960s is said to have been densely forested.

The Nauje lodge district. The first Sherpa lodge in Khumbu is the small wood-roofed structure just to the left of the new-style lodge in the center of the photograph.

The national park-built lodge at Lobuche (4928m) and its woodpile of high-altitude shrub juniper.

Matted shrub juniper near Tugla at 4600m. Juniper has been removed from the area in the center of the photograph and is being dried at the left for use as fuel wood.

Urang zopkio loaded for trekking, Nauje.

Cattle-trail terracettes in the lower Dudh Kosi valley near Nauje.

7

Subsistence, Adaptation, and Environmental Change

One of the achievements of cultural ecology has been the documentation of the adaptiveness, ingenuity, and creativity with which indigenous peoples have developed ways of life that are based on profound knowledge of the local environment and ecosystemic relationships and processes.[1] This perspective on indigenous peoples contrasts sharply with older views that denigrated the land management of "traditional" peoples in many parts of the world as unscientific, superstitious, ignorant, and even destructive. At the same time cultural ecology research has cautioned against a counter tendency to overromanticize indigenous ways of life and resource management. There was a period in cultural ecology research during the 1960s and early 1970s when much fieldwork and analysis reflected possibly premature assumptions about indigenous peoples' harmony with nature or, put another way, their development of a way of life that established a dynamic homeostasis between traditional land use and environment (Netting 1984; Rappaport 1968). Anthropological work especially tended to deemphasize human impact on environment and overemphasize stability and homeostasis, partly due to a lack of concern with historical change (Moran 1984:15–16). In recent decades, however, there has been increasing recognition that many indigenous peoples had profound impacts on the ecosystems of their homelands and that this took place in pre-capitalist, so-called traditional societies as well as in those whose former ways of life and social organization had been transformed by colonialism, frontier dynamics, increasing integration into national and global economies, and other processes. Cultural ecology has largely moved away from assumptions about the harmony or homeostasis of indigenous peoples with their physical habitats towards a more analytic consideration of the interaction of specific peoples and their ways of life with the environments of specific places.[2] This has been true in geography perhaps more than anthropology. The recognition of human impact on the earth throughout history has been important in geography for decades (Blaikie and Brookfield 1987; Sauer 1925, 1956; Thomas 1956).[3]

Khumbu Sherpas, as might be expected of a people who have lived for more than four hundred years in a few small valleys, possess considerable local environmental knowledge of their homeland. This has certainly influenced their land-use practices in many ways and has enabled them to develop relatively sophisticated local resource-management systems through which they have buffered to some degree the environmental impact of their forest use and pastoralism. Local spiritual and other cultural values and beliefs also contributed to channel resource use and moderate some environmental impact, especially through protecting spirit trees and sacred forests.

Sherpas' Buddhist ethics influence daily life and land use also in a number of other ways. Buddhism gives Sherpas a strong belief in nonviolence towards other forms of life. Sherpas value the consciousness within all life, which they believe extends even to the smallest blades of grass. All Sherpas consider the killing of animals a sin and some believe that cutting trees or uprooting plants is sinful (Fürer-Haimendorf 1979:276). Many people go to great efforts to avoid even the killing of insects. When unwelcome insects are discovered inside home they are carefully shepherded outdoors rather than exterminated.

There are limits, however, to the degree to which Sherpas adhere to these Buddhists beliefs in everyday life. It was not true, for example, that "green wood was never cut, mainly due to the maintenance of Buddhist practices" (Bjønness 1983:170). Villagers have long felled trees for timber and for fuel wood. And Sherpas have been know to exterminate wildlife when they believe that it endangers their crops and livestock.

Villagers, for instance, try to prevent pheasants from uprooting their crops of potatoes and grain by constructing scarecrows, mock bamboo traps, and other devices to attempt to frighten off the birds. Before the national park implemented strict wildlife-protection regulations, however, Sherpas sometimes as a last resort sent for the non-Buddhist blacksmiths of Nauje and asked them to shoot the pests. The blacksmiths were also called in to kill snow leopards when these were discovered near the main settlements of lower Khumbu. And Sherpas have so tenaciously trapped and beaten wolves that no more are alive in Khumbu.[4]

While the conservation qualities of the Sherpas' Buddhist beliefs are important influences on their lifestyles, they must not be over-romanticized. It would similarly be a mistake to simply ascribe to Sherpas, or any other indigenous people, a set of attitudes and practices that amount to an ethic of living lightly on the land in terms of conservation beliefs that limit household consumption of certain resources or provide ethical constraints on overexploiting resources from particular sites. Some indigenous societies have held such beliefs, or at least certain enlightened individuals have lived and taught these views among their people. But the degree to which these kinds of conservation ethics have influenced land-use practices and continue to shape them today must be carefully evaluated for each indigenous people and community. It would be easy to assume that people living in difficult environments would be very likely to have developed such adaptive sets of ethics, particularly toward key subsistence resources that are most limited, but this need not be the case. I have not found that Khumbu Sherpa cultural attitudes toward resource use are tempered by such scruples. Instead household decisions about the amount of resources to use and where to obtain them are largely made on the basis of lifestyle preferences, individual desires, and convenience. For all the role of religion in daily life, Khumbu economic life nonetheless seems to have operated primarily according to materialistic values which were themselves also culturally sanctioned and socially rewarded. The constant tension between the implications of these private goals and community efforts to enforce certain restrictions on seasonal resource use and particular resource-use practices has long been a significant feature of Khumbu Sherpa life and affected the relationships among individuals and families within villages, levels of support for pembu, and the regional environment.[5] And the goals and regulations of the community management institutions were themselves also shaped by cultural attitudes about respect for individual household economic freedom, cultural definitions of the good life, and the limits of local environmental ethics.

This view of Khumbu Sherpa resource use and management contrasts sharply with the more romantic "conventional wisdom" about Sherpa life during the pretourism era. Khumbu traditional resource-management institutions in particular have been widely depicted as a fine example of a local system based on communitywide participation and decision making which relatively successfully regulated the use of common property resources (and particularly forest and pasture use) to levels that conserved them. This depiction draws primarily from the description of the nawa system and shinggi nawa institutions as they existed in 1957 given by Fürer-Haimendorf (1964, 1975). Fürer-Haimendorf concluded that:

> Compared with the forests of lower and climatically more favoured regions where peasants of Chetri, Brahman, and Newar stock have in

recent generations wrought enormous devastation, the forests of Khumbu are on the whole in good condition. This is mainly due to an efficient system of checks and controls developed and administered by a society which combines strong civic sense with a system of investing individuals with authority without enabling them to tyrannize their fellow-villagers. (1964:112–113)

Many others have echoed Fürer-Haimendorf's evaluation of the shinggi nawa forest management system (eg. Bjønness 1980a, 1980b; Byers 1987b; Fisher 1990; Hagen 1980; McNeeley 1985; Rhoades and Thompson 1975; Schweinfurth 1983; Thompson and Warburton 1985). For twenty-five years it has been the standard interpretation of Sherpa forest management, and indeed the most commonly cited depiction of a Himalayan resource-management system. Yet these assertions about the effectiveness of traditional local resource use and management have been reached without much attention to historical change in forest (or pasture) use and management, analysis of the relationships among re-sources, resource-use patterns, and resource-use regulations, or atten-tion to the Sherpas' own observations of historical change in their local environment.[6]

This chapter examines evidence of historical environmental change in Khumbu and the ways in which this change reflected both the particular types of "traditional" Sherpa land use and "traditional" local resource management. I begin with a discussion of the possible role of early Sherpa and even pre-Sherpa Khumbu inhabitants in shaping the basic patterns of Khumbu distribution of forests, woodlands, grasslands, and shrublands. I then consider how the historical patterns of Khumbu Sherpa forest use and forest management influenced change in the extent and composition of Khumbu forest between the mid-nineteenth century and the 1960s. I focus here on the period before forest nationalization in the mid-1960s and the subsequent establishment of Sagarmatha National Park, both of which had profound impact on forest use. This chapter thus explores forest use and change during the time when Khumbu resource use and management was entirely in local hands. This is also the period before tourism began to have a major role in the regional economy and resource use. The second part of the chapter looks at pastoral regulation. I analyze the nawa system as a conservation institution and survey some of the historical impact that Sherpa herding practices and local management institutions may have had on Khumbu vegetation.

I take up the environmental impacts that resulted from the decline of some traditional local resource-management institutions after the mid-1960s in the next chapter and examine the environmental impact of tourism in chapter 10.

Fire, Forests, and Grasslands

Forests and woodlands now cover only approximately half of the area that may once have supported forest cover (Hardie et al. 1987:21, 49; Naylor 1970). Virtually all of the remainder of the area below 4,000 meters is grass and shrubland, in which occasional isolated trees are found.[7] These vast grass and shrub-covered slopes are especially prominent in the vicinity of settlements, each of which is at least partly encircled by a zone of grasslands. The proximity of the main villages to the most expansive, lower-altitude (below 4,000m) grasslands, the importance of these areas as rangelands, and the probability that local climate and soils can support forests and woodlands raise questions about the possible human role in the creation and maintenance of these grasslands. It is possible that Sherpas may have chosen to settle at the margins of already existing grasslands, preferring a settlement site that did not require clearing forest to establish fields and building sites and that was close to both good grazing and readily available timber and fuelwood resources. Alternatively they may have played a part in creating these grasslands. Whichever the case may be, there is no doubt that they have played a role in maintaining them (and in places expanding them) in recent centuries through their herding activities, forest use, and perhaps through the use of fire.

Oral history and oral traditions suggest that virtually all of the area that is now grassland has been so at least since the late nineteenth century. There are a few localities where small-scale forest and woodland clearing has occurred since the mid-nineteenth century, as will be discussed below. But there is no oral-history evidence to suggest that Sherpas during the past century or more had any part in the creation of the extensive grasslands on the eastern side of the lower Bhote Kosi river, the slopes immediately above Nauje, or the slopes above the Dudh Kosi river east of Khumjung and Nauje. Soil and pollen analyses also suggest that the grasslands are of considerable antiquity. Some of the grassland areas were established long enough ago to have developed a distinctive soil horizon above an earlier forest soil. Analysis of pollen from soil samples (unfortunately no pollen cores from lake sites are available) suggests that open fir/birch/alder forests were converted to grasslands on sunny, south-facing slopes at least 400–800 years ago (Byers 1987b:199–204). Charcoal found in this fifteen-centimeter-deep soil layer suggests that increased frequency and severity of fires could have been a factor in this vegetation change. People could have played a role in this transformation through deliberately setting fires and also by clearing forest.

Other pollen findings suggest that people may have played a role in the conversion of forest to grassland even if this took place eight-hundred years ago or earlier, well before the arrival of the Sherpas. Byers found cereal pollen (unfortunately of an undetermined variety) dating back perhaps as far as two thousand years (ibid.:199).[8] If correct, these dates suggest that people have inhabited Khumbu, or seasonally visited it, for a very long time indeed. The idea of ancient Khumbu settlement or resource use, while startling to those, including many Sherpas, who have believed that Khumbu must have been uninhabited when Sherpas arrived, is not entirely alien to Sherpa oral traditions. There are Khumbu oral traditions of pre-Sherpa settlement, such as the tradition mentioned in chapter 1 that the ancestors of today's Rais attempted to settle in Khumbu for some years before ultimately abandoning it as a permanent habitation site (although they may have continued to use Khumbu for summer grazing ground). It is also possible that earlier groups had crossed into Khumbu from Tibet just as the ancestors of the Sherpas ultimately did, later either moving on or being incorporated into the Sherpa population. Of this, however, there is no oral tradition except a possible reinterpretation of the story of a Sherpa who is said to have come to Khumbu at the very beginning of the settlement of the region via Rongshar and Rolwaling, rather than over the Nangpa La, and to have lived in the Bhote Kosi valley of Khumbu. It may be that this tradition in fact recognizes that there were migrations from Tibet into Khumbu even before the Sherpas arrived, perhaps long before they arrived, and that the time frame has simply been compressed in the oral tradition to become incorporated with the oral traditions of early Sherpa settlement of the Bhote Kosi valley.

The charcoal found in soil-sample layers, which corresponded to the early era of major Khumbu forest change and the establishment of the grasslands, may be evidence of the early use of fire in farming and pastoralism. According to Markgraf, who conducted the pollen analysis of the soil samples Byers collected, the pollen evidence and the associated charcoal in the samples suggest both fire and possibly logging as factors in forest change (V. Markgraf 1987, cited in Byers 1987*b*:199). The use of fire to improve grazing conditions and perhaps also for swidden cultivation may have had an impact on much more extensive areas of forest than would have the felling of trees to meet the timber and fuel-wood needs of a presumably rather small number of people. If the Rais or more ancient groups from the country south of Khumbu had settled long ago in the high valley or grown supplementary crops there while using the high alpine summer pastures it is probable that they, like most of the peoples of the hill region of eastern Nepal before 1850, practiced swidden cultivation.[9] Buckwheat and possibly other crops could have been successfully swidden cultivated in lower Khumbu.

Gurungs haved historically carried out swidden cultivation at comparable altitudes in central Nepal and as recently as the 1960s Pharak Sherpas used swidden fields to grow buckwheat. Early Khumbu settlers may also have used fire in order to improve grazing. It is a very common technique in much of the Himalaya to fire the forest and woodland floors in early spring in order to improve the growth of new grass. This is done by Pharak Sherpas.

It is impossible at this time to say whether or not the Sherpas themselves were involved in the early creation of the Khumbu grasslands. If Byers's estimate is correct that the key final phase of vegetation change took place between eight hundred and four hundred years ago (1987b:199), it is conceivable that Sherpa land use may have played a late role. Khumbu oral traditions, however, offer no insights into possible early Sherpa forest clearance or early Khumbu Sherpa use of swidden agriculture. In the past hundred years Khumbu Sherpas have had no tradition of setting fire to forests and woodlands to improve grazing or clear forest. But this does not rule out their earlier use of fire. The early Khumbu settlers might have already been familiar with swidden techniques from Kham, for several peoples of present-day Sichuan and adjacent Yunnan certainly make use of them. Or they could have learned them from their new neighbors in Nepal, as Sherpas have presumably done in Pharak, Shorung, and the Arun region. Sherpas could also once have used fire to improve pasture in a way that they have long since abandoned.[10] As of now, however, we simply do not know. Further pollen interpretation will be needed to refine our understanding of the dynamics and timing of early Khumbu vegetation change, and archaeological study is needed to clarify the early settlement history of Khumbu.

Traditional Forest Management: Strengths and Limitations

It is likely that the most extensive conversion of Khumbu forest to open woodland, shrubland, and grassland took place centuries ago, conceivably long before Sherpas arrived on the scene. Yet Sherpas have also had environmental impacts on forests over the past four centuries. The nature of these impacts was related not only to historical patterns of local forest use but also to the effectiveness of local forest management. Change in Khumbu forest during the period of Sherpa settlement reflects the goals and effectiveness of local Sherpa managment of different types of forests and the types of vegetation change associated with their use both of these protected forests and other unprotected Khumbu forests.

The goals set for the various types of protected forests generally appear to have been fairly well achieved. Sacred groves were kept from desecration. Administration of bridge forests, avalanche-protection areas, and rani ban generally met the objectives set for them. Residents of Nauje, Khumjung, Kunde, and Pare were able to continue to find beams in adjacent rani ban, Bachangchang continued to supply critical soluk.

Elderly Sherpas maintain that throughout their lifetimes both sacred forests and some rani ban were very well protected. The strength of local respect for sacred groves remains notable to the present day. Phurtse villagers, renowned in Khumbu for their strict protection of the local lama's forest, often observe that there was little for shinggi nawa to do because no one would think of violating the customs protecting the sacred forest. They say that in the especially sacred forest area north of the settlement people kept an eye even on their own relatives. The sacred forest at Yarin, the sacred junipers at Pangboche, and temple groves throughout Khumbu have also been relatively well protected from cutting and lopping throughout this century.

The regulations established for several of the rani ban were also well maintained. The Bachangchang rani ban, for example, was long exceptionally well administered during the early part of the century when Yulha Tarkia was pembu. Villagers declare that the rules were nearly as carefully observed during the years when his son Konchok Chombi had power. The fact that Bachangchang had been considered a sacred forest by local villagers before it was designated a rani ban must have made the enforcement of protective regulations there relatively easy. The Khumjung-Kunde rani ban and the rani ban near Nauje were also relatively well maintained during the course of several changes in pembu administration, and regulations were carefully enforced in both until the mid-1960s. The Bhotego rani ban, which was administered by a rotation of the shinggi nawa office among the residents of Pare, is also known for having long strictly enforced use regulations.

Control of some other protected forests, however, was less successful. Some of the rani ban were not strictly protected for long, as pembu or their successors seemed to lose concern with ensuring that regulations were enforced. This seems to have been the case at Samshing, Tesho, Gupchua, and Nakdingog, where any formal enforcement of regulations appears to have lapsed well before the 1950s. Even when pembu or local communities did maintain fairly good control over forest use it was not always possible to enforce the rules. Fürer-Haimendorf observed in Khumjung in 1957 that shinggi nawa had had to issue several fines that year for unauthorized tree felling (1979:112). In some formerly protected areas near Nauje people note that for years there were small-scale violations of the rules. Clandestine tree felling took place over a period of decades before the 1960s especially when dense summer mist or low clouds made detection difficult. One elderly villager, when describing what had happened to the trees in the Chorkem area near Nauje earlier in the century, said, "when the clouds came in the trees went away."

For the most part, however, Sherpas continued to respect the customs that placed sacred forests off limits to cutting and restricted the use of rani ban and some other protected forests. Violations of the rules, other than the cutting of much juniper at Shyangboche in the early and mid-1960s, were small scale and seem to have been relatively uncommon. Community cooperation, social pressure, and the continuing practice of appointing shinggi nawa and other forest-management officials appear to have kept forest-management systems functioning relatively effectively until the 1960s.

Yet local management had significant limitations in terms of how well it regulated local forest use to regionally sustainable levels. These shortcomings had nothing to do with how effectively regulations were enforced. They were instead the result of the rules themselves, or more accurately, of the forest management objectives they were devised to meet. These goals did not include regulating regional forest use to sustainable levels or maintaining, protected forests in a given state of composition or density. In all of the protected forests uses were allowed which were capable of transforming the density, composition, age-structure, and extent of those forests. Nowhere in Khumbu, for example, was any forest or woodland ever closed to grazing throughout the year, much less for a period of years. At most, protected forest areas in lower Khumbu were closed to grazing for three months during the summer, and during the rest of the year many of them endured considerable grazing pressure. This nearly unregulated livestock browsing and trampling may have had an adverse effect on forest regeneration, and the lack of more than brief seasonal restrictions on the collection of dead wood, leaves, and needles from the forest floor may have further contributed to this by removing valuable nutrients. In a number of protected forests, including nearly all of the rani ban and the bridge forests, trees could be felled with proper authorization, and it was up to the user to decide which trees to cut. This selection was often made with convenience in mind rather than any view to minimizing environmental impact and forest change. There were no cultural conservation ethics or community customs or regulations that limited the size of one's home or required that one fell trees only in areas where similar trees were plentiful or other Sherpas were not cutting. Each household chose which trees it wanted to fell in a rani ban and the key concern in this selection seems usually to have been to obtain appropriate trees as close as possible to the building site. There were thus several significant loopholes in protected forest management which could have affected both the long-term sustainability of use and the integrity and continuity of the forests themselves.

There were thus several serious limitations to local forest management as a system of conservation. There were even greater limitations to local forest management when it is viewed on a regional scale and all Khumbu forests are considered rather than just the protected forests. Only about half of the Khumbu's forested area was under one of the several local management systems. The remaining forests were affected by the patterns of use that the protected forest regulations established, however, because the relatively strict protection of sacred forests and, in the twentieth century at least, of rani ban, tended to shift the main demand for timber and fuel wood away from the protected forests and onto forest and woodland areas outside their boundaries (map 18). In these areas there were no rules whatsoever. Here villagers cut as much as they wished from wherever they chose. Time, energy, and demand governed the process which followed. Shinggi nawa or other administrators exercised no authority here, nor were there any cultural constraints on the scale of individual household fuel wood or timber use. At the regional level Khumbu forest management thus created a situation in which some forests were protected relatively well and pressure was instead focused on others in which a lack of management prepared the way for the classic "tragedy of the commons" scenario described by Garrett Hardin (1968) in which unregulated individual access to commons ultimately leads to environmental degradation.

Map 18 illustrates patterns of forest use by Nauje, Khumjung, and Kunde villagers after the establishment of the local rani ban and before Sagarmatha National Park regulations banned tree felling in Khumbu other than for house beams.[11] As can be seen in this map, dead wood and beams could be obtained from rani ban areas immediately adjacent to the settlements, and prior to the establishment of the rani ban timber was also obtained from these areas. People recall, for example, that the forest near the Khumjung village entry arch (*kani*) was a very good source of boards at the beginning of this century. Even at this time, however, more remote areas were also being exploited for timber, including forest on the far side of the Dudh Kosi. After the establishment of the rani ban, tree felling for timber, rafters, and fuel wood was channeled as shown into other nearby, unregulated areas. The resulting impact on those forests and woodlands is recorded in the landscape, in oral traditions, and in living memory.

Early Sherpa Impact on Forests

The uncertainty about the extent of forest cover in Khumbu at the time of the first Sherpas' arrival there and the lack of oral traditions about forest use during the early centuries of Sherpa settlement make it impos-

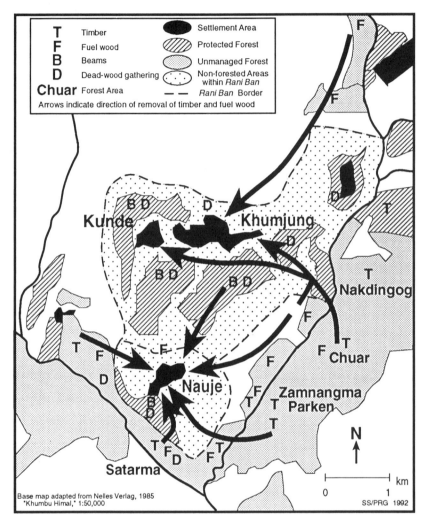

Map 18.   Village Forest Use Patterns, 1915–1976

sible to evaluate fully the scope of Sherpa impact on forests over the past four centuries. Oral traditions and oral history, however, do make it possible to say something of the scale and processes of forest change during the last hundred to hundred and fifty years.[12] Sherpa interpretations of place-names, family traditions about the sources from which trees were obtained generations ago for house beams, and elderly Sherpas' recollections about the forests of their youths all testify to historical deforestation.

Place-names contain several possible clues to the former location of forests. One widely cited example is Namche Bazar, the Nepali name for Nauje, which has been said to derive from the Sherpa for "big forest" (Byers 1987*b*:201, n. 9) or "dense forest" (Bjønness 1983:270). Sherpas are uncertain about the origin of the name Namche Bazar, but some speculate that the name Nauje may be derived from a phrase which would translate literally as "big corner forest," possibly a reference to the natural amphitheater shape of the place and its past vegetation. The upper part of the basin in which the village is set is now conspicuously bare of trees and only a few scattered lu-inhabited trees are found within the settlement area of Nauje itself. Elderly residents indicate that the slopes above Nauje have not been forested during their lifetimes and that they were not told of major deforestation there during the lifetimes of their parents and grandparents.[13] There are oral traditions, however, about smaller-scale forest change. Some people have heard that there was once fir on the south side of Nauje near the present site of the weekly market and that villagers formerly (probably before 1900) felled trees there for rafters. The name of one field in the central pair of Nauje, Tongbajen, is thought to possibly refer to large rhododendron, and here and in several other areas in the village, as well as on the slope immediately above it, many roots were found when establishing terraces half a century or more ago. It is also remembered that until the 1960s there were more juniper in one small area of the upper slopes near the crest of the basin where today several abandoned terraces can be seen. Formerly there was a hermitage at that site, and the juniper were cut down after it was abandoned as a retreat.

The stories told about house beams recall forest where there is none today. In Thami Teng and Thami Og, for example, there are houses that have been rebuilt several times and have immense juniper beams of a size unobtainable today in Khumbu. The prized standing beams often have a history, for the work of cutting and hauling them is long remembered, and beyond that they are regarded as one of the house's more sacred features. Stories about them are sometimes passed through generations. In Thami Og and Thami Teng such stories tell of juniper cut at sites which are now adjacent to the village or within the settlement area by the grandparents of villagers who are now in their seventies. In those places today there is not a tree standing. In Phurtse similar stories are told of the building of the village's early houses with juniper from now bare slopes, and in Nauje at the end of the nineteenth century it was possible to fell large juniper from an area now in the middle of the village.

Oral histories of forest change and deforestation are much richer, of course, from the experiences of living Sherpas. Here there is abundant recollection of the disappearance of trees from subsistence use areas. In the 1930s, for example, a primary wood-collecting region for Pangboche was a juniper woodland along the Imja Khola to the west of the village. A short distance downstream Phurtse villagers cut juniper for fuel wood on the slopes above the Imja Khola. Both these areas ceased to be major fuel-wood gathering areas forty or fifty years ago after wood had become scarce. Several areas along the Dudh Kosi between Nauje and Teshinga were cut by fuel-wood collectors over the past half-century. Woodland was thinned and cleared at Shyangboche and from several sites on the slopes above Khumjung. Tree felling for timber and fuel wood also affected birch forest near Thamo, birch and fir forest north of Samshing, birch and rhododendron forest above Thami Teng and near Kerok, and birch forest further north along the Bhote Kosi at Dokyo. Elderly Sherpas also note more gradual changes in the composition of some forests which took place over the course of many decades. The forests at Bachangchang and Tesho and the small grove south of Phurtse are all said, for example, to have gradually decreased in density and to have changed in composition, with fewer large birch and more young rhododendron today than before. Table 25 lists sites where pre-1965 forest change was reported, and these are indicated on map 19. Note the correlation here between unregulated forest and reported areas of forest change. Those protected forests where pre-1965 degradation occurred are usually either considered to have been laxly administered or were places that had lost their protected status some time earlier as a result of village boundary disputes.

The total amount of forest area lost during the twentieth century has been small, for each of the sites mentioned above as having experienced deforestation or other change is at most only a few hectares in size. Yet the cumulative effects are noticeable to those who remember the forests of early in the century. In 1985 an eighty-three-year-old Sherpa from Thami Teng told me that in his youth it was not possible to look from the western side of the Bhote Kosi valley and see people moving across the river along the forest paths between Thamo and Nauje because the trail was hidden within the forest. Now, he pointed out, there are long stretches of clearly visible trail in that area and people can be seen moving up and down valley.

When examined from local perspective and in historical context Khumbu forest management was a diverse set of institutions and practices that had a variety of economic, environmental, and spiritual concerns. Khumbu forest management was not a static, "traditional" institution. Instead forest management reflected the dynamic development of a number of different approaches to using and conserving forests which embodied cultural beliefs and assumptions about the natural and supernatural environment, natural resources, and the proper sphere of social

Table 25.   Oral History Accounts of Forest Change before 1965

Place	Period	Change
Thami Teng	mid-19th century	clearing of juniper
Bachangchang	1930s–1960s	conversion of birch to rhododendron
Dokyo	1930s	clearing of birch forest
Samshing	1930–1960s	clearing of fir-birch forest
Thami Og	before 1950	felling of juniper
Thamo (w. bank)	1940s	clearing of fir
Chanekpa	pre-1965	thinning of juniper
Chosero	pre-1965	thinning of birch, rhododendron
Nauje	19th century	clearing of juniper in lower village, clearing of fir
Chorkem	1960s	clearing of juniper
Mishilung	early 20th century	clearing of pine
Shyangboche	early 1960s	clearing of juniper forest
Komuche	early 20th century	clearing of fir, pine
Kenzuma	1950s–1960s	clearing of fir
Phurtse	1930–1960s	clearing of juniper
Phurtse	1930s–1960s	
above village		clearing
bridge forest area		clearing
east of village		conversion of birch to rhododendron
north of village		clearing of fir on north slope
Imja Khola near Phurtse	1930s–1960s	clearing
Mong	early 20th century	thinning of birch, rhododendron
Pangboche (n. bank Imja Khola)	1930–1960s	thinning juniper
Milingo	pre-1965	thinning and clearing of fir, birch

intervention in household economic decisions. Some aspects of forest use and management in Khumbu reflect an awareness of environmental change and local institutional responses, to it, yet the pattern of resource use and the nature of institutional and cultural regulation suggest that regional sustainability of forest use was not the primary orientation that outsiders have assumed it to be. Sherpas have had impacts on forests within the remembered past, and had not developed a regionally sustainable system of forest management before the establishment of Sagarmatha National Park.

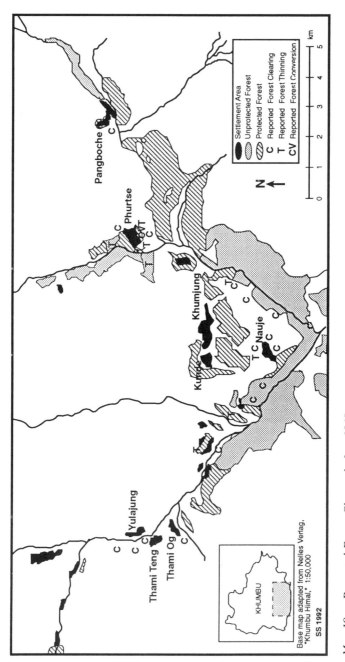

Base map adapted from Nelles Verlag,
"Khumbu Himal," 1:50,000

KHUMBU

Pangboche

Phurtse

Khumjung

Kunde

C Nauje

Yulajung

Thami Teng

Thami Og

N

| Settlement Area |
| Unprotected Forest |
| Protected Forest |
| C Reported Forest Clearing |
| T Reported Forest Thinning |
| CV Reported Forest Conversion |

0 1 2 3 4 5
km

Map 19.   Reported Forest Change before 1965

The continuing preservation of fine forest in the immediate vicinity of villages, however, *does* reflect an extraordinary continuing commitment in Khumbu to the protection of sacred places and the timely development in some areas of new institutions to protect key sources of certain forest products. The diversity of forest management systems in Khumbu and the relatively large area regulated by them is unequalled among the five peoples who inhabit the Dudh Kosi valley and unparalleled among the Sherpa groups with which I am familiar. Although earlier assumptions about local resource use and management in the Mount Everest region certainly require major revision, Sherpa forest management in Khumbu nevertheless may well remain an example of local resource regulation that is exceptional by Himalayan standards.

Traditional Pastoral Management: Strengths and Limitations

The evaluation of the degree to which past and present Sherpa pastoralism has represented an effective adaptation to local environmental conditions hinges on two questions. One is the degree to which Sherpas have been able to develop pastoral strategies, techniques, and social arrangements based on local knowledge of Khumbu environment which enable them to limit risks to levels they consider acceptable. The second is the degree to which regional livestock-keeping practices and grazing levels have been maintained within the carrying capacity of Khumbu rangelands. Here the issue is whether or not pastoralism has been carried out and regulated in a manner that is environmentally sustainable in that it has not diminished the productivity of regional rangelands. These questions are interlinked, for ultimately even small-scale and gradual diminishment of the productivity and carrying capacity of regional rangeland may lead to increased risks for stockkeepers unless new patterns of pastoralism are adopted or grazing intensity is reduced through changes in herd size, stock types, procurement of feed from outside the region, or increases in the productivity of hay cultivation in Khumbu itself.

In examining the success with which a people has coped with the challenge of adjusting grazing levels and intensity to local pasture conditions there are several basic factors to consider. Do cultural values, community institutions or outside authorities act to control grazing pressure through approaches such as controlling herd size and composition at household, community, and regional levels, rotational use of pastures, or limits on the number of stock grazing particular pastures? If communal systems regulate pasture use precisely how do these function and how effective are they? And is there evidence in the landscape and in local perceptions of changes in vegetation, pasture quality, fodder availability, or slope stability?

Khumbu Sherpas depend more on institutional regulation of grazing than do some peoples. Few cultural beliefs affect decisions about herd size and structure other than to encourage the keeping of cattle and yak rather than smaller stock and to equate large herds with high prestige.[14] There are no beliefs that it is unlucky to graze in a particular place more than a certain number of days, or that to have more than a particular number of animals in a given site is inauspicious. Decisions about herd size are entirely left up to the individual family as are decisions about the movement of herds during most of the year. It is only during the summer and early autumn, when the nawa system is in operation, that the community intervenes in any way in household pastoralism.

The operation of the nawa system thus becomes the pivotal place at which to assess the adaptiveness of Khumbu pastoralism in terms of its effectiveness in regulating pasture use to sustainable levels. Yet it is precisely this detailed knowledge of the functioning of the institution relative to environment which has been lacking in earlier work on Khumbu pastoralism. Earlier treatments of pastoralism have described the nawa system without mentioning the existence of zones of pasture management (Fürer-Haimendorf 1979), ignored community pastoral management altogether (Bjønness 1980a), or described the nawa system as a rotation system of grazing without critically assessing it (Brower 1987). Here I will examine Khumbu Sherpa pastoral management in more detail and then look at evidence for grazing-related environmental change.

Evaluating Sherpa pastoral management systems, or indeed any system of common-property resource management, requires attention to several different major features. First the effectiveness of the system must be evaluated in terms of its achievement of its own goals. How well, for example, are rules enforced that limit the group of resource users, such as by restricting use of commons to the members of a particular village? How well are they enforced within the community itself, and particularly against attempts to circumvent them by politically powerful families? On a second level the orientation of the rules themselves must be considered. What are the goals of the system? Is environmental sustainability one of them? What means are taken to achieve this? Are they adequate, or are there shortcomings inherent in the system itself? At a third level the flexibility of the system must be evaluated. How well can it be adapted to perceived changes in environmental conditions or to newly evolving patterns of land use and economic orientation? How well can it be maintained in the face of commercialization, redefinition of the membership of the resource-users' group, or challenges to its existence by government through the nationalization of land, the establishment of other systems of resource management, rejection of local efforts to limit resource use to a particular group, or challenges the legality of specific local use regulations?

In the first of these dimensions, the enforcement of regulations against both Sherpas and outsiders, the Khumbu nawa system has been relatively successful as a form of pastoral management. Throughout the region nawa continue to effectively administer the opening and closing of zones to livestock and hay cutting except in the Bhote Kosi valley and Nauje. Where nawa management has been maintained the system has successfully protected crops from livestock depredation and from fear that village activities might cause catastrophic blights. It has continued to reduce the risk from year-round pastoralism in Khumbu by protecting expansive areas of lower Khumbu from summer grazing, thus ensuring both that there will be considerable wild grass available for autumn hay making and that there will be good winter grazing in the main village areas. In eastern Khumbu the operation of the more complex zonal system of the upper Imja Khola and Lobuche Khola valleys has established a limited rotational grazing system in the late-summer high pastures. Until the undermining of the system in Nauje in 1979 and Thamicho by 1984 the regulations regarding pastoralism were consistently implemented throughout Khumbu despite a number of challenges by Sherpa villagers, non-Sherpa Gurung herders, and Tibetan refugees. Attempts by Gurung shepherds in the late ninteenth and early twentieth centuries to obtain exemptions from observing the zone closures were resisted. Until the late 1970s no major challenges by Sherpas had succeeded in eroding the authority of communities to administer nawa regulations or seriously compromised the enforcement of regulations. Several attempts by villagers from one community to deregulate another community's restrictive grazing zones by challenging village land boundaries came to naught, although to preserve the grazing controls in one case the men of Khumjung had to march to Nauje to back up their nawa's enforcement of the herding regulations with the threat of violence. A number of attempts by wealthy and powerful herders of several different villages to ignore the rules were thwarted by social pressure and in one case, in Phurtse, by a change in rules governing the selection of nawa to prevent political influence from undermining equitable enforcement. Even the widespread disobedience of nawa regulations at Phurtse during the 1940s, when many yak herders moved their stock into the village before it was authorized, did not lead to the collapse of the institution there. Instead the enforcement of the regulations was tightened successfully the following year and has been maintained ever since. The undermining of the Nauje nawa system in 1979 by massive civil disobedience and the parallel collapse of pastoral regulation in Thamicho in the 1980s are recent anomalies in what has otherwise been a long record of successful operation.

The Khumbu nawa system has a more mixed record at the second level of evaluation of communal management systems, where goals rather than merely enforcement are considered. The system clearly has a number of different goals: protecting crops from livestock, protecting winter grazing, protecting fodder through keeping stock out of hayfields and areas where wild grass is cut for hay, restraining the collection of wild grass until it has matured, and rotating grazing through a series of zones. Some of these are immediate, subsistance-related goals: protecting the harvest, ensuring that stock can survive the winter and spring, and maintaining good grazing through the summer through shifting the concentration of grazing. Some of these goals deliberately, or perhaps inadvertently, also support environmental sustainability. Rotational grazing limits the intensity of grazing of particular areas and decreases the risk of overgrazing. It is important, however, to recognize that the nawa system has systemic limitations in terms of conservation objectives. Although some dispersion of grazing is certainly accomplished through the opening and closing of various zones to livestock, in four of the five systems this rotational system only operates in limited lower-valley areas, leaving most Khumbu rangeland as unregulated, open-access commons.[15] There is no mechanism to control the intensity of grazing within a given zone by placing limits on the number of stock that can be grazed there or to set a ceiling on the number of stock herded in a village, valley, or in Khumbu as whole.[16] Herders may decide that poor grazing in one area is a reason to move elsewhere, but this is solely up to the household itself as are all decisions about household herd size and composition. The system does not include any mechanism for incorporating environmental monitoring and response. Nawa do not evaluate pasture conditions and have no authority to respond to observed increases in grazing intensity or deterioration in pasture by ordering or recommending special changes in zonal boundaries or the period when areas are open to grazing. Such changes could be made by a consensus of residents and herders, but there are no traditions of such issues having been brought up for consideration and acted on. The dates of zone closures and openings in some cases are relatively fixed and embedded in a socioreligious calendar and the border zones have not been adjusted for generations.[17] There is also no precedent for closing a degraded grassland or woodland area to grazing in order to allow it to recover from intensive use.

There thus remains within the nawa system itself the possibility that overgrazing can take place and cause the degradation of pastures and forests without there being any compensatory mechanism. To control localized overgrazing would require expanding the powers and concerns of the system beyond any historical precedent. The lack of strong conservation goals suggests that pasture deterioration might be allowed to proceed for some time without any institutional response, for to redefine zone boundaries or take more radical steps such as limiting the number of stock requires fundamental changes in the institution of a sort that may be socially difficult to arrive at. Recent Nauje and Thamicho history suggests that strong demands on resource use and increasingly scarce resources may lead to the abandonment of the system rather than attempts to modify and strengthen it, with grass cutters and herders choosing to compete with each other without community intervention or long-term environmental concerns.

The nawa system thus represents relatively constrained communal intervention in individual household herding decisions. In most areas Sherpas enforced little more than the minimal level of stock exclosure necessary to protect crops from livestock depredation and blight and to protect winter pasture and fodder resources. The continuing institutional capability to achieve these goals is now under severe testing in the region. For a century or more the nawa system has operated on the basis of community consensus and social pressure. This base has deteriorated over the past ten years in much of Khumbu. Communal management of livestock in the Bhote Kosi valley has been discontinued, is under pressure in Khumjung-Kunde and in the upper Imja Khola valley, and was abandoned in Nauje in 1979. This situation and what it suggests about the limits of the institutional adaptability and flexibility of the system are discussed in the context of the decline of local resource-management institutions in the next chapter.

Grazing and Historical Vegetation Change

Grazing has probably had a significant role in shaping Khumbu vegetation and landscape. Centuries of herding by Sherpas, Gurungs, and possibly Rais or other lower Dudh Kosi valley peoples have brought thousands of yak, cattle, crossbreeds, sheep, and goats to graze and browse in Khumbu grasslands, shrublands, woodlands, and forests. Historical changes in livestock numbers and types, patterns of transhumance, the regulation of community grazing, and the relative importance of hay and other fodder, all contributed to place different degrees of stress on different Khumbu rangeland areas. Centuries of intensive use of certain areas could be expected to substantially pattern vegetation. So too would herders' efforts to improve grazing by the use of fire which, as already mentioned, could even have shaped the relative distribution of forests and grasslands familiar from the past century. The vast expanses of grasslands below 4,000 meters that have provided such fine grazing as far back as oral history can illuminate could very well be a legacy of pastoralism.

Yet it is difficult as of now to analyze the historical extent of vegetation change in the region due to grazing. Oral traditions and oral history have little to say about declining pasture conditions before the past few decades. There is no testimony that any particular herding area has dramatically declined in productivity during the nineteenth and early twentieth centuries, much less that grazing has been abandoned in any part of Khumbu because of gullying, landslides, or heavy erosion related to pastoralism.

It seems probable that the composition of much of the rangeland vegetation in Khumbu reflects the historical effects of cattle grazing. In particular the relative role in Khumbu grassland and shrubland vegetation of a number of relatively unpalatable plants such as barberry, rhododendron, rose, ephedra, iris, and potentilla probably reflects the cumulative and continuing effects of pastoralism (Brower 1987:277). The great changes in Khumbu pastoralism in the past fifteen years would be expected to have had a particular role in shaping the vegetation in some areas, especially the lower Bhote Kosi valley and the Nauje area where large numbers of crossbreeds have been grazed year round for the first time in known Khumbu history. But the relatively high incidence in the vicinities of main villages and heavily used, higher-altitude herding areas of certain temperate and subalpine species that are poisonous or unpalatable to yak, cattle, and crossbreeds could be expected to be a general pattern that goes back long into the past. In some areas there may be more unpalatable, thorny rose and barberry shrubs today than decades ago, but Sherpas recall that concentrations of such shrubs near villages are not new in the Khumbu landscape.

The spectacular terracetting so ubiquitous in lower Khumbu and certain other areas such as Dingboche also has historical origins. Vast slopes are tightly patterned with tiny terraces two or three feet deep and several feet high, apparently the result of many generations of livestock crisscrossing the landscape in search of grazing. Some observers have taken these terracettes to be evidence of recent overgrazing and they undoubtedly are a reflection of continuing grazing pressure. Areas recently closed to grazing by the national park to establish forest plantations have begun to recover within a few years as grasses and annual plants recolonize and grow over the bare ground of the cattle trails. Yet while extensive networks of cattle trails reflect continuing patterns of high grazing intensity, they are a long-standing feature of the landscape. Sherpas report that the slopes near Nauje, Phurtse, Dingboche, Pangboche, and other sites have been terracetted in this fashion throughout their lifetimes.

The composition of forests, woodlands, and shrublands has also very likely been affected by grazing, particularly in areas near main settlements where seasonal herding patterns have long focused considerable grazing pressure. Browsing and trampling have probably inhibited seedling survival and maturation and may account for forest areas near some villages that have aging stands with little evidence of regeneration. In other areas there seems to be little evidence of regeneration other than by rhododendron, which is unpalatable to local livestock. Over time this has probably contributed to the twentieth-century transformation of birch and fir forests near Phurtse and Thami Og to rhododendron thickets.[18] Grazing is also very probably a major factor in the noticeable lack of tree colonization in shrubland and grasslands adjacent to forested areas.[19]

During the past half century the decline in the use of winter pastures in Tibet, the increasing emphasis on crossbreed and cow keeping, and the breakdown in effectiveness of communal pastoral management have all affected the use of Khumbu pastoral resources. There is some evidence that increased competition for grazing and grass for hay has led to a growing scarcity of critical winter pastoral resources in some areas and that the intensity of grazing may also be decreasing long-term range productivity. This is not entirely only a phenomenon of the late 1980s. Some herders in the Bhote Kosi valley, for example, are convinced that grass was much better in that area in the 1950s and 1960s before more Thamicho stock began wintering in Khumbu rather than in Tibet. As one large nak-herd owner noted, during the era when the herds went north to Tibet in winter, the spring in Khumba was a time of fattening for stock. Since the 1970s herders have had to feed nak hay all winter and when they really need it in spring none remains. A number of Khumjung herders also note that earlier in their lives both hay and wild-grass fodder was much more available than today even though the number of livestock then herded in the upper Dudh Kosi valley was much greater and the number of hayfields, on the western side of the valley at least, was about the same as today. They also relate this to the decline in the practice of herding nak and yak in Tibet during the winter as well as to the increased Khumjung herding of zopkio, zhum, and cows.

Herders also note that there was a major decline in the quality of grazing in some parts of Khumbu during the early 1960s as a result of the large numbers of Tibetan refugees who brought their yak, nak, and sheep with them. For a few years Khumbu summer pastures were especially intensely grazed and there were severe shortages of winter and spring grazing. Herders who did not have considerable fodder stored ran the risk of having their stock starve. The herds of the Tibetans were particularly at risk and the numbers of these stock that perished in the early 1960s were tremendous. After this, however, grazing pressure decreased and Sherpas do not note any long-term, adverse impact from this period on Khumbu pasture quality.

Adaptation, Land Management, and Environmental Change

Khumbu Sherpas are an excellent example of a people who during the course of many generations have come to know intimately the resources and environment of their homeland and who have developed strategies for making use of its opportunities and avoiding its risks. Their development of local resource-management institutions has been one component of this adaptive way of life. In Khumbu this institutional development has been the central way in which local demand on natural resources has been regulated and has been much more important than the development of cultural conservation ethics in regulating individual and household land use and their environmental impact. Yet the power of the community to intrude into household economic behavior was also limited by cultural and social beliefs about the limits of communal management and the community's right to influence household economic decisions.

There were also other limits on communities' powers of resource regulation and management. In many parts of Nepal, as in other regions of the world, villagers have limited use of common property resources to their own residents or even to particular groups of residents. In Khumbu, however, community resource management could not draw on such power. Since the mid-nineteenth century villages have not been able to restrict the use of common lands to their residents even though there were defined village land boundaries that were legally upheld and over which court battles were fought. The declaration of a particular area as the property of one village or another only influenced which community might decide if the area was to be treated as a protected forest or as a zone in which grazing and grass cutting might be regulated by nawa. But no village had the right to limit access to that land to its own residents or to enforce the resource-use rules differently for outsiders. Nauje people could cut trees in the Khumjung-Kunde rani ban as long as these were to be used for house beams and the shinggi nawa were consulted or could take their stock to graze anywhere in Khumbu as long as they respected the local nawa-enforced regulations. For many years even Gurung shepherds were welcome to use Khumbu resources provided they adhered to local community management regulations.

Khumbu resource-management institutions have also historically been unable to limit levels of resource use through enforcing household or community ceilings on the use of forest products, the amounts of wild grass cut for hay, or the number of stock pastured in particular areas. And while nawa, shinggi nawa, and pembu had the authority to issue fines, there have never been any use fees which could have been used to manage the level of resource demand and the ways it was met. Communal institutions thus were not a mechanism that limited the scale of household resource demand or influenced local standards of living.

No Khumbu resource-management institution was charged with monitoring the scale and intensity of use in particular places and associated environmental change, much less was given the power to respond to these observations by adjusting resource-use policy and regulations. Nor should it be assumed that the goals of Khumbu Sherpa traditional resource-management institutions were foremost and centrally concerned with environmental conservation. These institutions addressed a broad range of local concerns, not all of them by any means concerned with moderating the environmental impact of subsistence. Some central goals were concerned only with safeguarding subsistence resources in as short-term and specific a fashion as keeping livestock out of fields. Other goals concerned broader kinds of subsistence protection such as defending against disease and bad luck by protecting forests, trees, and springs which were the domains of powerful spirits. Such goals and the rotational grazing regulations and restrictions on tree felling and pollution through which they were achieved certainly had important ramifications for environmental quality and resource conservation. But these were by-products of economic and spiritual concerns, not the result of either core beliefs in preserving biological diversity or native vegetation communities or a desire to place regional resource use in general on an environmentally sustainable basis. Such cultural values were not translated into Khumbu-wide environmental and land-use policies nor coordinated into community or regional planning as explicit goals of local resource-management institutions.

Traditional pastoral, agricultural, and forest-management systems, however, were concerned in some ways with the adverse environmental effects of unmanaged, individual household resource use. This included concern with change in specific forests, sometimes reflecting a concern for protecting trees in order to spiritually protect the community and in other cases to ensure a sustained supply of timber and forest litter. Despite the shortcomings of local resource management from the standpoint of regional protection of native vegetation and biological diversity, the development of the nawa system of pastoral management and the rani ban and bridge-forest systems of forest use are evidence of local concern with moderating demand on certain resources to ensure their availability for at least the near future.

In Khumbu, as in many cases elsewhere in Nepal (Zurick 1990:32), perception of a real need (such as averting an environmental crisis or natural resource shortages that would affect local lifestyles) may have been the crucial factor in mobilizing community support for implementing such changes in resource-use patterns. In the case of Khumbu forest management it seems to have often been the case, as Gilmour (1989:3) suggests for other parts of Nepal, that "a perceived shortage of forest products is the stimulus that has been most important in causing individuals and communities to change their patterns of behavior toward their use of forests."[20] A perceived scarcity of important forest products does seem to have been a factor in the selection of some rani ban sites and their subsequent protection through community support. The pembu who selected forests near Nauje, Khumjung-Kunde, Pare, and Samshing (and perhaps also Nakdingog and Gupchua), and who instituted rules allowing only the cutting of trees for beams in those places, seem to have been reacting to concern about a threat to these resources.[21] Yet it should also be noted that the first Khumbu protected forests—and those which have remained the best conserved—were created not out of concern with the supply of forest products or avoiding immediate environmental calamities but out of a community dedication to honoring sacred places and trees.[22]

The history of the Khumbu protected forests does not fit neatly into a model such as the one Gilmour develops in which a direct relationship exists between the level of accessibility to forest resources and the degree of community resource management (ibid.:4).[23] It might be expected that there would be relatively little interest in protecting forests when forest resources are abundant near a community and that a community's concern with regulating individual resource use would rise as forests were depleted and their resources became scarce—especially as the gathering of fuel wood and other products began to require a full day's trip from home (ibid.). But the first and best-protected forests in Khumbu, the sacred forests of the Phurtse and Pangboche areas, were established adjacent to villages where there was not yet any shortage of forest products, whereas there was no community response in other areas such as Thami Teng, Thami Og, and Yulajung where nearby forest resources were depleted even in the nineteenth century. In this part of Khumbu, when the rani ban were established in the early twentieth century, there were no suitable forests left to be protected near the villages other than the sacred grove at Bachangchang. Indeed, rather than being an example of a place where there has been a simple relationship between increasing resource scarcity and local institutional response, Khumbu seems to be a much more complex set of different, special cases. No local community action had been taken anywhere in the region to establish protected forests in which only limited cutting was authorized until orders came from Kathmandu to establish the rani ban. Only this intrusion into local politics permitted several pembu to mobilize community support to conserve resources. Since then, however, perceived threats to resources have inspired other community efforts at forest protection, including efforts to revive local protection of some forests during the past decade.

Sherpa resource-management institutions thus both reflected adaptive responses to environmental degradation and increasing resource scarcity and illustrated a lack of a holistic, historical concern with regionally sustainable resource use and environmental protection in the sense it has now come to be regarded by ecologists and conservation planners. This should not detract, however, from recognition of the significance of the Khumbu Sherpa achievement in developing their systems of forest and pastoral management. These institutions were for the most part highly effective in meeting the goals for which Sherpas designed and maintained them. Such attempts to control and channel natural resource use through community and even valleywide systems are not universal in the Himalaya, and the Khumbu institutions are certainly more complex, sophisticated, and environmentally protective than those developed by their neighbors. These institutions reflect the application of considerable creativity, foresight, and social and political skill by both individual Sherpa leaders and their communities across several generations.

The preliminary reevaluation of Khumbu traditional resource-management institutions I have attempted here suggests the complexity of resource-management goals and means and the historical dynamism of Khumbu resource management. It underscores the caution with which the subject of the historical relationship between Sherpas and their environment must be approached. Sherpas have had an impact on their environment. They have fashioned a productive landscape of terraces and rangelands from what may have been a rather different environment at the time of their initial settlement of the region. In the past century and a half they have in some places rolled back the edges of forests and woodlands and altered their composition and density, have inadvertently encouraged subtle changes in grassland vegetation, and perhaps have increased the degree of livestock trailing on the heavily grazed slopes near villages. Their resource-management institutions probably have not regulated the use of regional forest resources within long-term, sustainable levels. Neither resource use nor resource management was explicitly geared as much toward minimizing environmental change as to achieving subsistence goals and minimizing risks. Short-term environmental change was not necessarily a concern if it did not threaten these goals and long-term environmental changes that might ultimately threaten standards of living (such as declining forest regeneration) were not necessarily perceived or responded to with alterations in land use or institutions. When communities did act to manage local resources it was either to protect sacred places or to meet immediate challenges such as preserving fodder and grass for getting the herds through the winter, maintaining sources of leaf litter and large beams, or guarding against crop disease and wildlife and livestock depredation. Protection of the wild diversity of flora and fauna and native vegetation communities, regional forest and woodland cover, or the environmental quality of rivers, lakes, and streams were not direct concerns except in those spheres of daily life where local religious belief touched land-use practice. This was most notably the Buddhist respect for life that extended to all species not perceived as threats to crops and livestock and a religious respect for springs, trees, and forests believed to be inhabited by spirits. At the same time Khumbu Sherpas have not, during the past century at least, precipitated any known environmental crisis. They have not despoiled Khumbu resources to the point that a deteriorating land base has created regional poverty and spurred large-scale out-migration. Nor have they created conditions fostering unstable slopes and massive erosion.

Khumbu Sherpa traditional resource-management institutions thus had both strengths and limitations even before 1960 when Khumbu land use and the economy had not yet been affected by major tourism development, government nationalization of forests, and the establishment of Sagarmatha National Park. How Sherpas responded to these new pressures and what became of local resource management and the Khumbu environment is the subject of the following chapters.

8

Local Resource Management: Decline and Persistence

One of the greatest changes since 1960 in Sherpa land use and management has been the decline in several parts of the region of traditional resource-management institutions. The abandonment of local forest regulation in some areas during the 1960s has been widely reported. Less well known is the abandonment of the nawa system of pastoral management in Nauje in 1979 and in the Bhote Kosi valley during the 1980s. These events have had great significance for Sherpa subsistence, the Khumbu environment, and current conservation efforts. This chapter examines the change in Khumbu resource-management institutions since 1960. I focus first on the processes and events that led Sherpas to abandon local management institutions in some places and to adapt and maintain them elsewhere, then explore the environmental changes that have accompanied the resulting changes in natural resource use.

Khumbu Forest Nationalization and Its Aftermath

Fürer-Haimendorf, revisiting Khumbu in 1971 after fourteen years away, was the first to report that there had been a major change in Khumbu forest use and management. He observed considerable tree felling near Nauje and Khumjung in areas where this had been tightly regulated when he had lived in the region in 1957. Here, he learned, the traditional shinggi nawa regulation of forests had been abandoned because of the government's nationalization of forests. Fürer-Haimendorf subsequently reported that all forests which are not on privately owned land have been declared state forests and the villagers have no more control over them. Persons who require timber for house building have now to apply to the office of the district panchayat which is at Salleri in Solu, at least four days' walk from Khumbu. The permits issued by that office specify the quantity of timber to be felled. The procedure is cumbersome, and as there are no officials of the forest department in Khumbu, unauthorized fellings cannot be controlled. The forests in the vicinity of villages have already been seriously depleted, and particularly near Namche Bazar [Nauje] whole hill-slopes which were densely forested in 1957 are now bare of tree growth and villagers have to go further and further even to collect dry firewood. In this case the replacement of an efficient and well-tried system of local control by a bureaucratic machinery has not been successful, and the Sherpas are conscious of the diminishment of local timber resources without being able to stop the inroads into forests which, being claimed by government as state property, are no longer under their control. (1975:97–98)

This analysis has become the standard description of the demise of Sherpa forest management in Khumbu and one of the classic Himalayan examples of how national government resource-management efforts can undermine previously effective local resource-management institutions to the detriment of local environment (Byers 1987*b*; McNeely 1985; Thompson and Warburton 1985:122). Thompson and Warburton, for example, used the Khumbu case as their sole example in a discussion of the shortcomings of Nepal's national bureaucratic resource-management efforts. Their summary of the events in Khumbu drew strongly on Fürer-Haimendorf's observations and analysis in concluding that

> in the 1950s the forests were nationalized. Management was taken out of local hands and transferred to central government—a reform which, we know, destroyed the indigenous management system (based on a rotating village office of forest guardian who managed the traditional village forest as a renewable resource) that had worked successfully for centuries and then did not work itself. (1985:122)

Khumbu has in this manner become a classic example of what has been supposed by a number of authors to be a Nepal-wide process through which post-1957 forest nationalization undermined traditional customs of local forest management (Acharya 1984; Arnold and Campbell 1986; Bajracharya 1983; Messerschmidt 1986, 1987:375).

In the past few years this thesis has been increasingly questioned. Foresters and anthropologists working in several parts of Nepal have found that greater familiarity with local oral history suggests that the stereotypical view of the impact of 1950s forest nationalization on local management institutions and Nepal's forests does not hold true in some areas and that it may be an exaggeration even in national terms (Archarya 1989; Gilmour and R. Fisher 1991:12–13). Gilmour and Fisher, who are very familiar with forest use and management in hill regions near Kathmandu write that:

> It is generally believed that widespread indiscriminate cutting of forests occurred during the years following 1957 because village people felt that their forest had been taken away by the government. . . . While this unintended effect of nationalisation has been widely reported, we believe that it has been exaggerated. In the course of fieldwork in Nepal we have obtained a great deal of oral evidence which calls this phenomenon into doubt. . . . We have never heard local claims that there was a crisis in forest management following the nationalisation of forests. . . . Contrary to the view that rural people reacted to the 1957 legislation by destroying forests is the observation that a great number of indigenous forest management systems (which were commonly set up by villagers to protect degrading forests) had their origins about 1960. (1991:13)

Closer oral history examination of Sherpas' response to forest nationalization also raises questions about early reports about the impact of the 1957 legislation on Khumbu forest management and use. On the one hand Sherpas present testimony that supports Fürer-Haimendorf's observations. Forest nationalization did indeed undermine some local Khumbu forest-management institutions and had an adverse impact on forests. Yet only some Khumbu forest-management systems were affected and then only in certain localities. Elsewhere local forest management was maintained. Sherpa institutions were not as fragile as Fürer-Haimendorf and Thompson and Warburton had assumed. From the Khumbu perspective the issue of nationalization and its impact is thus much more complex than either Fürer-Haimendorf and Thompson and Warburton, on the one hand, or Gilmour and R. Fisher, on the other, suggest.

Sherpas today verify that local forest management in several communities was indeed undermined by forest nationalization, although they disagree with some of the details of Fürer-Haimendorf's account and put the date of the abandonment of local management in the 1960s and 1970s rather than the 1950s. They also testify that increased tree felling took place as a result in some forest and woodland areas (particularly near the villages of Nauje and Khumjung) and note that the impact of this was visible in the landscape. They disagree, however, that villagers lost all control over the forests they had previously administered or that widespread deforestation took place throughout the region. Instead they testify that in much of the region they continued to maintain local forest management despite the nationalization of forests. They speak of a far longer, and still continuing, process of conflict and negotiation between local residents and government officials over the management of forests rather than of a simple, regionwide collapse in the 1950s of Sherpa efforts to maintain their regulation of protected forests. This difference in perception is not just a matter of historical interest, for it has enormous implications for national park management today and in the future. It is therefore worth reviewing the process of the nationalization of Khumbu forests and its effects on local institutions and the environment, paying particular attention to variation within the region.

The Nepal government has long taken an interest in the management of forests, including issuing occasional orders for forest conservation. In the nineteenth century both the Shah and Rana administrations sent out edicts to the hill regions of the country which specified restrictions on tree felling near water sources, along roadsides, and in religious places, and both governments authorized village headmen through numerous directives to regulate local forest and pasture use according to their local customs of management (Acharya 1989: 16, 17; 1990). We have already seen one example of this which had direct significance for Khumbu forest conservation, for Rana regime policies in the early twentieth century resulted in the creation of locally managed protected forests in eastern Nepal, including the eight Khumbu rani ban. After the fall of the Ranas with the 1950–1951 revolution, which restored the monarchy to greater power, the national government began to become much more involved with the nationwide administration of forests as national resources. One of the early measures in this process was the enactment of the Forest Nationalization Act of 1957, which returned to state ownership and management large-scale private forestlands and those areas of common forestlands that had not been already legally recognized as communal lands (*kipat*).[1] In the Khumbu context the Forest Nationalization Act of 1957 reasserted the government's legal authority over virtually all forest and woodlands in the region. Only a few, very small, private forests near Thami Teng were exempt from the act. Yet the passage of this legislation in Kathmandu in 1957 had no immediate impact on Khumbu. No government officials arrived in Khumbu ordering Sherpas to give up their former forms of forest management, and local residents continued to administer rani ban, lama's forests, and other protected forests exactly as they had before.

Khumbu forest use was more greatly affected by the government's continuing efforts to establish nationwide forest management based on a permit system of tree felling. This was to be the responsibility of district Forest Department offices.[2] Decisions about where tree felling would be allowed, which species could be cut, and on what scale felling could be carried out were to be made by professional foresters and the administration of these policies in the forests was to be the responsibility of trained government forest rangers. These rangers were to select the particular trees to be felled, collect timber fees, and monitor the operation. The first efforts to implement this style of government forest management reached Khumbu about 1965, when a branch (*sahayak anchyaladhish karyalaya*) of the Sagarmatha zone office, which coordinates the administration of Solu-Khumbu and neighboring districts, was opened in Nauje. The officials of this office were given the responsibility of administering tree-felling permits along with their other duties for several years until district forest offices could be established at Salleri and Lukla. Thus for the first five years of forest nationalization in Khumbu, until about 1970, officials based in Nauje itself rather than in distant Salleri issued permits.[3] This meant that the officials responsible for administering the new forest rules in Khumbu were not so far removed from the region as Fürer-Haimendorf had thought, and it made the procedure of obtaining permits for tree felling, before 1970 at least, somewhat less cumbersome than he had portrayed.[4] The permits issued in Nauje specified the area from which the trees were to be cut and the total volume of timber authorized. An important part of the permit process, one hitherto overlooked in the literature, was the requirement that Sherpas first obtain the approval of the *pradhan pancha,* the leading locally elected Sherpa official.[5] Only after this had been granted could one apply to the branch of the zone office for a felling permit. The system thus had some degree of local control built into it, a fact unrecognized by Fürer-Haimendorf.[6]

The Nauje branch office is said by Sherpas to have granted them permission to obtain any amount of timber which they wished, provided that the appropriate fee was paid. And it authorized the felling of trees for timber anywhere in the region, irrespective of earlier Sherpa protected forest boundary lines and regulations.[7] According to one of the pembu of that time, the Nepali official who implemented the system overrode the objections of Sherpa pembu who confronted him with the documents that had authorized the establishment of the rani ban and demanded recognition that they had legal jurisdiction over Khumbu's forests. The official's reply was that a paper issued by the Rana regime was not binding on the present government.[8] Some Sherpas say that the government's early interest did not seem to be forest management as much as collecting forest revenue, and they call the Nauje office the "wood-selling office" and say that it "gave away the forests."

Sherpa forest regulation did not automatically collapse throughout the region despite the refusal of national government officials in Nauje to recognize any local rights to forest management. Local forest protection might not have had any legal basis from Kathmandu's perspective, but the local forest-use regulations continued to meet local concerns. In a number of areas Sherpa villagers decided that even though they could now obtain permits to cut within protected forests they would simply not apply for them. Instead they would continue to administer their own regulations in their own forests. Phurtse villagers continued to choose shinggi nawa to administer sacred forests and Pare residents continued to select shinggi nawa for the Bhotego rani ban. In Pangboche the Yarin forest continued to be watched over by the temple caretaker and the nawa continued to enforce the bans on bringing freshly cut wood into Pangboche or Dingboche during the summer. Protection was maintained especially well for religiously safeguarded forests, and temple and lama's forests were very seldom transgressed even in places such as Khumjung which soon did abandon its other traditions of local forest protection. Local forest management continued to be enforced in all of these areas other than Yarin until the establishment of Sagarmatha National Park, when some communities reconsidered the relevancy of their forest management.

In Nauje, Khumjung, and Kunde, however, a different process took place. Here rani ban regulation was undermined. In these three settlements it became increasingly difficult to administer the old rules after the permit system was inaugurated. Villagers from these settlements applied for permits to cut in formerly protected forests and defied shinggi nawa and nawa by maintaining that they had obtained both the permission of the locally elected pradhan pancha and the representative of the Kathmandu government. Here people began to cut wood where it was easiest to get, which was often in the rani ban where timber was relatively plentiful close to the villages precisely because it had been so carefully safeguarded before.[9] Within a year Nauje nawa gave up efforts to enforce any of the old regulations, feeling that it was useless since, as Sonam Hishi observed, few people were paying any attention to the nawa and instead took advantage of the new system to obtain timber more easily than before by felling trees that the nawa would not allow them to cut. In Khumjung and Kunde there was an attempt to keep the system going, and locally elected village panchayat officials were given the shinggi nawa positions. But those villagers who had permits to fell trees were nevertheless allowed to fell them, and by the early 1970s both Khumjung and Kunde had ceased choosing shinggi nawa.

Clearly in some villages a process of abandonment of local regulation took place which did not occur elsewhere. Several factors might have been important. One could have been that in these three villages, all within a short walk of the Nauje branch office, the new forest-use regulations might have been more vigorously promoted and enforced. In Khumjung the fact that powerful and wealthy villagers, including pembu, took an early lead in felling trees in the former rani ban was probably a more important factor. Here the pradhan pancha himself

encouraged such tree felling and approved permit applications in the
rani ban, with special dispatch it is said, for friends and relatives.

The role of the Khumbu pradhan pancha of the 1960s and early 1970s
in undermining forest management in their home villages of Nauje and
Khumjung has not previously been explored. These men were certainly
placed in an awkward position and it is understandable that they did not
continue to confront central government officials in Nauje over forest
ownership and management. Yet they could have nevertheless supported
local management by simply refusing to forward permit applications that
requested permission to fell trees inside sacred forests or asked for trees
from inside rani ban for purposes other than house beams. In these cases,
where Sherpa rules were stricter than the new government ones, Sherpa
rules could have been supported. Shinggi nawa and other Sherpa officials
could have continued to enforce Sherpa rules in Sherpa protected forests.
At the same time government objectives would have been met. Tree
felling would have been put on a permit basis and the central government
would have received revenues on all timber cut in the area.

Khumbu pradhan panchas, however, apparently did not make any
use of their role in the permit system to uphold traditional rules in rani
ban. At least one pradhan pancha, however, made use of these new
responsibilities to play out old contests of power in new arenas through
denying permission to cut trees in certain areas near settlements (including former rani bans) to enemies while promoting the applications of
relatives and friends. In such struggles pride could win out over a sense
of community and environmental responsibility. One former pembu who
had been noted for his zealous enforcement of Khumbu rani ban regulations recalls triumphantly a victory over an old rival who had been
elected pradhan pancha which came at the expense of the formerly
protected forest at Khumjung. After he had been denied permission for
his permit to cut timber to rebuild his house he waited until the pradhan
was away from the area and then obtained the permit from a subordi-
nate official. He cut his timber from the slopes just above the village, an
area in which trees have long been scarce and that had been within the
rani ban.

The undermining of local management regulations and institutions
could, of course, have been easily avoided if the new approach to national
forest management had been implemented differently. A central government directive that preexisting local regulations were to be supported
when they did not conflict with the goals of Forest Department regulations would, for example, have made a considerable difference in
Khumbu. But Kathmandu did not send out such instructions and the local
officials in Nauje who implemented the new system showed little interest
in initiating such an approach themselves. Their efforts seem instead to

have been primarily aimed at the establishment of the government's claim to forest land and its sole power over forest regulation. There were occasional exceptions. One case is remembered, for example, when the branch office used its veto power over logging-permit applications to protect trees in a previously locally protected area. In this instance government officials denied a Nauje villager permission to cut trees in one area of the former rani ban on the grounds that it was necessary to conserve these trees for future use in bridge repairs. Such action by a central government officer, however, seems to have been highly unusual.

In other communities Sherpas integrated traditional, local institutions with the new permit procedures. In these places both local and national institutions were in place simultaneously for several years. Villagers obeyed the stricter local regulations maintained by local values, social pressure, and fines. At the same time they followed the new procedures of the national system, despite their redundancy and costs in time, trouble, and cash which the government enforced with the threat of arrest, prosecution, fines, and possible imprisonment. It was possible to continue to obey the local rules and to comply with the new Kathmandu bureaucratic regulations as well. In effect this established a two-tier system of forest regulation. A particular forest use first had to be locally legitimized through the local institutional procedures. Only then could one take the locally sanctioned request on to the government for it to certify to its satisfaction. In other communities no such two-tier system developed, and instead some individuals used the existence of the national institution to circumvent the local one. That the two-tier system did not develop in the 1960s in Nauje, Khumjung, and Kunde could be related to the fact that powerful local individuals, including pradhan pancha in some cases, took the lead in subverting initial community efforts to establish this kind of parallel set of institutions. The location of the national government office in Nauje itself—close at hand also to Kunde and Khumjung—moreover enabled villagers to appeal to the primacy of central government regulations as a justification for violating community customs. The prospect of ready support from Nauje government officials must have enhanced the ability of Sherpas who wanted to violate local customs to negotiate their actions with shinggi nawa, nawa, and their fellow villagers.

Forest Nationalization and Deforestation

It has been widely reported that the nationalization of Khumbu forests led to rapid and widespread deforestation prior to the establishment of Sagarmatha National Park. Fürer-Haimendorf's 1972 observations that "forests in the vicinity of villages have already been seriously depleted" and that "near Namche Bazar whole hill-slopes which were densely forested in 1957 are now bare of tree growth" (1975:98) often have been cited and assumed to have been an accurate accessment of conditions.[10] Ecologists, foresters, and park planners who visited the area in the early 1970s also reported a deforestation problem (Lucas, Hardie, and Hodder 1974; Mishra 1973) and this perception was one of the major factors that lent urgency to the proposal of national park status for the region and shaped early national park resource-management policy.

Sherpas do not agree that widespread deforestation took place between the time the government forest-use regulations were implemented in 1965 and the establishment of Sagarmatha National Park in 1976. They contend that descriptions of entire recently deforested slopes are exaggerated. They suggest instead that forest clearing was very localized. In contrast to the vague general impressions of extensive forest change offered by outside visitors their oral history accounts provide detailed insights into perceived change in particular forests and woodlands. Oral history also offers insights into the processes that Sherpas believe precipitated these forest and woodland changes and their assessment of the degree to which forest nationalization was the cause.

Sherpa testimony about forest change since 1960 indicates that in some areas small-scale clearing occurred prior to the establishment of the national park. In several other areas changes in forest density and composition became increasingly evident even though there was no major change in forest or woodland cover. Most of the areas where thinning or clearing took place are said to reflect tree felling for timber for house building, although vegetation change at a few sites (including some areas near Nauje on the slopes north of the Dudh Kosi) are thought to have primarily reflected the cutting of trees for fuel wood. The table below identifies sites of reported forest change and the types of forest use reported in those areas; these are shown in map 20.[11]

Most of the sites of forest change between 1960 and 1976 are in areas which were not traditionally protected forests. A few, however, reflect the abandonment after 1965 of earlier local restrictions on timber cutting. The most outstanding examples of forest change that were directly related to forest nationalization were the result of tree cutting in parts of what had been the Nauje and Khumjung-Kunde rani ban. This is presumably the logging that led Fürer-Haimendorf (1975:98) to remark, perhaps with some exaggeration, that considerable deforestation had taken place adjacent to Nauje between 1957 and 1971 and that forest had also been depleted near other villages. Although forest clearing near Nauje was not nearly as extensive as his account suggests, Sherpas testify to localized change after 1965 in two areas adjacent to the village:

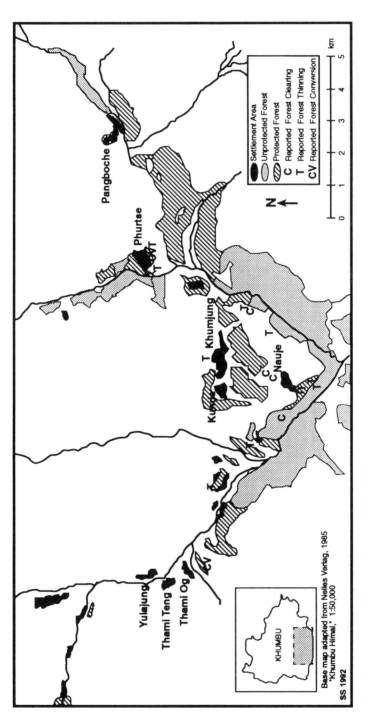

Map 20.   Reported Forest Change, 1965–1979

Table 26.   Reported Forest Change, 1965–1976

Site	Reported Change
Impacts associated with fuel wood gathering:	
Kenzuma	clearing and thinning
Lower Dudh Kosi valley, north slopes east of Nauje	clearing and thinning
Lower Bhote Kosi valley, eastern side	thinning
Phurtse lami nating	thinning and conversion to rhododendron
Shyangboche	clearing of woodland and shrub juniper
Bachangchang	thinning
Impacts associated with tree felling for timber:	
Lower Bhote Kosi, Samshing area	clearing
Nauje rani ban	thinning
Khumjung slope	thinning
Phurtse bridge forest	thinning
Shyangboche	clearing

the slopes just to the south of the settlement at Namdakserwa Nasa and the slope immediately below and west of the village. Both of these areas were previously within the Nauje rani ban. Villagers began applying for permits to cut house timber as well as beams in these areas soon after the Nauje branch of the Sagarmatha zone office implemented the new government forest-management system, even though there was far from unanimity in Nauje that such tree felling was a good idea. As far as I have heard no requests for cutting in these areas were turned down by central government officials or by the local pradhan pancha, and several Nauje families who built or rebuilt houses between 1965 and 1976 can trace their timber to trees cut in these two areas. The small platforms that were dug in the slope to support board-sawing stations can still be seen. The region below Namdakserwa Nasa, which is said to have been so densely forested that it was difficult to gather leaves there while carrying a basket, is now considerably more open, and today three soluk gatherers moving abreast would have no trouble negotiating this area. Sherpas testify, nonetheless, that there was no major change in tree cover on the slopes above Nauje during this period. Trees here, they say, had already long since been cut except for a few juniper that surrounded the site of a meditation retreat high above the village. These juniper were cut during the 1960s after the resident of the hermitage passed away.

In the late 1960s the slopes above Khumjung and Kunde also became an important timber source following the relaxation of rani ban regulation. Trees here had previously been protected against felling other than for use as beams. A large number of junipers were felled above Khumjung where today few trees of any stature are left.[12] The Khumjung slope had not been densely forested before 1965,[13] but people do recall some stands of denser, taller juniper thirty years ago as well as more scattered juniper. One of these denser stands was high on the slope in the vicinity of a hermitage. Another was situated lower on the slope and closer to Kunde village near where local Tibetan immigrants now maintain a prominent shrine. There are no large junipers in the latter site today and the trees around the hermitage are said to be fewer. Sherpas note that a great number of trees were cut in a long-regulated area and they no longer refer to the slopes above Khumjung as being forested.[14]

Other forest change during the 1960s and 1970s reflected continued cutting in areas that were outside the boundaries of Sherpa protected forests. Nauje and Khumjung residents, for example, recall that during this period there was much fuel-wood collection from several areas along the north slope of the lower Dudh Kosi above its confluence with the Bhote Khosi, including the Kenzuma area and slopes closer to Nauje. This is said to have led to a decrease in the tree cover of the open woodland there. Nauje villagers mention another area near Tesho and Samshing on the east side of the Bhote Kosi where the felling of fir for house rafters led to some small-scale clearing in the 1960s.

Some of the vegetation changes reported by Sherpas in the 1960s and 1970s can be attributed neither to the nationalization of forests nor to long-established Sherpa forest-use and management patterns. At Shyangboche, for example, considerable change in juniper woodland cover took place within a rani ban regulated area under extraordinary circumstances unrelated to the operation of the Nauje branch of the Sagarmatha zone office. Before the early 1960s, according to local memory, parts of Shyangboche were places of fifteen-foot-high junipers which were dense enough that it was easy for herders to loose track of yak there on foggy days. Now it is very open woodland and large areas have very few tree-sized junipers. Some of the juniper was cut by Tibetan refugees who camped in the area between 1959 and 1962 and some was cut during the construction in the early 1970s of the Everest View Hotel and its airstrip. Most, however, was cut by Sherpas during the 1960s following a land dispute. The area had been coveted by Nauje villagers for some time on account of its many junipers and rich pasture and, as mentioned previously, there had been arguments over it for years. In the 1960s Nauje villagers increased their timber and fuel-wood gathering in the area and one Nauje family tried to file a land claim there with district officials. This led to more widespread cutting in response by Khumjung and Kunde families, which in turn spurred more cutting as Nauje families who were determined to claim a share of any remaining Shyangboche juniper vied for resources.

Other special cases of forest change resulted from outsiders' violations of protected-forest regulations or from their being granted exemptions from the usual rules. Phurtse villagers, for example, say that they began felling trees for house building from the area that had been protected as a bridge forest after the village had decided to cut in that area for timber with which to build the Hillary school in the early 1960s. Thamicho villagers insist that they only began lopping and felling trees in the Bachangchang rani ban after Nepali members of the local police post began cutting trees there for fuel wood.

New Demands on Forests

Khumbu demand for ·forest products was changing at the same time as some Sherpa communities were loosening their control of forest management. During the 1960s and 1970s several new demands were placed on regional forest resources as a result of changes in subsistence forest use, the requirements of the growing tourist trade and the increasing numbers of Nepali officials and personnel stationed in the region, and the establishment of Sagarmatha National Park.

Increased Sherpa use of fuel wood, possibly the most widespread and significant of these changes, was related to a change in lifestyle. Until the mid-1960s many Sherpa families spent much of the winter away from Khumbu on trading expeditions and pilgrimages to lower-altitude regions as distant as northern India. It was common for families to set out from Khumbu in November and to be away for most of the winter, returning in March to begin preparations for spring planting. Families who had livestock either sent these to Tibet or left them in Khumbu in the care of a family member or a relative. A substantial part of the regional population thus spent four or five months outside of Khumbu, and some men were away as much as ten months out of the year on more extended trading journeys in Tibet and India.[15] Food was available more cheaply in the south, and the more time families spent subsisting there on grain grown at lower altitudes the less grain they needed to carry back on foot. This changed when Sherpas ceased barter trading salt for grain in the lower-altitude regions of Nepal. The declining availability of Tibetan salt in the early 1960s and the establishment a few years later of the weekly periodic market system in Solu-Khumbu rapidly undermined the former practice of winter trading in the south. More Sherpas began to spend the winter at home in Khumbu unless trekking work took them elsewhere.[16]

This change in life-style had a significant effect on forest use as well as prompting adjustments in agricultural practices.[17] A side benefit of the old way of life was that families were away from home during precisely the part of the year when their fuel-wood requirements would have been the highest. After 1965 when families began spending the entire year in Khumbu their fuel-wood requirements approximately doubled. A family that burned half a load of fuel wood per day in the winter would have found it necessary to obtain as many as seventy-five more loads (2.25 metric tons at 30kg per load) of fuel wood per year. If this family lived in an area where a half a day's work was required to collect a single load of wood, not an uncommon situation at that time, gathering a winter's supply of fuel wood would have required thirty-seven person days of hard labor.

The change in Khumbu seasonal settlement patterns in the 1960s could thus have created a huge new demand for fuel wood which was met from the forests and woodlands closest to the main villages. To this increased local demand for fuel wood were added several new sources of demand. Fuel wood was required by the increasing number of lowland Nepalis who began to be stationed in Khumbu during the 1960s and 1970s to staff army, police, and medical posts, schools, the post office and bank, government offices, the government yak farm, the airstrip at Shyangboche, and the national park. Tibetan refugees had significantly increased fuel-wood demands in the region for a few years during the early 1960s. Increasing numbers of mountaineering expeditions during the first half of the 1960s and the beginning of commercial trekking in 1964 also began to put new pressure on the fuel-wood resources of some areas, particularly along the main route to Mount Everest. Tourism has been widely reported to have had an especially severe impact on Khumbu forests. This issue deserves separate treatment, for many of these reports have considerably exaggerated and overgeneralized recent deforestation while overlooking tourism's role in more subtle vegetation change in the region. This will be discussed in detail in chapter 10 along with other impacts of tourism on Khumbu land use and environment.

One effect of these changes in fuel-wood demand and the presumed increasing scarcity of dead wood near main villages that accompanied it, may have been a new Sherpa concern with using fuel wood more frugally. Faced with a choice either of devoting considerably more time to fuel-wood gathering or economizing on wood use, it appears that many households became interested in the conservation of fuel wood.[18] It is precisely at this time that more fuel-efficient means of cooking began to be widely adopted. Until the mid-1960s virtually all families cooked over open fires, either setting the cooking pot on a tripod of standing stones or on a simple iron grate. Now many families began to install cheap and efficient stone and mud wood-burning stoves, copying an old Tibetan design. According to one estimate this type of stove increased the efficiency of fuel wood by five to thirteen times over simply cooking on an open fire (M.N. Sherpa "Conservation for Survival: A Conservation Strategy for Resource Self-Sufficiency in the Khumbu Region of Nepal," 1985, cited by Fisher 1990:64). Sherpas must have long been aware of this type of stove from their trading and pilgrimages in Tibet, and the several Khumbu families who owned houses in Ganggar may well have had them in their homes there.[19] During the 1970s some families also began experimenting with other fuel-saving measures including the use of pressure cookers, which many families now use routinely. In the 1960s large Chinese thermoses also became popular as a way of avoiding having to rekindle fires frequently during the day to brew more tea—a significant task in a region where tea is the main nonalcoholic beverage consumed and where it is common for adults to drink dozens of cups per day. More recently there has been considerable interest in a new stove design that adds a simple water-circulation and reservoir system (backboiler heater) to the wood-burning stove so that it heats water at the same time as it supplies energy for cooking. These conservation measures have been widely adopted despite the many beliefs about the sancity of the hearth and fire, both of which are inhabited by spirits who must be treated with special care since they are believed to affect the health and welfare of the household. There have also been recent changes in vernacular architecture that reflect a concern with the conservation of fuel. Until recently the interior design of Khumbu Sherpa houses had changed little for a number of generations, but today there is a marked trend in several villages towards remodeling older houses or building new ones in which the old style single, large, open-beamed room in which the hearth occupies a central place along one wall has been replaced with smaller, separate rooms, a wood stove in a separate kitchen, and a ceiling.[20]

Sagarmatha National Park and Forest Regulation

In the late 1960s and early 1970s there were increasing pressures on Khumbu forests from both new levels of Sherpa demand for fuel wood and timber and from tourism. In some areas, at least, local forest management had been abandoned. Highly visible tree felling was taking place in the Khumjung and especially in the Nauje area on a scale very different from previous decades. These changes troubled a number of visitors to the Khumbu region in the early 1970s, among them a series of government officials and advisors who had come to the area to evaluate its potential as a national park. These observers reported that deforestation was a serious regional environmental threat (Blower 1971; Lucas, Hardie, and Hodder 1974; Mishra 1973).

As early as 1971 John Blower, an official from the United Nation's Food and Agricultural Organization involved in the development of conservation and park programs in Nepal, urged the establishment of a national park in the Mount Everest area. Subsequent park-planning visits were made by Hemanta Mishra, a Nepalese ecologist, in 1972 and in 1974 by a New Zealand team sent to evaluate New Zealand's participation in the possible establishment of a national park. New Zealand, the homeland of Sir Edmund Hillary, ultimately signed a bilateral agreement with Nepal in which it took responsibility for training several Sherpas to be park administrators and offered to provide park-management personnel as advisors during the early years of the park's operation. Sagarmatha National Park was formally gazetted by the government of Nepal in 1976. For the first five years of its operation a series of New Zealand park managers, assisted by Nepalis from the Department of National Parks and Wildlife Conservation, oversaw the establishment of the national park headquarters complex on Mendelphu hill adjacent to Nauje, prepared the first park-management plan, and implemented new resource-use regulations.

The New Zealanders endeavored to make Sagarmatha a world-class national park and launched an ambitious program in which both the protection of the environment and concern for Sherpa welfare were central. Real efforts were made to consult with Sherpas about the design and management of the park. Village meetings, for example, were held as early as 1974 when the New Zealand planning mission drew up its recommendations for the establishment of the park. A local advisory committee was set up. And through New Zealand's efforts the first Khumbu Sherpas joined the Department of National Parks and Wildlife Conservation, completed university training in conservation and park management, and were stationed in national parks as administrators. This far-reaching concern for local involvement in future park management resulted in several Sherpas acting as chief or assistant administrators in Sagarmatha National Park for most of the 1980s.

Forests were an early concern. Nurseries were begun in 1979 to supply seedlings of indigenous conifers for a reforestation program. Plantations were established at Mendelphu, Khumjung, and Kunde, each protected either by stone walls or by barbed wire fencing. Furthermore, forest use throughout the region was placed under a system of regulation intended to curtail the adverse impact on forests of both local subsistence use and tourism. Regulations were implemented which banned fuel-wood use by mountaineering and trekking groups and new rules were enforced which outlawed much traditional Sherpa use of Khumbu forests.

Sherpas were suspicious of the idea of their home becoming a national park, as they made clear in the 1974 village meetings held by Nepali and New Zealand park planners to broach the concept of the park and solicit local input. Sherpas accused Hillary, who had been one of the main proponents of the park, of betraying them. Many Sherpas did not share the New Zealanders' vision of the national park as a way to buffer and ameliorate the adverse impact of tourism on both the local environment and culture, and were afraid that the establishment of the park would be a disaster for their way of life. Many feared that the entire population of the region would ultimately be evicted and resettled elsewhere in Nepal in order to make Khumbu a wilderness area. This fear was not entirely groundless, for this is what had already been done in Nepal's only previously established national park, Chitwan, and it was soon to occur again at Rara National Park. There were fears that even if Sherpas were allowed to remain in their homeland that park authorities would implement new regulations that would prevent them from continuing with the subsistence practices on which their settlement of the region had relied. Many Sherpas were worried, for example, that national park administrators would introduce new concerns about overgrazing into regional pastoral management and with them restrict the size of livestock herds and regulate access to pastures. Here too their concerns were not entirely misplaced, for the first national park management plan was soon to advocate these measures. And there was a widespread fear that the park would ban cutting trees even for subsistence purposes. Anticipating the establishment of the park, many Sherpas put up large stockpiles of fuel wood and timber while there was still the chance. Tree felling for both timber and fuel wood may thus have been at extraordinary high levels from the time the news of the coming establishment of the national park reached Khumbu in 1973 or 1974 until the late 1970s when the new forest regulations Sherpas had feared were indeed imposed.

The early management policies developed at Sagarmatha National Park were undoubtedly intended to be responsive to local Sherpa concerns as well as international environmental ones. An attempt was made to incorporate local consultation in park management in a way uncommon in that era in developing countries. By legally excluding the villages and herding settlements from the park boundaries, national park planners acknowledged Khumbu Sherpas' rights to remain in their homeland and choose for themselves their own paths of village development, a gesture that reflected a degree of sensitivity to the rights of indigenous peoples which was then extremely rare in the world national park movement.[21] Park administrators also recognized the importance of support for Sherpa subsistence lifestyles and culture even when these required use of natural resources located inside the park—allowing grazing, for example, even though domestic stock may have competed with the Himalayan tahr for forage and allowing trees and shrubs to be cut for fuel for cremations and ceremonial chotar and gotar poles. Ultimately park authorities decided against imposing any new grazing regulations in the region and did not even step into the management vacuum created by the abandonment of the nawa-system regulations in Nauje in 1979. But the national park did take a stand on forest use and it is this stand that so greatly colored Sherpa attitudes about the park in general for many years. Things might have gone differently if new forest regulations had been implemented more gradually and with a stronger foundation established first through programs designed to build good relations with local communities, develop local conservation education, win support from local political and religious leaders, evaluate current patterns of forest use and management, and find ways to integrate new conservation concerns with older, local forest-management institutions. But there was a perception that immediate action was vital.

After the establishment of Sagarmatha National Park, New Zealand and Nepali administrators instituted new forest-use regulations that were the strictest in Khumbu history and represented a fundamental change in resource-management goals. These were enforced by patrolling Khumbu with a Nepalese army "protection unit."[22] All tree felling was prohibited and lopping branches for fuel wood or fodder was banned. Only dead wood from the forest floor was to be gathered for fuel wood and all timber was to be brought in from outside the national park. In Nauje a permit system to regulate dead wood gathering was introduced on a trial basis. Free passes were given to gather wood in particular areas on particular days. This system of regulating dead wood collection was intended to rotate the gathering of dead wood and leaves from the forest floor in order to avoid overuse of the most accessible areas. Some Sherpas, however, feared that national park authorities intended to begin charging a fee for the passes and thus make gathering even dead wood costly as well as more difficult.

The approach that Sagarmatha National Park adopted for Khumbu forest management, unlike that followed for the management of grazing on park lands, thus adhered to the strict nature-protection philosophy that goes back to the establishment of the world's first national park and which is often referred to today as the "Yellowstone model." Here the idea is preservation uncompromised by human use, or at least uncompromised by commercial and subsistence land use other than tourism development, for the Yellowstone legacy also gave world national park planning a precedent that tourist access must be safeguarded and enhanced

(Stevens 1986*a*). This approach to protected area management led to the resettlement of indigenous peoples from newly created national parks in many parts of the world and greatly influenced early national park planning throughout Nepal. Sagarmatha, like Langtang National Park, was on one level an exception in that efforts were made to devise park planning that protected traditional settlement and land use. But in forest-use policy a rather different line was drawn. Here the national park directly confronted Sherpas over their traditional access to an important subsistence resource and assaulted their sense of land ownership and homeland. The unfortunate result of this clash of perspectives was continuing resentment and fear and a local lack of support for the national park which has remained widespread ever since.

The new park forest policies evoked anger and local protest. Park authorities, perhaps taken aback by the strength of this reaction, responded with efforts to meet local concerns at least part way. The permit system for dead wood collection was abandoned and there was some relaxation of the tree-felling prohibition. Under the new rules families were to be allowed to fell three trees for use as beams when building new houses. This was not entirely satisfactory to many Sherpas, some of whom noted that three trees were not sufficient even to provide all the necessary house beams, much less to compensate them for the expense and trouble to which they were now faced to go in order to have all other timber cut in Pharak and transported to Khumbu.[23] Some saw this as no better than declaring all of Khumbu a rani ban in which one was entitled to fell even fewer trees than the rani ban system had allowed.

Over the years an uneasy truce evolved. Sherpas came to accept the existence of the park, but many bitterly complained that all of Khumbu had been turned into a protected forest at the cost to Sherpas of making their lives more difficult. Some could see no reason for this or for the establishment of the Sagarmatha National Park forest plantations other than creating more forested scenery for the enjoyment of foreign tourists. Anger and bitterness were further fueled with fear when the army unit began to patrol forests and arrested Sherpas for "illegally" obtaining fuel wood and construction timber even though they were obtaining them in traditional ways for traditional uses from areas where they were traditionally procured. This kind of confrontation continued through the 1980s. The distrust and hostility created during the first few years of national park administration over forest use have handicapped other park efforts including its reforestation program. Many Sherpas suspect that the land walled in for reforestation is land permanently taken away from Sherpas, and some people breach walls or barbed wire fences in order to allow their stock to graze in the former rangeland.[24] Few Sherpas seem to believe that they will ever obtain much benefit from the reforested areas. They cannot be blamed for having reached this conclusion, since ten years after the plantations' establishment they have not yet obtained any natural resources from them and have not yet even been allowed to cut the grass for fodder. There has been no indication from the national park that Sherpas will ever be allowed to obtain fuel wood or timber from the mature plantations.

National park forest regulations and particularly their enforcement by soldiers were widely regarded by Sherpas as having "taken away" their forests and their responsibility for their management. Forests began to be referred to as "the national park's forests" and villagers felt that park planners and administrators were oblivious to their continuing efforts to maintain local forest-management institutions. Some park officials may indeed have believed that new approaches to Khumbu resource management were necessary and that existing Sherpa institutions were inadequate to meet changing conditions and new challenges. But early New Zealand park planners (Lucas, Hardie, and Hodder 1974:15) and administrators as well as the New Zealand-trained Sherpa administrators who succeeded them after 1981 have also continued to affirm that Sherpas must have a real role in national park management and that this is vital to its success. Many Sherpas have overlooked these good intentions, however, just as many have misunderstood the intent of conservation policies. A greater dialog between the park and its people is sorely needed.[25]

After the establishment of Sagarmatha National Park some of the communities that had managed to maintain effective local forest administration despite forest nationalization began to lose their ability to continue to enforce their own regulations. Soon after the establishment of the national park the residents of Pare decided to stop shinggi nawa administration of the Bhotego rani ban. Shinggi nawa continued to be chosen in Phurtse, but a village controversy erupted over whether national park regulations or traditional rules should be enforced. Some people began to lop branches for fuel wood from the lama's forest to the north of the village.[26] When confronted by angry fellow villagers they maintained that this was permitted under national park regulations. The controversy over the protection of the sacred forest was further increased by a respected Sherpa from Kunde who, when consulted about the situation, mocked the tight traditional control at Phurtse and asked villagers if they were just saving the forest to provide wood for their cremations. In other areas Sherpas complained that the national park was insensitive to local protected forest regulations, since the authorities permitted trees for beams to be cut even in areas such as the Yarin lama's forest where this was not allowed under local forest regulations. There were also charges that the national park gave permits to people who had not first obtained approval from shinggi nawa and other local officials.[27]

Sagarmatha National Park forest-use regulations instituted in the late 1970s remain in force today and have been enforced since 1981 by a series of three Khumbu Sherpa administrators as well as several Nepalis. These Sherpas were trained in New Zealand in national park management and have come to support many "Western" attitudes about resource management. They have, however, introduced significant new forest-use management approaches that are based on Sherpa shinggi nawa traditions. Such measures represent creative attempts to promote conservation and a synthesis of Western and Sherpa concepts of resource management by working with local residents and traditions. This may in time help to defuse tensions, but issues of resource and land rights will undoubtedly long remain a source of distrust between Sherpas and the outside authority of the national government and Sagarmatha National Park administrators.

Recent Forest Use and Management

Since the end of the 1970s Sagarmatha National Park regulations have greatly decreased the felling of trees in Khumbu for timber and other purposes and have also shifted patterns of fuel-wood procurement away from forests and woodlands adjacent to villages by requiring that only dead wood be used and by banning trees from being felled. For more than a decade all Khumbu timber other than a limited number of beams has come from outside of the region and new areas for village fuel-wood collection have developed at sites which in some cases require a full day's journey to obtain a single load of fuel wood. Nauje, Khumjung, and Kunde villagers have experienced the greatest change in their fuel-wood gathering. Before Sagarmatha National Park forest regulations were implemented the residents of these three settlements obtained fuel wood from relatively near forest and woodland and were able to fell trees for fuel wood (outside of the rani ban) rather than depend solely on dead wood. Most families obtained adequate household supplies from areas that were no more than half an hour from their houses. This changed in the 1970s with the introduction of new attitudes about forest use and preservation. With tree felling (and even lopping branches) banned, demand increased for dead wood to a point where nearby areas were picked clean of fallen branches. It became necessary to go further and further afield in order to find adequate supplies of fuel wood.[28] The main fuel-wood gathering areas of Nauje villagers shifted from forest and woodland north of the confluence of the Bhote Kosi and Dudh Kosi rivers to the western shore of the Bhote Kosi (especially the Satarma area) and the slopes of Tamserku on the far side of the Dudh Kosi. Khumjung and Kunde villagers began regularly making expeditions across the Dudh Kosi (using the bridge at Phunkitenga) to the Nakdingog and Chuar areas. These areas are shown in map 18. Most Nauje villagers today obtain fuel wood by hiking through the fir and pine forests near the village, descending three hundred meters to the Dudh Kosi, crossing the river on flimsy "woodcutter's bridges," hiking roughly another three hundred meters up the far slope, gathering fuel wood, and then carrying thirty- to forty-kilogram loads back down to the river and up the steep climb to the village. Many hours are spent getting to and from the fuel-wood gathering area whereas only a few minutes are actually taken to gather a full load at the site. Khumjung and Kunde villagers leave very early in the morning for the long trip to Nakdingog and Chuar. Travel time has also been increased for residents of the other villages.[29]

Most Sherpas obey national park regulations despite the inconvenience required to respect them. These regulations have without question radically altered local forest-use patterns and in general established a new level of forest protection in the region. Some people, however, revert to more traditional resource-use patterns when these are not likely to be observed. A good deal of fuel wood is burned which is not dead wood. Much of this is obtained by lopping branches and occasionally felling small trees in areas far from the villages and national park headquarters. But at dusk and at dawn one also comes across people carrying loads of green wood from the vicinity of settlements, most often elderly people or young children for whom the expedition to the now-distant sources of dead wood would be difficult or impossible. Trees are sometimes felled surreptitiously for building material. Other people abuse permits. I have known a case where a man who claimed that he had permission to fell three trees for beams had instead cut at least ten times that number in the Dole area from one of the highest-altitude forest regions, and a Nauje case where a man who had been authorized to fell three trees for beams had ended up with fourteen—the result, it was remarked wryly in the village, of these three having conveniently toppled the others when they fell. Still more ironically, in both cases much of the lumber thus obtained was not used for main beams at all but rather for joists and rafters. In one case the operation was carried out on a foggy day in a remote corner of the region, but the other took place quite openly with the trees hauled into the center of the village and sawn into joists on the main street on the day of the weekly market. These incidents as well as another much-discussed local incident in which a national park ranger apprehended a man in the act of illegally felling a large juniper only to have his superior release the unfined offender and return to him his axe and the firewood suggested to some Sherpas that the same national park rules did not apply to all Khumbu residents equally.

In some cases villagers have become concerned that the national park rules have not worked well enough. This is especially so when formerly tight local control of sacred forests or rani ban has eroded. In Kunde dissatisfaction with national park protection of the rani ban adjacent to the village led villagers to resume choosing shinggi nawa in 1979 and to double the number of guardians to four. In 1984 shinggi nawa were chosen once again in Khumjung and Pare for the first time in many years to watch over areas that had formerly been rani ban, and residents of the settlements of Mingbo and Chosero chose shinggi nawa to oversee the nearby lama's forest. In all of these cases the impetus to community action was a perception that tree felling had reached unacceptable levels and that self-policing by villagers would be more effective than national park regulations and infrequent army patrols.

During the past decade several Sagarmatha National Park administrators reached a similar conclusion. In 1983 Mingma Norbu introduced a national park–sponsored, Khumbu-wide system in which two village-elected shinggi nawa would be responsible in each settlement for keeping an eye on local forests. These guardians would be paid a small salary (100 rupees per month) by the national park and would enforce Sagarmatha National Park regulations, turning offenders over to the national park authorities for judgement. The park would also keep any fines collected. This new system built on the traditions of forest management of Khumjung, Kunde, Phurtse, and Pare, extending the shinggi nawa approach to forest management to all Khumbu communities for the first time. Yet while drawing on indigenous forms of management it did not in any way reverse the shift in decision making about how forests were to be administered which had passed from local communities to the central government during the 1960s. The new shinggi nawa were in effect nothing more than employees of Sagarmatha National Park. Perhaps because of this the approach did not have universal support and its early effectiveness in many communities was limited. It represented, however, the first attempt by government officials since forest nationalization to work with local people and to build on local institutions in developing regional resource management. In some places there were notable successes in building greater local support for forest protection. During 1986 and 1987, for example, Mingma Norbu's successor Lhakpa Norbu appointed young monks from the Thami monastery to be shinggi nawa in the Thamicho area. These new shinggi nawa took up their responsibilities with considerable enthusiasm and several times even apprehended police-post woodcutters at work in the former rani ban of

Bachangchang near Thami Og. Lhakpa Norbu also responded to local appeals to grant shinggi nawa more independent authority and to give their forest-protection role more standing in the community. The forest guardians were given the power to issue and collect fines, and the money from these fines could now be put to village projects rather than being turned over to national park authorities. Lhakpa Norbu also called a major meeting of Sherpa leaders and shinggi nawa from all over Khumbu which did much to enhance the pride and stature of the shinggi nawa.

In the past few years there have been more far-reaching changes in relations between Sherpas and the national park which have had ramifications for local resource management issues. During the late 1980s long-standing local unhappiness with national park goals and policies and dissatisfaction with the personal administrative styles of two consecutive Sagarmatha chief administrators (the first a locally born Sherpa and the second a non-Khumbu Nepali) precipitated local demands for their removal. This culminated in a 1989 petition to the Kathmandu-based director general of the Department of National Parks and Wildlife Conservation to transfer the Nepali administrator from Sagarmatha National Park. This petition was the first attempt by Khumbu Sherpas to mobilize public pressure in this way to influence park policy and administration. The document contained a long list of grievances, some of them complaints about the stringent enforcement of national park regulations, others complaints about unjust treatment. Perhaps the most compelling charge was that the army and national park rangers had harassed village woodcutters who were not guilty of having cut green wood and had confiscated their fuel wood and cutting implements. The petition demanded action within a month. When the park administrator was subsequently transferred to another assignment this was interpreted as a victory for local people.

A new chief warden, Surya Bahadur Pandey, arrived in Khumbu in the spring of 1990. That spring was a momentous time in Nepal. Mass demonstrations in Kathmandu and many other parts of the country were being staged which resulted in the end of the panchayat form of government and redefined the role of the king in Nepal's politics. A significant politicization of Nepali society took place through widespread grassroots political organization and street demonstrations, and with the April legalization of political parties and the acceptance by King Birendra of demands for a new constitution, a general election, and a constitutional monarchy, expectations were raised about a new level of local participation in decision making and local authority. In Khumbu, as in the rest of the country, the local panchayat form of government was dissolved in the spring, and two multisettlement village development committees (one for each of the two former "village" panchayats) were established to handle vital administrative functions until a new national form of local government could be adopted. The new atmosphere also led some Khumbu leaders to believe that the time was right to press again for recognition of Sherpa rights to land and resource management that had been lost since the mid-1960s. When the new warden of Sagarmatha National Park called a meeting of Khumbu leaders in the summer of 1990 to discuss forest management and the establishment of a new local forest-management advisory committee, Konchok Chombi of Khumjung took the podium and delivered an angry demand that the Khumbu forests be returned to Sherpa control and that greater access to forest resources be given to villagers.

Pandey was startled at first by the degree of outrage in Khumbu over forest issues, but proved to be sympathetic to Sherpa concerns. He believed that a new relationship between the local population and the park had to be developed and was shocked at the level of distrust that existed after fifteen years of park operation. He believed that conservation must be based on local development and participation in resource management in order to be effective and advocated a greater role for the park as an agent of development. He also had considerable respect for traditional Sherpa land-use knowledge and resource-management institutions. Out of the new dialogue that ensued between the park administration and local leaders a number of basic changes in policy came about. These increased local communities' roles in regulating local forest use and established a regional forest management committee with a representative from each village which would consult with Sagarmatha National Park officials on forest-use policy. The national park would continue to subsidize shinggi nawa, of which there are now twenty-two in Khumbu, but these local officials would now enforce regulations that were at least in part decided by their own communities. The villages would decide how many trees they would authorize to be cut for beams and where these could be cut, although final authorization would require approval from national park authorities. Each proposal now had to pass through a multilayer review. It first had to gain the approval of the elected village representative to the forest management committee. Only then could it be reviewed by the national park. This established throughout the region the kind of two-tier system that had been in existence in some parts of Khumbu in the 1960s. Villagers now met to decide which regulations they should require their forest management committee member to uphold in his permit review and policed these with their own shinggi nawa. Communities could now much more effectively ban tree felling in sacred forests and other areas as they saw fit, and the fees formerly paid as royalties to the park for trees authorized to be cut as beams (130 rupees per tree) were now to go to the villages themselves, which would keep these on deposit in Nauje bank accounts and use them for projects approved by the village development committees.[30] Responsibility for all regulations concerning forest use and wildlife protection and their enforcement in the area around the Tengboche monastery, from the Milingo Chu to the Phunki Chu, were to be returned to the monastery.

These moves were hailed by local Sherpas, including some men such as Konchok Chombi, who had been among the most outspoken critics of the national park. Sherpas felt that they had regained some measure of control over their forests. At least for a short time a new spirit of cooperation with the national park developed. Nauje leaders, for example, accepted a national park–imposed ban on all fuel-wood gathering in the Satarma area, a place on the western shore of the Bhote Kosi where there had been heavy demand for fuel wood by Nauje families for some years. This was the first time any total ban on local fuel-wood gathering had been proposed by the national park, much less agreed to by Sherpas.[31] The national park administration, for its part, agreed for the first time to build a bridge across the Dudh Kosi river to a previously remote forest area (Tougekok) which would serve as a source of fuel wood to replace Satarma, and Pandey also agreed that some tree felling for purposes other than house beams could now be done in certain areas. One remote area south of the Dudh Kosi, for example, will henceforth serve as a limited source of fir shakes for roofing. Here there will be a limit of two trees per household.[32] In the spring of 1991 some Sherpas believed that other changes might also be ahead and that limited felling of trees for rafters and other uses might eventually be allowed, with the national park maintaining the authority to designate sites and select the particular trees to be cut.

The proposed development of a new national park management plan may provide the occasion to formalize these new approaches to park resource use and management, although they remain very much debated within the Department of National Parks and Wildlife Management. Whether or not these new regulations will be a brief experiment or a lasting new order remains to be seen.

These changes raise several important issues. One is the parameters within which local management can be exercised within a national park. National parks in Nepal, as elsewhere in the world, are established because of concerns for the protection of environmental quality which may include the preservation of some areas in order to safeguard biological diversity and protect endangered species and habitats. Sustainable use is a goal that is compatible with national park management only if there is zoning to buffer core areas where stricter nature protection is a goal and where there is agreement that some traditional, local practices may have to be regulated in nontraditional ways such as imposing hunting seasons and limits or bans on the killing or collecting of endangered species of plants and animals. If a national park or protected area is to have zones in which local settlement and land-use activities are supported, the introduction of sustainable use principles may require local people to modify long-established land-use practices or resource-management institutions. There thus remains the likelihood of disagreement about resource management, for local goals may not always be compatible with park goals.

At Sagarmatha National Park there are already examples of such conflicts between resource-use regulations sanctioned by the community and those upheld by the national park authorities. Most Khumbu villages have proposed their own conservation-oriented regulations for forest use. Typically they have raised the number of trees their residents can fell in local forests for house beams from the national park's past policy of three to five to seven. Most of the villages are very serious about enforcing these limits and have selected leading residents to the offices of shinggi nawa and forest management committee representative. Konchok Chombi, the Khumjung representative on the forest management committee, has even advocated that his fellow villagers ban the cutting of trees for beams everywhere in the village vicinity and require that trees for beams be cut only on the far side of the Dudh Kosi river. Yet it is still true that the new limits represent regulations that make sense in terms of the economic goals of land use—enough trees to build a house—but which do not necessarily restrict demand to sustainable levels in terms of the ability of local forests to continue to supply trees suitable for house beams in sufficient numbers over the long term. Sherpas have discussed what they need from forests and the importance of restricting their use to only certain forest products and certain sites. They recognize that heavy demands from local forests will change those forests, but the villages have had to make decisions about how much tree felling should be allowed without the benefit of an analysis of the condition of particular areas of forest. Ecological study of the forests on which villagers rely and a program to monitor their condition would probably be welcomed by many local leaders as being in accord with their own conservation concerns. But this is not necessarily the case in all villages. In Nauje, for example, there has been less concern for conservation, and one villager gained approval at the village level for felling forty-five trees to use in building a large lodge adjacent to national park headquarters. In this case the national park intervened and blocked permission. After negotiation permission was ultimately granted for twenty-one trees. But if Nauje continues to pressure for large amounts of timber the concept of local sustainability of use could be called into question.

A master plan is sorely needed that will provide a mechanism for establishing local guidelines for the annual number of trees that can be felled in specific forest areas. To determine this will require forestry research of a type that has not yet been done in Khumbu, and to enforce it will require considerable organization, effort, and coordination between the national park and villagers. The issue of control of forests is thus very likely still far from settled, and the process of developing a new master plan will very probably involve much negotiation over the permissible range and intensity of subsistence uses of forests. The new master plan, if it does indeed support the new notion of sustainable forest use rather than preservation, will have to propose mechanisms to establish how this level of use will be determined and enforced, and it may have to consider developing different types of use regulations for different Khumbu forests. The establishment of special protection areas for the preservation of biological diversity or undisturbed habitats, for example, would require that little or no subsistence uses be allowed. Some areas may have to be closed to some uses, even grazing, for a period of years to allow forest regeneration. Other areas may be found to be suitable for limited and carefully supervised tree felling for local use or for creating fuel-wood plantations. Planning for local forest use, moreover, will have to be reconsidered if the Nepal-Austria hydroelectric project now under construction near Thami Og is successfully completed and has the considerable impact on fuel-wood use in Thamicho, Nauje, Kunde, and Khumjung that it is intended to have.

The current changes in forest management and national park policies in Khumbu, moreover, raise the prospect that the purposes and structure of protected area administration in the region may need to be reconsidered. It may be that all of Khumbu is not appropriate for national park management. This was recognized at the time that Sagarmatha National Park was founded and all of the villages, gunsa, secondary high-altitude agricultural sites, and high-altitude herding sites were left as islands outside the legal jurisdiction of the national park. This in effect established a rudimentary zoning system. In this arrangement, the park controlled forest, rangelands, and the high peaks with regulations that allowed grazing and some forest use whereas the Sherpas managed their settlements as crop- and hay-production places.[33] In the early 1970s, when Sagarmatha National Park was planned, this was an innovative step and there existed no other model in Nepal for multi-zoned, protected area management. Today other approaches are being tried and the lessons being learned may well be applicable in Khumbu. In both the Annapurna range and the Makulu-Barun area a major distinction has been made between the more strictly protected core area and the areas where villagers live and make use of adjacent forest and rangeland resources. In the Makalu-Barun case the high country, including summer herding areas, will be zoned as a national park in which strict nature-preservation principles will apply, whereas the lower-altitude areas where the main Rai, Sherpa, and Bhotia villages are located are to be administered as a conservation area. This conservation area will place much more emphasis on local management of development and conservation and on sustainable use of natural resources. The same approach would be appropriate in Khumbu. Below 4,000 meters the valleys could be declared a conservation area. The area above that as well as some representative and relatively undisturbed lower-altitude areas could remain a national park. This would mean greater Sherpa management authority in the lower valleys, especially over the forests, woodlands, and grasslands they use most intensively. In time such measures may become necessary, for Khumbu Sherpas are already becoming aware of the differences between national parks and the new conservation areas and the significance that this has for the way they can live their lives. The establishment of a Khumbu conservation area to complement Sagarmatha National Park would formalize government commitment to Khumbu conservation (in the sense of the wise use of resources) and local development as well as promoting tourism and preserving nature in a stricter sense. This complex set of concerns and goals has underlain the policies of Sagarmatha National Park since its establishment, but recognizing them in this formal fashion would greatly inspire Sherpa confidence and involvement in Khumbu protected area management, and in so doing greatly further the successful achievement of these goals.

Forest Nationalization and Impacts Reconsidered

The common view that forest nationalization caused the rapid abandonment of traditional Khumbu forest-management institutions and precipitated extensive deforestation is obviously in need of revision. I have suggested that the introduction of new forest-use regulations and Sherpa responses to them were more complex than has been allowed for, unfolding in different ways and at different paces in different parts of Khumbu and affecting different forests to varying degrees and in varying ways. Existing local institutions were *not* abandoned everywhere simply because Kathmandu declared the forests nationalized or an official in Nauje began to issue logging permits. In some parts of Khumbu local management has continued to be effective to the present, and although in other places local management was indeed undermined, this did not take place for some years after 1957, had less far-reaching environmental impact than had been thought, and was not necessarily permanent. In some places Sherpas revived local management after a few years and they are in the process today of reinvigorating it throughout the region. Yet at the same time the history of forest change since 1965 also suggests that there was a basis for the concern with which Fürer-Haimendorf and the various national park planning teams reported that local forest-management institutions had been abandoned and that deforestation was taking place. Although deforestation was far less extensive and there may not have been a regional crisis underway, there were particular forest areas where logging for timber had increased and Sherpas had noted a sudden change in forest density in several areas where rani ban regulations had previously been in effect. There has perhaps been too much concern among outsiders with deforestation in the sense of a decline in forest cover and too little with the environmental implications of the new, post-nationalization patterns of resource use in terms of changes in the density, composition, and stand age of forests, woodlands, and shrublands and changes in the supply and accessibility of specific species and sizes of trees and shrubs in local demand for particular uses. In respect to both changes in vegetation and resource availability, moreover, it is important to place more emphasis on evaluating specific sites and variation across the region than simply to offer Khumbu-wide generalizations. The same attention to regional variation is also necessary, of course, in considering institutional change. Localized forest change in Khumbu after 1965 was significant in social as well as environmental terms in the places where it occurred, for it testified to conflicts within and among Sherpa villages over how forests should be used and managed and to the ultimate decision to suspend any local community responsibility for intervening in individual households forest-use decisions.

Khumbu forest management in the 1960s and early 1970s thus has to be seen as a highly variable, local, institutional accommodation to forest nationalization which had varying regional impacts on the environment and reflected regionally varying social dynamics. In some parts of the region the new, outsiders' concepts of forest management were wholly adopted and local institutions were abandoned. But elsewhere Sherpas acquiesced in the government's demand for new procedures and payments without surrendering their own authority to choose and enforce local concepts of how forests should and should not be used. In these places villagers observed when necessary the rules of the new system while continuing to use and regulate forests according to their own needs, concerns, and goals. And even in communities that ultimately abandoned their institutional regulation of local forests there were often cultural understandings that significantly shaped local patterns of resource use. Khumjung, Kunde, and Nauje villagers, for example, all ceased to enforce community controls on the felling of rani ban trees without as individuals abandoning the belief that trees in the sacred groves around village temples and nearby hermitages must not be felled or lopped. Here no community meetings were held to reiterate regulations and no community officals were selected. Cultural values and conscience alone protected the groves. Thus even in these villages some woodland remained under traditional forest protection. While in some places the relatively recently adopted rani ban regulations were abandoned, far older customs of carefully protecting sacred trees and forests continued to be respected.

The literature on recent change in Khumbu forest management and forests has perhaps focused too much on the assumed impact of forest nationalization and not enough on the impact of the establishment of Sagarmatha National Park. The fundamental contradiction between long-standing Sherpa forest-use practices and Sherpa forest-management goals and early Sagarmatha National Park policies has had a far more significant effect on Sherpa forest use and management than did forest nationalization. Khumbu forest-use history from 1965 through 1976 demonstrates that forest nationalization is not necessarily incompatible with continued local resource use or even continuing local resource management. But the strict nature protection-oriented policies of Sagarmatha National Park created a very different relationship between national conservation policy and local resource use and management. The conflict between Sherpas and the park over the use of natural resources in Khumbu reflected apparently irreconcilable differences between Sherpa traditions of local resource use and management, which are based on a principle of strict protection of *some* areas and unrestricted local use of others, and the national park policy of strict protection of *all* forest. But the national park attitude has been tempered by some appreciation of the importance of forest use for Sherpa subsistence, and although regulations that have forced Sherpas to turn to Pharak for timber remain in place, the national park has continued to allow forest grazing, the collection of dead wood, and some limited tree-felling for beams. The process of developing protected area management policies that meet both national park goals and Sherpa concerns is still continuing.

It is now twenty-five years since forest nationalization was first implemented in Khumbu. Two different central government–directed systems of forest regulation have been tried in the region and several important lessons have been learned about the effectiveness of different approaches to forest management. One lesson has been that neither the permit system of tree felling in use from 1965 through 1976 nor the national park ban on tree felling has been as fully effective as Sherpa religious beliefs, self-restraint, and community vigilence in protecting sacred forests. Another lesson has been that it may have been wiser to build on local management institutions to begin with rather than to undermine them for nearly twenty years and then attempt to reverse direction. A third lesson has been that while a ban on tree felling enforced by the army may noticeably lessen the impact of local subsistence use on forests in the short run, it may not create a climate in which a national park and its conservation ideals are likely to win support from the indigenous people whose sovereignty in their homeland has been compromised. This support for new conservation ideals and the new kinds of resource-management measures needed to achieve them, however, can be critical in the long term, especially if future park management is to be carried out with the participation of local people. These lessons are not unique to Khumbu and similar conclusions drawn on the basis of experience in many parts of the world have led to a new global movement in the design and management of protected areas (Stevens 1986a). Yet increased understanding of the strengths and limitations of pre–national park Sherpa forest management also suggests that future resource management in a place that is now both a Sherpa homeland and a national park of international significance cannot be based simply on a return to "traditional" patterns of forest use and regulation.

The Decline of Communal Regulation of Herding

There have been a number of challenges to the Khumbu nawa system of local pastoral management since Sherpas developed it in the nineteenth century. These include dissension within communities of users (both due to disregard of regulations by wealthy and powerful families and as a result of more widespread civil disobedience), conflicts among communities over the boundaries of village lands and the question of who makes the regulations for particular areas, and attempts by outside herders to circumvent local pasture-use regulations. Until recently, however, none of these efforts to undermine local traditions of seasonally excluding livestock from the nawa-administered areas was successful. Early in the century in the Imja Khola valley, for example, Pangboche and Dingboche residents were successful in forcing Gurung shepherds to honor local grazing regulations. In the 1940s Phurtse villagers were able to stop the widespread violation of grazing bans precipitated by a few wealthy nak herders by implementing a system of rotating the office of nawa among all village households so that it became more difficult for any household to use its power and influence to subvert community regulations.[34] For decades Khumjung-Kunde and Thamicho villagers were able to persuade Nauje herders to comply with their nawa regulations in areas near Nauje despite Nauje residents' disagreement over the ownership and control of those areas. And in recent decades community pressure in Khumjung and Kunde has forced obedience from powerful families who had tried to ignore the rules.

In recent years, however, Sherpas have been less successful in maintaining the effectiveness of communal resource management against challenges from individuals seeking private gain. Since the late 1970s the nawa system has come under enormous pressure in most of Khumbu, and in some regions the former resiliency and strength of the institution has been lost. Nawa ceased to be chosen in Nauje after 1979, and after 1983 the system went into decline as well in Thamicho. It was abandoned in some areas in 1984 and 1985 and throughout the Bhote Kosi valley after 1988. Khumjung and Kunde villagers have maintained grazing restrictions in Teshinga and the immediate surroundings of the village, but not without some contention. There has been much disagreement on whether to continue to choose nawa and on how to deal with the frequent violations of Khumjung-Kunde herding regulations by Nauje livestock. Pangboche continued to administer livestock bans effectively through 1986, but in 1987 regulations in the Dingboche area were widely violated when livestock were moved early into some areas closer to Pangboche. Since then some villagers have been quite concerned that the institution may ultimately be undermined. Only in Phurtse has there been relatively little sign of recent challenge to nawa enforcement of regulations.

The events that precipitated the crisis in Nauje in 1979 are well remembered. According to one of the three nawa of that year, trouble began immediately after livestock had been excluded from the village after the Dumje celebration. A number of families were tardy in removing their stock from the village and had to be fined five rupees per head of stock. Offenders were given two to three days to remove their livestock and were warned that if they refused to do so they would be fined more heavily. In 1979, however, even the levying of increasingly severe fines did not induce all households to comply with the village rules. Several households still had not moved their livestock out of the closed area after several visits from the nawa and being fined first ten rupees and then twenty rupees per animal. At this point matters were made far worse when a number of other families brought their stock back to the village, announcing to the nawa that if some herders were going to refuse to adhere to the rules, they would also refuse to obey the nawa.[35] According to one of the nawa that year soon "everybody came down." The nawa gave up trying to enforce regulations in the face of mass civil disobedience. "This is not one person's job, it is everybody's job," one of the last nawa noted. "If they [the other villagers] will not listen, forget it." The next summer no new nawa were appointed for the first time in memory and none have been chosen since.[36]

Since 1979 Nauje decisions about where to herd livestock have been left entirely to each household's discretion. Some families still move their livestock outside the old restricted area each summer to higher pastures, but many families keep zopkio in Nauje for at least part of the summer and it is now common for zhum and cows to be grazed in the village environs year round.[37] The presence of livestock in the village during the summer months has led to increasing problems with crop losses, despite the efforts by many families to build more secure walls around their fields, and is perceived to have degraded the quality of winter grazing around the village.

The decline of nawa management in Nauje has also affected adjacent areas. During the summer Nauje stock commonly stray onto nearby Khumjung, Kunde, and Thamicho pastures that have been closed to grazing. The continuing violation of the Khumjung-Kunde herding restrictions in the Shyangboche area by Nauje livestock has been especially prominent.[38] Khumjung and Kunde nawa have come in for considerable censure from their fellow villagers for not better handling Nauje violations in this area. Individual nawa have made real efforts to control Nauje stock, trying to move dozens of animals back outside the restricted area, but have been ineffective. Several dozen head of Nauje stock at a time are often on Khumjung-Kunde land and as one Khumjung nawa put it, "daytime I push them [the livestock] out, but at night they come back. I cannot do duty all night." Some Khumjung villagers note that in the old days there would have been more of a confrontation and recall times when the entire village had marched to Nauje to demand compliance and threatened to fight for it if necessary. But today there is a lack of consensus over what kind of response should be made to Nauje violations, and this and uncertainty over legalities have stymied the effectiveness of the Khumjung and Kunde nawa in dealing with Nauje stockowners. To make matters worse some Khumjung-Kunde villagers are now losing their own dedication to observing the rules. As early as 1984 some Khumjung herders were complaining about Nauje violations and proclaiming that if Nauje stock were going to violate the regulations and graze Khumjung winter grass then they were going to bring their herds down early too. Thus far, however, only a few herders have begun to ignore the rules and these have experienced considerable social pressure from fellow villagers to comply with them. Prominent Khumjung villagers note, however, that people are beginning to tire of arguing and fighting over the herding regulations and that the community may well decide one year soon to abandon the effort of maintaining nawa regulation. Thamicho families also found Nauje stock violations of

the grazing regulations in the lower Bhote Kosi valley extremely annoy-
ing and some of them unsuccessfully asked the national park to enforce
these regulations against Nauje families.

In the Bhote Kosi valley nawa administration went into decline during the 1980s and by the end of the decade the system had been abandoned throughout Thamicho. The first signs of difficulty came in 1984
when no nawa were selected in Pare. In 1985 none were selected anywhere in Thamicho and it appeared that the institution had been abandoned. In 1986, however, Lhakpa Norbu, a Thami Teng Sherpa who was
at that time chief administrator of Sagarmatha National Park, inspired
Bhote Kosi residents to resume selecting nawa. That year nawa were
again chosen, but they had a great deal of difficulty enforcing the regulations against fellow villagers and Nauje livestock. The following year,
with little interest by the general public in accepting nawa duties, the six
Thamicho adekshe (ward officials in the panchayat government) were
given the responsibility by their fellow villagers.[39] It was further decided
that stiffer penalties should be enforced and the fine was set at fifty
rupees per head of stock. Violations, however, were again widespread
and the adekshe had trouble addressing them. Attempts by the adekshe,
for example, to force an old woman to move a cow out of the closed area
in midsummer were defied to general amusement, and a villager noted
that "six adekshe cannot control one old lady with a cow." By the end of
summer the grazing regulations were being widely violated. The following year it was decided to go back to electing nawa rather than simply
assigning the job to adekshe, but it remained difficult to enforce the
regulations. No nawa were chosen in 1989 or in 1990.

The nawa systems of eastern Khumbu (Phurtse and the Pangboche-
Dingboche areas) have not been challenged to the same degree, but
friction in the Imja Khola valley appears to be increasing. In 1987 the
nawa of Dingboche had trouble enforcing their decisions about the timing of the beginning of the barley harvest and the opening of the area a
few weeks later to grazing. After two unsuccessful attempts to halt early
harvesters the nawa gave up the effort and many head of stock were
brought into the area before it had been declared open for grazing.
Some Pangboche families who have land at Dingboche openly wondered
whether nawa regulation would be maintained there in the future. There
remains, however, a very strong conviction that blight prevention and
livestock-control regulations should be enforced, and the nawa were
much criticized for not acting effectively in 1987.

Phurtse is the only place where livestock exclosures currently seem to
be maintained with little village debate. One factor here may be that the
system is administered by a single village to regulate its own residents'
livestock. It may also make a difference that this is a relatively small

village where many families herd nak. Perhaps there is a greater consensus about resource regulation here, or more effective social control or local leadership.[40] We have already seen, however, that even in Phurtse the enforcement of grazing restrictions has not always been straightforward and that the system was under considerable pressure even forty years ago. Whether the community can continue to respond so successfully to its own internal challenges is by no means certain. The 1990 decision to change very old regulations regarding the collection of fuel wood in the local lama's forest is a sign that differences of opinion on how to manage local resources may be increasing. In this case middle-aged and older Phurtse residents found themselves for the first time in a major debate with young householders on an important issue that involved land use and religion, and they ultimately acquiesced to the changes demanded by young villagers.[41] Now dead wood will be collected in the sacred forest next to the village, and in local interpretation this also allows the lopping of dead branches.

It is difficult to pinpoint the underlying causes for the breakdown in nawa management at Nauje and in Thamicho. Nationalization was not the factor here that it was in the undermining of some local forest-management institutions. Although the national park has considered the implementation of rotational grazing schemes and the possible establishment of ceilings on livestock numbers (Bjønness 1980a; Garratt 1981), the central government and the national park have not yet moved to intervene in any way in either pasture use or herd composition other than in encouraging the banning of goats. Indeed, as already mentioned, there have instead been at least two cases in which officials have supported the nawa system or urged its reinstatement. The continuing tension between private interests and communal attempts to regulate the use of rangelands has, of course, been a common theme in all the areas where there have been difficulties in maintaining the nawa system. But this is a very long-standing conflict between cultural values and social institutions and does not go far in explaining what has tipped the balance against community regulation so strikingly in recent years in Nauje and Thamicho. In the Thamicho case competition for resources due to increased summer grazing on Thamicho-administered lands by Nauje stock was one important factor, as was the Nauje precedent of abandoning nawa enforcement. But Thamicho villagers themselves emphasize the increased number of zopkio and cows that have been kept in recent years by Thamicho herders, the irresponsibility of some Thamicho herders who have disregarded the communal regulations by bringing their pack stock to graze for days at a time in the lower Bhote Kosi valley while waiting for the arrival of mountaineering expeditions, the increasing loss of crops to zopkio, and the inability or unwillingness of the nawa to successfully police the area. There have been so many different points of pressure and loss of control that people have lost confidence in the system. They no longer feel that they can rely on it to protect their crops or to ensure that they have equal access to lower-valley pastoral resources. There is also the possibility that some new herders who only keep a few head of dairy stock or zopkio would prefer to remain based in the main village in the summer and not have to bother with the lifestyle demands of shifting to high-altitude resa and herding huts. This last factor may have been important in the erosion of community commitment in Nauje to maintain the nawa system. Many of the Nauje families who began keeping zopkio, cows, and zhum in the 1960s and 1970s had not previously practiced the annual routine of shifting base to a summer herding station outside of Nauje. They may not have been happy with the traditional regulations that required cows and zhum to be herded far away from Nauje in places such as the Pulubuk valley east of Tengboche in the summer and may have relished the opportunity to forgo the cost of hiring herders to watch the stock in these areas or the trouble of doing it themselves. The opportunity to remain all summer in Nauje, with stock grazing adjacent slopes, thus may have been welcomed by many families even though it meant poorer winter grazing. And once some families began defying the system and keeping their stock in Nauje there would have been a great temptation for other families to likewise opt for personal convenience and the opportunity to give their stock the same chance at grazing the rich summer grass near Nauje that their neighbors' stock had.

Environmental Consequences of the Abandonment of Local Pastoral Management

Nauje's abandonment of communal enforcement of summer livestock herding regulations and autumn grass-cutting rules and the abandonment of herding regulations in the Bhote Kosi valley have had repercussions on the Khumbu environment and pastoral resources. Grazing pressure has increased in areas where seasonal grazing restrictions were once enforced, especially in the vicinity of Nauje and the lower Bhote Kosi valley but also in the Khumjung-Kunde area where Nauje stock have increasingly transgressed areas within the zone that Khumjung-Kunde closes to summer grazing. Nauje's abandonment of the nawa system has also affected pasture conditions by ending the former practice of restricting wild grass harvest until the nawa declared in the autumn that this could begin. Villagers feel that the amount and quality of hay made from wild grass have both declined. These pressures have been further accentuated by the increasing trend in these settlements towards the keeping of greater numbers of zopkio, zhum, and cows, which have different grazing preferences and fodder requirements than yak and nak.

Nauje herders have observed that winter grazing near the village and in easily accessible areas of the lower Bhote Kosi and Dudh Kosi has declined since 1979, making it more important than ever to store large amounts of fodder for use in the winter and spring. In the 1980s, for example, some Nauje families began sending stock (dimzo and urang zopkio) across the Bhote Kosi river to the Satarma area in the late spring because the grass had become so poor nearer to Nauje.[42] This, they noted, had not been necessary in the 1970s. Yet at the same time that spring grazing becomes more difficult it has also become harder to store large amounts of fodder. Nauje herders have noticed a major decline in their ability to obtain wild grass for autumn hay making. Before the 1980s a day's work cutting grass on the slopes near the village supplied a large, forty-kilogram load of wild grass. No one bothered to go far afield to cut grass and it was rare for grass cutters to cross to the far side of the Dudh Kosi or the Bhote Kosi. Now grass cutters work the other side of the rivers and in a day people are content with a small load of ten to fifteen kilograms. Some Nauje wild-grass cutters are now even working in more remote areas of Khumbu, including Tengboche and the upper Imja Khola valley, where they store hay for use by their yak and zopkio along the mountaineering and trekking route to Mount Everest.[43]

Villagers give a number of reasons for the increasing scarcity of wild grass for hay making. These include the lack of enforcement of the summer livestock exclusion and the resulting heavier grazing of areas that formerly supplied wild grass, the increase in the number of head of village stock caused by the increased herding of zopkio and dairy stock, and premature cutting.[44] Since 1979 Nauje stock owners have begun cutting wild grass earlier than was previously normal in order to gain an edge on their competitors. In 1987 grass cutting began on August 18, several weeks before the nawa would have opened the area to cutting in the years before 1979. The result is said to be that grass is cut before its seed fully matures, making the resulting wild grass less nourishing fodder. This practice may also lead to a longer-term loss of range productivity, for as Brower notes, "an early ritsa harvest means wild hay grasses are cut before they have finished their annual cycle, presumably undermining their chances for persistence" (1987:299). Heavier summer and early-autumn grazing must accentuate this process. The decline in the enforcement of summer livestock grazing bans has also undoubtedly increased the amount of grazing pressure in woodland, forest, and shrub areas in the lower Bhote Kosi valley, lower Dudh Kosi valley areas near Nauje, and Shyangboche. Although it seems likely that this has contributed to poorer forest regeneration, studies necessary to establish this clearly have not yet been carried out.

It seems likely that continued year-round grazing by large numbers of livestock, and particularly by large numbers of zopkio, may gradually lead to a decrease in pasture carrying capacity. Declining grazing resources and availability of wild grass for hay may lead to increased risks to herders of winter stock losses, particularly in years of high snowfall when greater amounts of fodder than usual are required to save livestock from starvation. Herders who own insufficient private hayfields and those who lack the financial resources to purchase large amounts of fodder would be especially at risk since they would be less likely to set aside enough of a stock of fodder for such conditions and would instead hope for normal weather.[45] It is tempting to ascribe the major winter livestock losses that took place in Khumbu in 1985–1986 to this process of declining range carrying capacity. But the fact that this mainly affected yak and nak—whose upper valley grazing and fodder resources have been least affected by the increase in the number of zopkio and the decline in nawa regulation—suggests that weather and not grazing impacts was the key factor in the scarcity of grass and subsequent starvation of stock. This view is also supported by the fact that these losses did not take place only in Thamicho but also in the upper Imja Khola valley among Pangboche stock.[46] No such large-scale winter die-offs have taken place since, even though autumn snowfall was early and heavy in 1987 and this might have been expected to create a crisis. Nor have disastrous winters been simply a recent phenomenon. Extensive loss of stock in winter and spring was reported for at least one winter in the 1950s and another in the 1970s (Bjønness 1980a; Fürer-Haimendorf 1975).

Whether the situation will worsen is not clear. Fodder shortages and increasingly high fodder prices may begin to discourage families from investing in more zopkio or even lead to a decline in their total numbers. A change in the policies of trekking companies and expeditions that favored the use of porters rather than pack stock would certainly also dramatically effect regional stock numbers and grazing and fodder requirements. It is also likely that Sherpas will respond to the regional shortages of fodder by attempting to increase hay yields in their walled fields and through more cultivation of fodder crops. Experiments in Nauje with cultivating wheat and barley as a second crop following potato production, experiments with irrigating hay in Samde, and continuing interest in the introduction of new hayfield grasses all suggest that Sherpas may be able to raise regional stock-carrying capacity. There are also more efforts today to import hay from Pharak, even though this is considered to be less nourishing than Khumbu-grown and gathered hay. And some Sherpas remain interested in reviving nawa regulation in Nauje and Thamicho, although there is not very much optimism that this can be done. There also remains the possibility that concern over livestock-related environmental damage will lead to government intervention in the changing relationships between pastoral use and Khumbu range resources. Sagarmatha National Park administrators have shown an interest in encouraging the continuing operation of the nawa system, and over the years there have been a number of proposals that the national park should take action to control Khumbu herding. These proposals have included a ceiling on regional livestock numbers (Bjønness 1980a) and the implementation of a new rotational grazing system (Bjønness 1980a; Garratt 1981). The issue of limitations on the number of zopkio has also been raised and debated (Brower 1987:310). Park managers and advisors and Kathmandu officials agree, however, that government intervention in Sherpa livestock ownership or management would be highly resented by Sherpas and very difficult to implement. For the time being, at least, Sherpas continue to work out the dynamics of their pastoralism within their own economy, culture, and environment. Whether they will successfully cope with new and increasing pressures on pastoral resources and institutions remains to be seen.

9

From Tibet Trading to the Tourist Trade

Khumbu Sherpas, like many Himalayan high-altitude peoples and many mountain peoples all over the world, have long relied on trade as a basic and vital component of economic adaptation to the limitations of their environmental conditions.[1] Exchange provides access to the resources of other altitudinal zones making possible a broader lifestyle than the limited local repertoire of crops and livestock allows. The possibility of obtaining lower-altitude grain in particular decreases dependency on highly variable high-altitude crop yields. This has several important implications for household and regional economy and lifestyles. It lessens the risk of food shortfalls, permits a larger population to be supported in the higher reaches of the mountains than local resources alone would allow, and makes possible a greater specialization in the crops and livestock most suited to local environmental conditions. In places like Khumbu, where there is relatively little interest in commercial agriculture and only a few families are involved in commercial pastoralism, transit trade based on the products of other regions has enabled families to keep decisions about crop production and some kinds of pastoralism free of market considerations. This has had important implications for land use and lifestyles.

Khumbu is situated at the edge of Tibet, at the entry to the final, high-altitude crux of a major trans-Himalayan trade route linking Nepal and northern India with Tibet. Although the Nangpa La is one of the highest of the many passes that cross the Himalaya to Tibet, it is not a difficult passage for yak and for people who are properly equipped for snow and cold and knowledgeable about mountain weather and glacier crossings.[2] Khumbu Sherpas thus have been ideally situated to be middlemen on an important long-distance trade route, and since 1828 the Nepal government has granted them a monopoloy on part of this trade by banning Tibetan traders from proceeding any further south than Nauje and non-Khumbu Nepalis (including Pharak and Shorung Sherpas) from trading across the Nangpa La.[3]

Khumbu became a major exchange and transshipment point. Goods from the south carried by Rais and by Khumbu, Shorung, and Pharak Sherpas were exchanged here for goods from Tibet carried south by Tibetan and Khumbu traders. Khumbu Sherpas carried some of the southern goods on into Tibet and some of the products of Tibet on south into Nepal. In Tibet they traded in the villages of the adjacent Tingri region, in the regional centers of Ganggar and Shekar and in more distant Shigatse and Lhasa. A few Khumbu traders continued on further across Tibet, crossing west to the trans-Himalayan regions of Nupri, Manang, Thak Khola, and Mustang in north-central Nepal and east to Sikkim, Kalimpong, and Calcutta. Many Khumbu people traded southwards, bartering Tibetan salt and wool in lower-altitude Nepal for grain to be consumed back in Khumbu. Some continued as far as the Tarai town of Rajbiraj and the north Indian city of Jaynagar to obtain trade goods for resale in Tibet.

These routes were part of much vaster networks that moved goods across large portions of Nepal, Tibet, India, and western China from central Nepal to Calcutta and from the Ganges plain to Lhasa and the tea-producing regions of Yunnan and the Chang Jiang (Yangzi River). Khumbu Sherpas were one of a number of Himalayan peoples who played a small role in this trade, benefitting from the movement of goods but powerless to do more than respond to the trade opportunities created or terminated by political and economic actions taken by other peoples in distant places. Throughout the late nineteenth and early twentieth centuries Sherpas prospered through their increasing participation in international trade and moved freely among Nepal, Tibet, Sikkim, and India. But during the 1960s, when both political and economic change in Tibet, China, and India took place, this vast international trade network was severely disrupted by war, civil unrest, border policies, new trade and transit regulations, and development policies. Sherpas continued to participate in the new, limited forms of trans-Himalayan trade, but their attention began to shift in two new directions: their new role as consumers in a new intra-Nepal periodic market system and their quest for new opportunities for earning the money to play that role. Their ability to participate in the new interregional economy and ultimately to maintain the adaptive way of life that trade had supported, came in a very short time to depend on their ability to earn cash through sources other than farming and trading. The foremost new source of income proved to be tourism, and the phenomenonal speed with which Khumbu Sherpas embraced new opportunities in the tourist trade led by the mid-1970s to a new regional and household economy in which the former role of trade had been virtually entirely transferred to tourism development.

This chapter surveys historical change in Sherpa trade with Tibet and lowland regions, the decline of long-established trade networks in the 1960s, and the rise of tourism as a basic component of the Khumbu economy and Sherpa subsistence.

Trade Routes, Trade Goods, and Trade Systems

Khumbu Sherpas have conducted several different types of trade, each of which has varied historically in terms of scale, products exchanged, routes, and markets. Families from different villages and with different capital and labor resources, trading experience, connections, and lifestyle preferences were involved in various types of trade. Seven types of Sherpa trade can be identified as well as three types of trade conducted in Khumbu by non-Sherpas (table 27, and figs. 15 and 16).

The most important trade over the Nangpa La for most Sherpas was the salt trade. Salt was an immensely valuable commodity in Nepal as it has been in many parts of the world, and for centuries the major supply came from the great interior saline lakes of the Tibetan plateau. Nomads evaporated salt along the lakeshores and brought it by pack stock toward the settled heartland valleys of Tibet where they bartered it for barley tsampa from villagers who came there to meet them. These villagers in turn traded the salt to merchants in Ganggar and Kaprak. Sherpas traded in both of these nearby places for salt offering barter goods from Nepal and India in exchange. Both Sherpa and Tibetan traders then carried salt south across the Nangpa La and into Khumbu. From there it was traded across a vast region of eastern Nepal that reached as far as the Arun valley in the east, the confluence of the Dudh Kosi and Sun Kosi in the south, and the Tamba Kosi valley in the west. Salt bound from Tibet to eastern Nepal was traded south over only a few passes, and the Nangpa La was one of the most important in terms of the area it furnished with salt.

In most other Himalayan border areas similar trade for Tibetan salt was based on a barter exchange for Nepal-grown grain. This has also been assumed to have been the case for the Nangpa La trade. Khumbu Sherpas are usually described as hauling Nepal-grown rice and other grains north to Tibet to barter on good terms for salt and then profitably bartering this salt in lower-altitude areas of Nepal for grain (J. Fisher 1990; Fürer-Haimendorf 1964, 1975, 1984). This view of the basis of

Table 27.   Types of Trade

Type	Scale	Goods	Routes
Khumbu Sherpa Trade			
Tsongba	large	bulk and luxury goods*	Khumbu to Ganggar, Shigatse, Lhasa
Tsongba	very large	salt, iron	Khumbu to Kaprak, Ganggar
Tsongba	small	wool, sugar, Indian goods	Jaynagar and Rajbiraj to Ganggar
Nangzum	very small	salt, grain	Khumbu to Pharak, Shorung
Tsongba	medium to large	zopkio	Khumbu to Ganggar
Tsongba	medium	horses	Ganggar to Maini
Tsongba	small to large	nak	Chang, Tingri, Rongshar to Khumbu
Tsongba	small	herbs, dogs	Khumbu to Jaynagar, the Tarai
Other Trade to Khumbu			
Tibetan	small to large	salt, pastoral products	Tibet to Khumbu
Dongbu	small	grain, salt	Pharak and south to Khumbu
Shorung Sherpa	small to medium	butter	Shorung to Khumbu
Shorung Sherpa	medium to large	zopkio	Shorung to Khumbu
Shorung Sherpa and Rai	medium to large	paper	Kulunge, Shorung to Khumbu
Kulung Sherpa	small to medium	butter	Kulunge to Khumbu

* Butter, paper, shere, iron, Indian goods traded to Tibet, tea, silk, carpets, art, jewelry, silver goods, books, musk to Khumbu, Shorung and beyond.

Sherpa trans-Himalayan trade is partly mistaken in that a salt for grain barter was important only for the Nepal portion of the trans-Himalayan trade and was not the key to the entire trade both in Tibet and Nepal. Sherpas exported very little grain north over the Nangpa La, and the substantial stocks of rice and other grains that passed through Ganggar came instead from the Kuti and Rongshar routes farther to the west. The Sherpa barter trade in Tibet for salt was based not on grain but on butter, paper, and other goods obtained from Shorung and the lower Dudh Kosi valley. In the late nineteenth and early twentieth centuries Sherpas also exported a good deal of iron from the Newar mines at Those. Some Sherpa traders also purchased salt in Tibet using Nepalese, Indian, Tibetan, and even Chinese currency. It was only once Sherpas had trans-

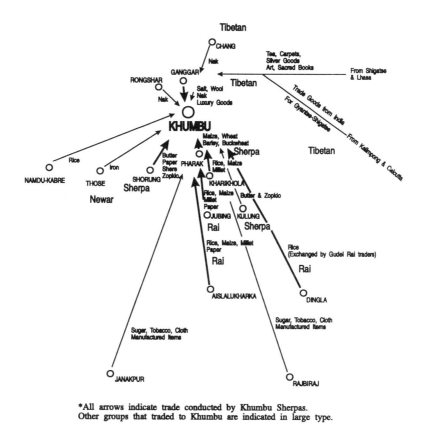

Figure 15.   Trade to Khumbu. All arrows indicate trade conducted by Khumbu Sherpas; other groups that traded to Khumbu are indicated in large type.

ported the Tibetan salt into Nepal that the grain for salt barter exchange became important. Sherpas traded salt for grain on a large scale with Rais, lower-altitude Sherpas, and other peoples both in Khumbu itself and in more southerly regions. Very little of the grain that Khumbu Sherpas thus obtained, however, was ever taken over the Nangpa La and into Tibet. It was bound instead for Khumbu where it became a fundamental component of the Sherpa diet, supplementing locally grown barley, buckwheat, and tubers. This trade freed families from relying entirely for subsistence on production from their own fields, and it was the demand in Khumbu for rice, maize, millet, wheat, and buckwheat from lower-altitude regions that spurred the widespread Khumbu Sherpa involvement in the trans-Himalayan salt trade. For most Sherpas participation in this trade was their only means to acquire grain.

Three very different types of traders were involved in the salt trade. Most of this business was in the hands of a very few big Sherpa and Tibetan traders (*tsongba*). For decades in the middle of the century four

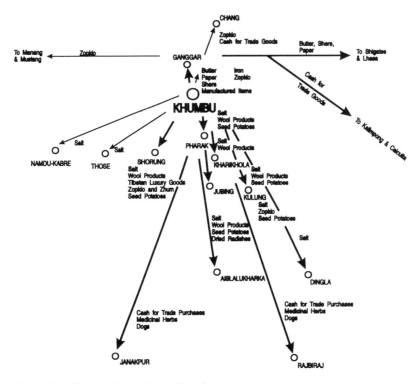

Figure 16.   Sherpa Trade from Khumbu

or five Nauje men dominated the trade who dealt in entire yak caravans of salt and imported tons each year from Ganggar (ten days north by yak) and Kaprak (six days north). One of the great Nauje traders of the 1940s and 1950s is said to have brought fifteen tons of salt into Khumbu per year (6,000 pathi), an amount that required 375 fully laden yak to transport.[4] These big traders did not bother transporting salt any further south than their Khumbu homes, from which they sold it to other Sherpas and Rais. More Sherpas imported salt on a smaller scale making use of their own few yaks or hiring a small number of pack stock or porters to bring it south over the Nangpa La to Khumbu. Still more families did not go to Tibet at all, but instead obtained small quantities of salt in Khumbu from larger traders. They then transported this salt south on foot by the basketload to trade with Rais and other residents of the lower altitudes. This form of trade, known as *nangzum,* required laboriously hauling loads of salt by tumpline in a series of shuttle trips to Rai, Sherpa, Hindu, and Newar communities where it would bring a better exchange than back in Khumbu. Families of moderate means and poor households participated only in this form of trade and often obtained their salt on loan from the big Sherpa traders. In the spring, after they had carried back to Khumbu the grain they had received in barter exchanges in the south, they repaid the loan of salt with a share of the grain. Such petty traders were the Sherpa equivalents of the many Rais who came by foot to Khumbu carrying baskets of grain to barter in Nauje and other places for salt. The better exchange rate they got from Sherpa traders in Khumbu compared to that which the Sherpas offered in their home communities made their time and physical effort worthwhile.

A very different type of trade revolved around the import of wool from Tibet. This was carried out by a number of Sherpa small traders who obtained wool (and also other products of the country such as dried sheep meat and fat) in Ganggar in the late spring and early summer in exchange for sugar, snuff, cloth, and other products of the Tarai and northern India. They then carried loads of wool back to Khumbu either on foot or with the help of a few yaks or porters. Back in their home village their families (and sometimes also hired helpers) worked the wool during the summer months and after the crops had been harvested in the autumn they went south with their entire families to trade these woolen goods to Rais and Shorung Sherpas. Woolen blankets and light rugs were traded to the Rais (along with heavy yak-hair tarps), but perhaps the most valuable wool products were *matils*, the striped aprons universally worn by adult Sherpa women. In Shorung they sold aprons for cash and also sometimes for other goods valued in the Tarai, such as a certain type of dog bred in Shorung. Then they moved on to the Tarai, and usually continued on all the way onto the Ganges plain to the town of Jaynagar, twelve to thirteen days on foot from Nauje.[5] Here they bought a variety of high-value, low-bulk goods ranging from Indian manufactured goods, dyes, and snuff to material for making prayer flags. Some traders also purchased quantities of brown and white sugar in large multikilogram cakes. They then moved these goods back toward Khumbu, sometimes hiring Tamang porters to help. On the way they often spent a few weeks in January and February in Shorung and the lower Dudh Kosi valley buying grain with cash, trade goods, or salt they had cached there on their way south. The rest of the late winter and early spring was devoted to shuttling grain back to Khumbu before family efforts had to be focused on preparing fields and planting crops. Once the crops were in it was time for another round of trading in Tibet. Here the sugar in particular turned a fine profit. In Jaynagar in the early 1950s two twenty-kilogram loaves of brown sugar could be purchased for five Indian rupees. Back in Nauje they would be worth five times that, whereas in Ganggar they brought enough wool for thirty-five aprons, each of which would bring seven or eight rupees back in Shorung.

There were also several different types of livestock trading. Most of these were linked to large-scale, international trade in dimzo. Khumbu was a major base for the crossbreeding of nak and Tibetan varieties of cattle, and virtually all male dimzo calves born in Khumbu were ultimately exported to Tibet. Female crossbreed dim zhum were in demand as dairy cattle in Shorung and most of these calves were also sold rather than being raised as part of Khumbu herds. Most of the dimzo calves also initially were dispatched to Shorung. Shorung traders purchased these calves from Khumbu stockbreeders and dealt them to Shorung herders who raised the calves for several years and then resold them to Khumbu men who traded the two-, three-, and four-year-old zopkio to Tibet. The Khumbu zopkio traders were primarily Nauje men, although a few men from other villages also engaged in this business. They either journeyed to Shorung to buy dimzo or else contracted in Khumbu for their purchase and delivery. In the latter case the young zopkio were delivered to them in Khumbu during the monsoon months and the traders herded them there until the autumn when they drove them over the Nangpa La to the Tingri area. Some of the larger stock traders took as many as a 140 head of stock north. In the 1950s a three-year-old zopkio worth 200–300 rupees in Nauje could be traded for two nak, forty-five kilograms of wool, or about eight loads of salt. Occasionally Khumbu traders drove zopkio beyond Ganggar, for there was also demand for them in the trans-Himalayan areas of central Nepal north of the Annapurna Range, particularly in Manang and Mustang. Groups of Khumbu herders sometimes drove scores of young zopkio across the grasslands of Tibet to this region.

There was also a trade in importing nak from Tibet to Khumbu, where they were sold to herders who were involved in the crossbreeding business. The nak trade also sometimes involved long-distance drives, for the best nak were obtained not in the Ganggar region but far beyond it on the high plateau country of Chang. Nak could be readily purchased for cash or obtained in exchange for zopkio calves in Ganggar or in the Tibetan villages closer to Khumbu, but sometimes Sherpas went as far afield as Chang to obtain better-quality nak at lower prices. This journey began with a six- or seven-day trip on foot from Ganggar to Lhartse (four days by horse). Here traders crossed the Tsangpo by boat (or took a different route to a suspension bridge) and then continued two more days on foot to Dokshum. Here nak could be bought for seven to nine Indian rupees in the 1940s. Mostly only small-scale herders made the effort to go this far, and in the spring or autumn groups of such Sherpas would set out for Chang to buy three nak or so each to build up their personal herds and others went each autumn (during the tenth and eleventh lunar months) with young dimzo to bargain for nak. Seven nak for two dimzo was the usual rate, and this included the cost of having the nak herded all winter by the Tibetans before the Sherpas returned to pick them up in the spring (in the fourth month). Sherpas note that for one's own use the best nak come from Pakrukok (beyond Lhartse) for that area has less grass than Khumbu and when the nak are brought to Khumbu they thrive on the richer pasture and breed better. It was considered best to import nak in the spring, for if they did not feed on Khumbu grass during the summer they were not considered likely to be healthy and strong for the winter ahead. If one was looking for nak to sell to other Khumbu herders it was considered best to buy them in the Sangsang area of Chang. These nak graze on excellent grass and their fine appearance brings a good price in Nepal. Knowledgeable stockmen, however, believe that such stock grow thin and weak on Khumbu grass and as a result breed poorly. Nak were also brought back from parts of Tibet closer to Khumbu, including the Tongnak, Gupchichutang, and Nyiring Gompa areas.

For twenty years one Nauje Sherpa conducted a rather different nak import business, keeping his own large herd of nak and yak in the Rongshar valley of Tibet, where hired Tibetan herders cared for them in exchange for a share of their milk and calves. He is said to have brought eighty or ninety nak per year to Khumbu for sale until 1962, when he discontinued his Tibetan operation for fear that it would be halted by the Chinese.

Horse trading was another special pursuit. The Nangpa La was considered to be too high in altitude and too cold for horses to cross, although there are stories of a few that survived the trip. Khumbu traders instead took horses that they purchased in Ganggar southwest over a low pass (the Pusi La) into Rongshar and then followed the valley of the Rongshar Chu south into Nepal. Some of these were sold in Shorung, others were brought all the way to Khumbu. Most were sold in the annual trade fair in southern Nepal at Maini, and subsequently became *tonga* horses pulling the carriages popular in the towns of the Tarai and northern India.

Some people also dealt in herbs. These families collected *hugling* and other medicinal mountain herbs in Khumbu and adjacent regions and took them south to the Tarai or India. Profits from their sale could be invested in grains for Khumbu consumption and also put into trade goods for sale in Tibet or bartered there for salt or wool.

Finally, a relatively small number of traders conducted long-distance trade in luxury goods. Most of them were Nauje men, although some Kunde and Khumjung men also were involved in this kind of trade that revolved around the export of luxury items from Tibet to Nepal, most of them intended for sale to Sherpas either in Khumbu or other nearby regions. Among the items that were important in various periods were "Tibetan" brick tea (actually tea grown in Yunnan, Sichuan or the middle Yangzi valley and specially prepared for the Tibetan market), carpets, silver goods such as teacup lids and stands and belt buckles, Chinese silk and silk brocade, jewelry and precious stones (turquoise, coral, and the prized zi agates), and a range of religious artifacts including paintings, statuary, prayer wheels and books. Such goods could be obtained in major Tibetan urban centers and Khumbu Sherpas sought them especially in Shigatse and Lhasa.[6]

Sherpa traders employed several different strategies for obtaining Tibetan luxury export goods in Sikkim, Darjeeling, Kalimpong, and Calcutta. The most common tactic involved a three-step process. First bulk goods from lower-altitude areas of Nepal or higher-value, less bulky items from northern Indian bazaar towns would be obtained during winter trading trips south. These would then be sold for cash in Ganggar and Shigatse. With this money traders would then purchase luxury goods in Shigatse or Lhasa to import to Nepal. Some traders with greater cash reserves chose instead to make the journey directly to Shigatse or Lhasa to purchase goods for import to Nepal. Others chose yet another tactic and began by traveling to Kalimpong or Calcutta. There they purchased goods which they resold in Lhasa or Gyantse for a profit which could then be used to buy Tibetan luxury goods for export home. Some traders left Khumbu with a caravan of yaks and porters carrying butter, paper, iron, and other bulk goods, sold these in the nearest Tibetan market centers, traveled across Tibet to Kalimpong on foot and then went by train to Calcutta, returned to Tibet and traveled north to Lhasa, Gyantse, and Shigatse to trade goods obtained in Kalimpong and Calcutta, and finally arrived back in Khumbu ten months or more later with a cargo of highly valuable Tibetan goods. Other traders spent years shuttling back and forth between Kalimpong and Tibet before ultimately calling an end to their trade activities and heading home with a few loads of immensely valuable trade goods.

Historical Changes in Trade

Although there are indications that Khumbu Sherpas have been trading with Tibet since at least the early nineteenth century, there is little evidence with which to analyze the origins of the trade, its scale, or its emphases prior to the late nineteenth century. It is clear that the Nangpa La was used as a trade route as early as the 1820s and that at that time Khumbu Sherpas had been given special trading privileges over the pass. Fürer-Haimendorf described an 1828 Nepal government document that he was shown in 1957 in Khumbu which granted Khumbu Sherpas alone the right among Nepal's citizens to trade across the Nangpa La and which restricted Tibetans from trading any further south than Nauje. I have seen later documents that refer to the 1828 edict and which indicate that Khumbu Sherpas repeatedly had to appeal to the government to continue to enforce their privileges.[7] The original decree probably only recognized and regulated already existing trade, but as of now there is no further evidence to establish how long this may have been underway. Khumbu oral traditions suggest that it predates the establishment of Nauje in the early nineteenth century, for there are accounts that the Chorkem area near Nauje was formerly a place where Sherpas and Rais met to exchange Tibetan salt and grain grown at lower altitudes. Ortner relates an oral tradition that describes an encounter by the early Sherpa settlers with Rais in Pharak in which they exchanged Tibetan salt for flour (1989:37–38). Detailed insights into early trade, however, are only possible after the mid-nineteenth century. Oral traditions and oral history offer a wealth of information about late-nineteenth- and early-twentieth-century trading practices.

Khumbu trade in the late nineteenth and early twentieth centuries revolved primarily around the Khumbu crossbreed trade and the exchange of Nepal iron, paper, and butter in Tibet for the Tibetan salt and wool which Sherpas needed in order to obtain lower-altitude-grown grain. In this era Khumbu traders apparently seldom ventured far into Tibet. Most primarily traded in Ganggar (fig. 17). Early in the century much iron was exported over the Nangpa La to Tibet and Khumbu traders obtained this either directly at the Those mines or from Shorung Sherpa middlemen who transported it as far as Nauje. Butter was primarily obtained from Shorung Sherpas who carried it to Khumbu in order to barter it for salt. These Shorung Sherpas used a now virtually abandoned summer trail that provides a direct route from the Junbesi area to Pharak via high summer pasture areas. Another important trade good was *shere,* a root gathered in Shorung that was used in the manufacture of incense. Paper came from the Hongu valley and other southern areas and was manufactured from the bark of the daphne shrub. Butter, paper, and shere were all largely exported for use in Tibet's monasteries.[8]

The degree to which Khumbu Sherpas were involved in trade to Kalimpong or Darjeeling in the nineteenth century is still unclear, but was very probably relatively minor. Sherpas began seeking income opportunities in Darjeeling by the 1880s and may have been frequenting the place still earlier. According to the 1901 Darjeeling district census, the first one taken, there were already 3,450 Sherpas living in Darjeeling (A. J. Dash, *Darjeeling,* 1947, cited in Ortner 1989:160), and they may have begun emigrating there any time after the British took control of the former Sikkimese village in 1835 and began developing it into a hill

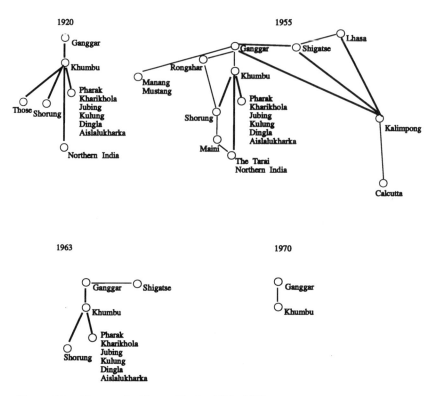

Figure 17.   Changes in Sherpa Trade, 1920–1970

station, a tea-producing region, and the summer capital of the government of India. How many of these people were Khumbu Sherpas is not
known.[9] It appears likely that most Sherpas who made the move worked
as laborers on road- and rail-building crews and in construction, porter,
and rickshaw work (Mason 1955:157).

The iron trade declined early in the century (Fürer-Haimendorf
1975:62), presumably eclipsed by British export of iron over the new
road from Kalimpong into southeastern Tibet. Although a small amount
of iron continued to be traded over the Nangpa La even into the 1950s,
primarily in the form of agricultural implements, trade had declined
considerably by the 1930s.[10] The livestock trade may also have declined
slightly by the 1930s, as is suggested by the end of the keeping of large
nak herds in Khumbu. The business of driving dimzo to Tibet, however,
was still lucrative, and a few of the really prosperous Nauje traders of
the 1940s and 1950s made their fortunes on dimzo. Salt also remained an
important commodity until the early 1960s. Here too, however, a decline in the scale of the business was apparent as early as the 1930s

(ibid.). Increased salt imports from India had apparently begun to cut some of the demand for Nangpa La salt in the southern edge of its former range. There remained considerable demand for Tibetan salt, however, and it was the main source of prosperity for the wealthiest Khumbu traders through the 1940s and into the 1950s. A few big salt traders, primarily Nauje men, still made fortunes based on salt during the 1940s.

Throughout this period and until the mid-1960s the small-scale barter trade of salt for grain continued to be the most important trade of all for most Khumbu Sherpas. These families lacked the resources to transport salt on the scale of entire caravans and often dealt in just one or two loads. Many earned the cash to purchase a load or two of Kaprak or Ganggar salt by working for the big salt traders as porters. Others obtained their salt in Khumbu itself on loan, either getting the money to buy it from the Tengpoche monastery or being advanced salt by the big Nauje traders who would take their repayment later in grain. Most Sherpa families spent the winter months transporting this salt south into lower-altitude areas of Nepal as far east as Dingla (ten days from Nauje), as far west as Those (six days west of Nauje) and as far south as the confluence of the Dudh Kosi with the Sun Kosi (ten days from Nauje). These lower-altitude areas were unsuitable for yak, and few Khumbu Sherpas had any means to move salt or grain other than by foot, hauling thirty- to sixty-kilogram loads up and down valleys in baskets. Entire families joined in the work.[11] Families made a number of shuttle trips to barter salt in the villages of Pharak, Katanga, Shorung, and the Rai country.[12] Usually salt was bartered for grain, especially the cheaper maize, wheat, and millet. Those who could afford rice bartered for it rather than for the less-preferred grains. Families often went to the same villages year after year to trade. There they stayed with friends and benefited from their help to trade with their fellow villagers. Often traders established bonds of ceremonial friendship with these southern families, and such bonds were sometimes made with non-Sherpa Rais as well as with Sherpas. The exchange rate in the south varied considerably during the past century and in general the amount of grain offered for salt declined through this period. Yet the relatively small amount of salt required to obtain a given amount of grain in Kharikhola or Aislalukharka as opposed to what had to be given up for that much grain in Nauje made all the difference for families who had very few resources. In the south in the 1950s one might get three pathi of rice for a forty-kilogram load of salt when the rate in Nauje was only two and a half pathi. This difference in comparative value also led many Rais to transport grain to Khumbu to sell to families who were well-off and who had no interest in the back-breaking work of hauling salt and grain on foot all winter.[13] Rais have been coming to Khumbu for a long time to trade grain for salt. There are many oral traditions from the late nineteenth century that attest to this and to the way in which particular Rai families dealt with the same Khumbu family every year. Some of these Rais went to Thamicho, Khumjung, and Kunde to trade their salt, but most dealt in Nauje. The Indian explorer Hari Ram noticed a large number of Rais headed to Khumbu when he was traveling up the Dudh Kosi valley in 1885 and commented that much rice from the rich rice-producing Ra Khola and lower Dudh Kosi valleys ended up in Khumbu (Rawat 1973:162, 164). Elderly Thamicho residents remember that early in the twentieth century the Rais were so interested in Tibetan salt that if a sufficient supply was unavailable in Thamicho the Rais would wait for Sherpas to go to Tibet for more, camping out in Khumbu in the meantime. Many Rais came to Khumbu for salt every autumn and it was not uncommon for them to establish a trading relationship with a particular Sherpa family that lasted for several generations.

It appears as if the overall volume of trade to Tibet over the Nangpa La may have been greater during the first forty years of the twentieth century than it was during the late nineteenth century, although there are no reliable records to support this impression gained from oral history accounts. It seems that during the twentieth century more Khumbu Sherpas began trading at the scale that involved transporting bulk goods north with caravans of yak. Ortner has also suggested that an increase in trade from Solu-Khumbu in general took place during this period and has linked it to economic growth in Nepal and Tibet sparked by British colonial development and trade activities and by independent economic growth in the Ganggar (Tingri district) region of Tibet (1989:105–108). I am not so certain that the impact of the British on Sherpa trade fortunes was significant. I believe that it would be extremely difficult to document that the British Raj's economic and political policies had an impact on the Tibetan demand in Ganggar and Shigatse for butter, paper, shere, dimzo, and other goods traded by Sherpas (although they may have had with iron) or that they affected the supply and cost of these items in the regions of Nepal where Khumbu Sherpas obtained them. It may be that increases in trade, if there were indeed increases, were not the result of sudden new Khumbu wealth from Darjeeling connections or a vast new demand in Tibet for Sherpa trade goods, but rather the result of a general, moderate growth in the population and economy of both Nepal and Tibet, Khumbu Sherpas' increasing interest in different types of trans-Himalayan trade, and the gradual (often multigenerational) family development of the requisite experience, knowledge, contacts, and capital. Certainly from the 1940s to 1960s there were many examples of traders who over the course of their careers went from very small-scale, relatively local trade to operating on the largest scale in the most lucrative and highly capitalized long-distance forms of trade.

In the 1950s a new generation of traders began to rise to prominence, many of them the sons of the great traders of the previous two decades. More families than ever seem to have begun trading over the Nangpa La, and in particular a high percentage of Nauje families who previously had just dealt in salt and grain between Khumbu and the lower regions of Nepal entered into the international trans-Himalayan trade. More traders seem to have been attracted to the import of luxury goods from Tibet and the large-scale export of paper, butter, and other goods not only to Ganggar but beyond to Shigatse. More traders went still farther afield as well to trade in Lhasa, Kalimpong, and Calcutta. The iron trade was over by this era and salt was not the business that it had been. There was still enough profit in salt and zopkio, however, that some traders who were familiar with these products and the necessary routes, markets, sources, buyers, and organization of transport remained content to continue with them. There was bigger profit, though, in trading butter, paper, and shere, particularly if one was willing to take it beyond Ganggar. The big traders of the 1950s set out from Khumbu with caravans of up to ninety yak loaded with these goods bound for Ganggar and Shigatse. Shigatse, second only to Lhasa as a Tibetan urban center, was reached either by a journey with pack yaks to Lhartse and then a five-day river journey by yak-skin and willow coracles or by a twelve-day mule-train journey by a more direct but slower land route. Both had their dangers: brigands, heat, wind, and poor grazing were problems on the overland route, whereas the river route ran the risk of a raft wreck. One Nauje trader remembers too well a cargo of twenty-six yak loads of paper lost on the river. But prices were better in Shigatse, double what were paid in Nauje for butter, paper, and shere delivered there on contract from the south.[14] Some traders who were unable to sell out in Shigatse or were as interested in pilgrimage as in business carried on to Lhasa, by the late 1950s only a day away by truck on the new Chinese road from Shigatse.

During the 1950s a few traders, mainly from Nauje and Khumjung, extended trade further still, some leaving Khumbu on trading trips that only brought them back to Khumbu many months later. This trade mainly focused on Kalimpong. Traders would go north out of Khumbu with yaks and porters, trade in Ganggar, continue on by mule or boat and sell out their Nepal goods in Shigatse, take their cash and travel by foot to Kalimpong, buy trade goods there and return for more business in Shigatse, and finally head back for Khumbu with a few loads of jewelry, carpets, porcelain, silver, Chinese brocades, tea, and other precious things. In Kalimpong they bought Bhutanese silk (*bure*), Indian red and green wool for Tibetan boots (*gonam*), incense, tobacco, turquoise, amber, zi stones, dyes, and cotton for trade in Tibet. Some traders went from Kalimpong to Calcutta by train in order to get lower prices and a better selection of some of these goods. A few Sherpas worked the Shigatse to Kalimpong trade for years at a time, returning only very infrequently to Khumbu. They took wool, horses, and other goods from Tibet south to sell in Sikkim and Kalimpong for cash to buy trade goods in demand in Tibet.[15]

The Transformation and Decline of the Tibet Trade   The place of trade in Khumbu economy declined dramatically during the 1960s. Several factors were involved. Probably the most important was the major political and economic change in Tibet following the flight of the Dalai Lama to India and the establishment of full-fledged Chinese administration. This affected the supply of trade goods and transport in Tibet, Tibetan demand for imports, Sherpas' freedom of movement within Tibet, and Sherpa traders' ability to maintain contacts and contracts with their Tibetan trade partners. But the war between India and China was also significant, as was the less dramatic but crucial undermining in Nepal of the demand for Tibetan salt due to the increasing availability of Indian salt and the gradual erosion of cultural biases against it. By the end of the 1960s trans-Himalayan trade was still being carried out by only a few small-scale traders. Today fewer than fifty men still go to Tibet each year, most of them either recent Tibetan immigrants, Bhote Kosi valley residents, or aging Nauje traders. Only one of the Nauje Sherpas who traded during the early 1960s was still going to Tibet in 1991 and he has not been able to interest his son in carrying on the business.

The transformation began in 1959 when the Dalai Lama and more than 80,000 Tibetans fled to India and Nepal. There was economic chaos in Tibet. Some Sherpa traders suffered considerable business losses when their goods were confiscated by Chinese soldiers or their Tibetan trading partners fled without fulfilling their contracts. As many as 5,000 Tibetans poured into Khumbu with their livestock, and for the next two years salt, sheep, and wool were so cheap in Khumbu that there was no need to make a journey to Tibet to obtain them. Many Sherpa built up substantial herds of yak overnight, so cheaply were these available, and gorged on one-rupee sheep. By 1962, however, the Tibetans had moved on to new refugee camps in Shorung, and Khumbu Sherpas were discovering that trade conditions in Tibet were considerably changed. Salt was scarce in Tibet for several years. The trade system that had brought salt from the Changtang lakes to the border region had apparently been weakened by the massive population movements and socioeconomic reorganization in Tibet. Even when small amounts of salt could be obtained it had become more expensive and difficult to arrange transport south from Ganggar, for Tibetan yaks were not available for hire as readily as they had once been. Meanwhile villagers in the country to the south of Khumbu were overcoming their initial resistance to using the Indian salt that was then becoming more easily and cheaply available.[16] By the time that Tibetan salt became more available there was no longer the market for it that there had been, although in Khumbu itself it remains preferred for many uses to this day. By the mid-1960s the salt trade to the Nepal midlands was finished.

At the same time long-established trade patterns were also being undermined in other ways. In Tibet the demand for some of the main Sherpa exports was also declining. During the 1960s and early 1970s thousands of Tibetan monasteries and temples were destroyed, and even before that many of them had ceased to function on their former scale after the mass exodus of 1959–1960. With the decline of monasticism there was a much diminished market for butter, paper, and shere. The war between China and India in 1962 led the Indian government to seal the border with Tibet and this abruptly blocked trade between Tibet and Darjeeling, Kalimpong, and Calcutta. And the terms of trade were beginning to change even as close to home as Ganggar. Sherpas were still free to come and go in Tibet, but the market for Khumbu livestock declined as collectivization spread and the Chinese encouraged local breeding of zopkio, banned the export of nak from Tibet, and limited wool exports.

Contrary to popular portrayals, however, Sherpa trade did not totally cease as a result of these difficulties. Some of the most prosperous Sherpa traders of recent times began their careers after 1959, whereas others who had been trading for many years continued to do so. Traders adjusted their scale of trade, product emphasis, and routes to new conditions. The eastern sphere of Sherpa trade, including Lhasa and Kalimpong, was abandoned, but Sherpas continued to trade in Shigatse until 1967. In 1962 a group of Sherpas drove more than 200 dimzo west across Tibet and into the Manang, Mustang, and Thak Khola regions north of the Annapurna range in central Nepal. Most trade in these days shifted from bulk goods to objects of great value and light weight. Some traders specialized in carrying artwork and religious objects out of Tibet. And in Tibet there was demand for certain luxury items, some of which may have been coveted by Chinese military personnel. Some Sherpas did a good business in Swiss watches.

The spread of the cultural revolution in Tibet, with the consequent suspension of normal law and order, made it dangerous as well as difficult to trade by the late 1960s. A few men tried to continue trading in Shigatse, traveling only under the cover of darkness. But they soon gave it up as too much of a risk. Many gave up trading altogether and few went beyond Ganggar. Some traders also experienced problems with the army and there were cases when goods were confiscated by border patrols. The Chinese had by now officially limited travel with a regulation that prevented Sherpas from going beyond the Ganggar area. Ganggar itself became off-limits once the Chinese had established a small trading post at Lungtang a few kilometers south of Ganggar. Sherpas who wanted to conduct business in the Tingri region had to spend their days and nights in this compound while they conducted business and to pay in the meantime for fodder for their pack stock. Business had to be carried out at fixed prices set by the Chinese, whose policies limited the goods available and those that were accepted in exchange. It was possible to trade with Tibetans, but only under Chinese supervision. If one wanted to sell a dimzo it was necessary to obtain permission first from an official in Ganggar. Prices were not what they had been nor was the quantity or quality of Tibetan goods. Yet still some money could be made by importing wool, Chinese tennis shoes, thermoses, flashlights, and other goods to Khumbu. Buffalo hides and some Indian goods were in demand in Tibet, and there was still profit in selling dimzo even when the only approved buyer was a government office. Only a few Khumbu men continued trading, but there was never a time when trade halted altogether.

China began easing trading restrictions between Tibet and Nepal in the early 1980s. Tibetan traders reappeared in Khumbu in 1983 for the first time in twenty years, and in December 1984 a group composed of Shato, Nejung, Penak and Kura men arrived in Khumbu with seventy-two yak laden with wool. Sherpa traders reported being able to barter now at the trade depot rather than having to accept fixed prices there, and they found that wool had become more available. Some had success in trading in the villages, although this was still technically not allowed. A few Khumbu Tibetan refugee traders began trading as far afield as Shigatse and a few Sherpas and Khumbu Tibetan traders began bringing a few more dimzo north. The refugee traders found it possible to sell crossbreeds again directly in the villages, although they were not clear on whether this was entirely lawful. One Nauje Sherpa family twice took more than 100 head of zopkio into Tibet for sale during the 1980s, but the total number of zopkio traded has remained far below 1950s' levels. Most of the current trade is in the hands of Tibetan families who settled in Nauje after 1959 or Bhote Kosi valley Sherpas who take advantage of their proximity to the border and large yak herds to import salt to Khumbu.

The scale of the Tibet trade today is very small. There is some demand still for Tibetan salt in Khumbu itself, but little is carried any further south. Wool is still good business and there is continuing demand in Tibet for zopkio, water buffalo skins, and some Indian and Kathmandu products including polyester, dye and snuff. The days of the big traders seem to be over, however. Few traders now are being succeeded by their sons, who find the money better and the life easier in trekking and mountaineering. Only Thamicho men still cross the Nangpa La in any numbers—fewer than thirty of them make the journey—and only one of the Nauje Sherpa traders is still going north. Unless new trade policies change current conditions in favor of trade over the remote passes such as the Nangpa La rather than via the main road between Kathmandu and Tibet, the long-vital Tibet trade seems destined to be nothing more than a dwindling relic in the Khumbu economy.

The New Regional Market System   As the old international system of trade declined during the 1960s a new form of economic interaction was developed in Nepal as a result of government initiative. This was based on the tradition of periodic markets. In some parts of eastern Nepal there were long-standing periodic market systems, in which a once-a-week, open-air market was held at certain designated sites in a number of localities on different days of the week. Traders circulated through a region selling their wares at a series of different periodic markets, whereas farmers and herders now had a convenient outlet at which to locally market their surplus and entrepreneurs found that shops and teashops could profitably cater to the market crowds.

There had not previously been a periodic market system in the Dudh Kosi valley or anywhere in the Solu-Khumbu district where traditionally trade had entirely been conducted by individual traders. In the mid-1960s, however, weekly markets were established in Solu-Khumbu at government instigation. These included weekly Saturday markets at Dorphu (later Naya Bazar), near the district center of Salleri, and at Nauje as well as additional weekly markets at Aislalukharka in nearby Khotang district and at Olkadunga in the Olkadunga district. The Nauje market was initiated in 1965 and was soon established at its present site at the southwest edge of the village. The operation of this market soon totally replaced the earlier salt for grain barter trade that had been such a fundamental part of the household economy of most Khumbu families. By the late 1960s no Sherpas were going south in the winter any longer to trade salt for grain. Rai farmers and traders, however, continued to carry grain to Nauje as they had for more than a century, and in larger quantities than ever before. But after the establishment of the market they now no longer bartered rice and other grains in Nauje for salt, but instead exchanged them for the rupees Sherpas had begun to earn in large amounts from tourism. The high prices at Nauje, where grain can be sold for triple what it would bring in Kathmandu, continue to bring many Rais and much rice into Khumbu. Today hundreds of hawkers converge on Nauje each Saturday with their wares, primarily rice and other grains, but also dairy products, eggs, fruit, vegetables, kerosene, cooking oil, and large amounts of candy, cookies, soft drinks, mineral water, and other goods destined for the tourist market. Goods are carried to Nauje from the road-ends at Jiri and Katari, respectively a week and ten days distance for a heavily laden porter, and rice is brought from places as far as ten days away.

Trade and Land Use   Trade long made a few Khumbu Sherpas rich and provided families throughout the region with important supplies of grain that freed them from any necessity to achieve self-sufficiency in crop production. This had important implications for Khumbu agriculture which could be specialized to a degree otherwise impossible. The ongoing Sherpa involvement in extraregional trade networks made possible the increasing reliance on monoculture potato cropping, which supported more people than any other use of Khumbu arable land could. The ability to gain grain through trade removed the necessity to devote large areas of cropland to buckwheat. It also rendered unnecessary subsistence strategies based on the cultivation of a combination of high-altitude and middle- or low-altitude fields. Khumbu Sherpa trade activities in lower-altitude regions made it unnecessary for them to acquire land there and grow winter crops.

Very little Khumbu-grown produce was ever traded outside the region. A few dried potatos (*shakpa*) were exported to Tibet and a few seed potatoes and dried radishes were sold to Rais in Nauje, but this was incidental. Lack of participation in a wider market economy has meant that the selection of crops and crop varieties has remained based on local preferences rather than on what Rais or Tibetans would buy. Khumbu crop production remained linked to local assessment of what is best to grow in terms of taste, yield, and risk in local environmental conditions.

Trade influenced pastoralism more profoundly. Pastoralism in Khumbu (or at least cattle raising) was not a matter of subsistence production for many owners. Certainly livestock products were valued for home use, and there was status and pleasure in keeping yak and nak. But the size and structure of the nak herds revolved around the sale of crossbreed calves to a significant degree, and changes in Khumbu pastoralism—including the major decline in nak keeping in the 1960s—may often have been related to changes in zopkio trade opportunities. The high degree of emphasis on nak raising and the characteristic herd size associated with it reflect the requirements of crossbreed production for trade. The striking historical lack of emphasis on zhum as compared to cows may also be linked to trade conditions, in this case to lucrative opportunities to sell zhum calves to Shorung herders. The importance of yak keeping was also historically linked to trade, for the main use of the yak as a pack animal was for Tibet trading. When owning large herds of yak for transport became unnecessary for trading activities early-century Nauje traders ceased to keep them. These trade-related changes in the scale of nak and yak herding have also affected use of pastures. Regional decline in the numbers of yak and nak probably decreased grazing pressure on some of Khumbu's high-altitude summer pasture areas.

Tourism in Nepal

Nepal was once one of the world's most remote and most difficult to visit countries. Until 1949 the Nepal government rigorously excluded foreigners from traveling outside of Kathmandu and other than by the most direct route to the capital from India. Beginning in 1949 a few mountaineering expeditions were allowed to visit remote parts of the country to attempt ascents of the high peaks, but until 1955 Nepal continued to adhere to a policy of excluding foreigners which dated back to the establishment of the kingdom nearly two centuries before.

With the coronation of King Mahendra in 1955 there came the beginning of a new attitude. In that year the first tourist visas were issued, the first hotel was opened, and the first visitors began to arrive. But tourism was slow to develop. In 1962, the first year in which a count was kept, only 6,179 tourists entered the country. Since then mass tourism has found Nepal and the Nepal government has responded with increasing interest in tourism development. During the 1970s tourism surpassed first rice exports and then Gurkha remittances to become the country's leading source of foreign exchange. Kathmandu's Tribhuvan airport was enlarged to accommodate jets, the government supported hotel development in the capital, and Nepal became a major tour destination. The pace of tourism development increased markedly during the 1980s with a profusion of new hotels opened in Kathmandu at every level from budget to luxury. The number of visitors climbed from 162,897 in 1980 to 265,943 in 1988 (Nepal. Ministry of Tourism, Department of Tourism [1989]:17), before political friction with India in 1989 and political unrest in Nepal itself in 1990 led to a slight decline in visitors. Twenty-seven percent of all the international travelers who visited Nepal in 1988 came from India, while the United States, the United Kingdom, West Germany, Japan, France, and Australia accounted for another forty-one percent (ibid.:13). The spectacular architecture, art, and living cultures of the Kathmandu valley continue to draw most visitors, but Nepal has also become a major adventure-travel destination during the past twenty years, offering mountain climbing and hiking, white-water rafting and kayaking, and wildlife viewing. Today the country is renowned especially for trekking, multiday hiking journeys on the extensive trail system that threads through what are still largely roadless mountains. Trekking offers a combination of adventure, closeup views of some of the most spectacular high country in the world, and direct contact with the land's diverse mountain peoples. These attractions have been skillfully marketed in Europe, North America, Japan, Australia, and New Zealand, and scores of trekking companies based in Kathmandu have prospered by catering to the international demand.

Tourism is today one of the leading sectors of Nepal's economy and a major focus of its development planning. Nepal continues to have one of the most pro-tourism policies of any country in Asia. The Ministry of Tourism favors further growth and to aid in this the government has recently relaxed its policies regarding foreign investment in hotels and transport. The official goal of current policy is to quadruple tourist arrivals to one million per year within the coming decade.

As Nepal tourism has increased in scale and diversified in scope, however, a number of potential problems have become visible. Thus far the economic gains from tourism have been highly localized, and in a large part investment and profit have remained largely restricted to the Kathmandu valley. The rise of mass tourism in Kathmandu has created small-scale "tourist ghettoes" of the "Freak Street" and Thamel areas and has supported the development of a smaller-scale "tourist trap" along scenic Phewa Tal lake in Pokhara. New social problems have developed in some areas, associated in part at least with tourism, including drug use, child labor, and professional begging. And a variety of economic, sociocultural and environmental problems have been reported in the scenic mountain hinterlands visited by trekkers. Only a relatively small number of Nepal's tourists take to the trails for multiday trekking holidays; in 1984 41,206 trekking permits were issued and the government estimated that about 15,000 tourists participated in trekking (Nepal. Ministry of Tourism 1985), and whereas by 1988 the number of trekking permits issued had climbed to 61,273, trekkers nonetheless constituted only 23 percent of the tourists who visited Nepal (Nepal. Ministry of Tourism [1989]:13, 50).[17] Even relatively small numbers of tourists, however, have had significant local impact on remote mountain areas and their inhabitants.

Khumbu has been one of Nepal's premier mountaineering and trekking centers ever since these forms of tourism developed, and today the Mount Everest region and its famous Sherpas attract about 20 percent of all trekkers and nearly half of all mountaineering expeditions.[18] The impact of tourism here has received more worldwide publicity than that in any other part of the country. At the same time Khumbu is the most striking Himalayan example of small-scale, locally controlled tourism development based on trekking tourism and mountaineering. Sherpas have responded to new opportunities in tourism with alacrity and adeptness and have prospered not only from earnings made from work as *sirdars* (responsible for managing porters, packstock, and camps), camping crews, cooks, and porters but also from their operation of strings of pack stock and owner-managed lodges, teashops, and shops. Sherpas have even begun establishing their own Kathmandu-based trekking agencies.[19] They have become one of the most affluent of all high Himalayan peoples as a result. The remainder of this chapter traces the increasing importance of tourism in the Khumbu Sherpa economy and describes the way in which it ultimately has replaced trans-Himalayan trade as one of the fundamental components of the regional economy and household subsistence strategies.

Early Sherpa Involvement in Mountaineering

Sherpas have been associated with tourism since 1907, when several were hired in Darjeeling as porters by mountaineers who were making attempts on peaks in Sikkim.[20] Within twenty years they had become world-renowned for their exploits on high mountains and particularly on Mount Everest. Sherpa high-altitude mountaineering porters soon developed a reputation for toughness, endurance, courage, and loyalty which made them preferred above all other Himalayan peoples as companions on the great peaks. They also began to earn considerable wealth as well as fame climbing on behalf of foreigners. Foreign mountaineering was forbidden within Nepal itself, but young Khumbu Sherpas could reach Darjeeling, the British hill station that was the major mountaineering center in the Himalaya until World War II, in only ten days of walking. It was from here that the British expeditions set out for Mount Everest, for in those days the only permissible approach to the mountain was via Tibet. Sometimes Khumbu Sherpas were able to make arrangements to sign on for these expeditions without going all the way to Darjeeling to do so. Several dozen Khumbu Sherpas, for example, simply crossed the Nangpa La and joined the 1933 Everest expedition at its Rongbuk (Rumbu) base camp. Some Sherpas lived this expedition life for only a few years, but others remained in Darjeeling for decades and settled there permanently.[21] Among these emigrants was former Bhote Kosi valley resident Tenzing Norgay, the Tibetan-born Sherpa who climbed Mount Everest with Sir Edmund Hillary in 1953. Those Darjeeling Sherpas who maintained ties with Khumbu sometimes returned money earned through mountaineering to Khumbu as remittances, and those mountaineers who ultimately returned to their native villages often did so as relatively wealthy men. This new money was used in old ways, to purchase land and livestock and to finance trade ventures. It appears to have had relatively little impact, however, on the regional economy before 1950. Trade remained the greatest source of Khumbu wealth.

Mountaineering Comes to Khumbu

In 1949 the Rana government allowed small teams of foreign mountaineers to explore several parts of Nepal and in 1950, the same year that a large French expedition climbed Annapurna, a small U.S.-British exploratory party received permission to travel to the Mount Everest area. Among the members of this group were two renowned mountaineers, Charles Houston, of K2 fame, and William Tilman. They reached Khumbu on foot from India, arriving in Nauje on October 1, and during the next two weeks camped their way across Khumbu to the Tengboche monastery and on toward the foot of Mount Everest. Tilman and Houston ascended a ridge of Pumori to a point at 5,545 meters which later became known as Kala Patar and thus became the first foreign visitors to enjoy what is now one of the most famous tourist vistas in the Himalaya (Tilman 1952). The route they had taken through Khumbu would one day become one of the major trekking thoroughfares in Nepal.

The following year a British team led by Eric Shipton (which included New Zealander Edmund Hillary) arrived for a detailed mountaineering reconnaissance of Mount Everest and made the first ascent of the dangerous Khumbu icefall. In 1952 Shipton returned, this time for an attempt on Cho Oyu, which was considered a training mission for the 1953 Everest expedition. The Swiss also were in Khumbu in 1952, sending large mountaineering teams for both a premonsoon and a postmonsoon attempt on Everest. The Swiss reached the south Col with the assistance of hundreds of Sherpa porters and a team of talented Sherpa high-altitude porters. Tenzing Norgay and Raymond Lambert very nearly reached the summit before abandoning the attempt. The British tried again in the spring of 1953 and, as all world soon learned, Edmund Hillary and Tenzing Norgay reached the summit on May 29, 1953. This achievement by no means ended foreign mountaineering interest in Khumbu, and for the next eleven years there were generally one or two

European, American, or Indian expeditions each year that ascended the valleys en route to Everest, Lhotse, Cho Oyu, Ama Dablam, or other peaks. The worldwide publicity these mountaineering expeditions focused on the area and its people brought other foreigners as well. Special permission had to be obtained to visit the area from the government of Nepal, but nonetheless a few anthropologists and journalists began finding their way to Khumbu along with several expeditions searching for the yeti. Tourism, however, in the sense of a stream of individual or group sightseers, had not yet reached Khumbu.

Most of the early expeditions were massive affairs that had dozens of climbers and employed hundreds of porters. The 1952 Swiss team hired 163 porters, the 1953 British Everest expedition 450 porters, and the 1963 U.S. expedition to Everest a full 900 (Dittert, Chevally, and Lambert; 1954; Hunt 1954; Unsworth 1981). Wages were high compared to the money offered for agricultural day labor. Base camp porters made as much as seven times the daily wage of field workers and the high-altitude porters who carried loads up onto the mountain itself were still better paid. Work, however, was limited. Expeditions were few, and for years they only came to the region for premonsoon spring climbing. Each required help only for a few weeks to establish and later abandon its base camp. Only the high-altitude porters who stayed with the expedition climbers throughout their time on the mountain had the opportunity to make a good deal of money. For other Sherpas the arrival of an expedition was simply a chance to make a little money more easily than could otherwise be done locally. Many families turned out every family member old enough and fit enough to carry a load. From Nauje to Everest base camp took four days, and if one walked all the way to Kathmandu to meet the expedition there was perhaps another twelve days' work. It was not enough money to live on or to finance a major trading venture, but the income might mean that one need not work as an agricultural day laborer that year or as a porter for a Tibet trader at low wages in order to be able to buy salt for the next winter's grain trading. Pangboche villagers note that during these years they ceased to serve as porters across the Nangpa La for Nauje traders as they had formerly done (Stevens 1983:25–28, 1988b:72–73).

The Trekking Era

The government of Nepal banned foreign mountaineering expeditions from 1965 until the spring of 1969. Ever since then, however, mountaineering has flourished. The Ministry of Tourism handles permits for 104 peaks that are open for summit attempts by foreign mountaineering expeditions and collects peak fees.[22] Between the autumn of 1987 and the spring of 1990 there were an average of 101 expeditions per year, of which 43 percent climbed in Khumbu according to records kept by long-time Reuter's Kathmandu correspondant Elizabeth Hawley (personal communication). In some years more than fifty expeditions have attempted Khumbu peaks. In the spring of 1990 alone there were twenty-eight expeditions in Khumbu, including six Mount Everest expeditions and four Pumori expeditions, with a total of 163 foreign mountaineers and 136 high-altitude porters. Some of the expeditions of recent years have been large-scale affairs like many of the expeditions of the 1950s and 1960s; for example, the autumn 1988 U.S.A.-U.K. Everest expedition had eleven mountaineers and twenty-eight high-altitude porters. Many expeditions, however, are considerably smaller and employ only a few high-altitude porters, many of whom are non-Khumbu Sherpas and Tamangs. Over the years many Khumbu Sherpas have died in mountaineering accidents, and many young men and their families have decided that less dangerous careers in trekking are preferable to the glory and financial rewards that can come from climbing. Most of the Khumbu Sherpas who climb today as high-altitude porters are from Phurtse and Pangboche. There has also been a major decline during the 1980s in the numbers of Khumbu Sherpas who work as porters for mountaineering expeditions. Most porters today are Rais or other lowlanders. There is also a general trend away from moving gear by porters to using pack stock. Khumbu Sherpas have profited considerably from this.

During the mid-1960s, when mountaineering was banned, the Nepal government opened several areas of the country's mountains, including Khumbu, to visits by ordinary tourists. In 1964 it became possible to obtain a permit to walk along certain designated routes into areas far from Kathmandu. A new tourism industry began to develop to offer hiking tourists multiday camping trips in the mountains, and Himalayan trekking was born. Trekking companies such as the Kathmandu-based Mountain Travel began to offer guided camping tours of the highlands to adventurous, middle-class Europeans and Americans, smoothing their way with trained staff who catered to them in a style long-since developed on Indian Army marches and on mountaineering expeditions in Tibet and the Himalaya. These trekking groups traveled about the land like small, self-contained armies, with their porters to carry tents and provisions and a local crew to cook and handle camp chores. Among the authorized routes was that leading from the Kathmandu valley to Mount Everest, a journey of two weeks or more in each direction across more than 200 kilometers of rugged hill country and a series of seven passes. The 1964 building of a small, STOL airstrip at Lukla by a team associated with Sir Edmund Hillary's Khumbu aid projects, however, had already provided an easier way to approach Khumbu. With Lukla only a thirty-five-minute flight from Kathmandu, one could be in Nauje the next morning. The first small commercial trekking group arrived in Khumbu in February 1965.

Trekking did not, however, immediately make the fortunes of its pioneering entrepreneurs. In the late 1960s it was an idea still a few years ahead of its time, and fewer than 300 of the nearly 35,000 tourists who visited Nepal in 1969 cited trekking as their purpose for visiting the country. But skillful promotion and a boom in "adventure travel" in the U.S.A. and Europe made trekking by the mid-1970s the big business by Nepalese standards which it has been ever since. By 1979 more than 25,000 trekking permits per year were being issued. Many of these tourists had the villages of Khumbu and a glimpse of Mount Everest as their goal. In 1964 only twenty tourists visited Khumbu (J. Fisher 1990:148). Seven years later tourism had increased to the point where during the fall season of 1971 and the spring season of 1972 a record 1,406 visitors arrived. The following year more than double that number came to Khumbu. Table 28 below illustrates the dramatic subsequent increase in the number of tourists.[23] By the end of the 1970s more than 4,000 visitors per year were touring the area and in 1980 the total surpassed 5,000 for the first time. Since then the number of tourists entering Sagarmatha National Park has risen to more than 8,000 per year. This is a small number relative to tourist levels in mountain regions in California, Colorado, the Alps, or Peru, but it is substantial relative to Khumbu's population. Most tourists, moreover, travel along a narrow corridor to Mount Everest passing through an area that is inhabited by fewer than 1,000 Sherpas.

In contrast to earlier mountaineering tourism the increasing number of trekkers made possible a fundamental change in the Khumbu economy. Khumbu Sherpas became very involved in the trekking industry. For a number of years they secured a virtual monopoly on jobs throughout Nepal as camp crews, cooks, and sirdar, while many more worked in Khumbu as porters. The work was very nearly as well paid as expedition work and was not only considerably safer than climbing but was more widely available. Although it was possible to work for most of the year if one was interested in doing so, most Sherpas worked only during the four or five peak months of the tourist year, especially during October-November and April-May. During the late 1960s and early 1970s many families took advantage of the new opportunities for income, and prestige began to become associated with employment with particular companies as well as with particular jobs. By the late 1970s a study of household involvement in tourism in four Khumbu villages (Nauje, Kunde, Khumjung, and Phurtse) found that in three of the villages three-quarters or more of all families were involved in tourism, and that even

Table 28.   Number of Khumbu
Trekking Tourists

1971/72	1,406
1973/74	3,503
1975/76	4,254
1979/80	4,348
1980/81	5,310
1981/82	5,092
1982/83	5,066
1983/84	5,130
1984/85	5,840
1985/86	6,909
1986/87	7,834
1987/88	8,430
1988/89	7,683
1989/90	8,290

SOURCE: From Hardie et al. (1987:25) and Sagarmatha National Park. Data through 1975/76 from Central
Immigration Office, Kathmandu. Data after 1975/76
from Sagarmatha National Park entrance records.

Table 29.   Sherpa Tourism Involvement, 1985 and 1991

Village	Percentage of Households with Direct Tourism Income*	
Nauje	86	91
Khumjung	85	—
Kunde	96	—
Phurtse	49	58
Pangboche and Milingo	66	77
Thamicho Villages	81	—

* Income is included from mountaineering and trekking work, lodges,
shops, and pack animals. Not included are wages from carpentry and
other construction work or sales of art, handicrafts, agricultural products, or other goods to lodges, expeditions, and trekking groups.

in relatively uninvolved Phurtse nearly half of all households had tourism income (J. Fisher 1990:115).[24] During the next few years still more
families became involved until by 1985 65 percent of all Khumbu families had income from trekking, and in all the settlement areas other than
Phurtse and Pangboche the percentage of families with direct tourism

income (from all sources including pack stock) was above 80 percent. This degree of involvement in tourism is unparalleled in highland Nepal where it is more typical for the villagers of areas that are trekking destinations to secure very few trekking jobs.

These jobs, it should be noted, were almost entirely limited to work for men. For many years there were very few opportunities for women in trekking and even today there are only an extremely small number of Sherpa women who work as sirdar, camp staff (often called "sherpas" regardless of ethnicity), or cooks in the trekking business. More work occasionally as porters. These are mostly women in their late teens or early twenties, and usually they come from households in which there are no other family members who can make wage incomes from tourism. Women tend to avoid portering work whenever possible, except for a few young women for whom it is a welcome source of cash and adventure. During the past few years more young unmarried women have begun driving pack stock on treks, often with groups for which their father or elder brothers are employed as sirdar or camp staff.

With the increase in the scale of trekking in Khumbu came new Sherpa interest in establishing tourism businesses. In Pangboche it became big business to supply expeditions and trekking groups with fuel wood. Villagers from Nauje, Khumjung, and Kunde began establishing small lodges along the route to Mount Everest. A number of Nauje families (especially Tibetan refugees) opened shops. Many families in the southern Khumbu villages, those closest to the entrance and approaches to the region, began to invest in pack animals, for a single yak or crossbreed could carry double the load of a porter, and several animals could earn a single herder a considerable daily wage (Stevens 1983).

The growth in shops was limited primarily to Nauje. The first tourist shop was established in 1967.[25] Today twenty-one of the twenty-nine shops in Khumbu are located in Nauje, flanking the two main streets of the village.[26] Three-fourths of the Nauje shops mainly stock goods aimed at the tourist market, although many also sell some staple goods and clothing to Sherpas and a few specialize in sales to local customers. The goods offered to tourists consist of a very similar array of mountaineering clothing, food bought from expeditions or imported from Kathmandu, and Tibetan and Sherpa curios. A few shops have specialized in selling Tibetan handicrafts or renting and selling mountaineering gear. Both men and women operate shops and a number of Nauje shops are owned and operated by Tibetan refugees.

More significant for local development has been the proliferation of small inns. Sherpas, unlike some other Himalayan peoples such as the Thakalis, had no tradition of innkeeping. The first lodges of the early

1970s were nothing more than Sherpa homes with a signboard out front welcoming guests to a bed on the floor and the chance to join the family for Sherpa-style meals. The first Sherpa lodge, the Sherpa Hotel, was opened in 1971 in Nauje. In 1974 one of the more successful of Nauje's traders opened the Tawa Lodge, the first lodge built to be an inn. By 1979 a network of lodges was developing along the trails that lead to Everest, with concentrations occurring at the spots that had become popular overnight stops from the time of the first expeditions.[27]

The total number of Sherpa lodges in Khumbu was relatively minor even in the late 1970s. In 1978, for example, there were nineteen in the region.[28] During the following five years the number of lodges more than doubled to forty-five. Most of these were built along the route to Kala Patar, and many at already popular sites on that path. Nine new lodges were opened, for example, in Nauje and three in adjacent Chorkem. Some lodges, however, began to be developed in new sites between the main overnight stopping places, and secondary development along the Gokyo route became more significant with six new lodges.[29] In the past few years both these trends have continued. The current regional pattern is indicated by map 21, which illustrates the location of the lodges opened by 1991 relative to the main trekking trails. Note the continuing small number in the Bhote Kosi valley north of Nauje and in the villages of Khumjung, Kunde, and Phurtse. Between 1986 and 1991 the number of lodges increased from sixty-two to eighty-one, many of them relatively simple lodges opened by Pangboche families in Pangboche, Dingboche, and Chukkung. The first Phurtse lodges were also opened. There were by this time also thirty teashops above Nauje which catered primarily to porters and trekking staff. In Nauje there was a score of such tea and beer shops, but many of them primarily did business with the Saturday market crowd and a number were only open for that event. Six more large Nauje lodges were under construction, two newly built ones were being operated, perhaps temporarily, as teashops, and two lodges were for the moment not being run due to extraordinary family circumstances. These ten establishments are not included in table 30 on page 365.

The growth of the local lodge system was supported by a new kind of tourism, independent trekking. The early lodges had a clientele of hardy souls interested in closer interaction with Sherpa families, a more authentic village experience, and freedom from organized group travel. Word of the possibility and delights of lodge-hopping through Khumbu (including the fact that even in 1990 it could be done for five dollars per day) spread by word of mouth, guidebook descriptions, and travel articles. A greater number of travelers came to the region on their own (among them many people who had originally toured the area with a trekking group) and virtually all of these independent trekkers traveled without

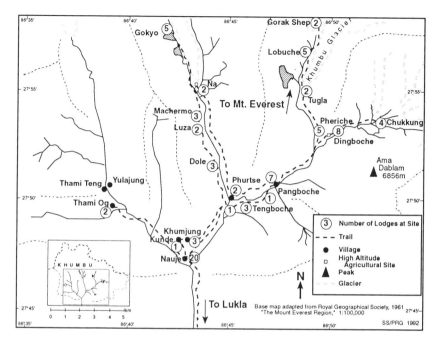

Map 21.   Sherpa Lodges, 1991

Table 30.   Sherpa Lodges, 1973–1991

	1973	1978	1983	1986	1991
All Khumbu	7	19	47	62	81
Everest route**	7	17	36	47	58
Gokyo route***	0	2	7	10	17
Other	0	0	1	4	6
Nauje and Chorkem	3	6	18	20	20

   * These figures include all lodges that primarily catered to tourists. Teashops that primarily offer food and drinks to Sherpa staff, porters, tourists, and local people, but that do not offer accommodation are not included.
   ** The Everest route here refers to all lodges along the main trail from Nauje to Gorak Shep.
   *** The Gokyo route here refers to all lodges above Nauje and Khumjung en route to Gokyo in the upper Dudh Kosi valley.

tents. By 1988 more trekking permits were issued to individual travelers to visit Khumbu (5,698) than were handled by Kathmandu trekking agencies (5,668) (Nepal. Ministry of Tourism ca. 1989:52–53).

    Increasing numbers of guests, income, familiarity with hotel-keeping practices in Kathmandu and elsewhere, and money and advice from foreign friends enabled Sherpas to begin developing more sophisticated lodges during the 1980s.[30] These have primarily been developed in Nauje where they have virtually captured all business from those earlier lodges that have not upgraded. The new lodges bear little resemblance to Sherpa houses—the main Sherpa element in the architecture, indeed, is merely the use of stone wall construction. Lodges now are multistoried, and three of the new Nauje lodges have four stories. Dormitory sleeping and dining rooms are standard and some lodges are beginning to add private double rooms. In 1983 the first lodge opened in Nauje with a top story dining room with large windows surveying the village and peaks. By 1986 there were seven, and one had been opened at the Tengboche monastery (Stevens 1983, 1988b). In the spring of 1991 there were eleven such lodges in Nauje, with two more currently being operated as local teashops and another two under construction. There were also two in Pangboche and one in Kunde as well as the earlier one at Tengboche. Some of these lodges were primarily financed with money saved from years of trekking work and smaller lodge operations, but most of them enjoyed substantial support from foreign benefactors in the form of gifts and loans. U.S., Japanese, and European money has helped build lodges in several Khumbu localities.

Another outstanding feature of the changing economy of Khumbu during the tourism era has been a tremendous increase in the number of livestock kept as pack animals. Until the 1970s yak were the most popular pack animals. In recent years, however, urang zopkio have become the principal means of moving expedition and trekking loads. Urang zopkio are preferred over yak for trekking and mountaineering work because they are able to haul gear on the Lukla to Nauje route as well as in the higher valleys. Sherpas believe that yak are not suited for the trip to Lukla because this requires them to be taken to altitudes below 3,000 meters and it is feared that the animals are susceptible to serious disease at such low elevations. Urang zopkio labor under no such handicaps and they are enormously profitable. A single urang zopkio carries two loads and earns more for his owner than a Sherpa earns per day as a porter. At 3,000 rupees (more than a months' salary for many sherpa), purchasing an urang zopkio is a substantial investment, but is a far more affordable one for most Khumbu families than establishing a shop or lodge. Urang zopkio keeping has become one of the most popular forms of investment in Nauje, Khumjung, Kunde, and the Thamicho villages. Families that have one or more members working as sirdar on Khumbu treks are especially likely to invest in urang zopkio. Since sirdar make the arrangements for pack animals and porters their own animals are assured of considerable work. By 1986 70 percent of Nauje families with sirdar positions owned urang zopkio.

The total increase in urang zopkio numbers in Nauje is astounding. In 1963 only five families kept urang zopkio pack stock totaling fewer than twenty head.[31] By 1986 there were more than 140 head owned by fifty-two households.

The Transition From Trade to Tourism

After visiting Khumbu in 1971 Fürer-Haimendorf came away with the impression that trade had more or less ceased after 1959 and that the resulting void in the Khumbu economy had fortunately been promptly filled by the rise of tourism. He suggested that:

> Had the Chinese stranglehold on the Sherpas' trade occurred even twenty years earlier the effect on their standard of living would have been catastrophic. Indeed it might well have caused a depopulation of the region of high altitude where farmers and pastoralists can only subsist if their income is supplemented by outside earnings. Fortunately for the Sherpas the checks on their trading activities imposed by political events in a neighbouring country coincided with the opening of Nepal to foreigners and the subsequent development of mountaineering and tourism which soon became major sources of income benefiting the inhabitants. (1975:3)

Yet, although this general process did occur, the transition was not nearly so timely and smooth. When the decline in trade culminated about 1967 there were a few years of regional doubt and concern. Mountaineering had not shown the capacity to be a major foundation of local economic growth during the previous seventeen years and was in any event at that time banned altogether with no assurances of when, if ever, it might be resumed. Trekking tourism had begun, but in the first years it was not the substantial phenomenon that it became by 1971. In the early days it must have seemed primarily a young man's field, work fit for the few Sherpas who had previously worked for the expeditions but hardly appealing to men who had spent their lives trading in Tibet and India. During the 1960s some men indeed began to abandon Khumbu to move to Kathmandu or to Darjeeling where they tried their luck at road contracting and other ventures. The rise of trekking tourism during the early 1970s slowed this emigration, although there remained a steady flow outward of many of the brightest and best educated to Kathmandu, where a number of Sherpas ended up working in trekking agency offices. With the increasing scale of Khumbu tourism in the 1970s and, more importantly, with the rise of both commercial group and individual trekking there were abundant jobs in the new service sector as well as opportunities for entrepreneurial activity that involved, in the early days at least, relatively minor commitments of cash. A new era in the local economy began as former traders opened lodges and shops, sirdar began to prosper, and most Khumbu households began making enough money from tourism to pay for the grain they had once earned through trade.

10

Tourism, Local Economy, and Environment

The rise of tourism as a major global phenomenon has made it one of the most important sources of international economic development and cultural exchange. It has also been associated in both the developed and the developing world with a wide range of adverse economic, sociocultural, and environmental impacts. Even extremely small-scale tourism can precipitate adverse impacts, especially in areas that have not been previously integrated into the global economy, have had little interaction with Western cultures, or are located in sensitive environments.[1] During the past few years international organizations, governments, citizens' groups, and local communities (as well as members of the tourism industry) have become increasingly concerned with the adverse effects of recreational, nature, and cultural tourism on the very ecosystems, peoples, and places whose special character has made them tourist attractions.

This chapter explores the role of tourism in recent change in Khumbu's economy and environment. I examine the past and present role of tourism in the Mount Everest region in changing local land and resource use and the concomitant environmental impact which this has had to date.[2] It is beyond the scope of this book to consider the wider cultural and social impacts of tourism, as significant as these are, to analyze the effectiveness of tourism in Khumbu thus far as a path for local economic development, or to present recommendations for more culturally and environmentally sensitive tourism and tourism planning.

Tourism and Subsistence

In the preceding chapter I suggested that for many Khumbu families tourism has taken the place of trade in their household economy. Like trade it has offered villagers the means to diversify their economic activities and to decrease their dependence on Khumbu resources for their basic sustenance. Tourism has provided opportunities for a range of different types of involvement. The equivalent in tourism to the common pre-1965 small-scale salt for grain barter trade is the widespread practice of working in trekking and mountaineering for a only a few months each year in order to make wages that are then primarily spent on grain and other food purchased at the periodic market.[3] But, like trade, tourism also offers more profitable avenues of activity to those interested in investing capital and running a more complex business operation. Here tourism has provided a range of options from the ownership of a few head of pack stock to the operation of shops and lodges, and even the establishment of Kathmandu-based trekking companies.[4]

Tourism is also similar to trade in its role in the Khumbu economy in that tourism employment and entrepreneurial enterprises can be integrated into lifestyles in which much time and attention are also given to subsistence household agropastoral production and small-scale, commercially oriented stockraising. Tourism has been fully integrated into long-standing Khumbu subsistence strategies. During these first decades of tourism development, at least, even families making considerable income from lodge operations have all continued to farm and to devote most of their land, as before, to achieving self-sufficiency in tubers and, if possible, in buckwheat and barley. This has also remained the practice of the many families who are involved in tourism through wage making only. Tourism has become the major source of regional income and probably directly or indirectly accounts for more than 90 percent of all the money earned in wages and business profits in the region.[5] Yet it remains, like trade before it, for nearly all families a supplementary activity which is integrated into a basic subsistence economy and which is less important in terms of household labor and time than crop production and pastoralism for household consumption. That this remains true reflects cultural values as well as economic conditions and is probably related to concepts of identity as well as to local perspectives on diversification, gender roles, and social interaction.

Tourism, Affluence, and Differentiation of Wealth

Differentiation of wealth is nothing new in Khumbu, as I discussed in chapter 2, and the region has long been relatively more wealthy than many other nearby areas of Nepal. Differences in regional, village, and household wealth made for differences in standards of living (shelter, diet, clothing, conspicuous consumption, leisure time, cultural pursuits), status, political power, and land use. The rise of tourism in the regional economy, however, has increased the overall affluence of Khumbu society and widened the gap between it and other regions. It has also accentuated disparities in wealth among Khumbu villages and among households within villages. This has contributed to new patterns of land and resource use and these in turn have also placed new pressures on some communal resource-management institutions.[6]

A variety of employment opportunities in tourism are possible in Khumbu today. Some are strictly day-labor wage work. Others are careers for which men receive year-round monthly salaries and benefits. There is a definite hierarchy of jobs in terms of both pay and status. The range of employment also often represents a hierarchy of experience with young men moving up a ladder beginning as porters or kitchen boys and ultimately rising to the better paying and more prestigious jobs of camp-crew sherpa, H. A. (high-altitude mountaineering porter or sherpa), cook, or sirdar. Today most men specialize in either mountaineering or trekking work, although a few pursue opportunities in both.

Mountaineering provides short-term day labor for large numbers of local porters who carry loads only as far as base camp. This was once an important source of income for both Khumbu men and women, but today it is very rare for Khumbu Sherpas to work as base-camp porters. Mountaineering also offers a relatively smaller number of other jobs, which range from the base-camp and climbing sirdar and the team of H. A.s who carry loads high on the peaks to cooks, cooks' assistants, and mail runners. Sirdar are the elite of mountaineering crews responsible during the approach for the logistics of the camps and the organization of the porters, for the efficient performance of the H. A.s on the mountain, and for the operation of base camp. On large expeditions there are often two sirdar, one in charge of base-camp operations and the other in charge of the high-altitude porters on the peak itself.

Trekking work provides a similar set of jobs with the exception that there is no need for mail runners or H. A.s whereas there are more opportunities for camp staff. Camp staff are today usually referred to as sherpas in the trekking industry, even though increasingly men of other ethnicities are filling this position. They are responsible for making and breaking camp and often accompany trekking group members as personal guides and companions. Typically there is one sherpa for every four trekking clients.

Pay varies with position both in terms of wages and other benefits. Porters are the least well paid in this larger sense, although in Khumbu they can make relatively good wages. In 1990 porters received from

120–250 rupees per day for work in Khumbu. This was without any additional allowance for obtaining food and shelter. The large range in pay rate reflects differences in supply and demand, tourists' itineraries, and the bargaining skill of the parties involved.[7] Individual trekkers in particular often pay quite high wages for porters since they tend to have little experience with hiring them, are unfamiliar with local rates, and often are given little choice by porters well aware of how short the supply of people willing to perform such work often is at the Lukla airstrip, the Jiri roadhead, and in Khumbu itself. The local rate Sherpas paid in autumn 1990 for porters was 120 rupees per day (up from one hundred rupees per day the previous year). Trekking groups paid 150–160 rupees per day without food. At these wage rates a porter can make a higher daily wage than men in the more prestigious trekking and mountaineering positions. They seldom, however, receive the substantial amount in tips and gifts that sherpas, cooks, and sirdar typically are given.

High-altitude mountaineering porters make between seventy and one hundred rupees per day whereas mountaineering sirdar probably average one hundred rupees per day. Both are also given an issue of equipment that can have a resale value which is much higher than the wages themselves.[8] Beyond this direct income many high-altitude porters and mountaineering sirdar who have attained the summits of major peaks have been awarded with long-term trips abroad. During these foreign trips they are often offered opportunities to work at wages which allow them to return to Nepal with considerable savings by local standards.

Trekking cooks typically made seventy to ninety rupees per day in wages in 1990 and cook-crew members fifty to seventy rupees per day. Trekking sherpas usually received slightly more than cook-crew members and slightly less than cooks. Pay varied with different trekking companies, the lowest-paying companies offering sixty rupees per day for sherpas in 1990 whereas the highest paying paid seventy to eighty rupees per day. Food and shelter were provided during the trek free of charge to all of the cook crew and sherpas as well as to the sirdar.

Trekking sirdar make different wage depending on whether they are "company sirdar" on contract to a particular trekking agency or freelance guides who work for different companies for short periods or even attempt to recruit trekkers more directly in Kathmandu or Khumbu lodges. Sirdar's pay also varies with their degree of experience, connections, and command of foreign languages. If they can perform as a group leader as well as sirdar, which requires considerable interaction with tourists as guide and counselor, they make considerably higher wages. Sirdar hired for an individual trip may be paid as little as seventy-five rupees per day plus food, whereas others on company contracts receive a monthly stipend of 1,700 to 3,000 rupees per month year round for four to six months of work per year.[9] A few of the finest group leaders may be paid by foreign salary standards and may make as much as fifty dollars per day, thus earning in two days the monthly salary of the highest-paid company sirdar.

Both trekking and mountaineering sirdar are also able to greatly augment their salary through money made from the recruitment of porters and pack stock and negotiations for campsites, fuel, and food. On all of these there are profits to be made from overreporting expenses to Kathmandu trekking companies and often also from kickbacks and other offerings. A sirdar might, for example, report to Kathmandu that he had hired porters for 130 rupees per day when he had actually obtained them for 100 rupees. This can very quickly add up to a substantial amount of money. Porters may also be expected to secure their jobs with a gift of some sort to the sirdar. All kinds of charges may be overreported, from fuel wood and camping fees to food expenses. These practices are widespread and can easily enable one to triple one's salary. When a trek or expedition takes place in Khumbu itself much more money can be made. One may choose to camp on one's own lands and those of relatives. Family members, household servants, and relatives and friends may be hired and one may use one's own pack stock rather than recruit porters. At 1990 rates of eighty rupees per load per day (160 rupees per zopkio), a man who could mobilize five or six of his own yak or zopkio could make as much money from this sideline in a month of trekking as an average sirdar's salary would bring him in a year.[10] Mountaineering sirdar who are in charge of large expeditions have the same opportunities on a far vaster scale, for they are responsible for arranging for hundreds of loads to be carried, for hiring dozens of high-altitude porters and a sizeable cooking staff, and purchasing enormous amounts of food and fuel. Some mountaineering sirdar have greatly increased their incomes by making advance preparations for expeditions. It is then possible to use their own livestock to shuttle tons of equipment and food to base camp. In one case a few years ago a Nauje sirdar who owned more than ten yak and zopkio transported more than four hundred loads to Mount Everest base camp from Lukla with the help of his son. At today's rates this would bring them over 200,000 rupees. Even in the mid-1980s, when prices were much lower, it was said that a sirdar on a major mountaineering expedition could make 100,000 rupees or more if he was skillful at the business side of his craft.[11]

A number of trekking sirdar and group leaders, like mountaineering sirdar and H. A.s, have been awarded with trips abroad by grateful clients and foreign friends. Some Sherpas have made repeated trips to Europe, Japan, and the United States in this way. The contacts developed over a trekking and mountaineering career also bring other benefits such as scholarships for one's children and low-interest or no-interest loans for building houses or starting businesses. Sherpas are very aware that the relationships they develop with tourists may change their lives and some carefully court possible patrons and benefactors.

There is also money to be made in tourist businesses from selling stocking hats at 100–120 rupees apiece to operating shops, lodges, and pack trains. The most lucrative business of all is owning and managing a lodge. All Khumbu lodges are run as family operations. The price of a bed is kept extremely low out of fear that anything more than minimal prices will scare away business. A bed in a dormitory runs from five to ten rupees, fifteen to thirty cents a night. Even a double room seldom exceeds fifty or sixty rupees, less than two dollars a night. Lodges do not make much money on rooms even by local standards of income. There are considerable profits in selling meals, however, since there are no restaurants in Khumbu and guests normally take all meals in their lodge (or pay a penalty of a higher lodging rate if they do not). It is common for guests to spend five dollars a day on food, and those with a beer or soft-drink habit may easily spend a great deal more at three or four dollars a bottle for beer and one to two dollars for a bottled soft drink. Prices are generally fairly similar among lodges in a given area, but increase with distance up valley from Nauje. Five dollars per day in basic food costs seems rather little, but it is nearly twice the daily wage of most trekking Sherpas. Many lodges serve at least ten tourists per day during the trekking season and more popular places average twenty tourists per night and still more for lunch.

All the cooking and other lodgekeeping tasks are generally done by family members if the business is of modest scale. If business booms families hire low-paid servants to help with gathering fuel wood and water, cooking, serving, and cleaning. Some of the largest lodges now have four servants working for them and in Nauje even small lodges have two. With these low labor costs it is not hard to understand why so many Khumbu families have been interested in going into the lodge-keeping business. Shopkeeping has been much less attractive. Most of the region's shops are operated by Tibetan refugees or as side businesses by families who keep lodges.

Since tourism arrived in Khumbu in 1950 wage labor rather than entrepreneurial activities has been the main source of Sherpa income from tourism. Over the years the nature of Khumbu Sherpa participation in tourism work has changed dramatically. Until the 1970s the majority of Khumbu Sherpas who worked in tourism held jobs as porters, either as local porters who carried loads to mountaineering base camps and accompanied trekking groups or as high-altitude porters on expeditions.

Only a few men had jobs as mountaineering sirdar and there were very few trekking sirdar before the trekking boom of the 1970s. During this era the most coveted position in tourism other than sirdar was undoubtedly that of the high-altitude porter. Mountaineering Sherpas risked death and debilitating injury, but the rewards were substantial. Until the 1980s Khumbu Sherpas filled almost all these positions on expeditions in Khumbu as well as other parts of Nepal. On expeditions and treks within Khumbu, moreover, nearly all mountaineering and trekking sirdar, cooks, and camp staff were Khumbu Sherpas. With increasing affluence and contacts in the mountaineering and trekking businesses, however, Khumbu Sherpas have begun to concentrate on better paying and less dangerous jobs, leaving the others for non-Khumbu people. Very few Khumbu people work as local porters anymore, and in Khumbu what is not transported by zopkio or yak is today mainly carried by Rais.[12] Fewer young men have been drawn to high-altitude-porter work as well in recent years. On the major U.S. Everest expedition of 1963 all thirty-two high-altitude porters were Sherpas (twenty-five of them from Khumbu, four from Darjeeling, and three from Solu). The 1988 French Everest expedition, by contrast, employed twenty-eight high-altitude porters, of whom twenty-six were Sherpas but only seven were Khumbu Sherpas (J. Fisher 1990:123). Most of the young Khumbu men who work today as high-altitude porters come from Phurtse.

The annual income Khumbu families make from tourism varies considerably as a result of differences in wages and other income made on treks and mountaineering expeditions and from tourism businesses. It also varies with the number of family members who work in tourism and the number of months in which wages are earned. The amount of time men commit to trekking and mountaineering work varies a great deal, and in general is rather less than the total length of the tourist season.[13] In Nauje it is typical to work only four or five months per year at trekking or mountaineering, although a few men work eight or even ten months per year. Some trekking sirdar who own no livestock or who mainly trek in areas outside of Khumbu make as much as 7,000 rupees per month between their salaries and other money made through their position when they are trekking, and 1,700 to 3,000 rupees per month for the rest of the year from their salary. This gives them a minimum annual income of more than 41,600 rupees (more than $1,400), a fortune by Nepal's national standards. Those who own zopkio or yak can make much more than this. A man who works as a sherpa, however, can expect to make as little as 7,200 rupees a year for four months of work. Income from lodgekeeping can also obviously vary tremendously. One Nauje lodge that today does a relatively small business takes in about 70,000 rupees per year. A lodge that averaged twenty guests per night over the entire eight months of the tourist season would very likely have gross receipts well over ten times that amount.

Khumbu Sherpas' relative affluence compared to other groups in the Solu-Khumbu region and adjacent districts has had important effects on regional labor migration and on the development of regional trade patterns. As already discussed, many young men and a few young women from other parts of the Solu-Khumbu district and neighboring areas have been drawn to Khumbu by the wages offered to household servants. And during the past fifteen years increasing number of Rai men and women have also come to the region for seasonal field and forest work. The money these migrant workers and servants can make in Khumbu is minor by Khumbu Sherpa trekking, mountaineering, and even agricultural day-labor standards, but this income is substantial in comparison to what can be made in their home regions. Many of the young men, moreover, are drawn by the possibility of beginning a trekking career and ultimately making larger incomes. They take positions in the households of Khumbu sirdar at low wages, or in some cases for no wages at all, in order to be able to accompany the head of the household on his trekking trips as a porter, pack-stock driver, or a member of the cook or camp crew. It is possible to earn substantial money by the standards of the Nepal midlands from such work even if one must share a percentage of it with one's benefactor and work the rest of the year as a household servant for nothing but food and a place to sleep.

Considerable differences in relative affluence exist within Khumbu itself. Some villages are getting steadily richer while others share more modestly in the new wealth from tourism. Nauje, Kunde, and Khumjung have far higher standards of living than the other villages. A half a day's walk in other directions leads to villages where rice is still only eaten on very special occasions and polyester dresses, stereos, and corrugated metal house roofs are only seen on market day trips to Nauje. And even within villages the contrast in lifestyles is tremendous between sirdar and lodgekeeping families and the quarter of Khumbu's Sherpa families who have little or no tourism income.

This regional differentiation in wealth is related to differences in the percentage of village households involved in tourism, the types of jobs they hold, the number of multicareer families, and the percentage of village families involved in keeping lodges and shops. Table 29 illustrated the contrast among villages in terms of the percentage of families who have income from tourism, including lodges, shops, and pack stock. The differences between Nauje, Khumjung, Kunde, and the other villages are striking. Further grounds for differences in village wealth become apparent from table 31 which illustrates the ways households from

Table 31.   Household Trekking Income by Percentage of Village Households, 1985

Village*	Sirdar	Sherpa	Cook	Porter	Pack Animals Only**
Nauje	46	9	9	1	11
Khumjung	34	22	11	5	11
Kunde	28	15	17	13	15
Thamicho	10	36	0	13	33
Pangboche	15	37	10	0	12
Phurtse	18	57	0	0	21
All Khumbu	25%	27%	8%	6%	18%

* Only Sherpa households are included. Each household was classified according to the highest status job performed by any household member in the following order: sirdar, sherpa, cook, porter. These figures do not reflect the fact that in many households more than one family member is employed in trekking.

** This category refers to those households that have no other direct income from trekking.

different villages are involved in trekking tourism. In Nauje, Kunde, and Khumjung a high percentage of those families who have trekking work have the highest-paying jobs available. Note especially the number of families having at least one member working as a sirdar and contrast this to the situation in Phurtse, Pangboche, and the Thamicho settlements. Here few families have a sirdar's income and many simply have what cash they make by hiring out some of their livestock for a few weeks per year as pack animals. By 1991 the contrast among villages had narrowed slightly as more Pangboche and Phurtse men improved their career positions to sherpa and sirdar. By 1991, though, there had also developed a contrast among Nauje, Khumjung, Kunde, and the other villages in the number of families in which two or more members were working in trekking.

The extent of ownership and operation of lodges has also long varied among villages. This was especially true in the 1970s and early 1980s, as can be seen in the figures for 1983 in table 32. Nauje families have for many years taken advantage of their village's site on the main route to Mount Everest and have also opened lodges further along the route in places such as Lobuche and Dingboche. The development of lodges in Kunde and Khumjung has been greatly inhibited by their location less than two kilometers from the major tourist center of Nauje and off the main trail to Mount Everest. As of spring 1991 there were only two lodges in Khumjung and one in Kunde, all opened only in the past few

Table 32. Khumbu Lodge Ownership, 1983 and 1991

Village	1983			1991		
	Number of Lodges Owned	% of Village's Sherpa Households	% of All Khumbu Lodges	Number of Lodges Owned	% of Village's Sherpa Households	% of All Khumbu Lodges
Nauje	16	18	33	24	27	29
Khumjung	22	21	45	19	15	23
Kunde	8	14	17	15	30	18
Pangboche	2	3	4	13	19	16
Phurtse	0	0	0	5	8	6
Thami Og	0	0	0	2	2	3
Thami Teng	0	0	0	0	0	0
Yulajung	0	0	0	0	0	0

years. Families from those two villages, however, have converted many herding huts on the high routes to Gokyo and Mount Everest into simple lodges. Villagers in Thamicho and Phurtse who own no land on the main tourist thoroughfares have had fewer opportunities. The two lodges in Thami Og have been opened relatively recently and do a light business. For many years there was no lodge in Phurtse. Today there are only two very basic ones in the village and Phurtse families have now opened a few humble teashop-lodges elsewhere in the upper Dudh Kosi valley. There were also few lodges in Pangboche until the late 1980s when several large, new lodges were built in the village and more local people also began to convert their herding huts in the upper Imja Khola and Lobuche Khola into lodges. By 1991 the relative involvement of the different villages had thus narrowed considerably.

The late involvement by Phurtse, Pangboche, and Thamicho villagers in the lodge business may be due to several factors. Location on the main route to Mount Everest may have been of especially great importance in the 1970s when far fewer trekkers visited the upper Dudh Kosi valley. The lack of a strong entrepreneurial tradition in Phurtse and Pangboche, where only a very few families traded with Tibet on even a small scale, may have been a factor. The relative lack of available capital in the poorer villages may well also have been important. It is also possible that villages in which few families had worked in trekking could also have been at a disadvantage because fewer families would have had the well-developed language, social, and culinary skills that are important for running a lodge. Few families in these villages, moreover, could have had the foreign contacts so often useful for obtaining loans to establish lodges. Village differences in the extent to which families set up lodges, like differences in types of tourism employment, could also reflect different levels of land and livestock wealth and contentment with agropastoral lifestyles. Pangboche, Phurtse, and Thamicho are all regions where a great deal of nak pastoralism is still practiced and where many families have relatively large land holdings. Families from these areas were not only slower than those from Nauje, Khumjung, and Kunde in opening lodges but were also somewhat slower in taking up work as sherpas and sirdar in the trekking trade. Villagers preferred, it seems, the more short-term time commitments of working as porters and H. A.s.

This regional differentiation in wealth seems likely to continue to deepen. It already has reached a level of inequality that probably surpasses the differences among villages during the trading era. In the pre-1960 era Nauje was noted as the wealthiest of the villages, for it was there that most of the big traders lived. There were many poor families in Nauje as well, however, and the overall contrasts between that village and the others were not as pronounced as they are today. Within Nauje the differences in wealth that once existed between the very few major trading families and the rest of the village have perhaps narrowed in the tourism era when a higher percentage village families have a major source of income. Families who own popular lodges, however, are becoming wealthy to an extraordinary degree in comparison to their fellow villagers. It remains to be seen what long-term repercussions intra- and intervillage economic differentiation will have on Sherpa society. There is a strong possibility that a new elite will develop in Khumbu, composed largely of families wealthy from sirdar work and lodge operations. There is not room for everyone to operate an lodge, and the costs of entering the business are becoming prohibitive for most families now that standards are rising everywhere and being competitive in Nauje requires a multistory lodge having a restaurant with a view. It is the sirdar and the lodgekeepers who have the money for new business enterprises and for sending their children to the upper classes of Khumjung's school or to boarding schools in Kathmandu or Darjeeling. These are the families, moreover, who gather most of the gifts and scholarships donated by philanthropic tourists who tend to be most generous toward the families of their guides and hosts. All of this will probably continue to translate into status and power as well as increasingly different lifestyles.

Monetarization, Social Relations, and Inflation

The rise of the tourist economy in Khumbu has brought a great deal more cash into the region than there had ever been in the past. To some extent this has fostered a shift toward cash-based economic exchanges. Grain and other goods from lower regions of Nepal that were once obtained through barter trade are now paid for in cash. And while labor in Khumbu was formerly paid primarily in butter, potatoes, and grain this has now given way to cash wages, although the custom of payment in food has not entirely disappeared. This increasing monetarization of Khumbu life, however, has not entirely permeated the local economy and social interaction, much less fundamentally altered cultural values. Reciprocal work groups are still, for example, very important in all the villages. Mutual aid groups continue to work alongside hired laborers in building houses and communities still maintain trails, temples, and ceremonies through the contributions of unpaid labor from all village families.

The degree of monetarization of the economy that has occurred has brought negative impacts, including inflation.[14] The price of rice has increased tremendously, from nine rupees per pathi in 1964 to twenty-six rupees ten years later, and thereafter rising more precipitously to thirty-five rupees in 1978 and ninety rupees in 1988 (J. Fisher 1990:116). In autumn 1990 rice cost ninety to one hundred rupees per pathi depending on quality. This is more than three times the price of rice in lower-altitude areas of the Solu-Khumbu district or Kathmandu. These prices are not the result of direct buying by mountaineering expeditions, trekking groups, or lodges, for tourists consume a relatively minor amount of rice, maize, millet, or buckwheat. They are an indirect impact of tourism, however, for they reflect Rai traders' perceptions that the now-affluent Sherpas can be made to pay high prices.[15]

Tourism has probably had some role in increasing the price of Khumbu-grown potatoes as well, although these have not inflated in price quite as much over the years as rice has. Tourists do eat a considerable amount of potatoes, and a relatively popular lodge may require as many potatoes during the eight-month tourist season for its guests as the family itself consumes in a year. Lodges that take in large numbers of guests may require double the family's subsistence requirements of potatoes in order to feed them. Across Khumbu, with its more than eighty lodges, the added demand for potatoes may thus be substantial. It is possible that tourists now account for over 10 percent of the total potato consumption in the region. The increasing quantities of potatoes required as tourist numbers rose over the past twenty years may well have contributed to the rise in the price of Khumbu-grown potatoes. The cost of potatoes rose from two rupees per eleven-kilogram tin in 1964 to fourteen rupees per tin by 1974 and twenty rupees per tin in 1978 (ibid.). By 1986–1987 potatoes sold in Nauje for twenty-five to thirty rupees per tin and in 1991 were selling there for thirty-five rupees per tin. Potato prices have been pushed much higher by tourism in some parts of Khumbu, particularly in the upper Imja Khola and Lobuche Khola valley. Here some lodgekeepers prefer to buy potatoes from Dingboche and Pangboche farmers as this makes transport much easier than from Nauje. Prices here reached fifty-five rupees per tin after the harvest in 1990 even though it had been a relatively good harvest in Dingboche, and by spring 1991 a tin cost seventy rupees.[16] Increases in potato prices over the past fifteen years would undoubtedly have been much greater but for the great increases in regional production due to the widespread cultivation of high-yielding varieties. Without this reliance on new varieties it is doubtful whether regional production would be sufficient for both feeding the tourist population and meeting Sherpa needs.

Tourist demand has had a much greater and more direct impact on the market price of important regional commodities such as kerosene, eggs, cooking oil, fruit, vegetables, and powdered milk, as well as goods of importance only in the tourist trade such as candy, cookies, beer, soft drinks, and bottled water. These products are bought by mountaineering groups, trekking parties, and lodges on a large scale at the weekly market in Nauje. Many Sherpas believe that these purchases not only greatly inflate prices but also create shortages, and this is one of the most widely cited local complaints about the adverse impacts of tourism.

Increases in the rates of pay for trekking and mountaineering work, however, have more than compensated for these commodity price increases. In 1971 rice cost less than twenty-five rupees per pathi (Fürer-Haimendorf 1975:77) and potatoes were fourteen rupees per tin, which means that rice quadrupled and potato prices increased by two-and-a-half times over the past twenty years. In 1971–1972 sirdar were paid twenty-five rupees per day while H. A.s and cooks received fifteen rupees and porters ten. In 1991 a sirdar's wages were seventy-five to one hundred rupees per day while cooks made seventy to ninety rupees, H. A.s seventy to one hundred rupees, and porters one hundred and twenty rupees. Thus while rice prices quadrupled over twenty years the pay of porters had gone up twelve times and that of cooks and H. A.s by five to six times. Sirdar, it is true, were being paid by the day at a rate only three or four times greater than twenty years earlier, but the relatively low pay given to them reflects the expectation that they would make much more money through the use of their position. Many individuals, moreover, have increased their incomes during this period through career advancement, moving from cook-crew work to positions as camp-crew members or cooks, or from camp-crew work to positions as sirdar and group leader. In many families total income has also risen as sons have taken up trekking and mountaineering work. The number of such multi-income families greatly increased during the 1980s.

Inflation thus has not been a serious problem in Khumbu. During the 1980s most Khumbu families who have been involved in tourism have probably become relatively beter off despite inflation. So too have families who perform skilled labor such as carpentry, for carpenters today make triple the wages that they did even ten years ago, and command day wages of eighty to one hundred rupees. Even Khumbu families who depend for their cash income entirely on wages from agricultural day labor have seen their earnings rise even faster than rice prices. Twenty years ago in Khumjung women were paid a maximum of three rupees per day when a local porter received ten rupees per day from mountaineering expeditions (Fürer-Haimendorf 1975:42, 89). Whereas rice prices have quadrupled and potato prices have increased by two-and-a-half times since 1971 agricultural day labor rates have gone up ten times to thirty rupees. The only households that have really suffered from inflation are those that have next to no cash income, such as elderly couples who are maintaining their own households. Sherpas who have been left out of the opportunity to make money from tourism work have become increasingly disadvantaged in comparison with the rising affluence of their neighbors.

Tourism and Agricultural Change

Tourism has been reported to have had a number of adverse impacts on local subsistence agriculture in Khumbu despite the fact that little Khumbu-grown food other than potatoes is served to tourists.[17] Tourism can have major indirect impacts on land use, labor, and lifestyles, and these are the factors which are thought to have transformed Khumbu agriculture. Among the impacts that have been reported are the development of a generational gap in farming (Bjønness 1983:268; J. Fisher 1990), a decline in the amount of land in crop production (Fürer-Haimendorf 1984:15; Bjønness 1983:269), a decline in crop yields (Bjønness 1983:269; J. Fisher 1990:122; Fürer-Haimendorf 1984:8), inflation of agricultural day-labor rates and shortages of agricultural labor (Fürer-Haimendorf 1984:8, 15, 17), and increased use of non-Khumbu migrant labor (J. Fisher 1990:122). According to one report, "tourism has replaced agriculture as the mainstay of the economy" (Karan and Mather 1985:93).

Whereas some of these reported impacts are indeed significant, others are rather less important than has been imagined. Inflation in agricultural day-labor rates has taken place, and there is indeed a greater influx of agricultural workers from outside the region than there was in the past,[18] but reports of labor shortages, declining yields, abandoned land, a generation gap in farming, and the eclipse of agriculture in the Khumbu economy may have been exaggerated. The production of food for household consumption continues to be the basis of Khumbu agriculture, and the role of crop growing in local economy, culture, and lifestyles has not yet eroded.

Tourism has not commercialized Khumbu crop growing to any significant degree. The need for larger amounts of potatoes has led some lodgekeeping families to acquire additional cropland, and the high regional demand for potatoes and the money to be made from the sale of surplus tubers may have been a factor both in the recent widespread adoption of high-yielding varieties and in the conversion of land from a two-year rotation of potatoes and buckwheat to continuous potato cropping. Yet an interest in production for sale to lodges probably has been a far less important factor in the recent regional intensification of potato production than continuing dynamics that go back a century in Khumbu. Recent intensification, like older intensification, probably primarily reflects a response to population growth and fragmentation of land holdings. Everyone has adopted the new potato varieties, not just families

who are interested in selling surplus potatoes. Likewise, buckwheat cultivation has been abandoned by the entire population of the Bhote Kosi
valley and not just by large landowners who have a surplus to market.
There has also been little evidence of local interest in the commercial
production of other crops for tourist consumption. Fields are not being
taken out of potatoes and put into cabbage, spinach, cauliflower, or
carrots, even though these bring exceptional prices at the weekly market. Some Nauje lodgekeeping families have shown great interest in
producing more vegetables and spices for use in their lodges and have
developed the finest household gardens in Khumbu.[19] But no one has
yet ventured to plant even a single field in vegetables for sale in the
market. In Pharak, however, the commercialization of farming in response to the new tourism market and increasing Khumbu Sherpa affluence has had an important effect on local land use and many Pharak
Sherpa families sell produce in the Nauje market.

Most families have not found that the time demands of trekking and
mountaineering work have compromised their ability to conduct field
work. It is true that the height of the spring trekking season overlaps with
the weeks when fields must be prepared and planted, and that the autumn mountaineering season and the beginning of the autumn trekking
season overlap with potato harvesting and with buckwheat and barley
harvesting and threshing. It has often been assumed by outsiders that the
absence of men during parts of the agricultural season must have a severe
effect on crop growing, yet this is not the problem it would be in a culture
in which men were responsible for most farming work. In Khumbu
women farm, and until now very few women have worked in mountaineering and trekking. Women still place a priority on farming and tailor
their other activities around its rhythms. Those who operate lodges, for
example, find relatives, neighbors, friends, servants, or children to run
things when they need to be in the fields or else arrange for servants,
hired help, or daughters to work the potatoes and buckwheat. If all else
fails they simply close their lodges for a few days. The preoccupation of
men with their trekking and mountaineering assignments may have some
small impact on those limited aspects of crop production in which they
lend a hand: plowing buckwheat and barley fields, carrying manure to
the fields and potatoes in from the fields, harvesting grain, and cutting
hay and wild grass. Their presence is missed in some of these tasks, and
as a result women do have to do more of the work than would otherwise
be the case. Probably the greatest additional pressure is put on women at
grain harvesting and threshing time, for in the pretrekking era in villages
like Pangboche men and women handled this work equally or men performed even a greater share of the work than did women. Now women in
most families are on their own, and for these tasks there is a element of

urgency in the harvest because it has to be completed before bad weather sets in and damages the crop. Yet while women's agricultural responsibilities are thus greater now than ever, the fact that they have long carried out most of the work of producing crops means that the loss of help from men is not as critical as it would otherwise be. The men's absence is also more than compensated for in field work by the increasing use of wage laborers from outside of Khumbu. Rather than a regional labor shortage tourism has created a situation of regional labor abundance through giving many more families the ability to hire agricultural day laborers. Such laborers, as we have seen, are paid less than half the daily wage of the lowest-paid trekking workers. Families who a decade ago either relied entirely on their own efforts or on their participation in reciprocal work groups today hire help for field preparation, planting, weeding, and harvesting. This has enabled Sherpas to cope with the increased labor involved in harvesting the large crops of yellow potatoes and development potatoes, made it possible for some families to expand the area they farm, and enabled more women to give their places in reciprocal work groups to hired helpers and to reduce the amount of work they do in their own fields.

Some Sherpas complain that the increasing reliance on help from non-Khumbu agricultural workers lowers harvests. They remark that outsiders do not have the knowledge and experience that local people have in farming and that hired people do not take the care in their work that they themselves do.[20] It is not clear how much credence should be attached to such remarks. In Shorung one hears similar comments about Khumbu Sherpa women being less skillful in farming than Shorung women. It may be that non-Khumbu people are less experienced in the planting and harvesting of potatoes, but digging fields and digging up potatoes are not tasks involving especially great skills and all the decisions about crop selection and the timing of agricultural tasks continue to be made by the families themselves. It is hard to imagine that field laborers' work could be too poor given that they are usually closely supervised by household members or work alongside Sherpa women as part of reciprocal labor groups.

There is no evidence that there has been a general decline in Khumbu agricultural productivity in the sense of lower regional crop yields. It is true that specific varieties of potatoes, especially kyuma and the red potato, are perceived to produce less than they formerly did—a change which could be related to a subtle decline in field fertility or to changes in characteristics such as disease resistance of the plants themselves.[21] Yet productivity per field and across the region has nevertheless increased since the 1970s due to the adoption of high-yield varieties. The universal perception among Khumbu villagers is that regional potato production is today at an all-time high due to the replacement of the red potato by the yellow potato and the development potato.

It is also not true that less land is in production than a few decades ago. As was discussed in chapter 6, the large numbers of abandoned fields that can be seen in some parts of Khumbu probably reflect emigration and intensification in the late nineteenth and early twentieth centuries, not the recent impacts of tourism. In the few cases where terraces have been taken out of production during the past twenty years these have mostly been associated with places where wildlife damage to crops has been particularly high, such as Tashilung and Nyeshe in the lower Bhote Kosi valley.[22] Fürer-Haimendorf appears to have been mistaken about recent abandonment of cropland at the gunsa of Teshinga (1984:15). The decline in nak pastoralism, however, may have led some families to deemphasize their production of hay in the high-altitude herding areas and even to abandon small potato fields in places where they no longer herd. There are a few examples of this in the upper Dudh Kosi valley. Neither this nor the general decline in nak herding itself, however, can be attributed to tourism.

The recent trend in Khumbu has been toward an expansion rather than retraction of the area under cultivation. During the late 1980s some long-abandoned terraces were clandestinely put back into production despite attempts by Sagarmatha National Park officials to discourage this, and since national park opposition declined in 1989 some Sherpas have begun registering and reclaiming substantial areas that have been out of crop production for half a century and more. Many old terraces in the Samde area, for example, are being put into hay and potato production. Other families are using income from tourism to buy cropland. There are a number of cases, for example, of land-poor Nauje families (many of them second or third generation immigrants from Tibet) who have made money in mountaineering and trekking and have used it to purchase crop and hay land in Nauje, Mishilung, Nyeshe, Langmoche, Chosero, and Tarnga. Some families who have opened lodges have also bought more crop land in these areas.

Many of the adverse impacts ascribed to tourism in Khumbu thus do not yet seem to have occurred. Land is not being abandoned for lack of workers with the time to cultivate it or declining dramatically in productivity. Crop production remains oriented to subsistence production rather than to market sale and tourism. Yet the issue of whether or not a generational gap in farming (Bjønness 1983; J. Fisher 1991, personal communication) is developing is nevertheless worthy of further analysis, for this could occur even without a decline in regional agricultural productivity.

I take generation gap in this context to be a change in attitudes towards lifestyle, customs, beliefs, and values between parents and their children which is so widespread a phenomenon as to seem commonplace or even characteristic of a region and an era. In some of the world's mountain regions, and particularly in the Alps, such a generational gap may indeed have been an important factor in a post-World War II decline in farming. The movement here of large numbers of young people to wage labor in the towns and cities has left the family farm in the care of aging parents and relatives. In the Alps this does seem to reflect value changes as well as responses to new economic opportunities in the lowlands. The Khumbu situation, however, is different. Here there has also been some out-migration to Kathmandu (and to other countries) and again this has primarily been by the young. Yet several points must be made about this. The scale of this migration, to begin with, is very different. Only a small percentage of Khumbu youth are emigrating from Khumbu. Many of those who do move to the city, moreover, are men who return to Khumbu a few years later to marry and establish a household there. Some young families have shifted their residence to Kathmandu for a few years, often so that their children can attend the good schools there as well as to enjoy a period of urban living and take advantage of job opportunities. Many of these families also return to Khumbu, although there is an increasing community of Khumbu families who appear to be permanently settling in Kathmandu. There are cases where Khumbu land is tended by aged parents whose children have all emigrated, and houses are occupied by renters or relatives of people who have gone to Kathmandu or abroad. But this remains the exception rather than the rule. Much research remains to be done to establish the scale of migration that has taken place and whether it is now increasing as it seems to be. I do not believe, however, that even the present level suggests that urban migration is characteristic of a generation of Khumbu Sherpas.

There also does not seem to be a decline in interest in farming as a way of life among Sherpas who remain in Khumbu. There is no generational discontinuity in lifestyles and household economy among Khumbu residents. Young Khumbu families living in the region have not declined their inheritance of main cropfields or sold, rented out, or abandoned the land. I do not believe that there is a single example today in Khumbu of any of these phenomena. Every Sherpa family in Khumbu today farms. No matter how well-to-do many Khumbu Sherpa families have become from tourism they continue to produce as much of their household food supplies as they can. This is as true of households that have just been established as of those that have farmed for decades.

Finally, in Khumbu a generational gap in farming means above all a change in attitudes and lifestyles between mothers and their daughters, for it is women who grow crops. I do not know of any cases in Khumbu today of daughters who have refused to work in the family fields or who on marrying and establishing their own households have chosen not to farm. There are probably no more than a few cases even of young women who rely entirely on hired agricultural labor or household servants to farm their land. Some may take advantage of their relative affluence to hire more assistance than their parents could afford to. And some certainly aspire to other careers, such as shopkeeping and running their own lodge. But for Khumbu women today as in the past nonfarm occupations supplement rather than replace crop production.

It is certainly conceivable that in future years Khumbu agriculture may change considerably and that tourism may be an important factor in this. It is very possible that agriculture may become more commercialized than it now is, with more land put to vegetables for sale to the lodges. It is possible that some households may begin to rely more on their income from tourism and less on crop cultivation and that as a result more land may be worked by hired labor, rented out, or even sold or abandoned. A generational gap may yet develop. That none of this has occurred thus far seems to be grounded in cultural attitudes more than regional conditions of land tenure, access to markets, or the particular forms tourism has taken. The high degree of land fragmentation, the inability of Khumbu families to be self-sufficient in food production from their own lands, and the resulting increasing reliance on purchasing food would seem to make the cultivation of high-value crops for market sale an appealing opportunity. Families throughout the region have ready access to markets, both to the weekly market in Nauje and directly to the lodges that are so numerous in much of Khumbu. There is considerable demand for fresh fruits and vegetables from both increasingly affluent Sherpas and from mountaineering and trekking groups and lodges. Yet Sherpas appear to be reluctant to give up a way of life that has proven reliable for so long. For many generations Sherpas have thrived on a balance of subsistence agropastoralism and nonfarm income and resources, and tourism may not yet seem a stable enough base to give up that diversity. Beyond that it may also be that many women continue to value crop growing as a way of life bound up with their sense of themselves as women and as Khumbu Sherpas. It is here that a generational gap in attitudes about farming could greatly affect Khumbu lifestyles and the economy. If greater opportunities for women in nonagricultural work should ever develop there may be more interest in scaling down farm operations.

Tourism and Pastoral Change

A number of tourism-related pastoral changes have taken place in recent years in Khumbu even though tourism has not yet resulted in the commercialization of either locally produced dairy products or meat.[23] Tourism has played a greater role in recent changes in pastoralism than it has in agricultural change and has been a factor not only in changing household economic practices but also in community resource management. These changes have primarily been related to the increased keeping of pack stock and particularly the keeping of urang zopkio. The zest with which Sherpas have purchased these crossbreeds from Pharak and Shorung breeders in recent years has, along with the less spectacular but also significant increase in numbers of cows and zhum, almost balanced the major decline since the 1960s in the number of nak being kept in Khumbu.

According to Sherpas zopkio were kept only in small numbers by a very small number of families before the mid-1970s. Despite their common use as plow animals in Tibet they were not widely used in this fashion in Khumbu before the 1950s (and even today only a very small number of teams are trained for this purpose). Yak were more valued as pack animals for the Tibet trade. Low altitude and poor bridges led Sherpas to pack loads south from Khumbu on foot rather than risk yak or dimzo, although a few families used sheep and urang zopkio to ferry loads.[24] In Nauje, where today zopkio are the predominate form of stock and more than 50 percent of all households own at least one, there are said to have been only five families who had zopkio in the early 1960s.

Since the 1960s, and particularly since 1975, urang zopkio have become increasingly important in Nauje, Khumjung, Kunde, and Thamicho.[25] By 1978 the total number of adult zopkio (urang and dimzo) had reached 80 (Bjønness 1980a:66). During the next few years this increased sharply, reaching 482 in 1984 (Brower 1987:189) when urang zopkio constituted 18 percent of all large stock in the region. Figures differentiating Nauje-owned urang zopkio and dimzo are not available for 1978 and 1984, but in 1987 there were 148 urang zopkio and 8 dimzo. These were kept by fifty-four families who owned an average of just under 3 head of stock per household. Fifty-two households, 58 percent of the Sherpa households in the village, owned urang zopkio and these constituted 65 percent of the village's 226 total head of large livestock (excluding calves). This is a substantial increase from just three years earlier when they were 48 percent of village stock.[26] By the spring of 1991 still more urang zopkio were being kept in Nauje. The village total had climbed another 15 percent to 171 urang zopkio, kept by fifty-one families. Nearly as many were kept by Khumjung (158) and in the Thamicho settlements (149), and the Thamicho count may well be incomplete. Across Khumbu there were at least 580 urang zopkio in the spring of 1991, more than any other form of cattle other than nak.

Urang zopkio are neither cheap to purchase (at approximately 3,000 rupees, an adult zopkio costs more than a month's wages for most trekking Sherpas) nor cheap to keep, since they require substantial amounts of fodder and are often owned by families who have no hayfields. But they can very quickly return the investment made in them. At 1990 pack rates of 80 rupees per load, 160 per zopkio, a pack animal can make back its purchase price in less than three weeks. Owning zopkio is particularly convenient for sirdar who are in charge of procuring porters and pack animals for their groups or expeditions and can ensure that their own livestock are kept fairly continuously in work. As mentioned earlier, 70 percent of those Nauje households where a family member works as a sirdar owned urang zopkio in the late 1980s. In the spring of 1991 twenty-five Nauje households having one or more members employed as sirdar had urang zopkio, representing 64 percent of all village households that had sirdar and 96 percent of sirdar households that owned pack stock. This link between sirdar and zopkio was not limited to Nauje. In Khumjung, for example, twenty-four of the village's thirty-eight sirdar households (63 percent) had urang zopkio.

The increase in the number of urang zopkio has brought more income for some of the already most well-to-do families and has increased the relative wealth of Nauje, Khumjung, Kunde, and Thamicho. This change in regional pastoralism has also put new livestock pressure on buckwheat crops in the Bhote Kosi valley and precipitated increasing conflict over summer range regulations in the villages and lower valleys. As I will discuss later in this chapter, zopkio have also placed new stresses on pasture and forestlands in certain areas with their very different grazing and browsing habits and different seasonal patterns of transhumance.

The other recent change in regional livestock keeping has been an increased emphasis on zhum and cows. Since the 1950s, and particularly during the past twenty years, the numbers of zhum and cows have increased considerably. Both are considered to be more effective milk producers than nak and are practical for households that want dairy products for their own use and are only interested in keeping a few head of stock without the commitment of time, energy, and lifestyle required to herd nak. Usually families keep only one or two zhum or cows each, enough to supply some milk for family use. Milk from such stock is only rarely sold and then only to fellow villagers in quantities of a few pints per week. Such arrangements are made directly among neighbors and friends when they are made at all; and zhum's and cow's milk and dairy products are not offered for sale at the weekly market in Nauje or sold to lodges.[27] Commercial dairy production for the tourist market thus is not a factor in this regional herding change, as Bjønness thought

(1980*a*:67). The increasingly widespread practice of keeping a cow or two, however, may well be indirectly related to tourism. It is the new income and affluence from tourism that now enables families who previously did not own cattle to purchase and keep cows and zhum.

The increasing involvement of so many Khumbu families in tourism, and particularly the employment of so many Khumbu men in trekking work, has also had an impact on herding. It is impossible today to find men, even those from the poorest households in the region, who are willing to work as hired herders. When men are often away from home for four or five months per year on treks and expeditions it becomes difficult for many households to maintain their former herding patterns using their own labor resources. Some families discover that it is difficult to maintain the size of herd they formerly cared for, and this is especially the case with nak herds. Families may lack a family member who has the free time and aptitude to take nak up into the high valleys in early spring or to cut and store the hay harvest and wild grass in the autumn when the men are away on treks. It is difficult for the women of the household to undertake these tasks in addition to their crop-production responsibilities, although there have been some cases when teenage daughters have taken on the job of nak herding. There is less of a problem with the care of cows, zhum, and zopkio, which can be allowed to graze freely in the village vicinity and are easily tended in the early morning and evening by women and by children who are still able to devote their main attention to crop-production tasks and school.

Families who have long herded nak have several options. They can forgo opportunities in tourism, arrange short-term, reciprocal labor exchanges with relatives and friends who may be able to watch stock while a herder is away on a trek, hire more day laborers to help with hay cutting and fodder gathering, ensure that a son or daughter can be entrusted with herding responsibilities, or decrease the scale of herding. It is very common to make use of cash income from tourism in order to hire workers to compensate for the lack of sufficient household labor at hay-cutting and fodder-collecting time in the autumn. Usually hired helpers are not entrusted with the care of nak, however, for many Khumbu herders are loathe to put their prized stock in the charge of people in whom they do not have full confidence, and it is widely believed that the lower-altitude migrant workers who make up so much of the wage-labor pool do not have the requisite experience and knowledge. There are households where either a father or a son deliberately forgoes tourism work when there is a scheduling conflict with herding. There are also cases where families have scaled down the size of their nak herd or even given it up entirely. This is particularly common when none of the sons shows any interest in continuing with nak herding and either their father's career in tourism or his increasing age lead him to decide that he cannot maintain the family herd in the same way any longer. Sometimes such families retain a few head of cows, zhum, or zopkio, but sell off their nak, high-altitude herding huts, and many of their hayfields. This is the one area of Khumbu agropastoralism where a generational decline of interest in "traditional" lifestyles has contributed to a shift in land use. It must be noted, however, that these situations were not responsible for the large-scale decline in nak keeping since the 1950s as has sometimes been suggested. The great decline in nak keeping occurred in the 1960s before tourism had a major impact on Khumbu pastoralism and, as discussed in chapters 6 and 9, was instead related to changes in international trade conditions.

Tourism and Changing Forest Use

Tourism has placed several types of new demands on Khumbu forests. For decades, until new Sagarmatha National Park regulations were implemented in 1979, mountaineering and trekking groups routinely cooked their meals and warmed themselves at night with campfires. Early teashops and lodges were built with timber from local forests until the national park put an end to most tree felling in Khumbu. And nearly all of today's lodges continue to rely solely on fuel wood.

Until the 1970s the major new demand for Khumbu forest products came from the mountaineering expeditions. Expeditions made use of kerosene and gas stoves at their camps on the high peaks, but made wood cookfires at base camp and during their approach marches through Khumbu. So did the hundreds of porters who carried their supplies. As expeditions made their way through Khumbu on the way to Mount Everest and nearby peaks they usually camped for a few days near Nauje, Tengboche, and Pheriche, scouring nearby forest for fuel wood. Once at base camp they established fuel-wood stockpiles large enough to supply the base-camp crew for six weeks or more.[28] It is difficult to estimate what the total fuel-wood requirement of an average expedition might have been. Nima Wangchu Sherpa (1979) has put this at 960 porter loads of fuel wood, or a minimum of about 28,000 kilograms of fuel wood over a two to three month period in Khumbu.

Much of a Mount Everest expedition's fuel-wood use came at base camp and thus was obtained from the region's highest-altitude forest and shrublands. The Swiss began the custom in 1951 of hiring local Sherpas (usually from Pangboche) for day wages to supply their base camp with fuel wood. Much of this was shrub juniper from the Tugla and Tsolo regions near the terminus of the Khumbu glacier. Some additional fuel wood was also carried up to base camps from the highest forest, the birch and rhododendron woodland northeast of Pangboche along the Imja Khola. During the early 1960s expeditions began to buy fuel wood by the load and a few years later it became customary to pay by the kilogram, a practice that encouraged the sale of freshly cut green wood rather than lighter dead wood. Since all of the areas that supplied fuel wood to the base camps for Mount Everest, Lhotse, Nuptse, Pumori, and Cho Oyu lay outside of the borders of locally protected forests, this new commercial activity did not compromise local values or practices. The sale of fuel wood became an important part of the Pangboche economy for a few years. Pangboche residents also benefited from the early expeditions' use of large numbers of tree trunks to bridge ice-fall crevasses. According to one report a 1974 Mount Everest expedition used 200–250 mature trees for ladders and crevasse bridges (Lucas, Hardie, and Hodder 1974).[29] These would also have come from the upper Imja Khola valley above Pangboche and good wages would have been made to cut and transport them to base camp.

Early trekking groups also required fuel wood, and with the rise in popularity of trekking during the 1970s this probably came to be a more significant demand on local forests than mountaineering. The large commercial groups almost invariably depended entirely on local wood supplies for their cooking fuel, and it was a popular custom to have an evening bonfire as well. Bjønness found in 1978, two years after the area had been declared a national park and just before tourists' campfires were made illegal, that only 7 percent of trekking groups carried fuel for cookstoves (1980b:126). Nima Wangchu Sherpa (1979) estimated that trekking groups used about three loads (approximately ninety or more kilograms) of fuel wood per day.

The Sherpa lodges that began to be opened in the 1970s also relied on fuel wood for cooking and in cold weather often kept fires going beyond mealtimes to warm guests. The amount of wood required by lodges varied both seasonally and with the number of guests they accommodated. In 1983 Nauje lodge owners estimated that their fuel wood use during the tourist season ranged from two to four loads per day and fell to half a load per day during the tourist off-season when it was only necessary to cook for their own household. Since then lodge fuel-wood use has increased in many cases due to the spread of the practice of selling hot showers to tourists.[30]

As new lodges were built beginning in 1974 there was also demand for construction timber. The first Sherpa lodge which was built for that purpose (rather than remodeled from a house), for instance, was constructed at Nauje from timber cut just below the village in an area that had been within the rani ban. Other lodges built or remodeled in the early 1970s also utilized Khumbu timber. This was a factor only in the

Nauje, Tengboche, and Pangboche (for timber for Pheriche lodges) areas, however. Most of the lodges opened elsewhere during the 1970s were simply remodeled from houses and herding huts and thus did not require the cutting of large amounts of new timber.

These early patterns of tourism-related forest use were greatly affected by the establishment of Sagarmatha National Park. Both the Nepali ecologist who scouted the region in 1973 (Mishra 1973) and the 1974 New Zealand mission that recommended New Zealand aid to establish Sagarmatha National Park (Lucas, Hardie, and Hodder 1974) reported that deforestation in the region was a critical problem and were concerned about new demands on forests because of tourism as well as uncontrolled Sherpa traditional forest use. The New Zealand team concluded that the impact of tourism on forests was so severe that, "There is no doubt that, in the present situation, the economic benefits tourism brings to the Khumbu [region] are largely off-set by attendant environmental pressures" (ibid.:11). A few years later the first draft national park management plan placed the blame for regional forest degradation primarily on tourism, noting that:

> The major problems stem from inroads made into the forests of Khumbu for structural timber for Sherpa "hotels" and other buildings to serve the growth of tourism and the increasing use of fuelwood by these tourist establishments and by the many trekking and other tourist parties which camp and cook their way up and down the valleys. (Croft 1976:22)

The measures that were implemented in 1979 in response to this perceived crisis were at first relatively effective. The regional ban on felling trees for timber (other than beams) averted further damage from the use of local forests for building material for lodges.[31] The regulations prohibiting the sale of fuel wood to foreigners and the use of fuel wood by mountaineering expeditions and trekking groups were effectively enforced and significantly lowered tourist fuel-wood use—although they did not entirely eliminate it since porters and staff were still permitted to obtain and use fuel wood for themselves.[32] A fuel depot was set up near the entrance to the park at Jorsale with assistance from the German Alpine Club to make kerosene and other fuels available to trekkers and mountaineers.

These early efforts to eliminate the impact of tourism on forests had one major loophole: no regulations were issued that prohibited or limited the use of fuel wood by teashops and lodges. During the 1970s when there were relatively few lodges and most tourists camped in tents this was not a critical failing. But during the 1980s the growing popularity of independent trekking and the rapid growth in the importance of local lodges made the early national park regulations increasingly inadequate. It is possible that that more fuel wood is now being burned in Khumbu due to tourist demand than ever before. Although well aware of the increasing lodge use of fuel wood for cooking and hot showers, Sagarmatha National Park administrators have felt that prohibiting fuelwood use by lodges would be too politically explosive a measure given the already-tenuous relationship between the national park and local residents. Such a measure, however, has been successfully implemented in the much-touristed upper Modi Khola valley of the Annapurna Conservation Area in central Nepal. Here a ban on fuel-wood use for cooking was enacted in 1987 by a local committee of lodge owners on the advice of conservation area administrators, and a kerosene depot was established (Stevens 1988a). Upper Modi Khola valley lodges have been cooking entirely on kerosene stoves for the past five years. The cost of the kerosene is passed along to tourists through slightly higher meal prices.

Across Khumbu, fuel-wood use by lodges and trekking-group and mountaineering-expedition porters may account for more than 20 percent of regional fuel-wood use (Hardie et al. 1987:38). But this demand for fuel wood is not spread evenly across the region, and tourism has increased fuel-wood demand on particular forests, woodlands, and alpine areas far above the level already exerted by Sherpas for subsistence purposes. The many lodges in Nauje and adjacent Chorkem, for example, probably require as much or more fuel wood during trekking season than the entire Sherpa population of the village.[33] Fuel wood for the lodges is obtained from the same areas where villagers collect fuel wood for household use, the slopes west of the Bhote Kosi river in the Satarma area, the area above the Dudh Kosi river just below Khumbu near Monjo, and the lower slopes of Tamserku. The pressure for fuel wood being put on the high-altitude juniper in the Tugla-Tsolo area in the Lobuche Khola valley region must exceed by many times the seasonal fuel requirements of local herders. The impact of this focused demand on local vegetation will be explored later in this chapter.

Tourism, Resource Use, and Land Use Change

Tourism has thus created several new demands on Khumbu natural resources, some of them through the direct requirements of tourists and their porters and pack stock and some of them through the indirect impacts of changing Sherpa subsistence practices as a result of new patterns of lifestyle and affluence. The direct demand tourism has had on resources in Khumbu thus far has largely added to preexisting patterns of

Sherpa resource use rather than introducing entirely new requirements. The relatively small scale of regional tourism to date and its basis in Sherpa-operated small lodges and camping trekking have kept new resource-use demands to relatively moderate levels. Tourism has not yet brought the major demands on local resources and land familiar in so many other major, international mountain resorts from roads, parking lots, and major hotels (even the Japanese Everest View Hotel has only twelve rooms) to ski facilities, swimming pools, tennis courts, and golf courses. Khumbu is not yet the Swiss, French, Italian, or Japanese Alps, nor is it the Yosemite valley or even an Indian Himalayan resort. Yet even the small scale of tourism development thus far in Khumbu has meant some increase in the use of fuel wood, timber, and building stone as well as more demand on local pastures, greater fodder production, and increased demand for water for drinking, cooking, and bathing.

The new regional and household income that tourism has created in Khumbu through opportunities for employment and business operations has also affected local resource demand and land-use practices by effecting changes in Sherpa standards of living, lifestyle goals, household labor allocation, and community resource management. These changes reflect not only the types and scales of Khumbu tourism but also Sherpas' responses to new opportunities and their decisions about how to use new wealth.

Affluence has led to new types and levels of resource demand, including desires to participate more fully in the "good life" as defined both in traditional terms and in new ones, which include the adoption of some Western tastes in clothing, diet, and electronics as well as new local fashions in architecture such as multiroom houses, wood paneling, ceilings instead of open-beam construction, glass windows rather than translucent paper ones, and corrugated iron roofing. Increasing aspirations for a higher quality of living in traditional terms and the wealth to meet them has meant some changes in land use. Land-poor families have displayed interest in acquiring new land, but have mostly been frustrated in this by the lack of sellers. More important has been the purchase of livestock. Here I would distinguish between the purchase of pack stock which is an investment in tourism and purchase of dairy stock which is intended solely to better family diets. Meeting new tastes and needs has almost entirely involved increasing the importation of goods from outside the region rather than changing resource demands and land use inside Khumbu itself.[34] The trend towards larger houses has had some implications for local forest use, but Sagarmatha National Park restrictions have diverted most of this impact to Pharak. The impacts on regional vegetation and pasture productivity of the new regional herd composition, however, may be more serious. In terms of their way of life affluence has not yet led people to decide that the effort to cultivate crops is no longer necessary, much less to conclude that it is incompatible with changing lifestyles and ideas of status or identity. With pastoralism the issue is again more complex and there has been some decline in interest in nak herding as a way of life.

The question remains of how these changing land uses and attitudes toward resources have affected the operation of local resource-management institutions. There is no doubt that tourism can have major impact on the organization, goals, and effectiveness of local resource-management institutions through introducing social, cultural, and environmental changes. Tourism can undermine the ability of communities to continue to enforce former resource-management regulations by changing the local household economy, patterns of labor allocation, and subsistence land use, commercializing natural resources, introducing new values, and creating new community patterns of wealth, class, and status. I have already noted that new stock-keeping practices, and specifically the keeping of larger numbers of urang zopkio, zhum, and cows, may have played some part in the undermining of the nawa system of livestock management in Nauje and Thamicho. Some observers have also argued that the commercialization of fuel wood played a similar role in undermining the shinggi nawa management of forests: "The traditional forest wardens (shing nawas) had ceased functioning by the early 1970s (although they were still active in Phortse) as the astronomical sums tourists paid for firewood had led to massive cutting that systematically undermined the forest wardens' authority" (J. Fisher 1990:142; see also Nima Wangchu Sherpa 1979).

Yet too much should not be made of the link between tourism development and the decay of local Sherpa resource-management institutions. It is clear that the nawa system of pastoral management has not collapsed throughout the region and that it is still maintained today even in areas such as Kunde and Khumjung which are heavily involved in tourism and where a shift in herd composition has taken place along the same lines of that in Nauje and Thamicho. It also seems incorrect to draw a connection between tourism development and the breakdown of traditional Khumbu forest-management institutions. I argued previously that the most striking example of the abandonment of local forest management, the decline in management of Nauje and Khumjung-Kunde rani ban, resulted from local responses to forest nationalization. The later abandonment of rani ban management at Pare was similarly related to changes in the regional control of forests, in this case by Sagarmatha National Park, and not to tourism. And even at Pangboche, where local use of forests and other fuel-wood resources was indeed greatly affected by tourists, it is not correct to ascribe a decline in local community management to tourism. Prior to the advent of tourism Pangboche villagers managed forest resources in two ways, by banning the importation of freshly cut wood to the village while crops were maturing and by banning all tree felling in the nearby lama's forest of Yarin. The village has never ceased to enforce its regulations against bringing fuel wood into the settlement during the late summer and village-chosen nawa have continued to fine violators. Community protection of the sacred forest at Yarin did indeed diminish for a few years in the 1980s, although it was ultimately restored, but this had nothing to do with the sale of fuel wood to mountaineering expeditions and trekking groups. The main sources of that fuel wood, as I have already remarked, were far beyond the borders of the Yarin forest. The traditional restraints on tree felling were instead relaxed because of national park intervention in granting permits to outsiders (Kunde and Khumjung villagers) to fell trees for beams there and the subsequent Pangboche reaction to this.[35] I do not believe that any Khumbu local forest-management systems ceased to function because the profitability of selling fuel wood to trekking groups or mountaineering expeditions led Sherpas to violate community resource-use regulations and defy shinggi nawa.

Tourism and Environmental Change

> *Nowhere in the Great Himalaya is concern for the environment more intense than in the Khumbu area. Processes of change have brought a plethora of environmental disruption to this formerly remote, unspoiled region. A major factor is tourism and the hordes of overseas tourists and trekkers. (Karan and Mather 1985:93)*

Tourism has both accentuated old pressures on Khumbu's natural resources and environment and introduced new ones. The impact of tourism on Khumbu forests and the increasing accumulation of trash along the trails and at camping and lodge sites have been among the most widely reported Himalayan environmental problems. Less heralded, but potentially perhaps even more serious, is the possibility that tourism development may be contributing to localized overgrazing.

Litter, Pollution, and the Garbage Trail to Mount Everest

The increasing amount of litter along Khumbu trails and the large dumps that have developed at some expedition base camps have shocked and offended many Khumbu visitors and generated some concern in Nepal that this may ultimately diminish the attractiveness of the region to tourists. During the 1980s the accumulations of garbage became a much more visible eyesore at such renowned tourist stops as Nauje, the Tengboche monastery, and Lobuche, and the mountains of discarded trash left behind at base camps by climbing expeditions became notorious. Cleanups organized by concerned foreigners, trekking companies, Nepalese organizations, the Sagarmatha Club (a Khumbu youth organization based in Nauje), and the recently founded Sherpa Pollution Control Committee have periodically rid the main routes through the region of litter and have consolidated, burned, buried, and carried out expedition rubbish from high-altitude base camps. One 1984 expedition collected sixteen tons of trash from the Khumbu glacier Mount Everest base camps alone. But these efforts have only temporarily alleviated conditions, despite Sagarmatha National Park regulations requiring mountaineering expeditions and trekking groups to remove their garbage. Expeditions and groups continue to flaunt this regulation and the national park has not yet developed an effective system of administering fines to violators. The situation has been compounded by the fact that Sherpas themselves contribute to the solid-waste problem by scattering discarded clothes, cans, bottles, and other refuse into village environs, including streams. There are no village dumps anywhere in the region. Much of the trash in the village vicinities is local rubbish, not tourist waste. But there is also a tourist component. Even those lodgekeepers who carefully place rubbish containers in their dining and sleeping rooms usually seem to dispose of their waste in the same haphazard way that villagers handle household garbage. There was also previously little concern with establishing facilities to deal with human waste in upper Khumbu. At such important tourist overnight sites as Gokyo, Gorak Shep, and Lobuche there were no pit toilets or other facilities until the late 1980s, and the construction of a number of new pit toilets at Lobuche may not have entirely solved the problem there, for one of these has been built immediately above the local spring.

Trekkers, trekking agencies, and mountaineering teams have begun to show some signs of taking increased responsibility for policing their own littering and trash disposal, but increased national park and local Sherpa enforcement activities are clearly required. The problem of expedition garbage could be dramatically reduced by requiring all expeditions to post a substantial bond that would not be returned until their camp had been certified clean. If it was not, the forfeited money could be used to fund Sagarmatha National Park cleanup campaigns. Nepal's association of trekking agencies (TAAN) could be required to organize a once- or twice-yearly cleanup of the trails. More trash receptacles could be set out for use by individual trekkers and a campaign could be organized to make them more aware of the importance of using them. Lodges could be required to collect their own trash and keep a waste pit for disposing of it. But it will require other types of programs to address the wider problems of tourist waste. Until a village-based program is put in place to deal with solid waste and to recycle glass, plastic, and metal no number of effective measures to force mountaineering and trekking groups to clean up their camps, or lodges to collect their trash, will entirely solve the current problem. New values must be promoted among Sherpas as well as tourists, for only this will make it possible to effectively establish and maintain a Khumbu-wide waste-disposal and recycling system. Dealing with trash in Khumbu will require procedures for regularly disposing of garbage in local dumps, packing recyclables out of the region, and burning combustible trash. These practices must be carefully developed both to avoid pollution and to avoid offending local religious beliefs. Burning bad-smelling things at expedition base camps, for example, would be considered dangerous by many Sherpas, for it might offend the spirits of the mountains on whose slopes they risk their lives.

A new effort initiated in Khumbu during the summer of 1991 offers some hope that the garbage situation may soon be alleviated. Mingma Norbu Sherpa (former chief administrator of Sagarmatha National Park and now director of World Wildlife Fund USA's South Asian and Himalayan program) approached Khumbu leaders with a proposal to establish a Sherpa Pollution Control Project with the assistance of funds provided by World Wildlife USA. Mingma Norbu suggested that an ongoing, regionwide, community-organized cleanup and recycling program be initiated based on a combination of volunteer efforts and hired labor. The idea won considerable support from influential Sherpas and at a public meeting held in Nauje. A local committee was formed to oversee the project and several different operations were initiated. The first Sherpa cleanup of Mount Everest base camp took place a few days later. More than eighty Sherpas participated and much use was made of hired yak to pack several tons of garbage out of the base-camp area to a site where it was burned and buried. Permanent rubbish dumps have been established at a number of places. The project also opened an office at Sagarmatha National Park headquarters near Nauje where all expeditions are required to leave a deposit based on the number of loads they carry into the region. Thus far there is no provision to inspect base camps to ensure that cleanups have been carried out, but expeditions have to show they are packing out garbage in order to reclaim the deposit left in Nauje. The project has enormous potential, but it remains to be seen whether or not it can be maintained over a period of years.

Tourism, Pastoralism, and the Environment

During the early years of international concern over environmental degradation in Khumbu there were a number of reports of overgrazing. The original United Nation's Food and Agricultural Organization's recommendation for the establishment of a national park cited overgrazing and severe, localized gullying among the environmental problems which made a park necessary (Blower 1971).[36] The New Zealand mission that laid the groundwork a few years later for the establishment of Sagarmatha National Park also reported overgrazing and gully formation (Lucas, Hardie, and Hodder 1974).

These early observations of overgrazing were supported by the first research on Khumbu herding patterns, grazing intensity, and pasture conditions. Bjønness concluded that overgrazing was a significant regional problem and reached this assessment on the basis of a livestock census, a study of the degree of grazing that took place in specific herding sites, and an analysis of pasture conditions on the basis of Landsat satellite imagery. She identified a number of overgrazed areas, including the immediate surroundings of the villages of Pangboche, Khumjung, Thami Og, and Thami Teng and the high-altitude settlements of Tarnga, Dingboche, and Gokyo and concluded:

> Today overgrazing has led to heavy depletion of vegetation in many areas, leaving a remnant cover less capable of holding the soil. Early stages of erosion are much in evidence; the result is not only due to overgrazing, but also to forest depletion partly through firewood collection and browsing. (Bjønness 1980a:63, map and 74)

These findings suggested to her that national park action would be needed to regulate grazing, and she proposed both the development of a rotational grazing system based on grazing permits and regulations to limit the size of household herds.

Subsequent researchers, however, have questioned whether Khumbu is indeed overgrazed (Brower 1987; Byers 1987b). In particular they have cast doubt on whether vegetative ground cover is as poor as has been reported or erosion as serious. Both Brower and Byers suggest that alarm over poor vegetative cover and presumably high erosive rates may have reflected an overreaction to the deceptively sparse vegetative condition of Khumbu pasturelands between late autumn and early spring by observers who were unfamiliar with the way in which Khumbu ground cover improves markedly before the onset of the summer monsoon rains in mid-June. They also note that early observers apparently failed to appreciate that the total summer precipitation in Khumbu is considerably less than in the central and eastern Nepal midlands and that there are few occasions when rainfall is heavy. These conditions of climate and vegetation make for lower erosion rates than had been expected (Brower 1987:278, 280–282; Byers 1987b:229–231, 245).[37] Byers' careful studies of soil erosion uncovered no evidence that serious erosion was taking place even in heavily grazed grassland areas in lower Khumbu. This led him to question whether even the extensive cattle-track terracettes on slopes in the vicinity of villages should be regarded as proof of overgrazing, for the shrubs that grow on terracette risers may act to stabilize the slope by impeding the down-slope progress of eroded material (ibid.:132,134). Byers suggests that "the hypothesis of overgrazed and eroded Khumbu pastures . . . is clearly in need of some revision" and that "the contemporary stability of the slopes and soil solum . . . indicates that they are and have been in balance with land use practices as observed in 1984" (ibid.:232). Brower, who conducted studies of Khumbu grasslands during the summers of 1984 and 1986, concurs, noting that although "Khumbu is a place that has been subject to human manipulation for a long time" and that "the result is a probably radical conversion of natural vegetation," this human impact on local environment "has enhanced its utility to people with livestock, without creating serious environmental degradation" (1987:303).

Yet while present levels of Khumbu grazing do not appear to be causing severe or extensive regionwide erosion, there still remain questions about whether or not localized overgrazing is taking place. Byers, for example, found that erosion from the slopes above Dingboche was much higher than in the sites he monitored in lower Khumbu (1987b). He suggested that the sparse groundcover here was due to heavy grazing and the cutting of juniper and other shrubs for fuel wood.[38] And if vegetation change rather than erosion alone is used as a measure of overgrazing, one might evaluate the impact of current grazing levels in lower Khumbu very differently. Here the issue, for example, might be whether or not current grazing practices are changing the composition of grassland and shrubland vegetation or lowering their productivity in terms of supporting cattle. Sherpa perceptions of changes in the availability and quality of grazing suggest that the productivity of some heavily used areas may indeed be declining. There remains a dearth of data, however, on the nature, pace, and severity of either pasture degradation or grazing-related inhibition of forest regeneration. In the absence of such data it is too early yet to judge whether or not subtle vegetation change is underway and what the role in it may be of the tourism-related practices of keeping larger numbers of urang zopkio, zhum, and cows. The issue of whether or not terracettes are an indication of overgrazing also requires further consideration, for whereas erosion levels may not be high even from these areas, the productivity of the heavily trailed pastures is very probably quite lower than other grassland areas and they may well have a reduced carrying capacity. Heavily terracetted areas appear to have a much higher percentage of bare ground than other grassland areas and a far higher percentage of shrub species not palatable to Khumbu cattle. Only multiyear monitoring of pasture and woodland areas and the establishment of permanent cattle-exclosure study plots is likely to yield data to evaluate the current environmental impact of Sherpa grazing practices and determine the direct and indirect roles of tourism in pasture degradation and poorer forest regeneration.

Tourism and Deforestation

There have been many reports of widespread and severe deforestation in Khumbu due to tourism development. Several observers who have been important in developing Sagarmatha National Park policies, for example, noted in an article titled "Saving Sagarmatha" that

> this mecca is fast becoming an environmental mess. The success of the Park over the past two decades brought more people (over 5000 a year) and with the masses has come the scourge of deforestation because of excessive cutting of forests for fuelwood. Since the watershed is now nearly destroyed, this in turn causes severe soil erosion and devastating spring floods in downstream areas from the Park.
>
> Of course, deforestation is a general malaise in the Nepalese Himalayas, but it is particularly acute in the Park. (Hinrichsen et al. 1983:203)

They concluded that:

> Sagarmatha has suffered more deforestation during the past two decades than in the preceding 200 years. Visitors to the Park are confronted with stark and denuded slopes in many areas, especially around Namche Bazar [Nauje]. Twenty years ago, foreign visitors described the area as rich in forest cover with lush stands of juniper. Today, these forests have been leveled, leaving only isolated clumps of shrub juniper. . . . Deforestation has, of course, been aggravated by tourism. (ibid.:204)

Another outsider familiar with the region, Sir Edmund Hillary, similarly contrasted the forests and high-altitude juniper shrub he saw in Khumbu in 1951 and 1953 with the much changed landscapes of the 1980's. He observed that

> the valley of the Dudh Kosi river was still very beautiful, but the forest was woefully thinned by the axes and saws of the Sherpas, cutting timber for

buildings. . . . The forests around Thyangboche [Tengboche] had lost many of their mighty trees, and the Pangboche area was almost bare. Up the Khumbu Glacier Valley [upper Lobuche Khola valley] there was hardly a juniper to be seen. (1982:698)

Amplifying this testimony, *Newsweek* not long ago reported that:

Dense alpine forests once covered the snowy slopes of Sagarmatha, as the Nepalese call Mount Everest, and the steep sides of its Khumbu valley were lush with a dark green carpet of junipers. But that was the Everest of 1953, when Sir Edumund Hillary and his Sherpa guide Tenzing Norgay, became the first men to conquer the highest peak on earth. Today the dense forest at Everest's base is 75 percent destroyed, replaced by a tumble of rocks interspersed with lonesome trees. (Begley and Moreau 1987:104)[39]

Bjønness reported that the "cutting of firewood is not only the most obvious [impact of tourism] but is also causing *the most critical impact* on the natural environment" and linked large cleared areas along the main trail to Mount Everest to trekking tourism (1980*b*:124,126). She and others have also suggested (incorrectly, as discussed in chapter 7) that the prominent, nearly treeless slopes above Nauje are another example of the destructive impact of tourism and forest nationalization (1983:270).

Such depictions of widespread Khumbu deforestation have been challenged in recent years by Charles Houston (1987) on the basis of his recollections and photographs of forest cover in 1950 and by several researchers (Brower 1987; Byers 1987*a*, 1987*b;* Pawson et al. 1984; Stevens 1983). After visiting Khumbu in 1981 he announced that he had observed little change in forest cover since his 1950 visit there as a member of the first group of Westerners ever to enter the region, claiming that there is "as much or more forest cover [today] than there was in 1950 and I have the pictures to prove it" (Charles Houston, "Return to Everest . . . A Sentimental Journey," 1982; cited in in Thompson and Warburton 1985:120). Byers has questioned earlier depictions of wide-scale deforestation on the basis of a comparison of 1984 conditions and those shown in photographs taken in the 1950s and early 1960s (1987*a:*77–80, 1987*b:*205–221), as has James Fisher (1990:144–145).[40]

It is obvious to anyone who visits Khumbu today that accounts of ongoing, extensive deforestation in Khumbu are considerably exaggerated. The new understanding of historical forest change in Khumbu that is emerging from the analysis of oral traditions, oral history, and pollen samples also questions whether tourism can be blamed for what forest change has taken place. This does not mean that tourism has not had any impact on Khumbu forests, but these have been considerably more minor and highly localized than has been depicted. To assess these impacts it is necessary to change scales from looking for highly visible, regionwide deforestation to exploring possible recent change at the particular sites where the demand for fuel wood by mountaineering expeditions, trekking groups, and lodges has added to local demands on specific areas of forest, woodland, shrubland, and the scattered shrubs of alpine regions. This focuses attention on the specific sites that supply fuel wood to popular tourist overnight or lunch stops on the major tourist routes through Khumbu. On the route to Mount Everest these places are Nauje, the Tengboche monastery, Pangboche, Pheriche, Dingboche, and Lobuche, whereas on the upper Dudh Kosi valley trekking route they are Dole, Machermo, and Gokyo. The fuel-wood gathering sites where the greatest amount of added pressure has been focused are shown in map 22, including the areas near Tugla and Tsola Tso where large amounts of high-altitude juniper are gathered for the lodges of Tugla, Lobuche, and Gorak Shep, and where early Mount Everest expeditions also obtained much of their fuel wood. At these places tourism has indeed added significantly to the demand of local seasonal residents, sometimes dramatically raising the total amount of fuel wood gathered. The sites vary in their capacity to absorb this pressure, for they differ considerably in type of vegetation cover, altitude, and microclimate. No study has yet been done to monitor change in these specific sites, but Sherpas insist that although no extensive areas of forest have been cleared there may have been some thinning of woodlands near Dole (a major source for fuel wood and construction material for lodges at Dole, Machermo, and Gokyo) and that there has been a decrease in the juniper shrub cover in the Tugla and Tsola Tso areas. It is this change in high-altitude juniper cover that Hillary alluded to in his account and which has been exaggerated by the overly dramatic *Newsweek* report. These and other post-1979 changes that Sherpas have noticed in forests and woodlands are shown in the table below and in map 23. Those areas where Sherpas associate vegetation change with tourism (as well as with local use) are indicated with an asterisk. They believe that forest change in the other sites is related solely to local subsistence use.

Yet while Sherpas suggest that in some places tourism has accentuated local use of fuel-wood resources and helped diminish the density of certain small areas of forest and woodland they do not believe that any widespread loss of cover has taken place in the region. They find puzzling the reports that tourism has caused the stripping of the slopes above Nauje and Khumjung, for while they have witnessed localized change in the vegetative cover on those slopes they attribute this to very different causes. They are also puzzled by reports that the Mount

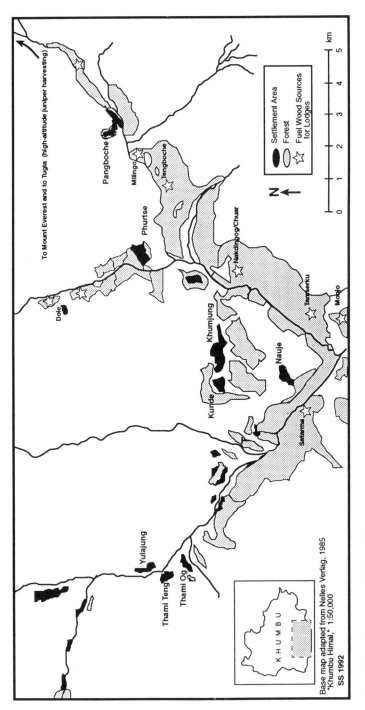

To Mount Everest and to Tugla (high-altitude juniper harvesting)

Pangboche

Milingo

Tengboche

Phurtse

Dole

Nakdingog/Chuar

Khumjung

Tengserku

Nauje

Mojo

Kunde

Setarma

Yulajung

Thami Teng

Thami Og

N

Settlement Area

Forest

Fuel Wood Sources
for Lodges

km
0   1   2   3   4   5

K H U M B U

Base map adapted from Nelles Verlag, 1985
"Khumbu Himal," 1:50,000

SS 1992

Map 22.   Tourism and Forest Use

Table 33.   Post-1979 Changes in Forest, Woodlands, and Alpine Shrub Vegetation

Bachangchang rani ban	thinning and conversion to rhododendron
Mingbo-Chosero lama's forest	clearing and thinning
Phurtse lama's forest	thinning and conversion
Yarin lama's forest*	thinning
Satarma**	clearing and thinning
Chuar**	clearing and thinning
Tugla-Tsola Tso areas**	clearing of high-altitude juniper
Pheriche**	clearing of high-altitude juniper
Dingboche**	clearing of high-altitude juniper

  * Felling of trees for beams for lodges a factor.
  **Tourism fuel-wood use a factor.

Everest base-camp area has been deforested during the past few years, for throughout their lifetimes that area has been located far above the highest trees. Local perspective suggests that many accounts of deforestation have been exaggerated by people either too ready to find a crisis or too quick to attribute to tourism what are really historical changes in the landscape. This assessment has also been reached by preliminary photographic research (Byers 1987*a*). It seems clear that lack of familiarity with the history of Sherpa forest use and the historical extent of Khumbu forest cover has led some people to confuse old and new change and link to tourism deforestation that took place long ago and as a result of other processes altogether. The impacts of local subsistence use on forests both before and after the nationalization of forests have also been underestimated or misattributed to tourism.

Yet it must also be kept in mind that even small-scale forest change may be highly significant for people who depend on forests for critical subsistence resources and have cultural concerns for the protection of sacred places. This seems especially important given that there has been little progress during the 1980s in slowing the demand for fuel wood by lodges. Even the Austrian-Nepal hydroelectric power project under construction near Thami Og in 1991 will alleviate the impact of tourism on forest use only in the Bhote Kosi valley and the Nauje, Khumjung, and Kunde areas, leaving current patterns of forest use unaffected along most of the routes to Everest and Gokyo. In the coming years tourism demand for fuel wood may well yet have an increasingly visible environmental impact, particularly if the government of Nepal succeeds in its plans to increase the scale of tourism.

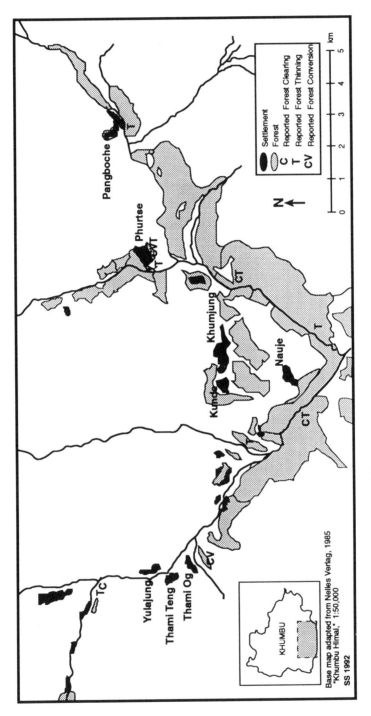

Base map adapted from Nelles Verlag, 1985
"Khumbu Himal," 1:50,000

SS 1992

Map 23.   Reported Forest Change after 1979

Tourism, Land Use, and
Environmental Planning

Tourism development in Khumbu has thus had a variety of impacts on Sherpa land use and the Khumbu environment. Thes have, however, mostly been less dramatic than has been widely assumed by outsiders unfamiliar with Khumbu history, culture, economy, and environment. The subtler environmental impacts of increased urang zopkio herding or fuel-wood use by Khumbu tourist lodges, however, should not be underestimated.

If further study does indeed establish that current Khumbu herding practices are leading to a poor forest regeneration in certain areas and a decline of pasture productivity in heavily grazed rangelands, there may be a need for new regional livestock-management planning. This will clearly be difficult for either national park officials or communities to address. The lack of local customs that would impose ceilings on livestock numbers either in terms of household ownership or the number of head of stock permitted to graze particular areas and the strong value placed on the freedom of individual households to decide which type of stock to herd and where to herd it make any intervention highly controversial and all too likely to misfire. Perhaps the place to start would be with an effort to convince Nauje and Thamicho villagers to reestablish their former nawa regulation of summer pasture use and autumn wild-grass cutting. This in itself would relieve stress on some of the most heavily grazed lower Khumbu areas as well as improve the quantity and quality of hay made from autumn wild grass. This might also have some effect on some families' decisions about the size and structure of their herds. Grazing pressure can also be relieved by developing more supplementary fodder. Many Sherpas are interested in increasing the productivity of their hayfields through better fertilization and the introduction of new grasses and some are closely following the results of pioneering efforts to increase hay production through irrigation. The introduction of new types of grasses should be very carefully considered, for it could well have a long-term impact on regional vegetation if new species become established outside of the walled fields. If this is indeed a threat then a policy about it will have to be developed by Sherpa communities, for national park officials lack the authority to issue orders about land use within settlements. Another possibility for increasing the fodder supply is to import more hay from Pharak, if this can be done without adverse impacts on that region. An increased fodder supply will be especially important if the number of stock grazing in lower Khumbu and the village vicinities continues to rise. Pack stock have become critical to supplying Khumbu's lodges and trekking groups and their numbers may well increase if the scale of tourism keeps growing. The recent arrival of a team of mules from the Annapurna region of central Nepal, owned and managed by people from that area, both suggests that the money to be made in Khumbu transport is enormously inviting and that this may continue to introduce new factors that will complicate regional pastoral management still further.[41] The restoration of communal management institutions in the Bhote Kosi valley should thus be a high priority for both the regional Sherpa government and for Sagarmatha National Park.

Decreasing or even eliminating the use of local fuel wood in the tourism trade seems much more politically feasible. Sherpas have shown much interest in adopting alternative cooking and heating technologies, and the main obstacle to alleviating fuel-wood use for tourists is the cost of developing alternative energy programs. Little real savings of fuel wood have been accomplished by the two small-scale hydroelectric facilities constructed in the 1980s at Nauje and the Tengboche monastery. The 600-kilowatt hydroelectric project now being built in the Bhote Kosi valley may, however, have a great impact on fuel-wood use in western Khumbu, and substantially lower the need for it not only in the lodges of Nauje, Khumjung, and Kunde but also in ordinary village households there and in Thamicho. It does not seem likely, however, to totally fill the role of fuel wood in these communities. All the lodges of eastern Khumbu, including most of those on the trail to Everest and more than two-thirds of the lodges in the national park, will continue to rely solely on fuel wood and dung for cooking. More thought and funding must be given to decreasing fuel-wood use by high-altitude lodges and to alternative energy for Pangboche and Phurtse. The use of more fuel-efficient wood stoves and cooking techniques, kerosene stoves, and small-scale hydroelectric and wind power should be explored, as should such a simple measure as banning the selling of wood-heated showers. Regional alternative energy planning requires careful reconsideration, beginning with a continuing discussion among Sherpas, Sagarmatha National Park officials, and donor agencies to clarify the types and scales of projects needed in the region. A regional energy development plan to coordinate the activities of the diverse organizations interested in Khumbu alternative energy and forestry projects should be another early step.

It would be unwise to relax concern over managing the diverse environmental impacts of toursim simply because these have not yet been catastrophic. The discovery that there is less of a crisis than has often been proclaimed does not mean that the measures taken thus far have been unnecessary. Clearly they have not yet been sufficient to cope with even the relatively small-scale environmental pressures experienced to date. There are no guarantees that future changes in the nature and scale of tourism will not have major new impacts on the regional economy, land use, and the environment. The development of more effective air transport or a road to the region would very likely rapidly swell tourist numbers, with major ramifications for the commercialization of regional natural resources, increased pressure on forests, local shortages of drinking water, and more serious pollution and solid waste problems. Even with existing access, Khumbu tourism shows no signs of leveling off and it can only be expected that if the government goal of quadrupling tourism within the next few years is realized this will have significant impacts on the Mount Everest region. One unfortunate corollary effect of the increasing scale of Khumbu tourism could well be greater interest by non-Sherpa Nepali and foreign entrepreneurs in the operation of Khumbu hotels and other tourist enterprises, a process all too likely to lead to the alienation of land and the loss of current opportunities for both Sherpa entrepreneurs and those employed in the trekking industry. This may already be underway. In the autumn of 1990 the Japanese Everest View Hotel reopened to great fanfare, including the helicoptering to Khumbu of the Japanese ambassador to Nepal and high-ranking officials from the Nepalese Ministry of Tourism. The airstrip at Shyangboche is back in operation, with several flights per day from Kathmandu of a Pilatus Porter aircraft that is leased to the hotel and also frequent helicopter service. Beyond renovating the earlier building the hotel operation has also begun expanding its facilities and may add more rooms. The autumn of 1990 also saw the arrival in Khumbu of the first hotel chain, Sherpa Guide Lodges, owned by a group of California investors in association with Kathmandu businessmen. No Khumbu Sherpas appear to be among the owners. The chain will operate thirteen lodges between the road end at Jiri and Lobuche, near Mount Everest. In the autumn of 1990 lodges were in operation as far as Nauje, and in the summer of 1991 construction was rumored to be about to begin at Dingboche and Lobuche. Reservations for rooms can be made in Kathmandu or booked through one of the foremost U.S. trekking companies. This chain may well begin to compete directly with Khumbu lodges for independent trekkers' business as well as shift some trekking groups from their tents into the chain's lodges. These two events raise major questions about the future of tourism in Khumbu as an avenue of local development, and they may preface an era of larger-scale tourism that may have have very different sociocultural, economic, and environmental impacts on Khumbu.

Sherpas and Sagarmatha National Park authorities have clearly not yet achieved an optimal integration of tourism with local land use and the environment. The challenge to do so only promises to become more difficult in the future.

Conclusions

The Sherpas of Khumbu have thrived for more than four hundred years in one of the highest-altitude regions on earth. Over the course of many generations in the valleys of the Mount Everest region they have devised and refined a set of different economic strategies (reflecting varying degrees of wealth and household labor resources), land-use practices, and resource-management institutions attuned to the opportunities and constraints of their high-altitude homeland. The mixed economy they have devised combines agropastoral exploitation of the natural resources of the highest reaches of the Himalaya with means of access to the agricultural resources of other altitudinal zones of the mountains beyond Khumbu. Sherpa subsistence strategies diffuse the risks of high-altitude agropastoralism and broaden the relatively narrow lifestyles inherent in an economy based on the agricultural possibilities of country above 3,000 meters. At the same time they enabled families to specialize in types of land use that were highly productive in local conditions.[1] Sherpas' historical success with these strategies has given them great confidence in their way of life. For Sherpas Khumbu is not a harsh land, but one which is fertile, bountiful, and valued. Intimate knowledge of local microclimates, crop and livestock varieties, and local pasture and forest resources, and a repertoire of risk-mimimizing practices and emergency fall-back strategies have enabled Sherpas to prosper despite the high variability of crop yields and the very real risks of major livestock loss following harsh winters.

In developing and refining these adaptive practices the Sherpas of Khumbu have demonstrated great creativity, resourcefulness, and adaptability over a period of generations. Their adaptive way of life is not simply a cultural heritage from remote ancestors but a living achievement of integrating knowledge and lifestyles. This creativity, resourcefulness, and adaptability has also long characterized their response to changing economic and political conditions. This too continues today. Sherpas have found ways to maintain environmentally adaptive practices amidst the major economic, resource-use, and resource-management changes of recent decades. In many respects Sherpas today continue to maintain central facets of long-standing Sherpa values, lifestyles, and subsistence practices. Outsiders unfamiliar with present Sherpa society have sometimes been too ready to see the Sherpas as victims of a transformation precipitated by outside forces beyond their experience, control, and ability to adapt to. Sherpas see not only change but also continuity in their culture and society, and recognize in much of the historical change in the regional economy and environment the results of their own choices and their own integration of new ideas, techniques, and technology within long-valued patterns of life. Much recent change, like change during the past century, reflects continuing Sherpa adoption of innovations and their own adaptability to changing circumstances and opportunities from within a context of cultural continuity that includes maintenance of cultural values, certain aspects of their identity and a grounding of land use in knowledge of the land.

This is not to say that there have been no changes in Sherpa life, for clearly there have been many during the past decades. Major social changes may well yet lie ahead as a result of increasing economic differentiation, affluence, education, and other factors. Yet, as of the early 1990s, change in Khumbu does not seem to represent the sharp break with the past which one might call transformation. The recent character of Sherpa land use and other facets of culture instead provide evidence of the type of continuing cultural change within shared values, beliefs, and identity that has long marked the Khumbu Sherpas as a living, dynamic society and culture.

In this book I have been concerned especially with a cultural ecological analysis of Sherpa life and in particular with the central issue of how a people's historical subsistence strategies relate to the ecosystems within and from which they gain their sustenance. This has required an exploration of the ways in which Sherpa land use and lifestyles reflect environmental perception and adaptation as well as other important factors such as cultural values and responses to outside political and economic conditions. Looking at land use in an ecosystemic context has also required an exploration of subsistence practices from the perspective of their environmental impact. In both spheres, and perhaps especially in the evaluation of local resource-management practices and local subsistence-related environmental change, the field of cultural ecology moves beyond other types of study of culture, economy, and land use and management. Here the evaluation of the balance, or lack of it, between resource use and resource availability and between land use and ecosystem dynamics becomes fundamental. Adaptation to the environment cannot be assessed without consideration of the degree to which particular land-use practices and other facets of lifestyle are ecologically sensitive and environmentally sustainable.

Khumbu Sherpas have historically demonstrated both the depth of their local environmental knowledge and the degree to which they have carefully patterned their land use and management practices on the basis of this knowledge. Yet the history of Khumbu resource use and environmental change also suggests that Sherpa forest and pasture management had shortcomings from an environmental standpoint. Sherpa land use both has been deliberately molded to the Khumbu environment and has transformed it. Thus while acknowledgment of the role of indigenous knowledge and ingenuity in Sherpa subsistence is basic to the understanding of Sherpa cultural ecology, so too is a recognition that in Khumbu (as in most other so-called traditional societies) there was no simple, innate "harmony with nature" nor even a dynamic, culturally controlled equilibrium in which villagers ultimately successfully respond to the increasing environmental impacts that may accompany changing patterns of resource use by developing more effective local resource management. Sherpas lived lighter on the land than many peoples of the Himalaya, but the relationship between Sherpas and their high valleys has not been a simple and romantically harmonious one. Instead there has been a complex historical interaction between Sherpa land-use practices, institutions, and values and Khumbu microenvironments. Sherpas' intimate familiarity with local resources and environments did not necessarily provide them with insights into all the processes involved in forest and grassland degradation. Nor did it necessarily furnish them with the ethical and institutional means and the individual and communal will to address perceived environmental changes. Environmental stability has not necessarily ever been a Khumbu Sherpa social goal—much less the preservation of wilderness or minimal interference with ecosystems. There have existed here, as in other parts of the world, stress, conflict, and compromise between awareness of the long-term impact of resource use and inability to address this in the short term. Here, as elsewhere among other indigenous peoples, there have been economic, social, political, and cultural constraints on action. And here too there have been inadvertent as well as deliberate impacts on local ecosystems. These new cultural ecological perspectives on past and present Sherpa subsistence raise important questions about adaptation and its role in regional economic and environmental change, a discussion which in turn has major implications for assessing future regional resource use and management planning and national park administration.

Adaptation

Incorporating history into the study of land use and adaptation raises a number of basic questions. If Khumbu patterns of land use were continually being readjusted during the past hundred years what does this say about the role of adaptation to the environment in Sherpa subsistence practices and in associated areas of culture? To what extent have these changes reflected an evolving adaptation and what processes have been involved?

One of the first lessons of Khumbu history is that diverse factors have been involved in economic change, and that many changes in land-use patterns and resource-management institutions represent Sherpa responses to factors other than physical environmental conditions. These changes may be adaptive, but only in the broader sense of defining adaptation as change that maintains cultural and social values while adjusting economic patterns to reflect new opportunities and constraints in the wider environment, which includes regional economic networks and political economy. Many historical changes in pastoralism, for example, appear to reflect responses to such factors as the market for crossbreed calves, the availability of porters and Tibetan pack stock for transporting goods, a changing investment climate in pastoralism as compared to other commercial ventures, increased costs and difficulties in winter herding in Tibet, and increased income opportunities from pack stock in mountaineering and trekking. In some cases individual and collective responses to these economic and political factors do not appear to have been developed with particular attention to their immediate impacts on local natural resources, much less to their long-term environmental sustainability. Outside factors have also figured in changes in Khumbu agriculture and forest use. Government restrictions on the expansion of cultivated land and reclamation of previously abandoned terraces have contributed to twentieth-century interest in the intensification of production through the adoption of higher-yielding potato varieties. The availability of relatively cheap buckwheat from Pharak has figured in decisions by many households to specialize in potato production. The increased demand for potatoes as a result of tourism may further inspire intensification not only by lodge owners but also by farmers who now may be able to more easily market their surplus locally. Government orders to establish protected forests in the early twentieth century, the implementation of forest nationalization in the 1960s, and the enforcement of Sagarmatha National Park forest-use regulations have all had major implications for regional patterns of forest use. Tourism has contributed to the increased demand for fuel wood from local forests and woodlands despite national park attempts to regulate this.

Some change, however, does seem to represent adaptation to the physical environment of Khumbu. Several Khumbu processes could be interpreted as adaptive in that they seem to promote a closer fit between land use and the perceived environment in terms of maximizing productivity and minimizing risk in the short term or contributing to long-term sustainability by moderating the environmental impacts of land-use practices. These practices reflect local knowledge and evaluation of environment, household strategies, and community institutional responses. The most important of these at the household level are the selection of types of crops and livestock and the timing of agricultural activities and herd movements where concerns over yield and risk significantly influence decisions. At the institutional level adaptation takes the form of local systems of forest and agropastoral management that include regulations to limit resource demands and protect the continued viability of the ecosystems that supply valued resources.

Land-use practices reflect Sherpa environmental awareness in several respects. Cropping decisions and the adoption of new varieties reflect local evaluations formulated from experimentation with crops and crop varieties in different microenvironmental conditions as well as attention to related variation in yield, taste, disease resistance and other qualities. The importance of such evaluations to crop selection is illustrated in the regional differences in the diffusion and emphasis on yellow and development potato varieties. The agricultural calendar also reflects environmental awareness. One of the primary Sherpa strategies for coping with climatic risks in specific places is a careful adjustment of the timing of crop-production practices. Special attention here is paid to planting times, and practices are based on local knowledge that is site- and crop-specific.

Pastoral practice is also shaped by knowledge and belief about environmental characteristics and resources, seasonality, and particular livestock types, sexes, and age classes. Grazing patterns reflect a perception of the requirements of different types of livestock for pasture, fodder, and protection from the elements at different altitudes and seasons. These perceptions influence decisions about herd size and structure as well as movement. Local environmental knowledge also greatly influences the timing of family pastoral movement and figures to some extent in the timing of closing zones to livestock and opening them to wild-grass cutting and grazing even though these are also scheduled to coincide with the local religious calendar. Environmental perception is far from the only factor in planning the seasonal movements of their herds, but it is a major factor in both day-to-day herding decisions and overall seasonal strategy.

The cultural criteria by which Sherpas evaluate crops may also be adaptative in the sense of fine-tuning practice to environmental conditions. Sherpas weigh some criteria more than others in making agricultural and pastoral decisions. Two particularly important concerns at many levels, for example, are the minimalization of risk and the maximization of yields. On the broadest scale the Khumbu Sherpa adoption of a mixed economic strategy including trade and tourism decreases the risks of relying on agriculture in a high-altitude environment where the crop repertoire is limited, the growing season is short, arable land is scarce, and where yields vary so tremendously from year to year. But beyond this, crop growing and pastoral practices reflect concern with anticipating and minimizing losses from a variety of environmental factors. Efforts to decrease risk directly depend on environmental perception and a combination of appeals to supernatural assistance, creating more household luck, and taking preventive and ameliorative measures through adjusting crop and herding practices.

In agriculture there are a number of areas in which risk minimization appears to be important: field siting in terms of soil, aspect, and altitude; the timing of planting and harvesting; selecting crops and specific potato varieties in different localities; raising multiple varieties of potatoes in individual fields; concern to avoid certain associations of crops in both individual fields and in settlements; measures to minimize frost damage and to discourage wildlife, bird, and livestock depredation; efforts to maintain and improve household luck; and, to some degree, the use of multielevational cropping strategies. These individual measures are also embedded in a number of institutional frameworks related to minimizing risk in agriculture, including regulations to prevent blight by controlling the times when fields may be worked and harvested, the spring circumambulation of the fields and other community rituals aimed at protecting the luck of crops, and the implementation of summer livestock bans in major crop areas.

Minimizing risks is also, of course, a fundamental part of herding strategies. Stock types are selected with hardiness for Khumbu conditions as a major concern. Less hardy stock are stabled in the winter and a good deal of trouble is taken to ensure that all stock has fodder to supplement grazing during the grass-poor winter and spring months. Pastoral movements are ordered and timed to diminish threats to the stock's health from vegetation and ground conditions, predators, and insufficient hay and fodder. At the institutional level risk is decreased by the nawa-enforced restrictions on grazing and on grass cutting that conserve the winter pasture and fodder supply and on a spiritual level by the communal herd-protection Yerchang ceremonies held in the high summer herding settlements.

A second major orientation of land use is towards maximizing yield. Khumbu Sherpas do not base their crop selection solely on maximizing yield, for obviously other factors are also involved in land-use decisions including cultural preferences and religious beliefs which may outweigh in some cases the value placed on high yields.[2] But yield is an important criterion for evaluating which crops and crop varieties are considered suitable for planting in given localities. And a general tendency toward individual household and regional maximization of food production seems to have underlain the increasing Khumbu specialization in potatoes and the adoption and diffusion (despite many individual reservations about taste) of new higher-yielding varieties of potatoes. Concern with maximizing yields through attention to microenvironmental and seasonal conditions reflects an awareness of the often extreme variability of Khumbu yields and a priority on achieving as far as possible household self-sufficiency in potato production. The adoption of higher-yielding varieties has been the major mechanism of agricultural intensification in the region, and one which thus far has kept pace both with the increase in Khumbu's population and with new requirements to feed tourists despite an apparent decrease in the area of land under cultivation since the early twentieth century.

There seems to be much less of a cultural emphasis on maximizing household pastoral production through introducing new breeds or increasing herd size. Although ownership of large numbers of yak and nak continues to confer considerable status, there is no cultural concept in Khumbu that subsistence, wealth, or status requires herding. Many families do not own livestock and most Khumbu households do not own enough stock to achieve self-sufficiency in milk, butter, meat, wool, and yak hair. Today there is little use of cooperative herding arrangements or hired herders to get around the difficulties of keeping herds of many more than thirty head of cattle while depending solely on household labor. Yet on a regional level it might be argued that Sherpas undertake a number of efforts to increase stock-carrying capacity, including their emphasis on gathering wild grass for hay, preserving crop residues for fodder, cultivating hay, and enforcing livestock exclusions from the lower-altitude areas of Khumbu. The earlier practice of using Tibetan winter pastures would also have increased total Khumbu livestock-carrying capacity and the twentieth-century exclusion of Gurung sheep from the region would have in effect allowed more Khumbu-owned stock to be herded sustainably. It is interesting that the final conflict with the Gurungs came in the early 1960s at the very time that Khumbu range resources were clearly incapable of supporting the large increase in grazing that had resulted from the arrival of thousands of Tibetans with their stock.

Conservation is a third way in which environmental awareness influences Khumbu land-use practices. Conservation of resources underlies several aspects of land use. Beliefs are widespread, for example, about the necessity to fertilize and rotate crops to maintain soil fertility and obtain good crop yields. Agronomic knowledge includes perceptions of the qualities of various soil additives as they relate to different types of soil and the cultivation of different crops. And although Khumbu resource-management institutions had shortcomings in terms of preserving regional environmental quality they also had limited conservation functions. These institutions enforced regulations that indicated a concern for maintaining land productivity and decreasing land degradation that might affect harvests, livestock health and winter survival, the ability to obtain house beams, and the availability of leaves for fertilizer in the immediate vicinity of settlements.[3] Although communal regulation of herding has been limited in some senses, there is enforced rotation use of some lower-altitude pastures which may conserve the productivity of these areas and protect crops, and it is possible that the temporary closures of some high-altitude pastures to grazing in the Imja Khola and Lobuche Khola valleys may increase summer pasture productivity. Khumbu forest use does not appear to be governed (other than in sacred forests) by the type of cultural conservation ethics sometimes assumed to be characteristic of the region. The establishment of some rani ban, the Langmoche avalanche-protection area and the Teshinga and Phurtse bridge forests, however, reflects a concern with the conservation of particular types of forest resources in particular locations and a response to perceived environmental change and increased resource scarcity.

Change in some aspects of Khumbu land use thus suggests that there are direct links between perception of the environment and efforts to maximize yields, minimize risk, and promote conservation. These adaptive responses to perceived environmental conditions have taken place both on the level of individual practice and that of establishing and adjusting community institutions. Change in economic practices has in some cases been based on change in shared knowledge and beliefs about the environment. In other cases economic change reflects increased social regulation of individual household economic activity in order to achieve more environmentally sound practices. This does not mean that Sherpa efforts to develop more finely tuned, safe, productive, and sustainable land use were the sole or even the most significant factor in economic or cultural change during the past century and a half. Much land-use change has reflected other priorities, and some changing

Sherpa economic practices have had adverse and maladaptive environmental impacts. There has also been historical tension in Khumbu between individual concern over maximizing personal convenience in land use and community attempts to regulate resources. This resulted in violations of communal agricultural and pastoral rules and protected forest regulations which were serious enough in the past to challenge some communities to reorganize their institutions for more effective enforcement, and which have led to the abandonment of local management in a number of places during the past two and a half decades. The long-term adaptiveness of Sherpa communal management institutions continues to be tested by changing socioeconomic and political conditions as well as environmental ones.

Resource Management, Environmental Change, and Adaptation

Khumbu resource management has also not been a tradition-bound custom but rather a set of attitudes and institutions which have been developed in new ways since the nineteenth century. One frequent theme of institutional change here has been a continuing effort to develop more effective enforcement by coming to terms with the continual challenge to communal regulations by individual resource users. This theme continues to be highly relevant today in local attempts to revive effective forest management and in the increasingly widespread problems with maintaining grazing controls. A second type of institutional change reflects several attempts to redefine management systems in order to expand their functions. The early-twentieth-century introduction of rani ban principles of forest management, for example, was a major conceptual breakthrough. So too was the development of the nawa system of agropastoral management which apparently came into its present form during the mid-nineteenth century. Khumbu resource management was a process of continuing experimentation with institutions that had multiple objectives and were conceived and operated within cultural and social constraints. Only when this dynamic nature of resource management is understood and assumptions about the ecological harmony of "traditional" indigenous land use are discarded can the actual functioning of these institutions be considered, adaptation discussed, and historical Khumbu Sherpa achievements appreciated. Adaptation in the sense of adjusting land-use practices to a long-term sustainability within local ecosystems thus becomes a possible individual and social goal rather than a basic trait of indigenous societies and a living process rather than a static state.[4]

Acknowledgement of the dynamic, historical nature of Khumbu resource management, however, also calls for a reexamination of what has previously been treated simply as a long period of traditional harmony with the environment. Khumbu land use and resource management have been shown to represent not only apparently adaptive responses to the physical environment but to other factors as well. The outcome of historically changing land use cannot be assumed to have necessarily been adaptive in terms of a dynamic balance between land use and the environment. Evidence of continuing change in Khumbu landscapes suggests that historical resource use has had environmental impacts. Khumbu resource-management institutions appear to have had limited jurisdiction in controlling individual household resource use and were not oriented toward a goal of regional sustainability in terms of adhering to any local, environmentally optimal ceiling on the number of livestock grazed, the amount of forest used, or the steepness of slopes cultivated. They also did not contain many mechanisms for facilitating flexible responses to local change in environmental conditions.

Historical Sherpa land use and management in the high valleys of Khumbu suggests that adaptation to a difficult environment is one important component of individual subsistence practices, community resource management, and some aspects of cultural change. Yet increased insight into resource use and management decisions and into historical change in the environment as well as land-use practices suggests also that adaptation to the environment is not the only significant factor in land use and that a harmonious accord between resource use and ecological conditions does not necessarily result from a people's long habitation of a particular place. The cultural ecology of Himalayan peoples such as the Sherpas of Khumbu constitutes a continuing adjustment not only to the valleys and slopes where they make their homes but also to wider political economic conditions and the increasing presence of the national government in local and regional affairs. And land use, like other aspects of cultural life, is also continually interwoven with diverse customs, beliefs, assumptions, and perceptions that are culturally shared and valued. Only by considering culture as well as the political economy and the environment in terms of historically dynamic processes rather than as simply "tradition" can the complexity and character of Sherpa land use and historical patterns of land use and environmental change be acknowledged and the creativity and continuity in the centuries-old processes of Sherpa adaptation to and simultaneous refashioning of Khumbu be appreciated. And only by considering environmental impacts as well as environmental adaptation can we recognize not only the role of cultural beliefs, attitudes, values, and shared environmental and land-use knowledge in subsistence practices and resource management but also the extent to which peoples have also historically fashioned their homelands.

In Khumbu this means the discovery of Sherpa perspectives on Khumbu history which testify to centuries of small-scale vegetation change, including changes in the extent, composition, and density of forests and woodlands and the extent and composition of grasslands and shrublands. Oral history also suggests the Sherpa role in these processes that transformed the Khumbu environment on a local scale and that may largely represent the cumulative impact of Sherpa forest use and grazing for subsistence purposes. Such change was not catastrophic, but neither was it controlled and sustainable. The realization of the antiquity of environmental change in Khumbu has overturned earlier assumptions about the impact of tourism as well as about the relative balance between Sherpa subsistence and the Khumbu environment. Old impacts on the environment have been mistakenly interpreted as new ones generated by outside rather than by local dynamics. Observers without a sense of history or local land-use practices have drawn incorrect conclusions from landscapes about the extent, pace, and severity of the impacts of the new demands on forests that tourism brought and have proclaimed a crisis that does not yet exist. The results of that perception have been as unfortunate for Sherpas as they have been for the success of efforts to achieve local support for and participation in national park management.

Continuity and Persistence

During the past century, and particularly during the past thirty years, there has been considerable land-use change in Khumbu. There was so much continuous change before 1960 when Sherpas supposedly simply followed a "traditional" way of life that it seems necessary to raise questions about some of the standard assumptions that outsiders have long made about the "traditional life" of indigenous peoples.

Recent changes have nonetheless been profound. The amount of time Sherpa men spend in earning tourism wages has no parallel in the earlier wage labor some Khumbu families sought as porters for traders or as agricultural day laborers. Most families today have incomes far beyond those of twenty or thirty years ago. This has created changes in lifestyle and land use. New affluence has affected diet, vernacular architecture, local economic differentiation, and local politics. New income has created new patterns of labor migration between lower-altitude regions and Khumbu and has increased the ability to buy land, livestock, labor, consumer goods, and low-altitude-grown grain. Affluence has enabled large numbers of families to invest in tourism ventures, from lodges to zopkio, and thereby become yet wealthier.

These are major changes in the local economy and they have significant social and environmental implications. The land-use changes that have accompanied them, however, have not yet been more significant in sociocultural terms than earlier changes in subsistence practices and the environment created by the series of Sherpa potato introductions, the shifts in the scale of pastoralism of the early century, the end of winter grazing in Tibet, or the introduction of the rani ban. Indeed, nearly forty years into the tourism era, it is striking how relatively little land use has changed. Khumbu families (at least the overwhelming majority of them who have chosen to stay on in Khumbu rather than emigrate to Kathmandu) have not given up crop growing, for example, to focus solely on wage labor or entrepreneurial pursuits. The great majority of fields are still cultivated by reciprocal work groups rather than by non-Sherpa immigrant labor. Crop varieties are not selected on the basis of their marketability. Agricultural technology and techniques are little changed. Only a few Sherpas have experimented with chemical fertilizer or pipe irrigation, and no one has invested in power tools. Nak herding remains the major pastoral emphasis. Communal protective ceremonies remain important in most villages, and in some villages blight-protection and herding regulations are still maintained which impose major inconveniences on agricultural and pastoral practices and daily life due to beliefs about the causes and communicability of crop diseases. Agriculture and pastoralism remain cultural enterprises, a way of life that is deeply entwined with attitudes toward nature, luck, religion, and being Khumbu Sherpa as well as being grounded in shared knowledge of local environmental conditions, risks, and resources.

It has been widely assumed, and not just in the case of Khumbu, that economic transformation means cultural transformation, as if traditional cultural beliefs and practices were too fragile to endure in the midst of modernization, Westernization, or incorporation into the global market system, and that economic practices necessarily condition social organization and culture. There have undoubtedly been places in the world where such transformations have occurred. Thus far, at least, Khumbu does not seem to be one of them.

In many respects Sherpas continue to maintain many basic features of their culture and social organization despite their increasing involvement in tourism and the other changes in Khumbu life during the past forty years. As Sherpas themselves note, for all the new money and lifestyle changes they are still Sherpa, and still distinctively Khumbu Sherpa. Maintaining this identity is not only important to them, it is fundamental to most aspects of their lives.[5] Definitions of what is and is not Sherpa vary among people and can and do change with time, but nevertheless key aspects of this are widely shared throughout the region and between generations. In Khumbu, now as before, social structure revolves around the family and the clan. Many distinctive life-cycle rites are maintained, including the three-stage marriage custom that requires men and women to continue living with their respective parents until their late twenties and usually until well after they have had one or two children. The major community celebrations of Dumje, Yerchang, Losar, and Pangyi are still the great events of the year. The regional Mani Rimdu festivals continue and a major new religious rite, *Bum Tso,* has been conducted several times at Pangboche and at the Tengboche monastery. At the household level families continue to perform the proper rituals every four months when replacing the prayer flags on the roofs of their houses, and make sure the household lu are properly honored and the luck of the household maintained. The religious rituals accompanying the moves up to the summer pasture are still faithfully observed, as is the custom of having lamas perform multiday rituals each year to bless the family house. Fundamental characteristics of Khumbu Sherpa cultural life from language and religion to the gender division of labor continue to be reaffirmed. Buddhism has particularly thrived during recent years as economic prosperity has increased. A great deal of money made in tourism is being used to support religion and visible signs of religious patronage are now evident on a scale that is probably unique in Khumbu history as Sherpas build chortens, prayer walls, and private shrine rooms, rebuild and expand temples and monasteries, and lavishly support monasteries and festivals. They also spend large amounts of cash to hire monks to officiate at life-cycle rites, house blessings, and at family exorcisms and other rites that may involve housing and feeding them for weeks and even months on end. This is not a society that has abandoned religion because of having been corrupted by the monetarization of life but rather a people for whom religion remains vital and whose affirmation of it in such spectacular fashion is a prominent sign of cultural persistence amidst economic change.

The Sherpas of Khumbu are a people of remarkable adaptability, ingenuity, and persistence who thus far at least have been anything but another example of the tragedy of the demise of a traditional people and their culture following the impact of tourism and other new outside cultural, economic, and political pressures. There has been change during the past thirty years, more change, perhaps, than during any comparable period for a century or more. Khumbu Sherpas have reorganized some facets of their economy, their local political system, and their customs. They grow their crops, herd, and harvest forest products somewhat differently than they did during the 1950s, just as they dress somewhat differently, build slightly different houses, and have a somewhat different view of the world and its natural and supernatural beings, forces, and processes. But they remain Khumbu Sherpas, upholding a set of values, institutions, and practices that are rooted in the legacy of their cultural past and still order the tenor of social and economic life in ways that give a distinctive cast to their way of life and set them apart as Khumbu Sherpas from the other peoples of highland Asia. Khumbu Sherpas have changed, but they have not been transformed. Instead they continue to adapt to new conditions, to experiment and innovate, and to reshape their land use and their culture with the same kind of inventive spirit, ingenuity, eye to opportunity, and reaffirmation of basic cultural values that has long characterized them. They have adopted new Nepal-wide systems of local government, participation in a new periodic market system, and work in the tourist trade with the same élan with which they have discovered down parkas, blue jeans, and stereos. They have not yet experienced any major identity crisis as a result. These things are now a part of being Sherpa and are not considered to be incompatible with the deeper outlook on life and customs that continue to define and identify Khumbu Sherpas. In the same way Sherpas have not yet found tending potatoes and herding nak to be incompatible with operating lodges and guiding tourists to the foot of Mount Everest.

The Sherpas' success in coping with the constraints of the Khumbu high country, like their success in coping with tourism, has reflected just this ability to adjust their land-use practices to new conditions and opportunities while continuing to ground them in their knowledge of the environmental realities of Khumbu and fundamental shared values. These historically dynamic land-use and management practices have created a complex, interwoven, regional pattern of economic, cultural, and environmental change. Sherpa adaptation to Khumbu environments has played a fundamental part in their past and present shaping of their way of life and their distinctiveness as a people. But Sherpa patterns of settlement and subsistence have in perhaps equal measure reshaped the incomparable highlands themselves where for four centuries they have with resourcefulness, faith, and celebration laid claim to the highest ground of all.

Appendix A
Major Himalayan Peaks

Eight-Thousand-Meter Peaks[1]

1.	Mount Everest*	8,848m (29,028')	Khumbu Himal	Nepal/Tibet
2.	Kachenjunga	8,586m (28,168')	Kachenjunga Himal	Nepal/Sikkim
3.	Lhoste*	8,516m (27,940')	Khumbu Himal	Nepal/Tibet
4.	Makalu	8,463m (27,766')	Mahalangur Himal	Nepal/Tibet
5.	Cho Oyu*	8,201m (26,906')	Khumbu Himal	Nepal/Tibet
6.	Dhaulagiri	8,167m (26,795')	Dhaulagiri Himal	Nepal
7.	Manaslu	8,163m (26,780')	Mansiri (Manaslu) Himal	Nepal
8.	Nanga Parbat	8,125m (26,660')	Kashmir	Pakistan[2]
9.	Annapurna	8,091m (26,545')	Annapurna Himal	Nepal
10.	Shishapangma	8,046m (26,398')		Tibet

Other Major Khumbu Peaks

1.	Lhotse Shar*	8,400m (27,559')
2.	Gyachungkang	7,952m (26,089')
3.	Nuptse	7,855m (25,771')

* Khumbu peaks

1. From H. Adams Carter 1985. All of the world's 8,000-meter (26,247') peaks are located in the Himalayan and Karakorum ranges. All Himalayan peaks of 8,000-meters or more are shown below. There are also four Karakorum peaks that exceed 8,000-meters; K2 or Chogiri (8,611m; 28,250'), the world's second-highest peak; Gasherbrum I or Hidden Peak (8,068m; 26,479'); Broad Peak (8,047m; 26,400'); and Gasherbrum II (8,035m; 26,360').

2. Nanga Parbat is located in an area of Kashmir that is claimed by both India and Pakistan and is currently administered by Pakistan.

Other Major Khumbu Peaks (*continued*)

4.	Pumori	7,161m (23,494')
5.	Baruntse	7,129m (23,390')
6.	Tengi Ragi Tau	6,941m (22,722')
7.	Ama Dablam	6,812m (22,350')
8.	Kantega	6,779m (22,241')
9.	Cho Polu	6,711m (22,017')
10.	Tamserku	6,623m (21,730')
11.	Tawache	6,501m (21,327')
12.	Arakamtse	6,423m (21,074')
13.	Kwangde (Shar)	6,091m (19,984')

Appendix B
The Sherpa/Tibetan Calendar

Months of 1987 Fire Rabbit Year[1]

1.	Dawa Tongbu	February 28–March 28
2.	Dawa Nyiwa	March 29–April 27
3.	Dawa Sumba	April 28–May 27
4.	Dawa Shiwa	May 28–June 26
5.	Dawa Nawa	June 27–July 25
6.	Dawa Tukpa	July 26–August 24
7.	Dawa Dimba	August 25–September 22
8.	Dawa Gepa	September 23–October 22
9.	Dawa Guwa	October 23–November 21
10.	Dawa Chuwa	November 22–December 20
11.	Dawa Chuchikpa	December 21–January 19
12.	Dawa Chuniwa	January 20–February 17

1. Sherpas follow the Tibetan lunar calendar, itself based on Chinese systems. The timing of New Year's Day and the beginning of the cycle of months varies as can be seen above in the contrast between New Year's Day on February 28, 1987, and February 18, 1988. The 1987 calendar was considered especially late in timing.

Appendix C
Chronology of Khumbu History

? pre-1500	Possible settlement or seasonal herding/hunting use by pre-Sherpa peoples.
ca. 1480–1500	Ancestors of Sherpas emigrate from Salmo Gang region of Kham.
ca. 1533	Kham emigrants cross Nangpa La and arrive in Khumbu.
pre-1667	Settlement in Bhote Kosi valley at Tarnga and other sites. Hermitages established near later site of Khumjung and Teshinga.
ca. 1615 to 1667–1672	Lama Sanga Dorje founds temple and community of devotees at Pangboche.
ca. 1667–1672	Temple above Thami Og founded.
ca. 1717	Sherpas defeated in battle south of later site of Nauje by Dongbu (Rai?) forces. Tribute possibly paid to Sen kingdom which controlled lower Dudh Kosi valley.
1772–1773	Gorkha conquest of eastern Nepal and establishment of authority over Khumbu.
1805	First known record of a Nepali government official visiting Khumbu. Sherpas fined for killing cattle.
1810	First documentary mention of gembu.
1828	First mention of Nauje in Nepali government documents. Proclamation by Nepali government of Khumbu monopoly on trade over Nangpa La.
1830	Founding of Khumjung temple and establishment of Dumje celebration there after quarrel at Thami Og.

429

1836	Tax document gives names of eight pembu and indicates a total of 169 tax-paying Khumbu households.
mid-nineteenth century	Nawa system established for controlling herd movements.
?1851–52	Assassination at Chorkem of pembu Nam Chumbi.
ca. 1850s–1860s	Introduction of potatoes.
1855–56	Nepal-Tibet war.
1860s	Small-scale emigration to Rolwaling from Thamicho.
1885–86	Hari Ram of the survey of India passes through Khumbu en route to Tibet.
1895	Tsepal of Golila-Gepchua (and Nauje) becomes gembu replacing Dorje of Thami Og.
1904	Khumbu contributes yak to Younghusband military expedition to Tibet.
1906–07	Nauje temple founded and consecrated. Thirty-four households from the village participate in establishing the maintenance fund. First Dumje held.
ca. 1912–15	Protection declared for eight rani ban established at order of Rana government.
1916	Founding of Tengboche monastery at direction of abbot of Rumbu (Rongbuk) monastery north of Mount Everest and through efforts of Khumjung Lama Chatang Chotar (Lama Gulu).
1919	Tengboche monastery consecrated by Rumbu abbot. The monastery is given a portion of the Nakdingog rani ban and other lands to administer.
ca. 1920s	Attempted assassination of gembu Tsepal and his flight to Tibet after committing a murder in Nauje.
1930s	Widespread adoption of the red potato (riki moru) and increasing reliance on potato as a staple crop.
1942	New land-revenue system introduced.
1950	First Westerners enter Khumbu.
1952	Spring and autumn Swiss attempts to climb Mount Everest.
1953	Ascent of Mount Everest by Edmund Hillary and Tenzing Norgay on expedition led by John Hunt.
1959–60	Arrival of as many as 5,000 Tibetan refugees in Khumbu.
1964	Lukla airstrip built and trekking tourism begins.
1965	Branch office of the Sagarmatha Zone office established in Nauje.
1965	Weekly market established in Nauje.

1966	Bhumi sudar, land reform and registration regulations.
1972	Construction completed of Shyangboche airstrip and the Everest View Hotel.
1973	Introduction of the yellow potato (*riki seru*) to Bhote Kosi valley from Darjeeling.
1974	Sagarmatha National Park announced in Bonn at an international conference of the World Wildlife Fund by Prince Gyanendra.
1974	New Zealand mission to evaluate New Zealand involvement in establishing Sagarmatha National Park holds village meetings about the national park.
1975	Introduction of the yellow potato to Khumjung from Singh Gompa in north-central Nepal.
1976	Sagarmatha National Park officially gazetted.
1978	District Forest Office established in Salleri.
1978	End of land registration opportunity under bhumi sudar.
1979	Nauje abandons the nawa system.
1981	Introduction of development potatoes (*riki bikasi*) from the agricultural station at Phaphlu in Shorung (southern Solu-Khumbu district).
1983	Nauje hydroelectric system completed.
1984	Tibetan traders resume trade to Khumbu.
1985	Langmoche flood destroys the still-unfinished Thamo hydroelectric project.
1988	Hydroelectric system installed at the Tengboche monastery.
1988	Thamicho abandons the nawa system.
1990	Forest-management committees established by Sagarmatha National Park which give village officials more authority over administering forest use.
1990	Control over forest and wildlife protection between the Phunkitenga Chu and the Mingbo Chu given to the Tengboche monastery by Sagarmatha National Park.

Dates are primarily estimated from oral traditions and oral history. Some early dates are from Fürer-Haimendorf (1979) and Ortner (1989).

Notes

Introduction

1. This includes recent books on Khumbu Sherpas by James Fisher (1990), Ortner (1989), and Fürer-Haimendorf (1984) as well as a wealth of other work. For further references to the literature on Sherpas and Khumbu see chapter 1.

2. Although Sherpas have inspired a considerable amount of research, Sir Edmund's evaluation may be exaggerated. The Thakalis, Tamangs, and Newars of Nepal have also attracted considerable study, and the work to date on Sherpas can hardly surpass anthropological efforts in parts of New Guinea, the Andes, the Amazon basin, Mexico, or Native American–inhabited regions of North America.

3. Here I refer to cultural ecology in a sense which also includes work sometimes referred to by its practitioners as human ecology and ecological anthropology as well as much work in geoecology. For surveys of the history of the field and its theoretical diversity consult (Moran 1979; Orlove 1980). Some of the differences between geographical and anthropological approaches to cultural ecology are discussed by Grossman (1977).

4. I include here the consideration of both land use and resource management which is adaptive to environment (in the sense of representing an effective and sustainable form of use) and that which becomes maladaptive in the sense of being environmentally nonsustainable. Both represent local adjustments of subsistence practices to local conditions. I am less concerned here with adaptation in the larger sense which includes physiological adaptability to environment and such cultural stratagems as vernacular architecture, clothing styles, diet, or patterns of work and rest, through which peoples may adjust to local environmental conditions. I define *subsistence system* for the purposes of this book more broadly than is often done. I emphasize noncommercial land use for household

sustenance, and this is what Khumbu economy is based on. But I also include the full set of economic practices (and particularly land-use practices) and social institutions by which peoples obtain and distribute natural resources. This especially requires attention to the role of trade and tourism in regional and household economies. Income and goods obtained from these activities influence the degree to which Sherpas rely on the resources of Khumbu itself to sustain themselves as well as placing commercial pressures on some Khumbu natural resources.

5. Brookfield called for cultural ecological focus on microregions as a way of working at a scale which includes individuals, stressing that only in this way could human geographers study process through exploring perception and decision making (1964). More recently he and Blaikie reemphasized this point in their stress on studying "land managers," the individuals and groups who make land-use decisions, while at the same time noting the need for attention to wider regional, historical, and political economic processes and conditions (Blaikie and Brookfield 1987).

6. The importance of linking the often local and microregional scale of cultural ecology to larger spatial contexts has been especially well argued by Blaikie and Brookfield (1987), who coined a new term—*regional political ecology*—for the larger regional, national, and international analysis.

7. This dissertation has recently been published (*Sherpa of Khumbu: People, Livestock, and Landscape,* Delhi: Oxford University Press, 1991).

8. In the nearby Sherpa-inhabited region of Shorung, by contrast, excellent work has been done by the German anthropologist Michael Oppitz (1968) and the Sherpa lama and historian Sangye Tenzing whose work is discussed in Ortner (1989). Both used written clan histories and oral traditions to reconstruct some details of the legend of the migration by the ancestors of the Sherpas from easternmost Tibet to Nepal in the sixteenth century and the history of Shorung.

9. Although it should be noted that Ortner (1989) has made an effort to consider the history of the common people and the ways in which this underlies and helps shape events that both Sherpas and outsiders often attribute simply to the actions of a few prominent individuals.

10. Chapter 5 discusses contemporary forest-use practices, but also provides background on how forests were used and locally managed before the establishment of Sagarmatha National Park.

11. These mapping activities were enriched by "walkabouts" with local experts such as Konchok Chombi and Sonam Hishi. Walking the country provided insights into the history and meaning of places. Particular views and sites often evoked memories and recitations of oral traditions, calling forth glimpses of the past and perspectives on change as it has occurred in particular places.

1: Sherpa Country

1. Geographically the Himalaya is usually distinguished from adjacent mountain ranges such as the Karakorum, the Zanskar and Ladakh ranges, and the Hengduan Shan. It is generally considered to extend from Nanga Parbat

(8,125 m) in the great bend of the Indus river to Namcha Barwa (7,775 m) in the great bend of the Tsangpo-Bramaputra river and is bordered on the south by the Ganges plain and on the north by the Indus-Tsangpo depression and the Tibetan (Qinghai-Xizang) Plateau. The name of the range is ancient Sanskrit and means the Abode of Snow.

2. All of the world's peaks of 8,000 meters or more in height are located in Asia. Ten of the fourteen 8,000-meter peaks are located in the Himalaya and eight of these are located wholly in Nepal or on its borders with Tibet and Sikkim.

3. In some regions of eastern Nepal, including the Mount Everest area, the Great Himalaya also forms the watershed between Nepal and Tibet, with southern drainage to the Ganges and northern drainage to the Tsangpo. Elsewhere the watershed lies well into Tibetan territory and transverse rivers such as the Sun Kosi and the Arun cross the Himalaya en route to the Ganges. In western Nepal the Great Himalaya runs far south of the Tibetan border and does not form the watershed.

4. The mountains of the Mount Everest region, like the rest of the Great Himalaya, primarily consist of highly metamorphized Precambrian gneisses that have been overthrust along the Main Central Thrust Fault upon older formations. In the highest reaches of the Khumbu Himal, and particularly on the upper heights of Lhotse and Mount Everest, the black gneisses (dark-colored paragneisses) are overlaid by sedimentary rock formed of sediments laid down in the Tethys Sea, the ocean that was displaced by the collision of India and Eurasia and the subduction of the Indian plate. These rocks are either unmetamorphized or only slightly metamorphized. The Everest metasediments include biotite schists, marbles and phyllites, and grey crystalline limestones (Vuichard 1986:44–47).

5. In Nepali the word *himal* may refer either to a mountain range, as it does here, or to an individual peak. The Mahalangur Himal could be translated as the Range of the Great Ape, possibly a reference to this as *yeti* country.

6. *Mount Everest* and *Lhotse* are foreigners' names for Khumbu peaks. So too are the names of a number of other famous peaks such as Nuptse and Pumori. In some cases these peaks had no traditional Sherpa names. Mount Everest was called Chomolungma by Khumbu Sherpas, as it was by Tibetans, and was considered to be the home of Miyolangsangma, a goddess of nourishment and wealth. The meaning of Chomolungma is not certain, though it is often translated as Goddess Mother of the World, or Goddess of the Wind (for a more detailed discussion see Bernbaum 1990:7). The Nepali name for the mountain, Sagarmatha, means Forehead in the Sky, or Forehead Touching the Sky. Sherpas often now refer to the mountain as Everest, a name which honors Sir George Everest, the early surveyor general of the Survey of India, whose work helped establish the baseline from which the height of the mountain was first determined in 1852.

The only 8,000-meter peak in the Mahalangur range located outside of Khumbu is Makalu (8,463m), just beyond the high ridge which defines the eastern border of Khumbu. Lhotse Shar (8,383m), a subsidiary peak of Lhotse, is sometimes counted as a fourth 8,000-meter Khumbu Peak.

7. This range was also uplifted by the same process which produced the Khumbu Himal crest to the north and consists mainly of migmatitic and granitic rocks (Vuichard 1986:48).

8. The northern stream is also known locally as the Pheriche Chu. I follow the Lobuche Khola usage established by the National Geographic Society in its 1988 map of the Mount Everest region. The names in Western languages and Nepali used today for Khumbu rivers were bestowed by outsiders. *Kosi* and *Khola, Bhote* and *Dudh* are Nepali not Sherpa words. *Kosi* and *Khola* are Nepali terms for rivers. *Dudh* is Nepali for *milk* and *Bhote* for *Tibetan*. How long ago these names were bestowed on local features is unknown. Two of the earliest European maps of Nepal, the 1811 Kirkpatrick map and the 1819 Buchanan-Hamilton map (Gurung 1983:18–21) show the "Dud Kosi." Khumbu Sherpas conceive of rivers in a rather different way than Western geographers. They do not refer to the entire length of a river as a single entity. Instead they refer to a river (*tsambu*) or stream (*chu*, which also means water or spring) in terms of particular reaches of water named according to the settlements and regions along the bank. An individual river may thus be referred to locally by many different names.

9. Hagen et al. (1963) and Vuichard (1986) suggest that the Bhote Kosi may predate the rise of the Great Himalaya. Hagen et al. note that the present valley appears to be far too large for the river flowing through it today to have fashioned it and suggest that previously it was the watercourse of the Dzakar Chu that now flows north from the Nangpa La into the Arun river in Tibet. Vuichard also suggests that this Bhote Kosi–Dzakar Chu system may have originally flowed northward, draining the slopes of a pre-Himalayan range located in the vicinity of the present Numbur-Kantega range.

10. There is some disagreement over the naming of the reach of river between the confluence of the Dudh Kosi and the Imja Khola and the confluence of this new stream with the Bhote Kosi. Some authorities refer to this as part of the Imja Khola (e.g., Schneider 1963). Sherpas, however, today refer to the same reach of river as part of the Dudh Kosi. I have adopted the Sherpa approach.

11. Sherpas did not name glaciers and the first cartographers to produce detailed Khumbu maps bestowed the names which are now in use (Schneider 1963). Often the names of local grazing areas and herding settlements were applied to adjacent glaciers. One of these, the Lungsampa Glacier in the upper Dudh Kosi valley, is misnamed, as the grazing area by that name is actually located in the next valley to the west alongside what is now known as the Sumna Glacier.

12. A more detailed treatment of altitudinal zonation of vegetation is provided in chapter 5, while soils are taken up in chapter 3.

13. Climatic data have been collected by the government of Nepal in Nauje since 1948 (Joshi 1982:399) and for shorter periods of time at Shyangboche, Khumjung, Tengboche monastery, and Lhajung (Byers 1987b; Muller 1959). Data for Nauje and Tengboche monastery, the two stations where they have been collected for the longest period, may have sometimes been based on haphazard measurement and recording.

14. The relatively dry conditions of these valleys enable agriculture to be carried out at higher altitudes than in wetter regions of the Himalaya. High rainfall and a high percentage of cloud cover during the summer encourages the development of cloud forest at altitudes above 2,500 meters on windward slopes in many areas of Nepal (Metz 1989:161).

15. In Nauje, for example, 89 percent of the annual mean rainfall (1949–1977) of 1,048 millimeters fell between May and October and 79 percent of it fell between June and September (Joshi 1982:400).

16. Occasionally autumn hurricanes in the Bay of Bengal cause torrential rains in the Himalaya which can fall as heavy snows. In October 1987 such a storm dropped more than 100 millimeters of rain on Nauje in less than twenty-four hours and heavy snowfalls occurred above 3,500 meters which blanketed much of Khumbu for many weeks.

17. The settlement of Milingo is also sometimes referred to as a village, although today only six families regard their dwellings there as their main house.

18. Occasionally Khumbu Sherpas say that Pharak begins at the first Pharak Sherpa village, Thumbug (Nepali Jorsale), rather than at the confluence slightly upstream. Similarly some people consider other Sherpa regions to begin with the first settlement that is culturally distinct from neighboring ones.

19. According to legend the Amphu Laptsa was crossed long ago by Rais from the upper Hinku valley to the southeast, who unsuccessfully attempted to settle Khumbu before the arrival of the Sherpas. There are no traditions of Khumbu Sherpas using the Amphu Laptsa or the Mingbo La other than when accompanying foreign exploring, surveying, mountaineering, and trekking tourist parties.

20. Another road has now reached a similar distance southwest of Khumbu and may well ultimately be extended to Salleri.

21. For an eyewitness account of the construction of the Lukla airstrip and a discussion of its consequences see James Fisher (1990). The Lukla airstrip has become the main conduit for mass tourism into the region. Shyangboche airstrip was used very little during the 1980s and is only suitable for the very small Pilateus Porter planes. In autumn 1990, however, the Everest View Hotel began running a regular Pilateus Porter air service to Shyangboche and helicopter charters to the hotel also became common. In recent years the Royal Nepal Air Corporation has kept Lukla airport open for limited service during the monsoon months.

22. The sacred peaks, however, are not places which are set aside from all resource use as has sometimes been suggested (Bjønness 1986). Khumbu Sherpas object to attempts to climb Khumbila and certain mountains sacred to particular clans, but they do not consider the slopes of these mountains off-limits to forest use, grazing, or agriculture.

23. See Berreman (1963b) and Karan and Mather (1987) for discussions and maps of Himalayan culture regions. The distinctively long-sleeved Tibetan and Sherpa cloak style, with sleeves extending a foot and more longer than the arm length, seems likely to have been adopted from Tang dynasty Chinese styles. *Zhi* stones may be a rare form of agate and are so valued that a single stone may cost three thousand dollars or more today in Khumbu.

24. The Bhotia peoples of the high-altitude Tibetan border regions of the Great Himalaya and the trans-Himalayan valleys of the Inner Himalaya (including the Sherpas) comprise less than 1 percent of Nepal's total population. According to the 1981 census (Nepal. Central Bureau of Statistics. 1984) there were 73,589 speakers of "Bhote-Sherpa" (a bureaucratic category rather than an actual language), accounting for only 0.5 percent of the total population of the country. If peoples such as the Tamangs (522,416) and Gurungs (174,464), some of whose ancestors according to legend came from Tibet (Fricke 1986:29–30; Holmberg 1989:14, n. 8; Messerschmidt 1976b), are included, the percentage of Nepal's population descended from immigrants from Tibet rises to at least 5 percent. It has been suggested that the 1981 census considerably underestimated the number of Tamangs and may also have underestimated Bhotias. The reported decline in the number of Sherpas in the Solu-Khumbu district between 1971 and 1981, for example, appears much too high.

25. The Nepali word *Bhote* and the anthropological term *Bhotia* are derived from *Bhot,* an old name for Tibet. It is used by a number of lower-altitude, nonethnically Tibetan peoples in India and Nepal to refer to the ethnically Tibetan peoples living on the border of Tibet.

26. While some of the inhabitants of the Taplejung region of northeastern Nepal (and particularly the village of Gunsa and other parts of the Kambachen valley near Kachenjunga) are often referred to as Sherpas (Bremer-Kamp 1987; Sagant 1976), their relationship culturally and historically to the other groups who call themselves Sherpa has not yet been established. The complex multicultural settlement pattern of Helambu and some other areas in which Sherpas form a component of the population is discussed below.

27. The Solu-Khumbu region (including Pharak) constitutes a large part of the district of Solu-Khumbu, a district shared also with substantial Rai and Nepali populations who actually outnumber the Sherpas. J. Fisher gives the number of Sherpas in the Solu-Khumbu district as 17,000 (1990:55), whereas the 1981 census puts the total at 15,166 out of a total district population of 88,245. Estimates of the number of Sherpas outside of Solu-Khumbu are far rougher. Hutt (1986) suggests that there may be approximately 20,000 Sherpas in Nepal. There may be another 7,000 in the West Bengal district of India, most of them in the Darjeeling area (J. Fisher 1990:55). Twenty thousand may be low for the total in Nepal. There may be as many as 25,000 or even 30,000 depending on how "Sherpa" is defined and whether or not the "Sherpa" population of northeasternmost Nepal in the Taplejung-Gunsa region or part of the population of Helambu are included. In Solu-Khumbu and adjacent areas (Olkadunga, Khotang, Ramechap, and Dolakha districts), however, it is likely that all or nearly all of the population classified "Bhote-Sherpa" by the Nepal government in recent censuses is Sherpa. The total number of people recorded in the 1981 census (Nepal. Central Bureau of Statistics. 1984) as native speakers of "Bhote-Sherpa" in these districts alone is 26,473. (In a number of other districts it is impossible to use census figures as even a rough guide to count Sherpas since these areas are inhabited by both Sherpas and other "Bhotia" groups).

28. It is somewhat ironic that Tenzing Norgay's achievement on Mount Everest made *Sherpa* a household word around the world. Tenzing Norgay grew up in the Khumbu village of Thami Og and married a woman from nearby Chanekpa before he emigrated to Darjeeling to seek his fortune in mountaineering. But he was not a Sherpa in the strictest sense, having been born in Tibet of non-Sherpa parents, and would be considered to be a "Khamba" by Khumbu Sherpas. This important social distinction is discussed later in this chapter.

29. All the pre–World War II Everest expeditions were forced to recruit their Sherpa high-altitude porters in Darjeeling and to approach the mountian via Tibet.

30. Among the most notable anthropological contributors to this still-continuing exploration of Khumbu society and culture have been Ortner (1978, 1989), Pawson, Stanford, and Adams (1984), Pawson et al. (1984), J. Fisher (1990), and Adams (1989). Fürer-Haimendorf's work (1964, 1979, 1984) is worth special note as still constituting the finest, broad ethnographic treatment of the Khumbu Sherpas. Ortner's analysis of the beliefs, values, and cultural patterns which underlie Sherpa social behavior, ritual, and history (1989) is also exceptional, although it should be read while keeping in mind that her earlier work (1978) is based on fieldwork in Shorung and that these observations and insights are not always fully applicable to Khumbu.

31. For a fuller review of the geographical literature and pertinent reports by physical scientists and national park administrators see Byers (1987*b*).

32. That these basic characteristics are widespread and have been relatively enduring, however, should not obscure historical and regional differences in Sherpa culture—a point which Ortner has also recently raised in discussing the difficulties of identifying a Sherpa cultural style or ethos (Ortner 1989:4–6).

33. I base these brief remarks on Sherpa identity and the discussion later in this chapter on Sherpa regions on my experiences discussing regional geography with Sherpas in the Solu-Khumbu district and adjacent areas. The term by which Sherpas refer to Nepalis, *rongba,* is also used by Tibetans and signifies people of the lower valleys.

34. In many cases there are clear differences between Sherpas and non-Sherpas on all four of these points. Tibetans and Tamangs share with Sherpas a faith in Tibetan Buddhism and are therefore considered, at least by Khumbu Sherpas, to be in a special class of peoples who are also "believers." But distinctions among Sherpas, Tibetans, and Tamangs are easily drawn on the basis of the other three as well as many additional points.

35. Many Khumbu people today speak of the Sherpas as a people who mountaineer and work for trekking companies and who raise yak and potatoes. Some of these features may indeed set contemporary Khumbu Sherpas apart from non-Sherpas or from some other Sherpa groups, but, like the characterization of some recently popular styles of dress or architecture as distinctively "Khumbu Sherpa," even such long-standing traits as mountaineering prowess and potato cultivation have not always set Khumbu Sherpas apart from other Sherpas.

36. If Sherpas did indeed migrate to Nepal from the eastern Tibetan area of

Kham (see below) it seems likely that their language may have even closer linguistic similarities to eastern Tibetan dialects.

37. Most Sherpa marriages are monogamous, although like Tibetans, Sherpas also tolerate polyandrous and polygynous arrangements. These, however, are very rare in Khumbu today and for at least the last several generations polyandry was not nearly as common in Khumbu as it was in Tibet.

38. Much of Tibetan society, including that of the central region of Tibet, is apparently not organized along a clan basis. Fürer-Haimendorf (1984:187), however, reports that Tibetans from the Kham region, the reputed original homeland of the Sherpas, do have clan affiliation.

39. The Nyingmapa is the oldest of the four major sects of Tibetan Buddhism (the name itself refers to "old ones"). It developed out of the introduction of Tantric practices by the Indian adept Padmasambhava, called Guru Rinpoche (Precious Teacher) by Tibetans and Sherpas. Nyingmapa is noted among the Tibetan sects for its rites of exorcism and other magical practices and for its married village lamas who are caretakers of village temples and preside over local life-cycle ceremonies.

40. Shrines and temples are constructed on Tibetan models as described in Tibetan texts. Some use may also be made of Tibetan geomantic principles, that are recorded in certain other texts. House styles differ in some important respects (particularly in Sherpa use of pitched roofs and noncourtyard design). Houses more similar to Sherpa design, however, are found in eastern and southeastern Tibet.

41. MacDonald (1980:141) also notes oral traditions of early administrative control of Khumbu by Tibet, Tibetan claims to which were given up after Nepali authorities sent troops to the region to rectify the situation. He relates this story, however, in the context of an era in Khumbu history which he dates to 400 years ago. This would put the Tibetan administrative presence in Khumbu long before the unification of Nepal in its current sense in the late eighteenth century. Suggestions of Tibetan administrative activity in Khumbu in the nineteenth century are still more startling, for Khumbu was supposedly incorporated into the Nepali state in the late eighteenth century and was visited as early as 1805 by a Nepalese official (Stiller 1973:265). If there is any substance to these legends it would testify not only to the ambiguous allegiances of Khumbu Sherpas of that era but also to the remoteness of the region from the centers of administration of both Nepal and Tibet. According to Khumbu traditions Sherpas paid tax simultaneously to both Nepal and Tibet in the early nineteenth century and discontinued this only after the defeat of the Tibetans in this area by Jung Bahadur Rana's forces in the war of 1855–1856.

42. G. Clarke (1980a, 1980b) has observed that many of the inhabitants of Helambu (Yelmo) who have often been called "Helambu Sherpas" or who call themselves Sherpas, are actually not related to the Sherpas of Solu-Khumbu. He traces their descent to intermarriage between Tibetan immigrants from the Kyirong area and Tamangs from the lower country just to the south. There are also a number of families, especially in the large village of Tarkyagang, who are descended even more directly from more recent Tibetan immigrants and in the

settlements of southern Helambu there are families who consider themselves fully members of Tamang clans. Yet in 1986 I met some families calling themselves Sherpa who trace their origins to migration from villages in the Balephi Khola valley just to the east of Helambu who refuse to intermarry with Tamangs. I have also been assured by Sonam Hishi, who has traveled in both the Balephi Khola and Helambu regions, that the Balephi Khola villages are indeed Sherpa settlements.

43. Some of the Sherpa settlers in the Arun region left Shorung around 1825 (Fürer-Haimendorf 1975:117). Others emigrated there from Shorung and Pharak later in the century. In the villages of Navagaon and Tashigaon there are families of several Solu-Khumbu clans (including Salaka, Chawa, Pinasa, Gole, Lamaserwa, and Goparma), and some elderly individuals trace their origins to their ancestors' migration from Solu-Khumbu settlements four or five generations ago. The Chyangma area was settled by a group of Shorung migrants between 1725 and 1750 (Oppitz 1968). Rolwaling was settled by families from western Khumbu around 1860 (Sacherer 1975; 1981:157). Sacherer notes that early Rolwaling settlers were poor families, some of which included people who were escaping from bad debts or fleeing after having committed crimes. Aziz (1978) has noted that some of the Tibetans who came to Khumbu in the nineteenth and early twentieth centuries were also people fleeing exactly the same kinds of situations. Similar stories go back as far as the tale of Dzongnangpa (see below) and the early days of Khumbu settlement. Some Sherpa emigrants to Darjeeling also may have migrated not so much because of the allure of career opportunities there as to leave behind awkward situations in Solu-Khumbu.

44. Sherpas began settling in Darjeeling in the nineteenth century, drawn by wage labor and trade opportunities in the then summer capital of the British Raj. By 1901 there were already 3,450 Sherpas in Darjeeling according to A. J. Dash (*Darjeeling* 1947, cited in Ortner 1989:160). Beginning in 1907 Sherpas became noted for high-altitude mountaineering and this became a major source of work during the period before World War II when Darjeeling was a major mountaineering center. By 1951 the number of Sherpas in the region had reached 7,539 (B. Miller, Ph.D. diss., University of Washington, 1958; cited in Ortner 1989: 217 n. 4). These Sherpas included emigrants from Khumbu as well as from Shorung and Pharak. The relative percentage of Khumbu migrants, the degree to which they hailed from different parts of the region, or the effect that this emigration had on Khumbu population density and land use are not known. In 1984 I was told by Sherpas in Darjeeling that their numbers had declined since the 1950s. A number of families were said to have moved back to Nepal following the eclipse of the mountaineering and tourism industry in Darjeeling by tourism development in Nepal.

45. Farther up the Arun are communities of non-Sherpa Bhotias known as Shingsawa.

46. Such cultural diversity within a population often considered to be a single "people" is not unique in the Himalaya. Berreman, for example, has noted considerable cultural variation among the Pahari of northern Uttar Pradesh (1960). The Rais are perhaps an even more striking example. Anthropologists

have discussed a dozen different Rai groups that are said to speak mutually unintelligible dialects (McDougal 1979) and among whom religious practices, architecture, and other facets of life are by no means uniform. Ethnic diversity among the Tamang has also been noted (Fricke 1986; Holmberg 1989).

47. Such marriages may have been forbidden, however, in Shorung.

48. In many other spheres, however, cultural variation within Khumbu is minor. Dialects, for example, do not differ nor do any customs of building houses or wearing a distinctive village style of dress.

49. Khamba is generally considered to mean "people of Kham", the region presently divided between the easternmost part of the Tibetan Autonomous Region and the Chinese province of Sichuan. In the sense in which Khumbu Sherpas use it the word means an immigrant from Tibet or from a Bhotia-inhabited region of the Himalaya who arrived in Khumbu before the great influx of refugees in 1959. Khumbu Sherpas even speak of Thakali, Nupri, and Lowi Khambas, Bhotia people from Thak Khola, Nupri, and Mustang. The very similar word Khyampa means "wandering" in the Tibetan dialect of the Humli-Khyampas of northwestern Nepal (H. Jasehko, *A Tibetan-English Dictionary* 1968; cited in Rauber 1980:59), and perhaps the Khumbu use of the word carries a little of that connotation.

50. Khumbu Sherpa society is much more ethnically varied than is the Shorung Sherpa population. There are very few Khambas in Shorung.

51. The number of lowland Nepalis stationed in Khumbu has increased from 84 in 1970 to 339 in 1982 (Fürer-Haimendorf 1984:32). Pawson counted 187 Tibetan refugees in Khumbu in 1982, a significant decline from the 274 counted in 1970 by the Langs (Pawson, Stanford, and Adams 1984:75). I know of fourteen Tibetan refugee families in Nauje, six in Khumjung, and one in Thamicho. There may be a few others.

52. Adams (1989:175–176) found in 1987 that only 45 percent (64 of 141) Khumjung household heads were considered to be Sherpa on the basis of their clan membership, while 43 percent (60) were Khamba or Khumbuwa. Presumably the "Sherpa" status here referred to "old" clan members. The remaining 12 percent of Khumjung households were Tibetans, a blacksmith family, a Tamang family, and several lowland Hindus who mostly came to the area as schoolteachers.

53. They may, not, however, have fully tallied the Bhote Kosi valley population since they reported an astonishingly high 32.5 percent of the houses there empty and presumably abandoned (Pawson et al. 1984:244).

54. The number of dwellings in most of these places is larger than the figures suggest since some families own houses in more than one village and I have tallied each family only once. Monks and nuns are not included in the count, nor are people who have retired to religious hermitages (*tsamkhang*). Some families who maintain their main residence in a community other than one of the main villages, such as the gunsa settlement of Thami Og in the Bhote Kosi valley or the secondary, high-altitude agricultural sites of Dingboche and Tarnga, may also have been missed in the count.

55. The full, multialtitudinal settlement pattern is most highly developed in

the Thamicho region. In this twenty-two-kilometer stretch of the Bhote Kosi valley and its tributaries there are forty-seven settlements situated at altitudes ranging from 3,400 to 5,000 meters. Only three of these are main villages.

56. *Yul* is used by Khumbu Sherpas to refer to village, valley, and region. Main villages are distinguished locally from other settlements in terms of function rather than simply by size, for some high-altitude secondary settlements (Dingboche, Tarnga) are larger than some main villages in terms of the number of houses in them. Main village residence defines one's participation in such important festivals as Dumje, Losar, and Pangyi; eligibility and responsibility to hold certain types of community offices; the site of ritual practices such as the placing of prayer flags on dwelling roofs three or four times per year; and the site of life-cycle ceremonies such as weddings (*zendi*). Main village houses are larger than those owned in secondary settlements and families usually spend more months per year living in them.

57. The sandy surfaces and level expanses of these three sites suggest that all three may once have been glacial lake beds.

58. The origins of both the names Nauje and Namche Bazar, the Nepali name, have been lost. Some Sherpas suggest that Nauje might be derived from the phrase *nating uk che,* which refers to a big forested "corner."

59. There were no bazaar or shops in Nauje prior to 1965. Tibetans and Rais instead visited Sherpas in their homes in order to trade.

60. Although lodges and shops are rare in these villages their economies today nevertheless are entwined with tourism. Most families depend on income from mountaineering and trekking work to purchase grain at the Nauje weekly market. Some also operate lodges in the high-altitude herding settlements on the approach to Mount Everest.

61. The origin of the name Khumjung is also uncertain. According to Ortner's reading of the nineteenth-century account of the Indian Survey pandit Hari Ram the name comes from Khumbu Dzong and was the home of the leading Khumbu administrator (Ortner 1989:23, 92–93). Fisher disagrees with this interpretation and suggests instead that the name signifies Khum Jung or Khum valley (J. Fisher 1990:xv–xvi). Perhaps this is short for Khumbu Jung, Khumbu valley. I can cast no further light on the question other than to suggest that Ortner has misinterpreted Hari Ram's account of his 1885 trip up the Bhote Kosi river and that his Khumbu Dzong is not at Khumjung but rather in Thamicho (the home of the *gembu,* the foremost officials of the nineteenth century), perhaps in the village of Thami Og. I have not heard of any oral tradition that makes direct reference to the site of the Khumbu Dzong which Hari Ram reported in 1885. This was the era when Shangup Dorje of Thami Og was gembu and it seems likely that the dzong was located in or nearby that village.

In a summary of Hari Ram's report Rawat (1973) describes the location of the dzong in the context of what appears to be the Bhote Kosi valley: "From a few miles north of Jubang [Jubing, a Rai village in the lower Dudh Kosi valley south of Pharak and Kharikhola] to Khumbu Dzong, the lower parts of the mountain sides are thickly wooded. . . . For the four or five miles beyond

Khumbu [Dzong], a solitary pine, rhododendron, or a Tibetan furze may be seen. After this not a tree is to be seen, and till the suburbs of the Ting-ri [Ganggar] are reached the only vegetation met with is a short grass." (Rawat 1973:165). This describes a trip from Jubing to Tibet which would have passed Nauje and then continued up the Bhote Kosi valley to the Nangpa La. Hari Ram notes that Khumbu Dzong is "about two miles north of Nabjia [Nauje] and on a flat part of a spur" (Survey of India, *Exploration in Tibet and Neighboring Regions, 1879–1892,* 1915 vol. 8, pt. 2; cited in Ortner 1989:23). Nabjia is clearly Nauje from his detailed description of the place. Two miles north might indicate the Thamo-Mende area, two major gunsa which might have been a seasonal home of the Thami Og gembu. Thami Og, the home of the gembu in those days, is about four miles north of Nauje and otherwise fits Hari Ram's description fairly well.

The only dzong about which legends still circulate in Khumbu, however, is the ruin of one building locally called a dzong that could be seen as recently as 1990 at Top Dara (Cannon Hill) not far from Nauje. This site is believed to have been a Tibetan post in the nineteenth century, but in spring 1991 a Magar work crew dismantled the ruins of the two-room structure to supply building stone for a nearby teashop. There was also said to be a large ruin early in the twentieth century at Tarnga which some people think may have been a dzong and attribute to Dzongnangpa.

62. Sherpas consider that this name was probably earlier Khum Te, or upper Khum in contrast to Khum Jung.

63. Upper village families do, however, have land in lower Pangboche, as do two families who are based in nearby Milingo.

64. Gunsa are often occupied only for a very few weeks each year during times when crops must be tended or when they serve as a herding base. In the Bhote Kosi valley, however, many families move to houses in settlements such as Pare, Thamo Og, Thamo Teng, Samde, and Mende for the winter. In the Thamicho region many families formerly only lived in their houses in the main villages for a few weeks each year, primarily in spring and autumn. These families often had quite large houses in the gunsa settlements and moved many of their possessions there with them in winter. In recent years these moves have been decreasing somewhat and the main villages are becoming more important winter bases.

65. The familiar name *Sherpa* is actually a distortion by outsiders.

The accuracy of the clan records that Oppitz discovered is by no means certain. Nor is it known when the texts were actually written (although Oppitz suggests that the most important, the *Ruyi* or *The Report on the Clans,* is "probably four hundred years old" (1968:143). MacDonald has challenged this claim and has even suggested that the apparently old text may be a recent prank (1987:58).

66. Both the date of the emigration from Kham and the arrival in Khumbu remain, however, very tentative.

67. It is entirely possible that the pass may already have been known as a route into the southern lands and it could conceivably have been used by earlier

migrants, pilgrims, and traders. This, however, is not clear from surviving legends. Ortner (1989:26) suggests that Tibetan hermits were already making use of some of the caves in eastern Khumbu before the Sherpa arrival. There is an oral tradition that early Sherpas may have arrived in the Bhote Kosi valley before the main group crossed the Nangpa La, and that they came to Khumbu via Rongshar valley, Rolwaling, and the Tashi Laptsa pass. But some Khumbu elders do not agree with this idea, and the version which I have stressed is the one most widely accepted today in Khumbu.

68. Some Khumbu Sherpa clans such as Mendewa, for example, trace origins not to Kham but to very nearby regions of Tibet. Presumably they came to Khumbu at some time after the first group of settlers. Other lineages trace their Khumbu ancestors to eighteenth- or early nineteenth-century Sherpa immigrants from Shorung and adjacent areas.

69. One variant Khumbu legend has it instead that the original Khumbu settlers were seven brothers, each of whose lines became a clan.

70. This is a common Sherpa interpretation of the Tibetan concept of *beyul* or hidden valley. Khumbu Sherpas believe that Guru Rinpoche, the great spiritual hero of Nyingmapa Buddhism and the man who did so much to establish Buddhism in Tibet in the eighth century, himself visited Khumbu and through his powers established it as a beyul. Guru Rinpoche is said to have flown north to Khumbu after having obtained special spiritual power during meditation at Mara Tika (Halashe), a cave in a ridge saddle near the confluence of the Dudh Kosi and Sun Kosi which is now both a Sherpa and a Hindu pilgrimage place. His stay at Arka Phuk, a cave on the slopes of Khumbila above Khumjung, is said to be marked by several of his handprints on the cave wall.

71. They could, however, also be relics of the activities of more recent Gurung shepherds or even former generations of Sherpa herders. These particular ruins predate the arrival of the Tibetan refugees who left behind many similar abandoned structures in the Bhote Kosi valley after their sojourn there in the early 1960s.

72. The Kulunge Rai inhabit the Hinku and Hongu Khola valleys southeast of Khumbu. One elderly Sherpa relates that Kulunge Rai have told him that they honor Ma Pe with special rites.

73. Unfortunately important details such as the type of grain pollen discovered are not given.

74. Arrangements between Sherpas and Rais in a number of areas appear to reflect Sherpa recognition of earlier Rai territorial rights. Some parts of Pharak, for example, paid a herding tax to Rais earlier in this century and families who use summer grazing grounds to the east of Pharak in the Mera area pay grazing taxes today. Arun Sherpas continue to pay a tax to nearby Rais in order to occupy their village lands and this may also be true for Sherpas in the Kulung and Salpa areas.

75. According to a Khumbu legend the Rais were asked to help restore to power the sons of a woman who had escaped the assassination of Dzongnangpa, an early Khumbu political leader and his associates (see below) and made her way south to settle in Dongbu country. The Rai king obliged and sent an army

up the Dudh Kosi. For six months they were stymied in upper Pharak below the confluence of the Bhote Kosi and Dudh Kosi and were unable to advance into Khumbu. The Rai forces were victorious, however, after the sons' mother helped them outflank the Sherpas by guiding them along a high route on the slopes west of the Bhote Kosi.

Ortner also discusses these events, and gives the Dongbu king's name (or the name of his descendant who was conquered by Prithivi Narayan Shah's forces) as Makwan Sher. She suggests that Makwan Sher was a Sen ruler (Ortner 1989:89), whereas Konchok Chombi testifies that he heard that Makwan Sher was the Kiranti ruler of Kirtipur in the Kathmandu Valley. He further notes that the troops who invaded Nepal were Rai from the lower Dudh Kosi valley region, not forces from the Kathmandu valley. But Karna Sena, ruler of Vijaypur (with a capital at that place in the hills north of Dharan in eastern Nepal) was in control of the lower Dudh Kosi valley at the time of Prithivi Narayan Shah's conquest of that region in 1772 and the Rai leaders there had recognized his sovereignty (Pradhan 1991:67, 113–115). Indeed, the current name Rai may derive from the Sen practice of giving the tital *raja* to the Rai leaders, which was then also conferred on them by Prithivi Narayan Shah. This was spelled *raya* and latter was changed to *Rai* (ibid.:52) for the groups that called themselves Khambus. The seventeenth- and eighteenth-century political status of the upper Dudh Kosi valley, including Khumbu, however, is still far from clear.

76. This conflict is said to have occurred because a Sherpa yak was killed in the Golila area (just south of the Lamjura pass) by Rais who were unfamiliar with yak and believed that it was a wild animal of a new type. The Sherpas and Rais agreed to meet on an appointed day and fight to determine territorial control. The Sherpas won and as a result came to exclusively settle the areas they inhabit today.

77. Ortner (1989:26) believes that the Pangboche and Dingboche areas of eastern Khumbu were the center of early settlement. It is conceivable that the arriving immigrants from Kham might have established themselves early on in this region, but from the oral traditions it appears that no permanent settlements developed there until several generations later following the establishment of a temple at Pangboche by Lama Sanga Dorje. Stories of the early settlement of Dingboche and the idea that Dingboche was the first Khumbu village are probably later Sherpa speculations based largely on observations of the area's suitability for yak and barley. The Bhote Kosi valley at the foot of the Nangpa La would have provided equally fine grazing and excellent opportunities at Tarnga and other sites for the cultivation of irrigated barley.

78. Several versions of the stories about Dzongnangpa's misrule and attempts on Lama Sanga Dorje's life are told. All accounts agree that Dzongnangpa was unsuccessful in having Lama Sanga Dorje killed and was eventually assassinated himself by Sherpas near Tarnga at Chakuparteng after he had killed the Zamde lama whose hermitage was above the Langmoche Chu just south of Tarnga. Dzongnangpa was killed during a party in a tent. Sherpas collapsed the tent and beat the trapped Dzongnangpa to death. Some people relate that a number of his followers were also killed.

79. Temples were established at Thami Og and Kerok in Lama Sanga Dorje's lifetime, apparently in the mid-seventeenth century. But it is unclear whether the Thami Teng and Thami Og areas were already settled or whether the temples were built in then isolated sites. Through much of the nineteenth century the gembu were members of a Thami Og lineage. The office passed to a Nauje resident (Tsepal, who was a Golila-Gepchua Sherpa who had married a Nauje woman) in 1895 following complaints against the previous gembu's (Shangup Dorje of Thami Og) handling of the office (Fürer-Haimendorf 1979:120–121).

80. Ortner describes this as a political conflict between two pembu (1989:51). My impression from the Khumbu accounts is that the pembu system had not yet developed at this time, and that while Dzongnangpa was a political leader who had built autocratic power the lama was a spiritual leader who was interested in pursuing his meditations in a series of hermitages and in instructing followers. From this perspective the conflict resulted from Dzongnangpa's jealousy over Lama Sanga Dorje's spiritual power and popularity among Sherpas and his disciples and his resentment of lamas, their disciples, and religion in general. This same attitude underlay Dzongnangpa's later murder of the Zamde lama (see Ortner 1989:85–86) which he accomplished by destroying his spiritual power. Sherpa horror at these excesses (and specifically the anger of some of the Zamde lama's followers) led them to assassinate Dzongnangpa and end his terrorizing of religion and the religious. There may also have been an element of political competition as well, however, if it is true that the Zamde lama was the local political leader in the Tarnga area at the time Dzongnangpa arrived. Lama Sanga Dorje may also have had some political aspirations. There is a tradition that he had been advised by a high Tibetan lama to be a spiritual and temporal leader in the style of powerful Tibetan lamas. This desire may have been behind his emigration from the Bhote Kosi valley with his followers and their establishment of a new community in the wilds of eastern Khumbu.

81. Sherpas sometimes talk of a four-hundred-year tradition and note that the temple has been rebuilt three times after the initial structure was destroyed by an avalanche.

82. This is also referred to in a 1919 document granting nearby local land to the Tengboche monastery (Ortner 1989, appen. 2).

83. MacDonald notes that most of the small, early Khumbu "temples" were "not within the limits of human settlements" (MacDonald 1980:141).

84. The settlement area of Khumjung and Kunde, a narrow valley between the cliffs of Khumbila and an area of tremendous boulders, is popularly likened to a horse. A rock which is said to be the horse's head is located on high ground southeast of Khumjung.

85. Ortner (1989:93) discusses the establishment of the Khumjung temple and its possible relation to increasing village political power, speculating that the office of the gembu may have become established in the village at that time. According to Khumbu oral traditions founding the temple was related instead to the conflicts at the Dumje festival between Khumjung and Thamicho villagers. It was also made possible by the increasing population of Khumjung and Kunde and the growing wealth and power of villagers such as Nam Chumbi, a man who

established several shrines (*chorten*) in the area and who was a powerful and at times feared and hated pembu prior to his assassination sometime after 1830 (and according to one account in about 1851–52). In any case it is incorrect to relate events in Khumjung to the gembu, since nineteenth-century gembu were Thami Og residents. There are no oral traditions of any Khumjung man ever having held the office.

86. They also had fields at Chorkem, a saddle just above Nauje, where today only Nauje families farm. Chorkem in that era was an important place in Khumbu, for here Sherpas from Khumjung, Kunde, and Thamicho met Rais hauling grain from the lower Dudh Kosi valley and bartered salt for it. There is some evidence of an old, direct trail from Chorkem to the upper Bhote Kosi valley and stretches of an old trail along what is said to have been the main route from Pharak to Thamicho and Tibet can still be followed along the Bhote Kosi valley slopes below the site of Nauje. This main route bypassed Nauje taking a more direct line up the valley. Chorkem is also infamous as the place where Nam Chumbi was assassinated.

87. According to some Sherpas Khumbu families in the early nineteenth century may have thus been paying taxes simultaneously to both Tibet and Kathmandu. One campaign of the 1855–56 conflict is said to have been conducted in the Dudh Kosi valley. Nepali troops outflanked the Tibetan fort at Top Dara near Dzong Chu Tok below Nauje and with the help of Sherpas temporarily pushed the Nepalese zone of control beyond the Nangpa La to the area of Kaprak. After the war the area north of the Nangpa La was returned to Tibetan administration, but Khumbu remained firmly in Nepali control.

88. It seems possible that this may have been an already existing local system of administration that was simply validated by the central government as a convenient way to implement its rule in a remote area. The Sherpa institution of gembu may well be derived from the similar Tibetan office of the Tingri region (Aziz 1978:199–200), as may have been the office of pembu. The gembu was the most powerful local official in the region until the early twentieth century when gembu Tsepal abandoned his office and fled to Tibet following a second failed attempt on his life during which he killed three men. (According to Ortner (1989:116) one of these was the former gembu, Dorje, but Fürer-Haimendorf (1964:121) and Konchok Chombi do not agree that Dorje was killed.) His son Pasang assumed the office and carried out the gembu's duties for some years, but eventually moved to the ancestral family home village in Golila-Gepchua and gave up Khumbu administration about 1933. The office thereafter lapsed.

89. Both the offices of gembu and pembu were regarded as hereditary, although either could be relieved of his authority by the Nepal government and the office given by the state to someone of its choosing. This, however, has occurred only twice in the last hundred years, once with a gembu and once with a pembu, both around the turn of the century.

90. A few years earlier, in 1810, there had been a royal decree which allowed Shorung traders to trade as far north and Khumbu traders as far south as they wished (Schrader 1988:245).

91. Khumbu Sherpas, for example, made use of Nepali courts for several

major cases in the nineteenth century, including a dispute between the gembu and villagers of Khumjung over herding regulations and a land dispute between Khumjung and Phurtse villagers which involved a sacred forest. In both these cases the government interestingly deferred the final decision to local Sherpa officials. In one twentieth-century case Khumbu-based Nepali officials were accused by Sherpas of hunting illegally in Sherpa protected forests.

92. A village panchayat often did not correspond to a single village, but was rather a governmental unit established to administer a population of approximately 2,000–4,000 people and might encompass a number of communities. This was the case in Khumbu. In some parts of Nepal lack of correspondence between the boundaries of the new local government's jurisdiction and the individual villages had important ramifications for resource-management policies, because forests and pastures previously owned and regulated by a single village sometimes now came under the joint control of a number of settlements.

93. The authority of the pembu remains substantial. Besides collecting taxes several pembu also continue to settle local disputes and carry out other roles that they formerly performed. Two have held the office of pradhan pancha as well.

94. In an effort to promote regional economic growth the central government encouraged the development of a country-wide system of periodic markets. Weekly markets were established in the mid-1960s in the Solu-Khumbu district near the district center of Salleri in Shorung (originally at Dorphu and later at Naya Bazar) and at Nauje as well as lower in the Dudh Kosi watershed at Olkadunga and Aislalukarka. Within a short time market transactions in Nauje for grain and other products virtually replaced the old barter exchanges that Sherpas had previously made with Rais and lower-altitude Sherpas. The establishment of the weekly market, however, has not transformed Khumbu subsistence crop production into a commercial activity.

2: A High-Altitude Economy

1. Altitudinal variation of climate and vegetation is certainly one of the most striking and fundamental characteristics of mountains and has rightly been highlighted in geoecological studies of land use ever since von Humboldt first called attention in the early nineteenth century to the altitudinal zonation of climate and natural vegetation and agriculture in the Ecuadorian Andes. But as Goldstein and Messerschmidt (1980) and Metz (1989) have ably noted it is important to examine other types of microenvironmental variation as well, and in the Nepal Himalaya land-use patterns are especially influenced by wide regional rain-shadow effects and local contrasts due to slope. Goldstein and Messerschmidt (1980:120) referred to these differences as "latitudinality" in contrast to altitudinal "verticality," highlighting the fact that the general west-to-east trend of Nepal's mountain ranges creates a south-to-north rain-shadow effect that some peoples exploit through seasonal movement of livestock and trading activities along longitudinal meridians.

2. I refer here to the difference between broad overall goals and plans (strategies) and individual techniques (tactics) used to implement these plans.

3. Murra's verticality thesis emphasized a strategy and cultural value of controlling the maximum number of such altitudinal zones, and he referred to the distant lowland jungle cultivation areas as archipelagos (Murra 1972, 1985*b*). Subsequent research has defined central and southern Peru and northern Argentina-Chile as the area where this subsistence pattern was practiced (Shimada 1985:xiv) and expanded attention to the role of trade, tribute, and other arrangements rather than direct cultivation for gaining access to the agricultural possibilities of diverse microenvironments (Salomon 1985). There has also been a realization, somewhat later than in Himalayan work, that the key factor is microenvironmental diversity rather than altitudinal zonal variation per se and as a result, by 1985, many Andean scholars had begun to refer to "ecological complementarity" rather than "verticality" (Murra 1985*a;* Salomon 1985; Shimada 1985:xiii–xiv; Yamamota 1985). I retain the term *verticality* here to emphasize not only the altitudinal dimension of microenvironmental variation but also the role of mountain topography, structure, and geomorphological processes in creating the conditions of local variation of climate, soil, and other conditions that make mountain regions places of such high microenvironmental density.

4. This remains a usable basic set of categories. Salomon (1985:512–516), in a more sophisticated analysis, had identified eight major strategies for gaining access to multiple ecosystem resources.

5. I prefer to refer to production systems rather than production zones to emphasize that these sets of practices have been developed for use in particular types of physical environmental conditions that in mountain regions are not found in simple zones but rather in complex mosaics of microlocalities whose environmental characteristics reflect a number of factors besides altitude. By using the word *system* I also mean to focus attention on the complex interlinkages between crops, livestock, and the natural environment. Individual production systems should be conceived of as agroecosystems.

6. These five strategies have also been important in other highland Asian regions, including adjacent areas of western China (Yunnan, Sichuan, Gansu, Qinghai, Xinjiang, and Tibet). Settled, mixed farming based on barley has been practiced at high altitudes in Tibet, at more moderate altitudes emphasizing wheat in the Hindu Kush by Pathans and others, and by the Bai (rice and maize) and Naxi (wheat, barley, maize) in northwestern Yunnan. Examples of peoples practicing middle-altitude agropastoralism can be found in the Hindu Kush, Karakorum, in parts of Tibet and Sichuan, and perhaps also in Yunnan. Swidden cultivation has historically been practiced by some of the peoples of mountainous northwestern Yunnan and western Sichuan such as the Yi. High-altitude agropastoralism is a familiar Tibetan strategy. Tibetan Drokbas, Mongolians, Kazaks, and Kirghiz (Kyrgyz) all practice nomadic pastoralism.

7. Other strategies based on commercial agricultural and pastoral production, circular migration and remittance economies (including mercenary service), cottage industrial production, trade, and income from tourism are, or have been, important for some rural mountain peoples. In rural Nepal these income-generating pursuits generally supplement rather than replace subsistence agriculture and pastoralism, although in certain regions they have become

increasingly important and now form the basis of the local economy (N. Bishop 1989; Fricke 1989; Fürer-Haimendorf 1979).

8. Government tax policies, inequalities of land ownership, widespread land tenancy, high land rents, rural indebtedness, and land fragmentation through inheritance have also undoubtedly contributed in some areas to interest in cultivating cash crops and the inability of some families to pursue multialtitudinal mixed farming and transhumance. Across Nepal, however, a relatively small number of families are today wholly involved in commercial farming. Metz (1989:156) estimates that 85 percent of Nepal's population engage in subsistence farming.

9. In distinguishing between integral and supplementary swidden (shifting agriculture) systems I follow Harold Conklin (1954, 1957). Integral swidden refers to the use of shifting field agriculture based on clearing and firing forest and planting in the fertile ash by peoples for whom it is the sole agricultural basis of subsistence. Supplementary swidden is carried out by peoples who rely on permanent fields for their main source of sustenance and for whom these swidden fields can be used for subsistence or commercial crop production.

10. The best-known Himalayan nomadic peoples are the Gujars and Gaddis, both of whom move herds to high-altitude (and rain-shadow) pastures in the summer, living in tents or herding huts, and move to winter herding grounds in lower-altitude mountain regions or the adjacent lowlands. Another such people are the Bakrwal, who primarily herd sheep in the Jammu and Kashmir region. Some of these groups, however, also cultivate some land. Gaddis in Himachal Pradesh, for example, combine long-distance pastoral migrations with fields and houses in main villages in middle-altitude areas. Some Bakrwal have acquired fields in the valley of Kashmir and the Jammu area since the mid-1960s (Casimir and Rao 1985:222). And the Kazaks of the Tian Shan, who follow a similar mountain transhumant strategy, today sometimes cultivate a few crops in their winter-quarter areas. A different type of pastoral nomadic herding strategy is followed by the Tibetan Drokbas, who remain on the Tibetan plateau year round at high altitudes (Ekvall 1968; Goldstein and Beall 1990; Goldstein, Beall, and Cincotta 1990). Both these styles of nomadism rely, as is common with so many other pastoral nomads, on trade with adjacent agricultural peoples to supplement the limited food resources available from herding. In westernmost Nepal there are also nomadic peoples such as the Humli Khyampas (Rauber 1982) who keep livestock primarily as pack stock and base their subsistence on an annual round of trade journeys between Tibet and the lowlands of southern Nepal and the Ganges plain.

11. Allan (1986) has argued that with the expansion of road networks and the commercialization of agriculture there is a fundamental change in the nature of mountain economies in which accessibility rather than altitudinal, environmental zonation becomes the key factor characterizing land use. While accessibility (and other market conditions) are obviously of great importance in those mountain regions that have become involved in commercial agricultural and pastoral production, this has been much less widespread thus far in the Himalaya (and especially in Nepal and the eastern Himalaya) than in the Alps or

Andes. And it should not be overlooked that altitudinal variation continues to play a fundamental role in land use even after commercialization, for environmental conditions are an important influence on the productivity and profitability of particular types of agriculture and pastoralism. Within commercial agropastoralism differences in emphasis can be noted which reflect not only access to markets, land rents, supply and demand, and state policies but also the limitations and possibilities of raising particular types of plants (e.g., jute, rice, tobacco, cardamon, ginger, tea, citrus fruits, apples, potatoes) and animals (e.g., dairy cattle, water buffalo, chickens) at different altitudes and in different microenvironmental conditions.

12. Swidden was once also much more common as a supplementary method of crop production among the Pahari.

13. Here there are parallels with Hindu hill-caste practices in India. In northern Uttar Pradesh swidden was practiced in the 1930s as a supplementary form of agriculture (Pant 1935).

14. Settled mixed farming has long been characteristic of the Newars of central Nepal as well as the Kashmiris of the western Himalaya and the Apa Tanis of Arunachal Pradesh in the eastern Himalaya. All of these peoples have agriculture based on irrigated rice cultivation. The Bai of the Dali region of Yunnan are another example of a rice-based permanent settlement at a similar altitude. Some Paharis, Magars, and Tamangs have adopted settled montane farming in the twentieth century, relying on adjacent forest and woodland for fodder and grazing rather than practicing transhumance. A second pattern of permanent settled agriculture is followed at much higher altitudes in Tibet, where it is based on barley and wheat rather than rice.

15. Not all families may practice this full round of transhumance, and there may be important variations among households that emphasize different degrees and types of herding. The patterns of transhumance followed by herders with sheep and goats and those with water buffalo and cattle are often quite different. Herd size, family labor resources, and the ability to hire wage labor can also be major factors.

16. The highest-altitude-dwelling peoples of Nepal typically inhabit rain-shadow regions. Many of these live to the north of the main crest of the Himalaya in environments similar to those of Tibet and Ladakh. The peoples of the Thak Khola and Mustang areas of the upper Kali Gandaki valley, the Manang valley north of the Annapurna range, the Dolpo region, Mugu and Humla all live in such country in north-central and northwestern Nepal. In eastern Nepal there are other regions which, while not north of the main Himalayan crest, have locations that shield them from the main impact of the summer southeast monsoon. These include the Sherpa-inhabited regions of Khumbu and Rolwaling. On the windward, high-precipitation, southern-aspect slopes of the main Himalaya main villages tend to be developed below 2,500 meters (Metz 1989:161). Above this high rainfall, a high degree of cloud cover, and wet soil conditions make crop production difficult and upper slopes and valleys tend to be heavily forested rather than settled and cultivated. In the Dudh Kosi region this difference can be seen quite strikingly in the contrast between altitudinal land use in

Khumbu and in the Dudh Kosi valley. In Pharak fields are not developed much higher than 2,800 meters. In Switzerland the highest-altitude agriculture also is conducted in rain-shadow regions (Netting 1971:133).

17. Current crop distribution does not reflect actual altitudinal limits of crops but rather historical patterns of production (chap. 6). Current crop distribution suggests an altitudinal limit for buckwheat, for example, of below 4,000 meters, but it was once commonly grown slightly higher as a major Tarnga crop. Wheat is now grown no higher than 2,800 meters in Pharak and is not grown at all in Khumbu, but a generation ago was grown in the lower Bhote Kosi valley at Tashilung (3,400m) near Nauje. Barley is certainly not physically restricted to the narrow, high-altitude range in which it has been grown in this century and could with irrigation be grown profitably in lower Khumbu as well. Some viable high-altitude crops such as turnip are not being grown to the edge of their range. Only potato appears to be pushed to its limits, and this on a very small scale in the uppermost crop-growing sites. The vast rolling slopes of the glacial, U-shaped valleys are not extensively planted to potato even at altitudes of 4,000 meters where they would certainly thrive.

18. It would also be useful to link production systems not only to altitude but also to other broad, zonal, mountain microenvironmental conditions such as types of terrain and levels of precipitation. A pioneering attempt at such a classification for central and western Nepal has recently been made (Metz 1989). Metz distinguishes ridge-valley sequences of production zones and places a set of five production types in both altitudinal and geographical contexts. These zones are the lower (below 1,500m), middle (1,200–2,000m), upper (1,000–2,400m), and high (above 2,500m in partial rain-shadow areas) elevation hill production types and the inner valley production type (1989:157–164). Metz relates these to locations in the midlands, high Himal, and trans-Himalayan valleys and describes their characteristic land-use features. This is a good beginning, although it would be useful to also examine how some peoples follow subsistence strategies that incorporate multiple production types and make use of the opportunities of several microenvironments. Metz's lower-altitude production type corresponds to the low-altitude, settled, mixed-farming category in my set of Himalayan subsistence strategies, his middle- and upper-production type to middle-altitude agropastoralism, and the high-elevation and inner-valley production types to the high-altitude agropastoral strategy.

19. In Switzerland potatoes were also grown year after year without rotation, and according to one report were cultivated in this way for generations without a sign of declining yield (Netting 1981:163). In the Andes, by contrast, potato fields were usually rotated and also fallowed for several years (in some areas for four to seven years) before being used again—in part to decrease crop losses from nematodes. Here many communities enforced this rotation and fallow system through communal regulations, a practice now known as sectoral fallow.

20. According to estimates offered by eleven Nauje families (average size 4.5 adults, including 1.1 servants and .7 children, counting teenagers as adults), a household required twenty-five loads per year of potatoes. In Nauje loads of

potatoes are often calculated at four tins per load, with a tin nearly 12 kilograms. This gives 1,200 kilograms of potatoes per year, or .72 kilograms per adult per day. Families from other villages, however, seem to consume a good deal more potatoes than do Nauje villagers.

21. The eleven families interviewed about family food requirements estimated their grain consumption at 15.4 muri per year (a muri is equal to 2.4 bushels or 90.9 liters and varies in weight with the type of grain), a little more than a metric ton of grain. One-third to one-half of this grain requirement was rice (70 kg/muri). Nauje villagers, however, probably tend to consume more grain than villagers elsewhere in Khumbu.

22. On my return to Khumbu in 1990, after having been away since 1987, I was told of a great tasting new sauce that could be used on potato pancakes. This proved to be mayonnaise! Families who own lodges may be particularly inclined to incorporate foreign foods into the family diet on a regular basis.

23. It may be that families for whom selling some surplus is an important component of their household economy are relatively quick to adopt higher-yielding varieties of potatoes, although further research will be needed to establish this. But it is certainly also true that many families with no interest in the sale of surplus are also often quite interested in new higher-yielding varieties. For example, such varieties have tended to be very rapidly adopted in Nauje, a settlement where almost no one raises a surplus. The interest in higher-yielding varieties seems to be widespread in a region in which land is scarce relative to food requirements. The price of potatoes at the weekly market in Nauje is the same for all varieties of potatoes, regardless of the local value put on one over another in terms of taste.

24. Formerly some goats were also kept until they were regionally banned at the request of Sagarmatha National Park administrators.

25. Apparently hay is not grown in walled fields in Rolwaling and only a little wild grass is cut for hay from steep slopes.

26. Gunsa holdings (or land in Nauje) enable families to eat fresh potatoes two or three weeks earlier than the main village harvest, a much appreciated benefit in August when the year's potato stockpile has been exhausted.

27. This practice also common at lower altitudes in Nepal (Metz 1989:160).

28. In a few cases, however, the women of wealthy families simply supervise the fieldwork of hired agricultural laborers.

29. The importance of demographic change, and particularly population growth, in mountain cultural ecology was highlighted by Netting (1981) and MacFarlane (1976). The role of the demographic cycle in household land use, however, was largely neglected in Himalayan studies until Fricke's recent work on the Tamangs (1986).

30. Care must be taken not to overgeneralize from Chayanov's model, however. Fricke (1986) notes that Chayanov's analysis of Russian peasant family dynamics was not entirely applicable to Nepal's Tamang population with its different cultural concepts of marriage, household economy, kinship, mutual aid, and other socioeconomic institutions, practices, and values. The same is certainly true of Khumbu Sherpas.

31. Ortner reports that youngest sons sometimes refuse to care for their parents (1978). While this may occur in Shorung, it is very rare in Khumbu.

32. An adult woman who is part of a reciprocal labor group can easily plant and harvest enough potatoes and buckwheat to meet household requirements in these staples, provided that the household owns sufficient land. Supporting young children may not be as great a burden as might be supposed. Women take their babies and young children to the fields with them when childcare is not available.

33. The prospective son-in-law of a woman whose parents have no sons works for his in-laws as a son once the engagement begins. He also tries to help out his own family until the time of the marriage ceremony and his formal move into his in-laws' home.

3: Farming on the Roof of the World

1. Such poor harvests are not disastrous since families have long been able to make up shortcomings through obtaining extra stocks of low-altitude grown grain as well as by trading for or purchasing surplus potatoes from regions of Khumbu which may have been less affected by poor weather or disease. These added expenses (and in former days the greater work they required to haul bartered grain from the Nepal midlands) are an inconvenience to all, and poorer families sometimes have to lessen the amount of grain in their diets during years when they have had a poor buckwheat harvest. But there is no memory of a Khumbu famine.

2. More land could be put to crops than is currently cultivated and there are even substantial areas of abandoned terraces in some parts of Khumbu. The wide floors of the upper valleys between 3,800 and 4,500 meters in particular are used more for hay and grazing than for growing crops even though potatoes and probably barley will flourish throughout this range and buckwheat can be grown to 4,000 meters. Terraces could also easily be extended in many gunsa areas and a number of main villages although the best land is already in cultivation.

3. Both terrace risers and the walls that protect the fields are similarly constructed with unmortared stone. Some terrace risers are very carefully built and in Phurtse some old terraces are even anchored into the slope with timbers. The degree of care taken in constructing field walls varies tremendously reflecting differing assessments of the degree of risk from livestock. In Phurtse, for example, walls may be only a single stone high, serving more as a boundary marker than a barrier, and there are excellent, large fields in the central part of the village having no walls at all.

4. There is evidence that some of these areas have podzolic palaeosol spodosols buried below their brown grassland soils. This suggests that forest once covered those regions and was subsequently modified by climatic change and/or human impacts (Byers 1987c:212; Hardie et al. 1987:19).

5. Several other soil conditions are also considered to be poor for crop growing. In the few areas where they occur in Khumbu waterlogged soils, for example, are regarded as a problem and special care is thought necessary in order to grow crops there.

6. Autumn frost is not considered a problem for potatoes.

7. There was considerable spring frost damage in some settlements in 1981 (Thamicho, Khumjung, Kunde) and in 1984 (Pangboche, Phurtse, Khumjung, Kunde, Dingboche). In 1986 frost was a problem throughout the region.

8. Radish is considered to be much less vulnerable to frost damage.

9. In 1986, for example, Phurtse villagers reported five or six frost days during April, and three or four were reported for Khumjung.

10. In this area the circumambulation is performed by monks from Thami monastery. That year the head lama of the monastery had chosen to make the circumambulation accompanied by two foreign tourists. Some villagers concluded after the frost that the presence of the foreigners was the cause for the failure of the protective rites.

11. Too much rain, however, is rarely a problem in the drier, high-altitude crop sites.

12. In Phurtse one Sherpa noted a connection between the two unusually heavy rains, observing that heavy tenju rains only occur in years when there has been a major yerchu rain. The heavy rain and snowfalls of autumn 1985 and 1987 were not considered to be tenju, although in 1987 more than 125 millimeters of rain fell within a twenty-four-hour period in Nauje.

13. Himalayan tahr are a problem, for example, in the potato fields at Rhokumbinan, adjacent to Nauje, and Nyeshe and Tashilung in the lower Bhote Kosi valley, and in some gunsa near Phurtse where they are reported to dig up potatoes with their hooves. Pheasants are a problem in Phurtse, the easternmost fields of Khumjung, Sonasa, Samde, and other areas. Fifteen or twenty birds at a time may pick over a field, rooting up potatoes and eating buckwheat seed. Choughs can make such an impact on freshly planted barley and buckweed seed that many farmers plant extra seed to make up for the share taken by the birds. Farmers try to scare off birds and wildlife. Some make replicas of bamboo traps and put these around their fields in the hope that it will keep pheasants out. In Nauje several families have begun to suspend lines of audio cassette tape across their fields, where they wave, glint in the sun, and hopefully discourage choughs. Earlier in the century, when wildlife or birds became too much of a problem, farmers called in the Nauje blacksmiths to shoot them. Sherpas themselves do not hunt, but the Hindu blacksmiths formerly kept guns and were willing to kill wildlife.

14. Sherpas note that blight affects buckwheat at the time it flowers in mid-summer. Cases are described in which blight-infected buckwheat formed linear patterns across fields, corresponding, Sherpas believe, to the track taken by livestock that had strayed into the field during the summer.

15. Sherpas are not sure about the origins of the name shimbak or the meaning of the word. Some Sherpas consider that it must refer to *shi,* death. An anonymous reviewer of this manuscript suggested that the word might be derived from the Tibetan *shing bak,* or "field pollution."

16. Late blight was first reported in the Himalaya in 1883 at Darjeeling. It spread rapidly into adjacent eastern highland areas of Bhutan and Nepal. By 1897 it was also being reported in northern Uttar Pradesh in the Indian Himalaya west of Nepal. Since 1900 blight has been prevalent throughout the Himal-

ayan potato-cultivating region. In middle-altitude eastern Himalayan areas, including Darjeeling, it occurs annually (Dutt 1964:70–71). Khumbu cases are remembered from before 1920.

17. In 1987 blight first appeared in Nauje in June and by mid-July most of the village's red and yellow potato plants had died. By that time there was also blight in Kunde, Khumjung, Teshinga, and Thamicho. Even in August, however, the infestation was minor in Pangboche and Phurtse. The spread of blight in 1987 might have been associated with the wetter than normal conditions, and local farmers also attributed it to violations of the community regulations believed to prevent the spread of blight.

18. People are still restricted from working in summer fields in some communities, including Pangboche, Phurtse, and Dingboche. Livestock are still excluded from Khumjung, Kunde, Phurtse, Pangboche, and Dingboche during the critical months of blight risk.

19. Warts have not led to major crop failures in Khumbu, but apparently have in some other Sherpa-inhabited regions. Sherpas in the Salpa area blame this disease for a major disaster about 1980 when entire crops of red potatoes and *kyuma* varieties failed. After this many families abandoned cultivating those two varieties and instead have relied on the yellow potato, which like the other varieties was originally obtained from Khumbu.

20. High rainfall in the upper Dudh Kosi valley, for example, is considered by some Phurtse people to be beneficial for hay and potato crops on the eastern side of the river, while the shadier fields on the other side of the river produce good hay but very poor potato harvests.

21. The main advantage here is that *gunsa* fields can be prepared, planted, and also harvested earlier than the main village fields that are slightly higher in altitude. A family can thus cultivate substantially more land through its own labor efforts than would otherwise be possible. Land in the high-altitude, secondary agricultural sites and the high-herding areas offers less of an advantage, for the tasks of the agricultural cycle often overlap with those in the main village, stretching the ability of households to conduct both simultaneously and sometimes forcing them to delay practices that optimally might be conducted slightly earlier. In terms of labor scheduling alone a family would be better off with more land in the main village rather than plots scattered at several locations in the upper valley.

22. Pangboche, Phurtse, and Nauje families do not own very much gunsa land, although in the Nauje case the main village fields are situated at an altitude comparable to gunsa elsewhere. Nauje households also do not own much high-altitude crop land. Khumjung and Kunde families own relatively little high-altitude crop land in comparison to the holdings of Phurtse, Pangboche and the Thamicho villages.

23. Maize (Sherpa *litsi*) was grown by one Nauje family thirty years ago. They grew it, however, not in Nauje itself but rather in a set of gunsa terraces at Jangdingma, a place along the shore of the Bhote Kosi at an altitude of approximately 2,800 meters. Wheat is no longer grown in Khumbu as a food crop, although it is raised on a very small scale for fodder. A generation ago, however,

it was raised for grain by Nauje families at Tashilung (3,400m). This raises questions about the reason for its relative lack of importance regionally both today and in the past.

24. Very little Khumbu land is put to vegetables. Mustard is grown by a few families in Nauje as a second crop following potatoes, which must be harvested slightly early to allow this. A few other families intercrop potatoes and mustard in fields adjacent to the house, broadcasting the mustard seed and covering it with earth scooped with a weeding tool. The mustard ripens earlier than the potato and is picked as it ripens. Mustard leaves are pickled (*shotzi*) and this is virtually the sole vegetable available in winter when it is very commonly eaten in noodle soups. No other vegetables are ever grown as field crops, although a few are raised in household gardens and window boxes. The largest gardens are found in Nauje where several dozen households raise plots, mostly pocket-sized patches on odd corners of ground between houses and adjacent trails and at the edge of terraces. Fewer than 10 percent of households even here, however, keep a garden as large as 100 square meters. Elsewhere in Khumbu gardens even of this size are very rare. Special care is taken with gardens, which are well-fertilized and watered. In March and April some people also protect vegetable seedbeds by covering them or by standing boughs of dwarf rhododendron (*kisur*) around the bed as a sort of tiny hedge or erecting these all through the plot. This is said to protect the seedlings from wind, heavy rain, frost, and strong sunlight.

25. The twentieth century decline of buckwheat cultivation is discussed in chapter 6.

26. The main barley grown in the adjacent Tingri region of Tibet is a naked white barley that is distinct from the Pharak variety. In the Tingri area the black Dingboche variety occurs only rarely, interspersed in white barley fields. Sherpas who have traveled through the area during the growing season note that only a few black barley plants per field is usual. The white Tibetan variety was grown in Khumbu up until the early twentieth century, but only in the Bhote Kosi valley, an area in which no barley cultivation whatsoever is any longer attempted. The custom of cultivating only black barley at Dingboche may have origins in local environmental and agronomic knowledge. One man who experimented recently with cultivating white Tibetan barley at Dingboche reported that the grain had not matured well and required not only a longer growing season but also more irrigation.

27. Barley responds to nitrogenous fertilizers by producing higher yields as a result of the increased photosynthesizing area it develops. Fertilizer use can also encourage a higher concentration of protein in the grain (Langer and Hill 1982:54). The benefits of higher-protein grains and increased grain yields may be especially high with a short-stem variety such as that raised in Khumbu, where the additional nutrients are channeled less to increasing stem height than to other parts of the plant.

28. Buckwheat, by contrast, is considered very inauspicious. Wheat and millet are also considered ritually impure. No millet, buckwheat, or wheat can be offered to the gods. In the case of millet its black color is the offensive quality.

29. It is also noted for having the ability to soak up much larger amounts of tea than other tsampa, a quality to which much importance is attached in Khumbu.

30. One Pangboche field, however, yields at best just under a one-to-four-ratio, and at worst yields only two to three pathi of grain from four to five pathi planted.

31. To is never planted, but instead grows as a volunteer (*to yem*) in potato and buckwheat fields and is harvested for its edible root. Care is taken to leave some plants in the fields to seed in order to ensure a harvest the following year. There is no oral history record of it having been grown as a crop in Khumbu. It is interesting, however, that it today and in the remembered past this plant only grows in fields, possibly reflecting original planting or at least transplanting. It is distributed from Nauje to high-altitude settlements such as Dingboche. The tuber is quite astringent and elaborate preparations are necessary to render it edible. It must be washed, dried in the sun, and then ground into flour (*to pe*) that is then mashed into sen mush, cooked into soup, and then finally rolled by hand into noodles. These noodles have to be eaten without chewing, for if chewed they burn the throat. To is occasionally eaten in a porridge, but if this is done it is also considered important to bolt it down without chewing.

32. The Sherpa custom of not eating potato skins unfortunately deprives them of a significant source of nutrition. They do feed them, however, to live-stock along with other table scraps.

33. Potatoes were introduced to one area in southern Dolpo, for example, only forty or fifty years ago by a lama who brought them from an area two passes to the west towards Jumla. They were adopted quickly, and made a major difference in the subsistence of poorer families who were able to become self-sufficient in their food production as a result (J. Fisher 1986b:58). Potatoes conceivably also could have been introduced relatively late to higher-altitude areas of Dolpo and to the far northwest of Nepal. The rate of cultural adoption of a new crop can also vary enormously after it has been locally introduced. In the case mentioned above the rate of adoption was rapid. Potatoes were ac-cepted only very slowly in Europe, however, after their initial introduction around 1570. Even in Ireland, where they became a staple earlier than else-where, potatoes had replaced porridge and wheat bread in only a small part of the region by the 1770s and were really only widely incorporated in local diets after 1790 (Salaman 1985:494–507). Its importance in Germany and the Alps came only in the eighteenth century after a series of famines spurred efforts by government, clergy, and learned landowners to promote it. Local cultural biases against the new crop were overcome only slowly (Netting 1981:160–161).

34. Today only eighty-one varieties of potato are reported for Nepal (Khanal 1988:28), a meagre range in comparison with the more than 12,000 named varieties known in Peru (Brush 1987:276). All the Nepal varieties belong to a single species (*Solanum tuberosum*), while in the Andes seven species were domesticated by Neolithic farmers (ibid.).

35. There may be several different varieties of development potato cur-rently being grown in Khumbu. The one that is being adopted for food and fodder use has a pink or purple flower. There is another variety, however, that

has a white flower and which some people say has a quite bad taste and is unfit for consumption.

36. Relatively high yields have also been reported for Swiss villages in which potatoes were a major summer crop. Netting reports 1960s' yields of nineteen metric tons per hectare (Netting 1981:163). This, of course, is at a much lower altitude (1,500m) than Khumbu, but also is from fields situated near the uppermost limits of potato cultivation in Switzerland (ibid.:191). The Swiss national average production of forty-two tons per hectare is the highest in the world (Langer and Hill 1982:281).

37. The finest yields of all are considered to come from the middle and upper Bhote Kosi valley, a region of Khumbu that is also noted for being relatively arid by Khumbu standards. The ability of potatoes to thrive in areas of moderate rainfall is also recognized in the Alps, where one local proverb in a village located in the lowest-rainfall region of the country proclaims "the drier the mountain, the better the potatoes" (F. G. Stebler, *Sonnige Halden am Lötschberg* 1914, cited in Netting 1981:163).

38. Inability to maintain viable seed potatoes is a factor in potato cultivation at middle and low altitudes in Nepal and as a result seed potatoes are often obtained by these farmers from high-altitude areas. Khumbu Sherpas supply some seed potatoes, for example, to Rais and middle-altitude Sherpas who come to Khumbu to obtain them.

39. Potato yields similar to those in Khumbu have been reported for traditional varieties cultivated with traditional technologies at the horticultural farm at Lumle (5,500m) in the hill country south of the Annapurna Himal, where twenty metric tons per hectare have been produced (Khanal 1988:27).

40. Potatoes were weighed in monocropped, yellow potato fields at Nauje (four fields), Pangboche (six fields), Dingboche (six fields), Thami Og (three fields), and Tarnga (two fields). Monocropped, red-potato-field yields were weighed at Nauje (four fields), Pangboche (two fields), and Dingboche (two fields). Tarnga yellow potatoes produced far better than Dingboche ones did, yielding 4.75 kilograms per square meter.

41. Small amounts of brown potatoes are also grown at lower altitudes and are planted, for example, by a few households in Thami Og and in Nauje.

42. The higher yellow potato yields may well be related to the larger leaf area of this variety, for leaf area is the key factor in determining yield once adequate moisture and soil nutrients are available (Langer and Hill 1982:281). The variety's large leaf area also acts to shade out weeds, which farmers consider to be less of a problem with yellow potatoes than with red ones.

43. Site as well as altitude, however, can also be a factor. Yellow potatoes grown at Tarnga (4,000m) are considered to be very good tasting (although not as good as red potatoes from the same place), far superior to potatoes grown at similar heights in the Dudh Kosi or Imja Khola valleys. Farmers in the Bhote Kosi valley, however, feel that the yellow potato changes in taste above Tarnga and only a few grow it at Arye and Apsona.

44. Some people recall that kyuma and koru 2 stored even better than the yellow potato does.

45. All crop and hay land is owned by nuclear households with the exception of a small number of fields that are owned by the Tengboche monastery in the upper Imja Khola valley and Lobuche Khola region and rented to individual Sherpa farming families.

46. There is certainly much interest in land purchase, for many families do not own enough land to produce all of their household's requirements of potatoes and buckwheat and few families own barley land. In recent years Nauje families have been especially active in purchasing land, most of it from Thamicho families. Many of the purchasers have been Khamba families who previously owned little land and have now acquired the means to do so from earnings from tourism. They find, however, that it is difficult to find sellers in most of the region. Today land is primarily sold for cash, with prices varying not only with the quality, size, and location of a field but also with the social context of the exchange, for relatives and friends may be given generous terms. Land prices vary greatly between villages. Ten to fifteen thousand rupees ($333–$500) will buy a 300-square-meter field in the famous potato-growing settlement of Tarnga, but the same size field would cost 25,000 to 30,000 rupees in Nauje, where its yield would be between ten and eighteen loads at best. A field with possible commercial value as a site for a lodge, however, could well command a price of over 100,000 rupees in Nauje.

47. In practice the division of land and livestock among sons may not always be equal (see Ortner 1989). Women normally do not inherit land, but receive instead household goods, jewelry, and cash at the time of their marriage. Occasionally, however, the daughters of wealthy families or of families without male heirs may inherit fields (*gyashing*). Such inheritances sometimes underlie anomalous land-holding patterns. The ownership by some Khumjung families of land in Phurtse and in otherwise solely Phurtse family-inhabited high-altitude herding settlements can be traced to a Phurtse woman who brought title to these lands to the Khumjung household into which she married.

48. The Tengboche monastery owns more than twenty cropfields and another twenty hayfields in a number of widely scattered areas in the Imja Khola valley. These include fourteen fields at Dingboche and seven or eight at Pangboche, plus hayfields at Pheriche (seven), Phulungkarpo (three), Yarin (two), Omoka (one), Ralha (one) and others at Tsadorji, Lobuche, Tugla and Kuma. These lands were bequethed to the institution by Sherpa families. Rents for these lands are very nominal and are used to finance prayers on behalf of their former owners. Giving land (or animals) to the monastery gains one good merit for future rebirths. The monastery above Thami Og, the other major Khumbu institutions, owns no hayfields and only two or three fields at Thami Teng and Tarnga.

49. Rent for Tengboche monastery–owned fields at Dingboche, for example, can be paid in barley, the amount being calculated for each field by the monks on the basis of its size and usual productivity. A rent of twenty to thirty pathi of barley, between a fourth and a half of the usual harvest, is typical. Renters must provide all their own inputs, including seed, fertilizer, and labor. This contrasts with tenant farming where the harvest is divided equally, but the landowner is obligated to provide seed and other inputs.

50. These men received this labor as payment for their services and were required to submit all the tax receipts themselves to the central government. As far as people today are aware the gembu did not receive any taxes. According to Hari Ram's nineteenth-century account, however, they may have claimed 15 percent of the net revenue of the region as their pay (Ortner 1989:23). During the twentieth century, at least, no labor had to be given to the gembu.

51. Although one twentieth-century Khumjung pembu acquired lands both in the Khumjung-Kunde region and in other areas through adroit use of the powers of his office, it does not appear that in general Khumbu pembu were able or interested in doing so. Those who had extensive lands by Khumbu standards—a hectare or two of fields—usually came by these through inheritance and often had family fortunes based on trade and livestock rather than on crop production. The degree to which their offices contributed to their accruing cash for trade enterprises is a question that requires further exploration. So too does the issue of whether or not the government in effect created a wealthy elite through conveying pembu privileges as Ortner suggests (Ortner 1989:92, 109–111) or only recognized already wealthy and locally powerful individuals and lineages.

52. There is also a house tax. Half a century ago this was six percent of a rupee per floor at a time when the day labor rate was one rupee per day. Today the rate is 2.5 rupees per floor.

53. Exceptions to the typical gender-based division of labor sometimes reflect household shortages of labor as well as cases of individual preference. Some women who herd, for example, do not do this by choice. In discussing two Thami Og families and one Thami Teng family in which daughters tend nak all year in the remote herding settlements one Sherpa observed they were doing so only because their families had made that decision for them.

In the adjacent Tibetan area of Tingri a similar gender-based division of labor exists. Here women do most agricultural work and men tend to herd, but women when necessary take up herding duties (Aziz 1978:108).

54. Separate groups are organized for each activity, and usually a single group of women does not remain together for all the activities of an agricultural season. Groups contain women of various ages, who may or may not be relatives or neighbors. The composition of groups also often changes from year to year. Ngalok are also organized for gathering fuel wood, and these can be mixed-sex or even all-male groups.

55. Wages can also be paid in food. This was formerly more common, but it remains customary in some places, particularly for haycutting. Wages may be paid in potatoes, grain, or butter. A *ghar* (about half a kilogram) of butter, for instance, might be paid for six days of labor, or a price might be set by the field for haycutting, which might range from one to three ghar depending on the size of the field. Thamicho laborers may also be paid in potatoes for some field tasks. A day's work weeding fields, for example, paid ten or eleven kilograms of potatoes in 1987.

56. Most of these migrants remain in Khumbu for two to five years, working for a Sherpa family on a year-round basis as a kind of household servant. A few come only for the harvest season. Those who stay on year round as hired hands

are given responsibilities that range from childcare and kitchen work to field labor, woodcutting and water hauling. Some are also given the opportunity to work with the men of the household as porters or camp staff on trekking tours, thus trading months of relatively low-paid labor for the chance to make a few months of good wages in tourism. After several years young migrant workers take their accumulated savings and return to their home regions, the money often being put towards dowries, wedding costs, and land. Only one has thus far stayed on to settle permanently in Khumbu. In 1984 house servants in Nauje could expect a wage of 2,000 rupees per year if they were considered good workers. By 1987 this had doubled. Household servants who join family members for seasonal trekking work may choose to keep their full earnings from this work and in return for this wage-labor opportunity they work the rest of the year in the household with their upkeep as their only pay.

57. There are a few household servants in Khumjung and Kunde, one in Phurtse, and none in Pangboche and Thamicho. In Nauje, however, in the autumn of 1990 at least fifty households had roughly ninety-two people employed as household servants. Of these forty-nine were non-Khumbu Sherpas, twenty were Tamangs, and seven were Rais. Most servants were young men, but thirteen were women.

According to James Fisher (1990:122) a majority of the households of Khumjung and Kunde had at least one servant as early as 1978, when they were given six rupees per day in wages as well as their food and a place to sleep. Many of these workers, however, may have been seasonal agricultural laborers rather than year-round servants. Even today only about a sixth of Kunde households have servants.

58. This differential between agricultural day wages and portering wages has been characteristic for at least thirty-five years. In 1957 a wage of three-quarters of a rupee to a rupee a day for women was reported, with two rupees a day paid to men who pulled a plow (draft plowing had not yet become widespread in Khumbu) (Fürer-Haimendorf 1975:42). Early mountaineering expeditions, by contrast, paid 7.5 rupees per day for portering work between Nauje and Everest base camp (ibid.:87).

59. Men are generally paid slightly more than women for ordinary field work. Cutting hay and wild grass, which are mainly handled by men, also pays much better than other agricultural work. In 1987, when agricultural day wages of less than twenty rupees per day were common, wages for cutting wild grass were twenty-five to thirty rupees per day in Nauje, thirty-five to fifty rupees in Phurtse, and forty-five rupees per day in Kunde. In the upper Imja Khola valley haycutting wages jumped from thirty rupees in 1986 for Pangboche men to forty rupees per day in 1987 when it became difficult to find men willing to forgo mountaineering-porter wages. In 1990 the wage for haycutting at Pangboche was eighty rupees per day, double the wage in that region for agricultural day labor.

60. The iron heads of these hoes are today obtained from Kathmandu. Formerly they came from the iron-mining center of Those.

61. Buckwheat is nearly always grown in a rotation with potatoes, and it is felt that if a field is manured every other year when potatoes are planted that soil

fertility will be sufficient for buckwheat cultivation the next year even without further manuring. In the rare instances when buckwheat is planted several years in succession in the same field it is considered important to manure the field well the second year.

62. This is considered a relatively meager haul, and lower than in earlier years. Increasing competition, partly sparked by interest in gathering dung for local sale, has made it increasingly hard to find nearby the settlement. Dung is in far more demand commercially for fertilizer than for fuel.

63. In 1987 some Nauje families completed their planting by March 19 while others were only then beginning. The earliest planting was carried out on March 8. In 1991, however, many families began planting a week or two earlier.

64. Occasionally there are a few men in these groups, usually hired agricultural laborers. It is more common for one or two men to continue to carry loads of fertilizer to the fields in which women are planting, dumping the loads and then returning for more while the women quickly spread the fertilizer over unplanted portions of the field and begin working it in as they plant. Not all fields are planted by group labor, and it is most common to see men working when families attend entirely to their own planting. These are usually households with so little land that mobilizing the entire family for a few days sees the work through.

65. A few families in Nauje and Khumjung who have relatively large potatofields now find it expedient to have them plowed by hired crossbreed teams rather than dug by hand. A large field that might require two days to dig by a crew of ten or more women can be plowed in a day. Substantial savings can be achieved by hiring a plow team rather than agricultural day laborers to prepare a field. Paying for plowing, however, may be more expensive than simply providing food and drink for a reciprocal work group. Harvests from plowed potato fields appear to be satisfactory, but some Sherpas believe that hoe cultivation is essential to achieve the best yields.

66. Seed potatoes are not seeds at all, but tubers that send out new stems, roots, and rhizomes from "eyes." Each stem arising from an eye becomes a new plant and there is no physical connection between them once the mother tuber dies (Langer and Hill 1982:278–279).

67. Seed potatoes that are small reduce the chances of disease (Langer and Hill 1982:281). Using cut pieces of potato increases the chances of loss of planting material due to disease.

68. The diversity of potato varieties planted per field and per family in some parts of the Andes is much higher. In eastern slope areas of Peru with fertile, volcanic soils and a comparable 3,000 to 3,800-meter altitudinal range a hundred or more varieties of potatoes may be grown in a single village. Individual families may have a repertoire of fifty varieties. The more meagre level of diversity found in Khumbu, however, is similar to that found in other areas of highland Peru where microclimatic conditions are drier. Villages in the western-slope Andean valleys at altitudes of more than 3,800 meters may cultivate fewer than twenty varieties and individual families may grow only five to ten varieties (Brush 1987:277).

69. In Nauje radishes are planted in late April. In Khumjung radish planting is done in early May, well before buckwheat is planted. In Pangboche, by contrast, radish may be planted several weeks later as the last event of the planting season.

70. In the Dudh Kosi valley potato planting is well underway at Phurtse by April 14 and in early May families are planting in Na, Tsom, and Charchung in the upper valley. Planting schedules in the Imja Khola valley are more complicated due to the requirements of barley sowing there (see below). In 1991 mild spring weather encouraged earlier planting; Tarnga was being planted in mid-April and by May 8 potato planting was completed in the Bhote Kosi valley.

71. This low-to-high sequence of planting is followed by most families in Khumbu, but in Pangboche a more complex sequence is common, as described later in this chapter.

72. According to one learned Tibetan monk living in Pangboche Sherpas do not follow the proper procedures for consulting horoscopes about planting times. Both the horoscopes of the man directing the plow and the woman broadcasting the seed should be consulted, he notes, and an astrological recommendation should also be taken concerning the direction in which the first furrow should be plowed.

73. Kunde and Khumjung families with fields at Dingboche mostly adopt a contrasting pattern. After barley planting they tend to return to Khumjung and Kunde for potato planting and then make another trip to Dingboche to plant potatoes in mid-May before the buckwheat in the main village requires sowing.

74. In 1987 the barleyfields of Dingboche were plowed by five teams of crossbreeds and a single yak. Most families relied on hiring the plow-team and its owner. Four of the teams were composed of two *urang zopkio* (male crossbreeds of yak and lower-altitude *Bos indicus* cows) each and one consisted of a pair of *dimzo* (male crossbreeds of nak and Tibetan *Bos taurus* bulls)

75. In the past few years there have been several experiments with irrigating other crops and there is some interest in irrigating hayfields.

76. In Dingboche this task is rotated among all the resident households, with a fixed order that is much more formalized than rotations of office elsewhere in Khumbu. One nawa is chosen from the houses above the main irrigation channel, and one from those below. In both cases families serve in a sequence that corresponds to their upstream position, beginning with the furthest upstream house and proceeding down through the settlement. All settlement residents are included in the rotation, which involves families from four villages.

77. A similar, special ceremony known as *sa yang* ("earth luck") is held in parts of Tibet.

78. They admit, however, that the opposite was said to be true that year in Pangboche. Such differing results may explain the persistence of such totally different responses and the lack of any single, accepted strategy for dealing with frost damage to buckwheat. There is also no universal agreement on which of two possible replanting techniques to use. The more common method is to use a small weeding tool to dig in the seed. People who did this after the 1986 frost

believe that they had better harvests than did people who instead broadcasted seed after scratching furrows with a thorn branch.

79. The more internationally renowned Mani Rimdu is not regarded by Sherpas in the same way, although it has often been depicted as the most characteristic Sherpa festival. Mani Rimdu, held at the Tengboche and Thami monasteries, consists of a multiday series of rituals and a day-long, masked dance drama celebrating the triumph of Buddhism over Bon in Tibet. It has none of Dumje's significance for community life, does not celebrate local history or religious heroes, and was introduced into the region only in the 1920s by the head lama of the Rumbu monastery on the Tibetan side of Everest.

80. Many Sherpas believe that the celebration was originated by Lama Sanga Dorje or his followers, who introduced it to Thami Og and Thami Teng, Pangboche, and to the temple at Rimijung in Pharak. Many Khumbu Sherpas regard the ceremony as a kind of memorial service celebrating him. Before 1830 Khumjung villagers celebrated Dumje in Thami Teng, whereas Kunde villagers went to Pangboche. In 1830, after arguments in Thamicho over Dumje administration, Khumjung villagers decided to build a new temple in their own village and organized their own Dumje rites. Kunde villagers also began to celebrate the rites there, as did Nauje families until a quarrel led to their own building of a temple in 1908. Only in Nauje is the festival today purely a single-village celebration. Elsewhere in Khumbu two or more communities come together each day for the rites, spectacles, and feasts and share in the responsibility for staging and administering them. It is the one festival of the year that absolutely everyone tries to attend. Sherpas even return from Kathmandu specifically for Dumje and herders who seldom come down to the villages try to catch at least a few days. Entire herds are often moved down from the high country for a few weeks so that families can be together with their fellow villagers to carry out the preparations and celebration.

81. Sherpas in other regions also have Dumje. Dumje is also held in Pharak (at Rimijung), Shorung (Junbesi), Golila-Gepchua (Golila) and the Likhu Khola region (Sete). Pharak holds the ceremony in summer as Khumbu does, but in the other areas it is a spring rite that contains some dances having clear fertility symbolism. An attempt some years ago to introduce these dances to Nauje ended in a fiasco when an outraged Sherpa woman grabbed a whip and chased off the participants.

82. In 1986 the festival was held on June 14–20, for example, whereas in 1987 it took place on July 4–10.

83. Fuel wood cut during the summer months is collected and stored just outside an agreed-on boundary line beyond the village and its fields. This can be seen today in the Phurtse area where large woodpiles appear each August on the western bank of the Dudh Kosi. Pangboche also bans freshly cut fuel wood from being brought into the village after crop planting is completed in May.

84. Some Sherpas also note that it is extremely unfortunate for villagers to die during the height of the dangerous blight season, because their deaths can trigger the outbreak of blight.

85. A few Nauje families also harvest some fields at this time in order to

clear land for a second crop of mustard or barley. Both are harvested by October. The barley does not produce mature grain, but is useful as fodder.

86. In 1987 the Nauje harvest was begun on August 25 in Thami Og, Thami Teng, and Yulajung at the beginning of September, in Khumjung and Kunde on September 7, and in Pangboche on September 21. In Nauje and Thami Og all but a few fields had been harvested by September 10, but elsewhere the work required from ten days to nearly three weeks longer. In Pangboche, for example, many families were still harvesting potatoes on October 1. The slower pace there, however, was related to the need to suspend work in the main village for part of September in order to harvest barley at Dingboche and cut hay in the upper Imja Khola valley.

87. Hot temperatures are also considered to lead to storage rot, and occasionally shelters are constructed to shade pit areas.

88. In 1987 Phurtse families, for example, began harvesting buckwheat on September 23 and completed work about October 15. In 1987 barley harvest began on September 12 and the final field was harvested on September 21.

4: Good Country for Yak

1. There is a complex local categorization of yak and nak, with different names to distinguish animals by age, color, size, and horn shape and size. Some types are specially prized. White is considered the finest of coat colors, and handlebar-shaped horns are highly valued. See Brower (1987:247–248) for more detail.

2. The original range of the wild yak may have been equally vast, and in the nineteenth century herds of as many as 2,000 head were reported (Perry 1981:123). This has been tremendously reduced, however, by hunting and by competition with domestic stock. Although there are still reports of wild yak in remote parts of the Changtang plateau of Tibet (Goldstein and Beall 1990:41), the wild yak may be extinct in Nepal and endangered in Tibet. The extinction of the wild yak would be a tragic loss, for it is one of the great mammals of Asia. Males stand as tall as five-and-a-half to six feet at the shoulder and have three-foot-long horns (Perry 1981:122–123).

3. Some tourists, however, mistake yak-cattle crossbreeds for yak. This greatly amuses Sherpas, who also find extremely funny tourists' references to yak's milk and butter. Yak, they point out, are male stock and do not provide milk or dairy products.

4. Yak tails were among a number of precious goods whose export or sale to foreigners was forbidden by the Chinese emperor in an edict in 714 A.D. (Schaffer 1963:24, 74).

5. For detail on the economic use of livestock in Khumbu see Brower 1987:177–187, as well as Bjønness 1980a, Fürer-Haimendorf 1975:48–52, and Palmieri 1976.

6. In Nepal the killing of females of all species of cattle and related animals is forbidden by law. Killing yak is also forbidden and Hindu Nepalis find the eating of yak abhorrent. According to legend Khumbu Sherpas were long ago

given a legal exemption from the ban on yak killing and eating, in token of which they were presented with a royal edict inscribed on a yak-head-shaped copper plaque.

7. Lack of culling has affected the age structure of Khumbu herds and may have been a factor in the great livestock losses suffered by some herders during recent severe winters. The end of culling in Khumbu probably also contributed to the increased importation of meat (primarily water buffalo) from the lower Dudh Kosi valley.

8. Yak are more often used as pack stock than nak, but *kama* nak, nak that are not lactating and are not considered likely to calve, were used as pack stock not only in Khumbu but also on trading journeys to Tibet.

9. Crossbreeds bred from yak sires and kirkong pamu dams are also known as *dim zopkio* or *dimzo* (male) and *dim zhum* (female).

10. In the Pangboche region zhum give milk six or seven months out of the year. They provide two liters of milk per day in the summer when they are milked twice per day. By autumn they are milked only once per day and only provide a quarter to a half a liter of milk each. After December or January they give no milk.

11. Breeding does not occur along entirely natural lines, and herders manipulate stock in order to accomplish the mating. Nak, for example, are reluctant to mate with *lang* (Tibetan bulls), which may require encouragement to proceed. Details of the skills and techniques involved are discussed in Brower (1987:245–247).

12. Palang pamu are not kept in Khumbu, where climatic conditions are believed to be too severe for them.

13. Livestock census figures for 1984 that categorize zopkio by type are not available.

14. Sheep dung is not used for fuel in Khumbu as it is in Tibet and Mongolia. Nor are sheep or goats employed today as pack animals as they are in some other high-altitude Himalayan regions, although a few Khumbu families used sheep as recently as the 1960s to haul salt and grain between Khumbu and lower-altitude regions.

15. I refer here to Sherpa-owned stock. The number of sheep that grazed in Khumbu from 1959 until the early 1960s was much higher due to the flocks that Tibetan refugees brought with them. These sheep severely overgrazed Khumbu pastures and many of them were lost to starvation. In the late nineteenth century and up until the 1960s hundreds and in some years well over a thousand Gurung sheep also grazed Khumbu each summer. Fürer-Haimendorf (1975:59) tallied 1,230 sheep and goats in the region in 1971, but unfortunately included no figure for Nauje nor did he give separate totals for sheep and goats for several villages. In 1978 Bjønness found that only forty-eight Khumbu households kept sheep or goats (12 percent of all livestock-owning households) and that only three families owned more than thirty sheep (1980a:69). A partial livestock census in 1984 found sheep kept by approximately 10 percent of Pangboche households, 8–9 percent of Kunde households, and 12 percent of those Thamicho households that were surveyed (Brower 1987:203 n. 6).

16. In Qinghai in 1984 the number of sheep was 14,392,000 whereas the combined total for cattle, horses, camels, and donkeys was 6,009,000. In Tibet these totals were 15,988,000 sheep to 5,344,000 head of other livestock and in Inner Mongolia 23,774,000 sheep to 6,972,000 head of other stock (Yan 1986:241). Goldstein, Beall, and Cincotta report that the Phala nomads of western Tibet keep herds composed of 87 percent sheep and goats and 13 percent yak (1990:141).

17. The impetus for the ban on goat keeping came from Sagarmatha National Park administrators, but the actual ban was enacted by the local Khumbu panchayat governments after a community meeting on the issue. The measure was controversial among Sherpas, some of whom believed it to be an unwarranted intrusion into traditional practices and economic freedom. Goats, like sheep and yak, are considered to be animals under the protection of the local god Khumbu Yul Lha and their effigies are among the offerings made to him at the annual Yerchang summer rites that protect the herds.

18. Lack of regionwide household data on livestock ownership by village makes it impossible to compare the numbers of households in each village following the three patterns of cattle keeping.

19. In 1987 six Kunde families and twelve Khumjung families herded nak. Kunde herd sizes varied from seventeen to twenty-six head. Half of the Khumjung families herded more than twenty head.

20. These patterns are typical only of recent years. Contrasts in herding styles among most of the villages were not so prominent in the 1950s when nak keeping was more characteristic of cattle keeping throughout Khumbu other than in Nauje. Nauje families have not been much involved in nak keeping since the early decades of the century.

21. This contrasts with the system of exclusive pasture rights for nomad encampments in part of western Tibet, which were allocated by outside authorities according to herd size until 1959 (Goldstein, Beall, and Cincotta 1990:148–149).

22. It should be noted however, that they are not village lands in the sense of the community's exclusive rights to resources. Villagers cross the boundary lines to cut timber and herd livestock in other areas without being required to own land there or pay fees. Community ownership was important, however, in decisions about how land was managed. Disputes over land have resulted in the violation of sacred and other protected forests when the particular villages enforcing protective regulations lost ownership or effective control of the land.

23. There are exceptions to this, however, which often reflect intervillage marriages and the inheritance of herding huts and hayfields from the wife's parents.

24. Bjønness was the first to identify and map village patterns of pastoral movements (1980a:71), although her effort was marred by the assumption that Kunde and Khumjung patterns were identical.

25. A few Phurtse families have herding huts at Machermo, however, and the several descendents of one Khumjung household have herding huts at Charchung, Tsom Teng, and Tarnak. These reflect land inherited through sev-

eral generations from a Phurtse woman who had brought it into the household of her Khumjung husband.

26. Two Kunde families, however, herd nak in the upper Dudh Kosi and the Bhote Kosi valleys.

27. The possible origins of the Khumbu system of communal agropastoral management are discussed in chapter 6.

28. In Pangboche (and formerly in Nauje), nawa were also in charge of enforcing some forest-use regulations as will be discussed further in chapter 5.

29. There was no rotation at Kunde, for example, where a resident pembu selected the nawa, and both Phurtse and Dingboche began rotating the office during the twentieth century (in Phurtse during the 1940s). In Khumjung, Phurtse, and perhaps also in Pangboche not all families were included in the rotation.

30. When a zone is closed to livestock usually an announcement is made publicly and a notice or barrier is set up at the border. Sometimes small sections of stone wall are set up to symbolically close trails or bridges that lead into the zone.

31. Some Khumbu families also grow radish specifically for use as fodder. In Nauje, Khumjung, and Thami Og a few families are now experimenting with growing wheat and barley for fodder in fields that are either planted at altitudes too high for the grain to mature for human food or are planted as second crops following potatoes and consequently have no chance to develop more than an edible stalk.

Forest fodder is extremely important in many middle-altitude regions of the Himalaya, and some excellent studies have been done of the role of forest grazing and fodder collection in local economies in the Indian Himalaya (Moench 1985) and in Nepal (Fox 1983; Metz 1990).

32. Tibetans in the Tingri area rely on the vast winter grazing resources of the Ganggar plain and adjacent areas. Certain areas of rich wild grass that grow in wet areas (*nama* or *na tsa*) are considered private property in the Ganggar region and are cut in autumn as hay. Several wild grasses are also harvested from nonprivately owned rangelands, including grasses known as *jap* and *ke,* which are cut, dried, and stored in homes as winter feed.

33. In the Bhote Kosi valley the additional step may be taken of manuring the area well for several years in advance by corralling livestock at the site. One Nauje family has made use of urea fertilizer available today only from Phaphlu in Shorung and then only at a relatively high cost.

34. Payment for cutting hayfields varies somewhat regionally. Khumjung-owned hayfields in the upper Dudh Kosi valley might bring one to three ghar of butter per field in 1986. A field that might bring one ghar of butter in the upper Dudh Kosi might bring double that (or 100 rupees cash) in Thamicho.

35. The fodder shortage in Nauje is particularly acute and grass cutters from that village now go as far afield as Tarnga in the Bhote Kosi valley and Dingboche in the Imja Khola valley.

36. Most Nauje families do not have crop residues for use as fodder due to the lack of grain cultivation in the village. A few families there, however, are now growing barley and wheat for use as fodder rather than as food.

37. Usually one or two Nauje families also lease grass-cutting rights to part of the large protected meadow at the Tengboche monastery.

38. Fodder requirements here may reflect the depletion of winter grazing in the area following the abandonment of summer livestock exclosure in 1979.

39. The amount of fodder required by an urang zopkio varies, of course, with age and with the amount of free time stock have to graze. A zopkio being used for tourism work in the spring has little time to feed and must be supplied with hay.

Urang zopkio may be fed grain (e.g., dry cornmeal) as well as turnips and potatoes. Potatoes are believed to supply good energy but, as one Sherpa put it, will not fill animals' stomachs the way that hay does. Earlier in the century, before regional population growth and tourism development increased regional demand for tubers, more of these may have been fed to livestock, especially to crossbreeds and calves. Surplus tubers are still commonly used as fodder on a small scale.

40. This procurement of different types of fodder from different sources can be seen in the following examples. In the winter of 1986 and spring of 1987 one Nauje family with six urang zopkio, one cow and a horse required approximately eighty-three loads of one *man* (forty kilograms) each, not counting the considerable amount of bamboo cut and fed to the horse or grain given to the horse and cattle. Of these twenty-five loads were dried wild grass, half of it collected in the Nauje area and half at Tengboche. Seven loads were *shruku tsi*, a wild fodder plant. Of the remaining 63 percent, fifty-two loads were hay. Of this about twelve loads were harvested from the family's own hayfields at Samde. Four were purchased from another Nauje family. A total of fourteen loads were purchased from Thamicho Sherpas from three settlements (Thami Og, Thomde, and Samde). And sixteen loads were purchased from Pharak Sherpas including hay from Monjo, Phakding, Benkar, Lukla, and Yulning, a small place just north of Ghat. In 1987 the same family purchased similar amounts of hay, part of it bought from seven Bhote Kosi valley families in Pare, Yulajung, and Samde. Another Nauje family with two dimzo and a cow harvested twenty-one loads of their own hay and bought sixteen, fourteen of them from Pharak at ninety rupees per load undelivered. Wild-grass hay figures for this family were unavailable. This level of investment in hay purchases suggests how great the demand for fodder is, how inadequate local supplies are, and how profitable urang zopkio must be to be worth this much cash outlay and trouble.

41. The effects of the sale of large amounts of Bhote Kosi valley hay to Nauje livestock owners is uncertain. It would be particularly interesting to investigate whether any Thamicho households put their own herds at greater risk by selling large amounts of hay.

42. In a few cases hay-growing sites are quite close to main villages and transport is more practical. Konar, for example, is only a few minutes walk from Phurtse. The amount of time herds are kept at Konar in the spring depends on the amount of manure herders need to collect there in order to fertilize their hayfields. If one already has plenty of manure on hand then it is possible to

move the herd more directly up to the high-herding settlements and utilize the hay stored there. In this case Konar hay can be shifted down to Phurtse and fed to stock there enabling herders to put off their move out of the main house for a few days.

43. In western Tibet, for example, nomad yak- and nak-herding camps are shifted seasonally, but within a rather small range of altitude. Here stock is herded year round at altitudes no lower than 5,000 meters (Goldstein, Beall, and Cincotta 1990:141).

44. Straw is not used for this purpose as it is feared that grass seed might thus be introduced into the crop fields.

45. Virtually all families today live in two-story houses. This was not as common earlier in this century and was very rare before 1900. This may have affected herding patterns for winter-sensitive stock. It is conceivable that for these stock more use was formerly made of winter herding bases in the gunsa where temperatures are somewhat milder than in the main villages. Some cross-breeds were also taken south to lower Dudh Kosi regions in mid-winter in former times.

46. Some Thamicho families herd in Gyajo in late winter if there has not been much snowfall in order to save their hay supplies. Others go in the spring during the worst years when the grass is poor everywhere else. Nauje families use the remote, unsettled, narrow valley during the summer for urang zopkio pasture. Gyajo was used for Nauje zopkio long before the recent increase in urang zopkio ownership there. Traders bringing young dimzo from Shorung to Tibet often summered the stock in Gyajo until the Nangpa La became easily crossable in autumn. During the 1950s some individual Nauje traders herded more than 100 dimzo at Gyajo. Gyajo was also much used by Nauje families herding yak during the 1960s after many families bought Tibetan yak cheaply from immigrants. Other important grazing areas for young dimzo were Pulubuk and the upper Imja Khola valley.

47. Yerchang is not a factor, however, for Thamicho families, for whom the celebration is not as important as it is for the other villages.

48. Residents of Dingboche do not consider the still-earlier ban on livestock in the Dingboche area to be a formal nawa-enforced exclosure but rather an informal understanding. The later nawa-enforced restrictions at Dingboche to guard against the outbreak of blight take effect on the fourth day of the sixth month, Dawa Tukpa. The entire area is also included in the post-Yerchang Dawa Tukpa livestock ban. Tengur in Pangboche and Phurtse is conducted in May-June, Dawa Shiwa. In both places regulations to guard against blight are not enforced until the end of the following month.

49. Formerly the other zones in the upper Bhote Kosi valley were closed on the fifteenth of Dawa Nawa (more recently this was set back to the twenty-fifth of that month).

50. Pasture areas in the lower Imja Khola valley near Pangboche are also closed about that time, within seven to ten days of the end of Dumje.

51. Besides the already-noted popularity of the Mong area for Khumjung herders, a number of Thamicho families have summer bases at Chosero,

Langmoche, and Mingbo, just outside the northern boundary of the Thamicho restricted area.

52. Thamicho is an exception. There the rites are conducted by individual households and are not held in such regard as they are elsewhere.

53. Such animals may be used as pack stock, however. Only prized animals with a particular coat color and other marks deemed important are dedicated. Sheep and goats can also be dedicated, but zopkio cannot because the infertile males are not considered suitable gifts to the gods.

54. This ceremony resembles one conducted on the second day of Dumje as well as the quarterly rite of erecting prayer flags on house rooftops.

55. Sometimes a family decides to move into a different social set, or young village friends decide to start their own group at a different settlement. This occurred in 1987, for example, when five young Pangboche men decided to begin holding Yerchang at Chukkung.

56. It must be noted, however, that not all Pangboche and Kunde herders elect to herd in those two upper valleys. Some families with zopkio, zhum, and cows prefer to remain in the relatively low-altitude Pulubuk and Ralha areas, both of which continue to be open to grazing throughout the summer.

57. This is the one area in Khumbu for which there are grass- and hay-cutting controls but no restrictions on grazing.

58. On September 14 Dingboche nawa reopened the settlement for cooking fires and the barley harvest got underway. Several families, however, had violated the community regulations and began harvesting on September 12.

59. The suggestion that the timing of the opening of Dingboche for harvest is based on the ripening of the barley there, and that this also serves as an indicator for when the wild grass and hay of adjacent areas is mature and cutting should begin (Brower 1987:228, 231–232) is incorrect.

60. Later I was told that this region and the Yarin area were not then officially opened, and that grazing was not supposed to take place there until October 2. People were upset about the widespread violation of the restriction on livestock and felt that the effectiveness of the management system was being undermined.

61. In many years Yerchang takes place in mid- to late July. In 1987 the ceremony was held late as a result of the same calendrical situation that resulted in the late observance of Dumje a month earlier, and the family spent only two weeks at Dusa in early August.

62. Our family owns land at Dingboche and cultivates barley there. Family members were harvesting there during the week when their nak were grazing in the Pheriche area. The amount of time that stock are kept in the Pheriche area following the opening of that zone to grazing varies from year to year by a few days. In some years when the weather is sunnier the hay and wild grass are cut, dried, and stored more quickly allowing the nawa to open the area to livestock a few days earlier.

63. The nak are not pastured at Teshinga during this time, although the family does have a house there and uses it as a herding base for their ten zopkio, zhum, and cows. These stock are based in Kunde for the winter and spring and in summer are herded with the nak.

64. The family does not have a house at either Pulubuk or Mingbo. At Mingbo they use a tarpaulin to set up a resa and at Pulubuk make use of the resa of Kunde families who do not arrive there until after Dumje.

65. Some go to Dingboche briefly en route. For several weeks in midsummer Dingboche can be used as a herding base. Although the field area itself is closed to grazing the surrounding area is not. Once the ban on building fires has been implemented on the fourth day of the sixth month, however, herders move out from the settlement and disperse with their nak to Shangyo, Pheriche, and Bibre.

66. Thirty or forty years ago the family did not go to Chukkung, but instead based at Do Ong Ma, a place north of Bibre on the way to the Kongma pass, where they stayed in caves.

5: Sacred forests and fuel wood

1. Fürer-Haimendorf and others refer to *shingo nawa*. I will use the pronunciation used in Khumbu, *shinggi nawa*, Khumbu Sherpa for "*nawa* of wood."

2. Lower temperate montane forests are more extensive in Pharak where there is considerable land below 3,000 meters and where there are oak forests (especially *Quercus semecarpifolia*) as well as large areas of *Pinus wallichiana*. The oaks are absent from even the lowest-altitude areas of Khumbu, perhaps reflecting human use rather than natural vegetative patterns.

3. By forest line I refer to the upper limit of contiguous forest. The tree line is considered to be the upper limit of tree-sized (2m) individuals, which may grow at sites above the upper limit of contiguous forest (Byers 1987*b*:55 n. 22; Price 1981:271). Juniper trees (*Juniperus recurva*) have been observed as high as 4,238 meters (Byers 1987*b*:64).

4. In Tibet the chotar is supposed to be erected only in front of houses in which a certain collection of religious texts is stored. But in Khumbu some Tibetan customs are more liberally interpreted.

5. Position relative to the fire is a matter of great concern in Khumbu and is determined by customs of status ranking. Men sit on the right of the fire with the highest-ranking males the closest to it. Women sit on the floor in front of the fire. The fire is always on the second floor of two-story houses, and both fire and hearth are considered to be the homes of respected spirits. Generally the amount of heat given off by the stone stoves is minor and is compensated for by those sitting at a distance of more than a few inches from it by wearing heavy clothes. Wealthy households sometimes burn oak charcoal in small braziers for the benefit of people sitting at a distance from the hearth. Traditionally Sherpas have worn heavy wool garments, but these are now being replaced by polyester and cotton and by down and wool mountaineering and trekking clothing. It is extremely common for men and women to wear multiple layers of clothing all year, with long underwear or jogging suits the preferred undergarments today.

6. Wood remains the primary fuel of many families at sites above the forest line such as Tarnga, Dingboche, Luza, Tugla, and Lobuche. Families in these places either make journeys down into forested country for fuel wood or depend

on high-altitude juniper (*pom*). Dried cattle dung, however, is much used in the high country as well as to a lesser but still significant degree in many of the main villages. Yak and nak dung burns with little odor and with a hotter flame than fuel wood. Sherpas consider that dung collected in the autumn makes the best fuel for it is the product of the best-fed and healthiest livestock. Winter dung is considered to be quite poor as a fuel and spring-deposited dung the worst of all. During the autumn many families go to great effort to collect large amounts of dung from slopes in the village vicinity and even at distances of up to several hours' walk. At this season people set out before sunrise and hunt by flashlight in order to maximize their search time. Such competition leads to well-picked-over slopes. People maintain, however, that the use of dung as fuel has not been at the expense of its use as fertilizer.

7. Oak is ranked above birch by Sherpas familiar with it from travels in Pharak and Shorung, but is unavailable in Khumbu.

8. Tibetan refugees who camped in the Nauje area in large numbers during the early 1960s are said to have dug *Cotoneaster microphyllus* roots and also harvested a great deal of juniper. Some Sherpas believe the immigrants relied on shrubs for fuel because they were unfamiliar with forests and lacked axes or other implements for felling or lopping trees. Nauje villagers asked a Nauje pembu to keep the Tibetans from digging up *Cotoneaster microphyllus*.

9. Wealth becomes a factor because well-to-do households may hire people to supply them with fuel wood or assign this task to household servants. This is particularly a factor in the villages of Nauje, Khumjung, and Kunde. In the mid-1980s, moreover, Rais began coming to Khumbu for periods of several weeks in the autumn and spring to cut fuel wood for Nauje villagers. This was sold by the load, and through very hard labor a man or woman could gather two loads per day. At sixty rupees per load in the spring of 1991 this was more than the usual daily wage of a porter for a trekking or mountaineering group. In April 1991 it was common to see groups of thirty or forty Rai cutters returning from the Satarma area.

10. Nima Wangchu Sherpa uses a figure of twenty-eight kilograms for a basket load of fuel wood. The weight of these loads, however, varies considerably and can be a good deal heavier. Some people consider a load to be forty kilograms.

11. Average use of a full load of fuel wood per day was reported by 36 percent of Kunde and Thamicho respondents and 40 percent of those from Khumjung (Sherpa 1979:19). Whether or not this actually reflects greater fuel use is uncertain. Villagers may have claimed higher needs out of fear that the national park–sponsored survey was a prelude to an imposition of wood-use limits and a desire to be on record as requiring a large amount in case of rationing.

12. The idea that Sherpas had a ban on green or wet wood (*lemba*) may have been derived from the Khumbu custom of prohibiting the importation of freshly-cut wood into some settlements during the height of the agricultural growing season. This, though, as already discussed earlier, was a measure to protect crops, not a forest protection measure and there were no ethical bans on felling trees.

13. Some residents also lop (and even fell) trees in very visible sites near their village when they consider this safe. There are some local differences in opinion on whether or not the national park regulations ban lopping.

14. It must be emphasized that these long distances between home and forest are not (except in the cases of Thami Teng and Thami Og) the result of forest and woodland adjacent to villages having been lost to deforestation, but only the result of more accessible forests no longer being places where it is possible to fell trees without risk of a fine or imprisonment. Some people, among them elderly, infirm, and young people, sometimes risk punishment and lop or fell trees closer to the settlement. Often such now-illegal forest use is done under the cover of darkness or foggy conditions.

15. Tree felling for Pangboche and Phurtse house construction, the building and rebuilding of the monastery, the construction of the nearby Devuche nunnery, and the building of houses for monks and nuns is very likely responsible for the large number of stumps in the forest to the east of the monastery. It seems likely that the present birch and rhododendron forest was once a mixed fir, birch, and rhododendron forest in which fir was much more common than today.

16. *Kyak shing* is used synonymously with the Sherpa terms for several different types of protected forests, including lama's forests and *rani ban* (from the Nepali for queen's forest). There is no separate term for the types I discuss below as temple forests, bridge forests, and avalanche-protection areas. Most Sherpas are unaware of the origins of some protected forests, almost universally referring, for example, to the rani ban simply as kyak shing. Young Sherpas from some parts of Khumbu would not even recognize the term *kyak shing,* having grown up in an era in which those near their villages have ceased to be protected. There is no term for *village forest* as such, probably because no forests were considered simply as the property of particular villages.

17. Sacred trees and groves have been a part of Buddhism since the origins of the faith. The groves associated with the great events of the Buddha's own life very early became important pilgrimage sites, particularly his birthplace in the Lumbini grove at Kapilavastu, the deer park at Sarnath near Benares (Varanasi) where he began his teaching, and the grove of Kusinagara in which he died. The pipal tree (*Ficus religiosa*), which became known as the "Bodhi Tree" (*bodhi* referring to the enlightened mind), became one of the symbols of the faith, celebrated for having sheltered the Buddha during the last crucial hours of meditation which culminated in his enlightenment. The pipal had already been regarded as a sacred tree in India, but it now acquired still-more-exalted status, and "the original Bodhi Tree of Gaya, under which the Buddha sat, became an object of pilgrimage, and cuttings from it were carried as far as Ceylon" (Basham 1985:263). The pipal today remains the great sacred tree of South Asia long after Buddhism has ceased to be the great religion of the subcontinent.

Although sacred trees figure significantly in the Buddhist traditions of India and Southeast Asia there has been little commentary on their importance among peoples following the Mahayana Buddhist sects of Vajrayana or Tibetan Buddhism. Sacred forests do, however, exist in some Tibetan-settled areas. In

Gansu, for instance, there is a protected forest near the monastery of Labrang, and a sacred tree associated with the founder of the Gelugpa sect of Tibetan Buddhism, Tsong Kapa, grows at Kum Bum (Taer Si) monastery in Qinghai.

18. Sherpa and Tibetan lu with the body of a woman and the tail of a serpent bear some similarities to Hindu *naga* (Basham 1985:317). These half-human, half-snake spirits are important in the Newar culture of the Kathmandu valley, playing a prominent role in religious belief and art through their connection with rain and fertility (Slusser 1982:353–361). Unlike the Newar naga, however, lu are not associated with clouds and rain, do not live in the earth, and are not usually identified with actual snakes. Sherpa lu are invisible except when seen in a shaman's vision when they appear as women and can be manifest in various ages and colors—black being angry and quite dangerous for people—depending on their current disposition (Fürer-Haimendorf 1979:266–269; Ortner 1978:279). I do know of one case, however, when a snake discovered in a tree near Kunde by a man who had been about to fell that tree was considered to be a lu. He abandoned his cutting of the tree.

19. It is believed that these lu migrate each year to Tibet. This is said to explain the dynamics of a Nauje spring whose flow decreases during the summer months in the midst of the monsoon rains and increases in the autumn after the rains ends. The spring usually only has a reduced flow during the absence of the lu, but has been known to go dry. About thirty years ago, it is said, the Nauje spring lu was offended by Nepali officials slaughtering goats beside it and went to Tibet prematurely. The spring went totally dry until Sherpa villagers performed special offerings to beg the spirit to return.

20. The rites for the household lu are also carried out by the women of the family. This is contrary to most Sherpa practice, for men generally conduct household rites. But women are believed to have a special affinity with the lu, so much so that there is danger that the lu may leave with a marrying daughter. It is said that the lu may envy the finely adorned bride and the attention she receives from the wedding party, and follow the procession to the bride's new house to take up residence there. To guard against this it is the custom for children to be dressed up grandly and given the task of dancing about the lu shrine to distract the spirit from observing the exit of the bridal party (Fürer-Haimendorf 1979:267).

21. The Junbesi grove is revered particularly by the Lama clan, which holds annual rites there. A particularly huge group of oak in Pharak near Lukla is also regarded as sacred by a particular Sherpa clan, in this case the Chawa.

22. At times of lunar eclipses, however, some people believe that it is safe to lop branches from lu trees since the lu at these times goes to the moon's assistance. One old juniper in Nauje had a great many branches lopped from it on such an occasion some years ago and died soon thereafter.

23. Groves adjacent to temples which are respected because of the sanctity of the place are different from another type of temple-owned forests found in many parts of Nepal which are used as a source of funds for the maintenance of the temple and its operations. These revenue-producing forests need not be immediately adjacent to the temple. There are none in Khumbu.

24. There are only two other private forests in Khumbu, both in the Thami Teng area and neither even half a hectare in size.

25. Khumbu oral traditions about the establishment of these protected forests were substantiated by documents I was shown in Shorung in 1987. The details presented here concerning the forest regulations and procedures decreed by Kathmandu are taken from the Shorung documents and traditions. I have not yet located surviving Khumbu documents. It is very likely that the Khumbu instructions were quite similar to those for Shorung, although this is not certain. Similar orders directing local village headmen to designate protected forests have been reported from the Jiri area, and documents discovered there suggest that such orders were issued in at least four districts of eastern Nepal beginning in January-February 1908 (Archarya 1990:130–133, 422–426). These "forest-protection circulars" outlined general regulations and procedures for forest management and specified the boundaries of the forest under the care of a given headman. In Shorung such orders were being received in 1911–1912.

26. Konchok Chombi recalls that his father, who was a pembu at this time, told him that he received these orders when Konchok Chombi was about two years old (1915). I have been shown similar documents in Shorung and Golila-Gepchua which were dated 1911–1912. It is not known whether these directives were sent out to all village headmen (*talukdar*) and pembu in what is now the Solu-Khumbu district at the same time. The Rana government began to issue such orders in areas slightly to the west of Solu-Khumbu such as Jiri as early as 1908 (Acharya 1990:130–134).

27. Petitions for permission to fell trees for use as beams had to be made with offerings of beer and the presentation of ceremonial scarves, but they were merely formalities since they were rarely denied. Pembu and other officials did not select the particular trees to be cut.

28. It is interesting that these first Khumbu efforts to establish a new type of local forest management came about only after Kathmandu's forest-protection directives. The high value that Khumbu Sherpas place on individual decisions about household economic activity and resource use may have hindered the earlier development of regulation of household forest use other than in the sacred forests. The Kathmandu government's action provided Khumbu pembu with power and legitimization to intervene in forest use and helped create a social context in which certain pembu were able to institute a new approach to local resource use that ran counter to the local spirit of individualism and customary rights to resources.

29. Some people say that the protection of forest near (and above) Thamo Og and Thamo Teng was in part to protect those places against avalanches. Avalanche protection is also given as a reason for the protection of an oak forest in Pharak near Lukla.

30. Neither these fines nor the grain that was collected each spring by the nawa from every community household went to these officials themselves as payment. The grain instead financed the ceremonies, celebrations, and village religious rituals that accompanied the annual installation of new nawa. In some other parts of Nepal, however, a *mana-pathi* system is followed in which each

household must give a *mana* (approximately one pint) or pathi (eight mana) of grain towards the salary of the forest watchers.

31. Phurtse changed this rule in recent decades to allow incumbent shinggi nawa to be rechosen.

32. Enforcing shimbak restrictions on the importation of freshly cut wood to the village was apparently the main forest duty of the nawa in both Nauje and Pangboche as well as the major enforcement activity of the Phurtse shinggi nawa.

6: Four Centuries of Agropastoral Change

1. This view differs from previous depictions (Bjønness 1980a, 1983; Brower 1987; Byers 1987b; Fisher 1990; Fürer-Haimendorf 1975, 1979, 1984) which have portrayed Khumbu agriculture from 1850 until the present as relatively unchanging except for the introduction of the potato in the mid-nineteenth century. Earlier treatments only noted very minor twentieth-century change, most often the adoption of draft plowing and changes in the availability of wage labor (Fürer-Haimendorf 1975:41–42).

2. Some high-altitude herding settlements in which potatoes are grown today, including Luza and Dole, were apparently not farmed in the nineteenth century. They then served only as herding bases or hay-producing areas.

3. Brower suggests that formerly to, the small root which today is a semi-cultivate, might have been an early, pre-potato Khumbu tuber crop (1987:99). Sherpas today, however, do not remember to as a crop and there is nothing in the oral tradition about it once having been the region's staple.

4. Brower notes that barley was previously grown at Chosero, a site near Tarnga in a tributary valley of the Bhote Kosi (1987:99). Barley may once have also been grown at Na, a Dudh Kosi site at an altitude similar to the Bhote Kosi and Imja Khola valley barley-growing sites and with similar irrigation potential. This may be commemorated in its name, which is the Sherpa word for barley. Residents of the area, however, have no oral tradition that the grain was ever grown there.

5. Barley may even have been grown on a small scale at Khumjung, where an eighty-two-year-old Sherpa recalled in 1985 that when he was very young he saw a barley *yusa* (threshing place) in his village. He has no recollection of seeing barley fields then, but he surmises that there must have been some barley grown nearby.

6. This is not the small ditch that today leads water from an out-take further down on the creek. The present system is only intended to provide enough water to supply some families with their household needs.

7. No oral traditions survive of any early concern at Tarnga, as there was at Dingboche, that introducing potatoes there would affect barley yields.

8. Sherpas certainly did not bring the potato with them from Kham, for the crop was only introduced there early in the twentieth century by French missionaries (Guibaut 1987:54).

9. A search for evidence of the introduction of the potato to Shorung and adjacent areas might cast much light on the history of the crop's diffusion across northeastern Nepal.

10. Different people sometimes use different names for these two early varieties. The long potato is known to elderly people as kyuma, hati, and belati and the round potato as koru, kyuma, and belati. I will refer to the long potato as kyuma, the round one as koru. And there appears to have been more than one koru. Sun Tenzing characterized koru as white in color, low-yielding (much lower yielding than kyuma), and producing a distinctively high number of tubers per plant. None of these characteristics matches those of the later koru. He also noted that the Phurtse koru produced very inconsistent yields. In good years plants produced an extraordinary number of tubers that attained very respectable size. In other years the tubers were extremely tiny, hardly worth harvesting.

11. No one in Khumbu today knows any tradition about the introduction of this early, round potato. It is possible that it may have survived longer in the Phurtse area than elsewhere. Elderly Sherpas in other parts of Khumbu do not remember it, only kyuma (which was grown in their villages before their births) and a later, round potato introduced during their lifetimes which had different characteristics than the round potato described by Sun Tenzing. Phurtse may thus have obtained kyuma later than did some other Khumbu villages, an example of multiple introduction of varieties to different parts of Khumbu at different times. If this did occur it raises the question of why Phurtse families did not adopt kyuma earlier from Khumbu sources. There is also the question of why it was first planted in a gunsa and was apparently only subsequently introduced into the main village.

12. Potatoes have been monocropped in Nauje as far back as memory goes. Buckwheat has been grown only on a very few, marginal, upper terraces during villagers' lifetimes. There are stories, however, of buckwheat being grown there on a larger scale long ago. One large and very productive potato field in the center of the settlement area, for instance, is said to have been famous many years ago for producing good buckwheat.

13. As already mentioned, there are several difficulties in interpreting the number and sequence of varieties that were introduced during the early twentieth century. There may have been even more introductions, some of which never had more than very narrow, local distribution. Names that were used for one variety, such as kyuma, may have later been applied to new varieties with a similar morphology. There may have been, for example, yet another introduction of a long potato that was locally called kyuma (but that had a darker skin than the earlier kyuma variety), which was brought back from the Lazhen area of Sikkim to Thami Teng about sixty-five years ago.

14. Today kyuma is still grown by a very few families in Khumbu, especially in Dingboche (where it is also known as hati). An early koru (perhaps koru 2) is also still grown on a very minor scale by a family or two. Both tubers will probably soon cease to be grown. After three years of searching for kyuma and koru potatoes I finally found a family that still had some in 1987. They were phasing the koru out by separating it from the other seed potatoes and cooking it

up in curry rather than planting it! In 1987 I also found kyuma in the Salpa region which had been grown by a woman who noted that her stock was descended from tubers that had originally come from Khumbu.

15. Families from Khumjung, Kunde, and Thamicho exported dried potatoes to Tibet, as did a few families from Phurtse. Earlier in the century dried potatoes were also sold by Dingboche farmers to Thamicho villagers for trading to Tibet.

16. MacFarlane suggests that the population of Nepal "more than trebled in the years between 1850 and 1960, from a base of between three and four million" (1976:292). In the national context potato cultivation was not a significant factor in population growth, although other agricultural change was, including the adoption of maize and the diffusion of irrigated, terrace agriculture into areas previously cultivated only by swidden farming.

17. It is not clear whether other tubers were planted in Dingboche during the period before potatoes were accepted.

18. Gembu Tsepal was one of the wealthiest men in Khumbu. He owned considerable land, the largest yak and nak herd in the region, and several houses and herding huts. Besides his Nauje land and the Dingboche fields he also, as mentioned earlier, had land at Tarnga.

19. In another version of the story it is Ang Sani, Ang Chumbi's wife, who defied the gembu. Once potatoes were introduced it was found that barley continued to yield good harvests and that the potato harvests were good enough to merit a biannual rotation with barley.

20. The most recent rebuilding of the Tengboche monastery in 1990–1991 was an exception to the earlier traditions of building religious structures with donated labor. For this project Sherpas raised donations in Khumbu and abroad and used them to hire workers, mostly Magars from the lower Dudh Kosi valley.

21. A few cases might be cited as examples. In Nauje the temple was supported through the donation by each family of a small amount of grain. The big prayer wheels that were the original centerpiece of the place, however, were contributed by wealthy traders and a famous lama. The several large private chapels in the village all belonged to trading families. The five large prayer wheels in the village stream and the two in separate shrines were built by traders. The building of the Tengboche monastery was made possible by the inspiration of the head lama of Tibet's Rumbu monastery who convinced a Khumjung man to raise funds to build it, but it was the wealth of four Sherpas (two of them from Shorung) that launched the project.

22. It is not possible to determine the total amount of land abandoned in various periods, nor to estimate how much of this was later reclaimed. The lack of land records or accurate tax lists provides no basis on which to attempt such a historical reconstruction. The relatively high percentage of old, abandoned terraces in the Bhote Kosi valley could either represent a greater degree of historical abandonment there than in other valleys or simply a lower degree of subsequent reclamation. At Nauje, Phurtse, and Khumjung some expansion of the agricultural area took place during the first half of the twentieth century.

23. The dates of terrace abandonment shown in table 21 are tentative in

many cases and further research will be required to pinpoint the processes and time period involved. The preliminary treatment given here suggests, however, the large number of sites involved and the relatively small number of areas where terraces have been abandoned since 1960.

24. The reason for the abandonment of some other extensive areas at Nyeshe in an earlier period, however, is not known.

25. It is said that once families from all over Thamicho had houses at Leve. Elderly Thamicho Sherpas remember it as a place of ruins throughout their lives, inhabited early in the century only by a few people such as Leve Karsong and Leve Pasang. The latter is the grandmother of a Thamicho man who is now about fifty years of age. Leve was apparently a secondary crop-growing site for some families and the main dwelling for others. The last families to leave either shifted their operations entirely to other villages in Thamicho or moved out of the region to Rolwaling and Darjeeling. It had developed a reputation as a place of bad luck and this belief may have been a major cause of its demise, along with a lack of water. Two explanations of the inauspiciousness of the site are given. One is that Leve is situated at a place that has bad luck due to its topographic situation. The second and more common view is that Leve's decline began with a contest of powers between a lama living there and a female nun who dwelt at Samde. The lama succeeded in causing a large landslide below Samde, the nun in causing the spring to dry up in Leve. In some versions she also caused the collapse of the Leve chorten, but other people ascribe this to the 1934 earthquake.

26. In the nineteenth and early twentieth centuries there was a greater emphasis in the Bhote Kosi valley on cultivating buckwheat (and perhaps barley) than there has been since the 1920s. The gradual and ultimately total abandonment of grain growing in favor of obtaining grain grown outside of Khumbu has greatly changed land-use dynamics in the valley.

27. Thamicho villagers, however, do not recall any major expansion in field area during this period despite population increases.

28. It will be extremely interesting to observe whether there is more widespread experimentation with plowing potato terraces in future years and whether a change in values and perceptions accompanies such a change.

29. He himself had forgotten the exact date by 1985, but a 1976 date was furnished by his neighbor Konchok Chombi, who obtained seed potatoes from him the following year.

30. Pemba Tenzing did not ask how Singh Gompa had obtained the variety. It is possible that it was introduced from Kathmandu agricultural experimentation projects, for in the mid-1970s a number of new varieties were introduced at Singh Gompa from Kathmandu (J. D. Sakya, National Potato Development Program, personal communication 1987).

31. Konchok Chombi himself makes no secret of having originally obtained the potato from Pemba Tenzing's wife, but a great many Khumbu Sherpas are astonished to learn about this. It is probably apt, however, that people continue to associate the yellow potato above all else with Konchok Chombi. His realization of the importance of a new, high-yielding potato variety, his lack of concern with personally profiting from his discovery of the variety, and his

ability to administrate his sales and gifts in order to maximize the early wide-spread diffusion of the seed potatoes all were a major contribution to regional welfare. He was clearly proud of his ability to do all of this and enjoyed the acclaim it brought him. Yet his motives seem to have arisen primarily from the particularly keen sense of civic responsibility that has characterized his career as a pembu.

32. It is not unlikely that some Thamicho families actually obtained their first yellow potatoes from Konchok Chombi, who has had long ties with many Thamicho families due to his having inherited responsibility as pembu for that area from his father.

33. In the last few years the adoption of the yellow potato has been given further impetus by the increasingly common perception that red potato yields are declining.

34. It is unclear which of the four varieties being promoted at Phaphlu is the one currently gaining acceptance in Khumbu. There may be more than one variety of "development potato" being grown today in Khumbu.

35. An initial 1984 experiment by one elderly Yulajung resident was very encouraging, with yields of twelve times the volume of seed potato. He was enthusiastic enough to consider putting as much of his land into development potatoes as he could obtain seed potatoes to plant, but poorer harvests during the next two years dampened his initial excitement and he has also had problems with the potato rotting in storage.

36. Some Sherpas consider the black potato to be identical to the brown variety. In the Pangboche-Dingboche area, however, the black potato was introduced even before the yellow potato. Further research on the introduction and diffusion of these minor varieties is required.

37. Buckwheat was grown on several sites near Nauje in the 1960s, but in the main crop-growing area of the village it has been virtually uncultivated since early in the century. Today some buckwheat is grown in gunsa such as Teshinga and Mende. Formerly, however, it was probably grown more widely in gunsa and was also grown in some high-altitude areas such as Tarnga, where it was a major crop early in this century.

38. Another reflection of this situation was the effort by some Nauje families to obtain more land. After 1975 at least ten Nauje families acquired fields at Tarnga and other families obtained land at Langmoche and Chosero. The Langmoche area was highly regarded because it was outside of the livestock-closure areas during the entire summer and also was close to forest. This meant that Nauje families could base their zopkio there for the summer in order to have sufficient manure supplies available for the following year and could also easily collect leaves for fertilizer from nearby woodland. Land was more readily available at Tarnga and more than a dozen Nauje families bought fields there. Fertilizing fields there was more difficult, however, for during the 1970s and until 1985 the summer livestock closure on this area was well enforced. Nauje families who only had Tarnga fields had difficulty pasturing their livestock in the area long enough to accumulate sufficient fertilizer and had to buy manure at relatively high prices from Thamicho households. In recent years manure has

been so much in demand at Tarnga that it is virtually impossible to purchase unless arrangements are made half a year or more in advance.

39. At Thami Og some long-abandoned land has also been reclaimed for hay production.

40. Konchok Chombi believes that the use of sheep manure indicates that the family had no yak because no one who had access to cattle manure would use sheep manure. He further speculates that at this time sheep keeping may have been common in Khumbu, and that many families were too poor to have yak.

41. Some Sherpas believe that the ban on goats initiated in 1983 by the national park may be offensive to Khumbu Yul Lha and that this has been responsible for various manifestations of poor luck in the region since the ban.

42. The Dzongnangba legend also refers to large tents, perhaps similar to the gaily decorated tents familiar in Tibet for summer use by the elite.

43. The fact that Sherpas were using black tents in the nineteenth century does not mean that they were more nomadic at that time, for clearly villages and gunsa settlements had been established much earlier. There were also high-altitude herding settlements with permanent huts at least as early as the late nineteenth century. Konchok Chombi's father and grandfather, for example, owned herding huts at Luza as well as the black tents that they used at Tugla, Tsola, and elsewhere.

44. The last time this occurred was in the 1940s.

45. Pharak Sherpas from Lukla, for example, use summer pastures on the upper Hinku Khola, an area that is also summer pasture for the sheep, water buffalo (and in at least one case yak) of Rais from Bung and Gudel as well as for Gurung sheep from the lower Dudh Kosi valley and its tributaries. In the upper Arun-Barun region Sherpas, Rais, and Gurungs share summer pasture at Khembalung in the Barun drainage, a place also regarded as a pilgimage site by all three groups, and Sherpas from Navagaon send their water buffalo to Rai-inhabited lower-altitude areas in winter. In these cases arrangements are made for the use of different areas and grazing fees may be collected. In the past Sherpas in the Salpa, Arun, and Katanga regions, moreover, paid taxes to the Rais for the right to inhabit their main village sites (and in the Arun pay them still), and in the early twentieth century Pharak Sherpas still paid Rais for the right to use some Dudh Kosi pasture as well as that in the Hinku Khola. It is not impossible that Khumbu Sherpas may have once had to pay similar taxes to the Rais.

46. Gurungs could have been coming to Khumbu for well over a century. They settled in the lower Dudh Kosi valley in the early nineteenth century.

47. Similarly some Gurung herders bound for the Bhotego area on the west bank of the Bhote Kosi river near Pare drove their flocks up the valley on the western side, choosing a difficult route rather than pass through major Sherpa settlement areas.

48. The powerful Khumjung Sherpa who had encouraged Dimal to come to Dingboche and apparently taken money from him was also the target of considerable anger and is said to have been seriously hurt when attacked near Khumjung by rock-throwing boys. This took place during the time of Nepal's Rana regime (e.g., before 1950).

49. It is conceivable that this unusually fierce resistance to Gurung grazing may have been linked to especially grave concerns about overgrazing. In 1959 and 1960 approximately 5,000 Tibetan refugees settled in Khumbu, bringing with them huge herds of yak and sheep. Many of them established themselves in the Nauje area and a serious grazing shortage soon developed.

50. And in the early 1960s a sheep could be purchased for a single rupee, compared to 200 rupees today for the dried meat of half a Tibetan sheep.

7: Subsistence, Adaptation, and Environmental Change

1. The recognition of the ecological wisdom of indigenous peoples has since become a basic theme of the radical environmental movement and agroecology as well as a factor in global conservation and protected area management planning.

2. Some cultural ecologists may still subscribe to the view that "in order to adequately conceptualize the ecological relationships of human groups, it may be necessary to treat them as if they were parts of a functionally integrated, persisting, homeostatic, isolatable ecosystem" (Netting 1984:231). Netting, however, cautions against a simple assumption that local ecosystems "epitomize a well articulated, self-sustaining interdependence of physical environment, subsistence techniques, and human population" (ibid.:227).

3. In particular this perspective has been a hallmark of the Berkeley school of geography since the 1920s, where it has underlain both the concept of the cultural landscape as a historically developed artifact distinct from the natural landscape and a continuing focus on studies of the past and present human transformation of the earth.

4. The last Khumbu wolf met his end in 1986. When villagers discovered in 1987 that a pair of wolves had apparently crossed from Tibet into the upper Bhote Kosi valley and produced a litter there they asked the national park to exterminate the animals. When this request was refused they took matters into their own hands. These activities bear a certain resemblance to the pragmatic Sherpa (and Tibetan) attitude toward the killing of livestock. Sherpas will not kill livestock suffering from disease, injury, or old age, nor will they kill unwanted crossbreed calves whose parentage makes them of little economic value. They will, however, allow livestock to starve to death or otherwise indirectly assist in their deaths. Similarly animals cannot be killed for meat, but meat can be eaten from stock that has "accidentally" fallen from trails.

5. This tension between communal institutions and cultural emphasis on household economic freedom has not been emphasized in the anthropological literature. Fürer-Haimendorf stressed the high degree of civic responsibility in resource-management institutions (1984:50–51) and the lack of village factionalism (1975:98). This is certainly characteristic of Khumbu community resource management, although there has long been conflict within and among villages over rights to resources, the power to administer them, and the enforcement of regulations. At a deeper level, however, the rules themselves are very limited in

some ways that suggest a long-shared cultural attitude about the proper borders of social intervention in household economic life.

6. Fürer-Haimendorf (1979:110–113) primarily focused on Sherpa forest management as a social institution in the villages of Khumjung and Kunde. He was not interested in the actual boundaries of protected forests, regional patterns of forest use, or historical change in forests or management institutions. Fürer-Haimendorf assumed that the system he observed in Khumjung and Kunde was a Khumbu-wide practice of some antiquity, that it regulated forests adjacent to communities that were used for subsistence purposes, and that it functioned effectively in conserving forests (1975:97–98; 1979:112; 1984:57). We have seen how erroneous a number of these assumptions are and the irony that the rani ban, including the very one Fürer-Haimendorf was describing, had in 1957 only been protected for less than fifty years. Brower (1987) describes the nawa system as a de facto rotational system of grazing that effectively distributes grazing pressure regionally, but does so without delimiting the areas regulated, analyzing the intensity of use of particular places, considering the constraints on the system as a result of sociocultural values and conditions, or exploring historical change in herding and pastoral management. Byers (1987b) finds that Sherpa land use is in relative harmony with the local environment but considers only slope stability as a criterion.

7. Isolated trees are especially common in gullies and near the edges of forests and woodlands. Scattered juniper and rhododendron shrub are more common and are found throughout the altitudinal range of the grasslands. Barberries and rose bushes are more common in the vicinity of settlements whereas *Cotoneaster microphyllus* is especially prominent on the slopes that have been heavily tracked by livestock.

8. Cereal pollen was found in soil layers for which radioactive carbon dating of charcoal provides dates of 1,480 years plus or minus 360 years and 2,170 years plus or minus 330 years (Byers 1987b:199).

9. For an account of the gradual adoption of permanent agriculture by the Rais and the decline of earlier integral swidden agriculture see English (1982, 1985). This process involved not only agricultural intensification but also migration, for the techniques were apparently brought to the region by Pahari immigrants from the central and western midlands of Nepal.

10. Twentieth-century Sherpas attest that they do not intentionally use fire as a tool for pasture improvement. Large fires such as the one that burned the grasslands and woodlands along the Dudh Kosi river just east of Nauje during the 1950s have resulted from herders' and woodcutters' cook fires accidentally getting out of hand. Many of these accidental fires seem to take place in the higher reaches of Khumbu, above the forests in the high grazing country. Here in any given spring or early summer one may come across blackened slopes very reminiscent of the intentionally burned high-pasture country in similar areas in the Annapurna range, where Gurung shepherds in the Modi Khola valley apparently deliberately encourage better grass growth in the alpine country above 4,000 meters. In the Khumbu case these fires are always said to be accidental, although many occur on slopes covered with rhododendron and other unpalatable shrubs and probably improve grazing there through encouraging grass growth.

11. This map also makes clear that the actual rani ban boundaries included a considerable area that is now neither forest nor woodland. This does not, however, represent post-1915 deforestation within the protected forest boundaries. Instead it reflects the original definition of rani ban and nawa agropastoral management zones in terms of topographic features such as ridges, streams, and trails. All trees within an area defined by such features were protected, regardless of whether they were within forests, and substantial rangeland areas were included in some rani ban.

12. Beyond a century and a half ago, however, there are no oral traditions about forest use and forest change. The earlier history of regional vegetation change will be sketched only with the help of pollen analysis.

13. This slope is sometimes incorrectly assumed to be the area near Nauje which Fürer-Haimendorf (1975:98) observed had been much affected by tree felling between his 1957 and 1972 visits.

14. There is a belief among Khumbu Sherpas that it is inauspicious for a household to own more than 100 head of yak and nak, and there is an oral tradition about an early Sherpa household that ran into misfortune when it exceeded this limit. The fact that no household in the past fifty years has amassed a herd of even half that size, however, makes this injunction less relevant than it might otherwise be.

15. In the Bhote Kosi and Dudh Kosi valleys almost no pasture area was closed above 3,800 meters.

16. This contrasts with the system in use in areas grazed by Phala nomads in western Tibet before 1959, where pasture allocations were based on herd size and each defined pasture area was considered suitable for a certain number of head of livestock calculated in terms of the type of stock (one yak was considered to be the equivalent of six sheep or seven goats). There was a limit of thirteen yak (or seventy-eight sheep) per standard pasture area. Every three years a stock census was conducted and pasture-use rights were reallocated (Goldstein and Beall 1990:69–71). This, however, was not a locally developed institutional arrangement. It was instead imposed by the landowner, the Panchen Lama of Tashilumpo monastery in distant Shigatse. It should not be assumed that this type of range-management system was typical of Drokba-inhabited areas in Tibet.

17. This has meant, for example, that when Dumje came late in the lunar calendar in 1987 that livestock exclusions were nevertheless not enforced until after the festival. Despite the perception that closing off the winter pasture areas to livestock a month late invited trouble, the stock were not sent out of the village until late July. Changes are very seldom made to the customary rules about the closure of zones. Nawa in some areas, however, do exercise some discretion in deciding when to open areas to grass cutting and livestock.

18. A lack of concern with the possible impact of grazing on forest regeneration is a feature of Khumbu protected-forest regulation as well as the operation of the nawa system. Grazing has never been prohibited in any Khumbu forest, not even in sacred forests, nor have there have been any regulations that restricted forest grazing by any particular type of livestock, such as zopkio or

goats, which might be expected to have a particularly heavy impact on forest regeneration. It may be that the absence of this kind of concern reflects the belief of many Sherpas that grazing pressure does not affect forest density or composition.

19. Similarly, grazing may be responsible for the relatively poor rate of seedling survival in some national park plantations that have not been adequately protected from livestock intrusions. These plantations are surrounded by stone walls or barbed-wire fences, but stockowners find ways to enable their animals to enter them. The forest plantations are extremely unpopular with some local herders, who complain that the government is taking away some of their most important grazing land.

20. The equivalent incentive for pastoral management may well have been concern over maintaining enough fodder and winter pasture to successfully winter stock.

21. On the basis of his experience in central and western Nepal Gilmour suggests that two products whose scarcity is of special concern are leaf litter and large trees for construction purposes. Concern with leaf litter has been mentioned by Khumbu Sherpas as a possible factor in the protection of the Bachangchang rani ban near Thami Og and the Khumjung-Kunde rani ban, although Sherpas did not develop rules such as those Gilmour reported in western Nepal where local users limited leaf-litter gathering to a set period (1989:5). There was even more concern in Khumbu over safeguarding the supply of trees large enough to serve as beams, and this may well have been a factor in the establishment of the rani ban at Nauje, Pare, Khumjung-Kunde, and Samshing.

22. The protection of sacred trees and forests in Khumbu, however, is connected with fear as well as with faith. Villagers perceive real risks of loss of health and good fortune if they do not take effective community management action.

23. Evaluating the accessibility of forest resources should include examining not only the physical abundance and location of the types of forests needed for subsistence, but also the effects of religion and other cultural beliefs, local resource-management systems, and differences between households in wealth, land, and labor resources.

8: Local Resource Management

1. The exclusion of kipat forest lands from nationalization suggests that the Forest Nationalization Act of 1957 was not intended to undermine all community management of forests. (Kipat refers to a particular type of communal land tenure once common in eastern Nepal, often in areas where subsistence was based on swidden cultivation.) Acharya thus notes that "the Act neither affected nor was meant to affect communally managed forests (an estimated 6 percent of the total forest area at that time) or private forests smaller than 1.3 hectares in the hills" (1989:17). The act was rather intended to reclaim for the state the estimated 17 percent of forestland that had been granted to individuals by the Shah and Rana governments as gifts to relatives and in lieu of payment or reward to employees

(ibid.). In many areas, however, communally managed forests and other common lands were not registered as kipat and no provision seems to have been made to legally recognize these at the time of forest nationalization.

2. Not all nonprivate forests are administered directly by the Forest Department, for with the Forest Act of 1961 the government created a mechanism by which local governments could apply for the right to manage local forests under Forest Department supervision as panchayat forests. Community forest-management plans, however, must be approved by the Forest Department and thus far a rather small amount of forestland is actually being administered by communities. There are no panchayat forests in Khumbu and none can be created under current Sagarmatha National Park policies under which all forestland is under park ownership.

3. The responsibility for issuing permits was transferred to Salleri in 1970 and the present Forest Department office was established there in 1978.

4. The closer involvement of central government officials in the Khumbu area and the relative ease of obtaining permits, however, may have only increased the speed with which local forest administration in some communities was undermined and the scale at which tree felling took place in some previously locally protected forests. By 1971, however, a Forest Department office had been set up in Shorung and Sherpas were required to obtain permits there rather than from the Nauje branch of the Sagarmatha zone office. In most years it has not been necessary to go in person to Salleri to obtain a permit, as usually a pradhan pancha or other Sherpa with business in the area can handle the application.

5. At the time that the Nauje branch of the zone office was established all of Khumbu was under the jurisdiction of a single pradhan pancha, but by 1970 the area had been divided into two panchayats, one containing Nauje and the Bhote Kosi valley population and the other the rest of Khumbu, each area with a separate pradhan pancha. Since the panchayat system was established in Khumbu all pradhan panchas have been residents of either Khumjung or Nauje.

6. Recently it has become necessary to gain the approval also of the Chaunrikharka pradhan pancha (who administers the area of Pharak directly south of Khumbu), following Pharak complaints that Khumbu Sherpas were cutting large amounts of timber in that area without any local role in the process of approval and control.

7. The office apparently discouraged the traditional Sherpa practice of felling trees for fuel wood. Permits were only available for obtaining timber.

8. The official also kept the documents. These were apparently the forest protection circulars from Chandra Shamshere Rana's government which assigned responsibilities for particular, defined, forest areas to each pembu. It is possible that some copies of these will yet be found in Khumbu. I have had the opportunity to inspect a number of rani ban documents in Shorung and Golila-Gepchua.

9. Some villagers also took advantage of the lack of careful monitoring of the amounts of timber cut and exceeded the quantities authorized on their permits. Some Sherpas reportedly chose to interpret permits in ways that granted them rather more wood than was intended. One strategy was to cut many more trees than authorized, arguing that the number of cubic feet speci-

fied for cutting on the permit referred to the trees' basal area only, and that the rest of the timber obtained from the tree was free for the taking (P. H. C. Lucas, personal communication, 1990).

10. Fürer-Haimendorf (1975) did not mention tourism as a factor in this change in forest cover and presumably he assumed that the tree felling had taken place as a result of Sherpas' own demands for timber. Even in the early 1970s, however, some observers were concerned about the impacts of mountaineering and trekking tourism on Khumbu forests, a topic which will be taken up in chapter 10.

11. It should be possible to corroborate these accounts of forest change between 1965 and the 1979 enforcement of new forest regulations in Khumbu by comparison of photographs taken before, during, and after this period. Comparison of a 1962 photograph made by Erwin Schneider and another taken by Alton Byers (1987a:78–79; 1987b:207–208) from approximately the same site in 1984, for example, substantiates Sherpa accounts of post-1965 forest and woodland change at Shyangboche and near Khumjung and Kenzuma. Two of Charles Houston's photographs (J. Fisher 1990:8, 144) similarly support Sherpa oral history testimony in that they indicate that the area adjacent to the northwest corner of Nauje was not forested in 1950 and that large areas on the slope immediately west of the Tengboche monastery were open woodland at that time. Much more could be done with historical photographs beyond the analysis of repeat photographs or individual historical "slice in time" photographs. What is really needed to augment oral history testimony are sequential comparisons of the conditions of particular sites over a period of years, with photographs from as many different times as possible between 1950 and the present.

12. That considerable logging occurred in the area has been confirmed by a ground survey which found large numbers of stumps (Byers 1987b:211). Byers contrasts the density of juniper in a less-disturbed site near Shyangboche with that of juniper on the slope above Khumjung. At the less-disturbed site there was a "general absence of stumps" and a density of tree-sized individuals (i.e., greater than four centimeters in diameter at breast height) of 725 per hectare. Above Khumjung the density of juniper was only 533 per hectare. A full 36.5 percent of the Khumjung juniper that Byers tallied as stems, moreover, were merely stumps. He further notes that "it could also be argued that considerably more juniper existed in the mid-1950's and, because of the Sherpa practice of removing all woody material, this was not reflected in the 1984 sampling quadrants" (ibid.:212).

It could be further noted that there are now very few junipers more than a meter in height on the slope immediately above Khumjung. This contrasts strikingly with the stature of the protected juniper at the Khumjung village temple or the *lu*-inhabited juniper elsewhere in the settlement.

13. Interestingly, Byers reports researchers' impressions that Sherpas have insisted that the Khumjung slope was densely forested in the recent past:

It is of note, however, that most Sherpa informants interviewed by researchers since the early 1970's insist that the Khumjung slopes were covered by "thick forests" 20–

30 years ago, although the specific Sherpa landscape terminology used and translated is not known. (1987*b*:212, n. 13)

Descriptions of the Khumjung slope as densely forested only thirty years ago, however, are considered by some elders to be extremely exaggerated. Local residents instead remember the vegetation there as having been between *nating tukpu* (thick forest) and *nating shreme* (thin forest) in density during their lifetimes. Certain areas had more trees than others.

This confusion may have several causes. Sherpas may have misunderstood the degree of detail demanded by researchers and offered simple generalizations rather than detailed accounts. Researchers or other investigators may have misinterpreted Sherpa responses or asked imprecise questions.

14. The difference between the tree cover above Khumjung and that above Kunde only a few hundred meters away along the same slope is tremendous. At Kunde there are big junipers growing only a few dozen meters above the highest houses and an open juniper woodland extends unbroken up the slope until it feathers into fir woodlands. This may reflect better protection of this area since 1915 as well as possibly different forest conditions at the time of the establishment of the rani ban.

15. Not all families spent the entire winter outside of Khumbu and some made a series of shorter trips to Pharak, Katanga and other areas to trade salt for grain. Fürer-Haimendorf (1979), for example, noted that Dorje Ngundu of Khumjung made a number of such trips during the winter of 1956–57.

16. By the 1980s, however, it had become common for entire families to spend the winter outside of Khumbu again. Destinations by this time had changed and most people wintered in Kathmandu or made pilgrimages to Buddhist sites in India or Tibet.

17. Villagers made several adjustments to these new conditions. One common strategy was an effort to produce more potatoes. This became relatively easy to do once the higher-yielding yellow potato became available in the mid-1970s, but during the years before that many families had difficulty in growing enough tubers for their new household requirements or finding other farmers who had a surplus to sell. It is possible that some families may have shifted their cropping emphasis at this time toward a greater reliance on potatoes and less production of grains. Another strategy was to purchase more grain. Income from tourism enabled many families to pay relatively high prices to purchase this grain from lower-valley traders at the Nauje weekly market.

18. According to Thompson and Warburton (1985:122) Fürer-Haimendorf noted a change in fuel-wood use between his 1957 and 1971 visits. Whereas in 1957 Sherpas had kept fires going all day, in 1971 they used wood only when it was necessary for cooking. Thompson and Warburton use this simple observation to estimate that Sherpas may have been using a third less fuel wood in the 1970s than previously. This may be an overestimate, for they did not take into account the increased numbers of months that families were now spending at home in the region using fuel wood. It is very probable that Sherpas were making more thrifty use of fuel wood by 1971, but several factors ignored in Fürer-Haimendorf's comparison are very important in evaluating the amount of

conservation being practiced. It is not clear whether or not Fürer-Haimendorf took seasonal variations in fuel-wood use into account. He may have been observing differences between late autumn or winter practices and summer ones. It is also not clear how much time would have been spent cooking. Today Sherpas cook four meals, all of which require substantial time. The preparation of dinner in particular, especially in winter when the evening warmth is welcome, can occupy several hours. Additional fires may be lit to prepare tea.

19. It is possible that Tibetan refugees may also have made stove use more visible in Khumbu in the early 1960s (J. Fisher 1990:64). It seems, however, that stoves were mainly adopted in the mid- and late 1960s rather than at the beginning of the decade.

20. Sherpas do not try to heat their homes as a rule, but rather cluster close around the fire when they are cold. The new architectural fashions may in part be a response to the lower heating abilities of the stone stoves compared to open fires. Not all the new fashions are concerned with conserving forest resources in the larger sense, for multiple rooms, ceilings, and wood paneling have increased the amount of timber required. There has, however, been some conservation of timber as well, including the use of smaller beams and rafters, a change that also reflects the conversion to lighter metal roofs from the older styles of slate slabs or heavy fir shakes.

21. Perhaps the fact that New Zealand advising was so influential in the establishment and early operation of Sagarmatha National Park is partly to be credited with this attitude. New Zealand national parks are managed with considerable emphasis on local consultation, which there is ensured by having local representatives as members on the councils that administer individual parks.

22. Under the regulations that established Nepal's national park system the parks are required to pay for the posting of troops who operate under the independent command of their own officers. At times the army has been involved in efforts to halt musk deer poaching by Tamangs from outside of Khumbu who have been active even in recent years in the Dudh Kosi valley.

23. The new cost of timber has led to increasing use of sheet-metal roofing in some Khumbu villages. Sherpas note that now the cost of having metal roofing material carried in from Kathmandu is competitive with the cost of having fir shakes cut and transported from Pharak. Timber use has also been influenced. Since the cost of having timber cut and hauled to house sites increases with distance from Pharak, it has become extremely expensive in the villages of eastern Khumbu. This has affected the lives of residents of Pangboche and Phurtse, many of whom are among the families least involved in the tourism industry and who can least afford high prices for what was formerly a resource free for the taking.

24. This has probably contributed to the apparently rather poor seedling survival rate in some plantations.

25. This could be established at many levels. One would be the reestablishment of the long-defunct national park advisory committee with regular meetings and legal minutes that would be submitted to the director general of the Department of National Parks and Wildlife Conservation in Kathmandu as well

as to park authorities in Khumbu. Annual open meetings should be held between the chief administrator of the national park and villagers. Conservation education programs could be developed in the schools. Greater mutual involvement of local people and national park staff in village development and Khumbu conservation activities could be encouraged. And at another level Sherpas could become a more intrinsic part of park planning and management through a reorganization of national resource management to grant more authority to village committees.

26. Some Sherpas consider that national park prohibitions on the use of green wood refer only to tree felling and not to the lopping of branches.

27. In the Pangboche area villagers complained that the park gave permits to Kunde and Khumjung villagers who cut beams to build tourist lodges in Pheriche and other sites along the trail to Mount Everest. These villagers apparently felt no need to respect Pangboche villagers' religious sentiments about the Yarin forest. Tree felling at Yarin by nonvillagers is said to have led some Pangboche people to obtain trees there as well. During the mid-1980s there was also a village controversy in Phurtse, this time precipitated by two villagers who felled trees for beams with national park approval without having obtained authorization from the shinggi nawa. In both cases park administrators could have been more aware of local institutions and procedures and more careful not to undermine them.

28. In theory government logging-permit policies after the nationalization of the forest would also have discouraged the felling of trees for fuel wood by awarding permits only for timber use. But there was no effort to enforce such a policy, and felling trees for fuel wood continued to be a common practice until the national park began to enforce its regulations.

29. Pangboche families have the most accessible fuel-wood supply today, with abundant dead wood available within an hour's journey across the Imja Khola.

30. The collection of these permit fees, however, was seen by park authorities as a reward rather than as a right and not all villages immediately won the authority to collect these fees for their own use.

31. This interpretation of the agreement, however, may not be shared by all Sherpas. In 1990 the bridge to Satarma was dismantled by national park workers. Much fuel wood continues to be gathered in the area, however, by Sherpas and also by Rai woodcutters who now come to Khumbu each autumn and spring to work for a few weeks selling fuel wood to lodges, government offices, and individual families. Two makeshift bridges have been established to provide access to the area. Some cutting also continues to take place without authorization on the Nauje side of the Bhote Kosi across from Satarma. It should be noted that many woodcutters at Satarma do exercise some care in selecting fuel wood. During a spring 1990 visit to the site I met Rai woodcutters who had selected a large birch tree that had been uprooted and fallen rather than fell a standing tree. Other Rais and Sherpas were climbing the steep slope above Satarma for up to an hour in order to reach remote areas of birch forest before gathering fuel wood.

32. Two large trees were considered to be enough to provide roof shakes for a modest-sized house. In Phurtse and Pangboche it was believed that areas closer to their settlements would also be opened for tree felling for roof shakes and in Phurtse, where slate slabs for roofing are popular, there was an understanding that families who were not interested in shake roofs could cut trees for use in window and door framing instead.

33. The New Zealand mission of 1974 recommended that three zones be established within Sagarmatha National Park, a human settlement zone, a pastoral zone, and a wilderness zone. Although these zones were not implemented the principles underlying them were embodied in park regulations.

34. It is remembered that about 1943 a large number of yak were brought down early in the autumn before the village had been opened. Until this time the custom in the village had been to choose two nawa annually but to place no limit on the number of successive terms a nawa could serve. As part of an effort to restore the effectiveness of the institution it was then decided that the offices must rotate each year.

35. The difficulty the nawa had in collecting fines is summed up in one incident where a Nauje family lavished beer (*chang*) and rum on the nawa as they are supposed to when they are being fined for having violated community regulations, but then charged the nawa for the alcohol they had drunk. This incident did not help other enforcement efforts.

36. The next spring the customary system of holding a village gathering and electing new nawa broke down. Usually the previous year's office holders are responsible for collecting a small amount of cash and grain (half a rupee or a rupee and a very small amount of rice) from each village family to make a barrel of beer and pay lamas for conducting a ceremony at the village temple in the third lunar month (Dawa Shiwa). At that gathering of the village prayers are offered to Khumbu Yul Lha and the three new nawa for the coming summer are chosen. In 1980 two of the incumbent nawa were away from the village in the spring. Some community leaders discouraged the mens' wives from performing the collection in their absence, and given the lack of community will to carry on communal pastoral regulation no one came forward to organize the ceremonies.

37. Many zopkio, however, are taken to the nearby (and traditionally unregulated) valley of Gyajo where they can be left for weeks on their own. Nauje families who have bought land in Langmoche and Chosero base zopkio there for at least part of the summer. A few Najue families during the last few years have begun taking zopkio still further afield, letting them summer on their own in Tengbo in westernmost Khumbu and in the Gokyo area of the upper Dudh Kosi valley. The two main Nauje yak-owning households also take their livestock to the uppermost Imja Khola valley and the high elevations of the Bhote Kosi valley.

38. Grazing disputes between Nauje and Khumjung-Kunde are not new. Nearly twenty years ago there was a major argument as a result of Nauje livestock (especially zhum) being herded at Shyangboche in summer when that area is closed by Khumjung-Kunde–administered herding regulations. There was discussion of ending nawa regulation. Villagers from all three settlements took their complaints to the Nepali official of the Nauje branch of the Sagarmatha

zone office. He supported continuing the nawa system and asked them what they wanted to do. Nauje and Khumjung-Kunde villagers alike then spoke up in favor of restoring the administration of the grazing ban while blaming each other for its violation.

39. The Thamicho action of turning the task of enforcing livestock exclusion rules over to the adekshe is very reminiscent of the last approach taken by Khumjung and Kunde when community support of shinggi nawa regulation of the rani ban was evaporating. It may reflect a lack of willingness by villagers to voluntarily continue to accept their turn at a rotated office rather than an effort to increase the effectiveness of enforcement.

40. In this regard it might be significant that Phurtse also has had a history of enforcing forest-management regulations unusually effectively.

41. This cautions also against assumptions that because sacred forests have long been carefully protected by a community that they will continue to be because of the strength of cultural values and beliefs.

42. Before sending the zopkio to graze on its own the family conducted a ceremony that included the reading of protective prayers such as are read before people go on trading trips to Tibet and other long journeys. There is some risk involved in sending the stock out to these areas, for some have died in falls on steep slopes and others have been drowned in the Bhote Kosi.

43. Many Nauje families now hire people to cut wild grass, and many young non-Sherpa family household "servants" (particularly young men) are assigned this task. It is said to be increasingly risky to grow hay in some lower Bhote Kosi sites near Nauje because these wild-grass cutters may "mistake" unattended hayfields for wild grass when the mist is thick and no one is watching.

44. Some Nauje residents also note that the local population of Himalayan tahr has risen markedly and also competes for grass on local slopes. A possible connection between the end of goat keeping in the area and poorer grass was also suggested, on the grounds that the lack of goats means that nearby slopes where they formerly grazed are now less effectively manured.

45. This aspect of herding management requires further investigation.

46. In the autumn of 1985 there was a poor hay harvest in much of Khumbu and wild grass for hay was also scarce. That year it snowed in late autumn and the snow stayed on the ground. Households began feeding hay and other fodder in the late autumn and had to continue feeding more fodder than usual to stock as the winter continued. By spring many families' fodder stores were exhausted. The grass was poorer than usual in the spring, and there was further snow as well. Hay was so rare in the spring that its price doubled and soon it became totally unavailable. Livestock began dying in late April and early May. Some people lost a great deal of stock (in some cases half their herd) and many families lost at least some stock. Thamicho herders especially lost a great deal of stock. One Thami Teng herder lost ten head of nak and yak and a Thami Og stock owner lost eight nak. Many herders lost five or six animals. Such disasters were not unique to Thamicho in 1986. Pangboche and Kunde nak herders in the upper Imja Khola lost unusual numbers of stock as well. One Pangboche family had eleven nak die at Dingboche. Stock owners herding in the upper Dudh Kosi

valley also reported losses, in one case of eight nak and yak. In all these cases the stock that died were yak and nak, always in the high herding settlements and often the oldest and least mobile stock.

9: From Tibet Trading to the Tourist Trade

1. The role of multialtitudinal and multienvironmental exchange has also been important in Andean subsistence strategies since pre-Columbian times (Brush 1976; Salomon 1985).

2. Livestocks and men have, however, been killed while attempting the crossing. Bad weather, crevasses, and frozen glacial lakes are all hazards, and there are reports of traders having lost more than a hundred head of dimzo on the pass.

3. There may have been one exception to this monopoly, for in Shorung there is a tradition that four powerful families of that region had government authorization to trade across the Nangpa La. Other Shorung families gained access by marrying into Khumbu families, and there was some illicit trading by both Shorung and Pharak Sherpas. Marriages into Khumbu families have boosted the trading careers of some Shorung men since the late nineteenth century. Among the early examples of this phenomenon was gembu Tsepal, who left his first wife in Golila to marry a Sherpa woman of Nauje. Ortner (1989:109) notes two other examples, Karma of Junbesi, who late in his life moved to Khumjung, and Kusang, Karma's son-in-law, who had earlier made the same move. In the latter cases, however, it is not clear that marriages to Khumbu Sherpa women were involved. Marriages between the sons of prominent Khumbu traders (especially Nauje traders) and the daughters of politically pow-erful and wealthy Shorung Sherpas continued through the 1950s, but the degree to which this kinship link was exploited by Shorung Sherpas for trading purposes is not known. Shorung Sherpas, it should also be remembered, had trading options other than maneuvering for access to the Nangpa La route. The route through Rongshar, which was far closer to Shorung and which was open year round, was utilized by Shorung Sherpas as well as Tibetan traders. This was indeed a much more important conduit for grain exports from Nepal than was the Nangpa La, and it was also the major horse-trading route.

4. Besides the Khumbu traders a number of Tibetans also carried salt south and traded it in Nauje to Sherpas (they were forbidden to deal it directly to Rais). Some of these Tibetan traders came south with fifty or sixty yak, and the total number of Tibetan yak arriving in Khumbu during trading season has been estimated at five hundred to a thousand. Many of these yak returned north lightly laden, and it was possible for Khumbu traders to hire them as pack stock paying a daily fee of eight or nine rupees per yak in the 1950s.

5. Traders who operated east of the Dudh Kosi in Rai country went south as far as Rajbiraj (Hanuman Nagar) in the Nepalese Tarai. A yak-hair chara that was worth four rupees in Nauje in the early 1950s brought twenty pathi of husked rice in places like Lokhim, a seven- or eight-day journey south of Nauje.

6. Sherpas did not buy, sell, and exchange goods in Ganggar, Kaprak, Nauje, or even Shigatse in market center bazaars. Instead all trade was carried out in households or camps. When Khumbu traders traded in Shigatse, for example, they rented rooms or a house from acquaintances and when word spread of what they had to offer, clients dealt with them directly on an individual basis. Similarly when Sherpas traded in lower-altitude regions of Nepal they stayed with trading partners, relatives, friends, or acquaintances in the villages and traded from their homes or made rounds of the various houses in the villages. Permanent marketplaces were scarce in eastern Nepal and much of Tibet until recent decades, although in lower-altitude Nepal and Tibet there were great annual trade fairs known as *melas*, often associated with religious celebrations.

7. In 1947, for example, 248 Khumbu men brought a court case against 3 Pharak Sherpas from Ghat whom they accused of going to Tibet and bringing back a flock of sheep over the Nangpa La.

8. Paper made in Nepal from daphne bark was the major paper used in Tibet to print religious texts and government documents before 1960. Butter was used on a vast scale by monasteries for keeping alight millions of devotional lamps, for warming monks with endless cups of salt tea, and for creating the large butter sculptures for which some monasteries constructed special buildings.

9. A few prospered and returned to Solu-Khumbu. One of these was the son of a Shorung pembu and trader who later became one of the greatest political figures and wealthiest men in Solu-Khumbu history, Sangye Tenzing Lama. He had already been active in trading with Tibet from Shorung (probably via the Rongshar route) before he moved to Darjeeling in the late 1880s. He did not go to Darjeeling to trade, however, but rather to be a labor contractor on two road-construction projects. Sangye Tenzing Lama took these earnings back to Shorung and only then invested them in trade in Those iron and paper (Ortner 1989:104).

10. Iron from Those was traded both over the Nangpa La and via the Rongshar route to the west. Most of the iron passing through Khumbu was in the form of agricultural tools, whereas the Rongshar iron was primarily round balls of raw metal. Both the trade via Rongshar and iron trade from Those to Khumbu were in the hands of Shorung Sherpas.

11. A few families used crossbreeds or sheep as pack animals.

12. From those places other middlemen such as the Rais of Gudel traded Tibetan salt still farther.

13. Rais brought grain each autumn to Nauje and other villages, exchanging it directly with particular families with whom they had been accustomed to trading. Rais have been coming to Khumbu in large numbers to trade since at least the nineteenth century.

14. Big traders arranged for large quantities of these goods to be supplied to them in Khumbu. A load of paper in Nauje cost 100 rupees delivered in the 1950s and big traders might take 300 loads a year north to Shigatse. Shere was 230 rupees a load in the mid-1950s and butter 400 rupees. Bigger profits were possible with butter. A trader might anticipate doubling his investment by selling

paper or shere in Shigatse, but butter was worth double the Nauje price in Ganggar and double the Ganggar price in Shigatse.

15. Khumbu trade to Kalimpong in the twentieth century may have reached its peak during the 1950s.

16. Lower-altitude Sherpas and Rais shared the common Khumbu belief that Indian salt is unsuitable for livestock and for many human uses. Khumbu Sherpas believe that feeding female stock Indian salt can result in miscarriage. It is also believed to be less effective as a preservative and to make poor salt-butter tea.

17. The number of trekking tourists is certainly well below the number of permits issued, since some tourists go on more than one trek and others never set out on the journeys for which they have obtained a permit.

18. Over a three-year period from autumn 1987 through spring 1990 43 percent (131) of the 303 expeditions that climbed in Nepal attempted Khumbu peaks.

19. In 1978 Sherpas had a majority financial interest in four-trekking companies. By 1988, however, Sherpas controlled twenty-six of the fifty-six trekking agencies registered with the Nepal Trekking Association (Kunwar 1989). All but a few of these Sherpa agencies were operated by Khumbu Sherpas and all had obtained their ownership of these businesses with support from non-Sherpa benefactors (Adams 1989:18).

20. The first recorded occasions on which Sherpas were employed as porters on a mountaineering expedition date to the 1907 climbing holidays of the Scottish doctor and mountaineer A. M. Kellas and the Norwegians Rubenson and Monrad Aas (Mason 1955:127). Both groups were impressed by the Sherpas' abilities as porters and praised not only their strength but also their good spirits, behavior, and bravery (Cameron 1984:154, 161). These were but the first of many accolades the Sherpas would soon receive from mountaineers.

21. It was also possible to engage in seasonal circular migration, journeying to Darjeeling each spring for mountaineering work and returning for the monsoon and the rest of the year to Khumbu. It is not yet possible, however, to reconstruct how important seasonal migration was in comparison to longer-term settlement in Darjeeling.

22. There are also eighteen peaks that were designated in 1978 as "trekking peaks." Permits to climb these are issued by the Nepal Mountaineering Association, and there are fewer conditions to be met and lower fees for these peaks. Several Khumbu peaks are approved as trekking peaks for climbing by tourists: Island Peak (Imja Tse, 6,183m), Lobuche East (6,119m), Kongma Tse (5,849m), Pokalde (5,806m), and Kwangde (6,011m).

23. These figures also include tourists staying at the Japanese-built Everest View Hotel. The number of tourists staying at this hotel, however, peaked during the mid-1970s and the hotel was not operated for most of the 1980s.

24. By village the percentage of households involved in tourism in 1978 was Nauje 84 percent, Kunde 85 percent, Khumjung 76 percent, and Phurtse 47 percent. In Nauje and Kunde more than a quarter of the entire population worked in tourism. In Nauje 150 individuals worked in tourism out of a village

population of 540, whereas in Kunde 68 people of a village population of 227 had tourism jobs (J. Fisher 1990:115).

25. By tourist shop I mean one considered by local residents to cater primarily to tourists and emphasizing the sale of mountaineering and trekking equipment, foreign foods, and souvenirs. Many such shops also sell some staple foods, kerosene, and clothing to Sherpa customers. A few shops in Nauje and the shops in Kunde and Khumjung primarily cater to local demands.

26. These two narrow lanes are flanked by continuous lines of multistory lodges. Until the late 1970s they were only paths fronted by a few houses and they then followed slightly different routes. With very few exceptions all the structures that line them today have been built since 1973.

27. These places were Nauje, Tengboche, Pheriche, and Lobuche. These sites formed a set of overnight stops that were located an easy day's hike apart and at altitudes that related well to optimal schedules for acclimatization. There were no early lodges in the other Khumbu valleys. During the 1980s there has been a process of developing lodges at intermediate sites along the main trail and establishing them on increasingly popular secondary routes (Stevens 1983).

28. All but two of these were located along the main route to Mount Everest.

29. The small mountain (Gokyo Ri) adjacent to the herding settlement of Gokyo in the upper Dudh Kosi valley also offered a view of Everest, and the route up to it gained a reputation for being both very scenic and less crowded than the main trail to Everest. By 1986 as many as one-fourth of Khumbu visitors were including Gokyo on their itinerary, usually combining it with a trip to Kala Patar.

30. Virtually all the big Nauje lodges, for example, have been built with financial assistance from foreign sponsors. Some Sherpas have received gifts of several thousand dollars, equivalent to several years' wages. Others have been granted long-term loans with low rates of interest. In some cases sums of more than $20,000 have been involved.

31. The higher zopkio figure given by Fürer-Haimendorf (1975:44) for 1957 includes calves temporarily in Nauje in transit to Tibet.

10: Tourism, Local Economy, and Environment

1. Now classic studies of the impact of tourism with particular focus on societies in developing countries include Smith (1989) and De Kadt (1976). Sociocultural impacts on host societies vary among communities as a result of their internal dynamics as well as with regional political economy and the nature and scale of tourism. Environmental impacts also vary enormously as different types of tourist activities, infrastructural development, and accompanying change in local economies and land use interact in diverse and complex ways with local ecosystems.

2. An assessment of the sociocultural impacts of tourism alone, for example, would need to examine tourism's role in change in local identity, beliefs, social hierarchy, conceptions of status, intergenerational relationships, religion, art, reciprocity, politics, demography, and emigration. Some of these topics have

been discussed by Fürer-Haimendorf (1975, 1984), J. Fisher (1990) and Stevens (1983, 1989). In 1990–1991 I conducted further fieldwork on recent sociocultural change in the Khumbu region.

3. Sherpas often say that they "eat" these wages. In the Khumbu Sherpa sense "eating" income, however, also includes spending it on consumer goods, house improvements, and pleasure and pilgrimage journeys to Kathmandu, India, and Tibet. But for households with a relatively small income from tourism, the main use of tourism money is for purchasing food.

4. Investment in tourism is also possible through the purchase of materials for producing handicrafts such as woolen caps and mittens.

5. Precise figures on income and profits are unavailable, but only a few Sherpas perform relatively low-paying day labor or trade with Tibet. Day-labor income, like income from the sale of agricultural surpluses, moreover, is indirectly linked to tourism because tourism furnishes the money used for payment. The most important source of regional income that is entirely outside of the direct and indirect economic ramifications of tourism is probably the sale of crossbreed calves to Sherpas from Shorung and other regions.

6. Other factors, however, have also been at work, and some change that has often been attributed to tourism has had other roots altogether.

7. Porters often expect foreigners to pay more than local residents for their services, and the amount they demand rises if the route involves high-altitude conditions, cold weather, danger, or transit through regions known for expensive lodging and food. The daily wage is based on a load of twenty-five to thirty kilograms, with a load of twenty-five kilograms becoming more common. More pay is expected for greater loads and longer carrying days. It should be noted that in Khumbu porter rates today are high by standards for other areas of Nepal, including regions that have become tourist destinations, and that most porters in Khumbu are not Khumbu Sherpas but Rais, Tamangs, and Pharak Sherpas.

8. Sherpas who already have a full set of the necessary gear from previous expeditions are often given a cash payment in lieu of equipment, which in some cases amounts to a thousand dollars.

9. The company which pays its sirdar 3,000 rupees per month offers cooks 2,800 rupees, both quite high wages by Kathmandu trekking company standards.

10. The profits can be considerable even if a zopkio driver must be hired, in which case the income is split evenly. But it is common for a family member (usually a son but sometimes a wife or daughter) to carry out this task. During the late autumn, winter, and early spring, however, profits are much less if hay has to be purchased.

11. Sherpa sirdar, though, often pay a price for their high income. Both mountaineering and trekking sirdar work is often enormously stressful, requiring the shouldering of much responsibility on a round-the-clock basis and a great deal of close interaction with foreigners. Several Sherpa sirdar have developed ulcers, and others have taken to drink.

12. Thamicho people, including a few women, continue to do some of this work, as occasionally do a few Phurtse and Pangboche men. More Pharak Sherpas work seasonally as trekking porters, including groups of young women.

13. Until the mid-1980s the autumn tourist season was by far the largest, and October and November the busiest months. There was a second peak in the spring, but this was formerly much less important. The number of tourists during the winter months has increased somewhat in recent years, and although there are still two peak seasons the tourist season as a whole now lasts from late September until the first of June.

14. Price rises also took place in the pretourism economy, as reflected not only in the rising cash costs of purchasing nak, salt, and other goods in Tibet but also in the relative barter exchange rates between salt and grains.

15. Most of the rice sold in Nauje is not grown by the men and women selling it but has rather come from the Dingla and Aislalukharka areas or the Tarai and has reached Khumbu through a series of exchanges. Prices in Nauje are reached by direct bargaining between sellers and individual Sherpas.

16. One popular Dingboche lodge, for example, that relied entirely on purchased potatoes, bought approximately twenty-seven loads of potatoes for the autumn 1990–spring 1991 season and ran short. This is equivalent to the annual consumption of many Khumbu families.

17. Mountaineering expeditions and trekking groups bring most of their food and supplies with them from overseas and from Kathmandu. The vegetables, fruit, dairy products, and grain they buy at the Nauje weekly market are all grown outside of Khumbu.

18. The rise in day-labor rates I have described, however, has not been a problem regionally since the higher wages from tourism have more than compensated farmers for the added expense. More families than ever today can afford to hire agricultural wage workers. The one place where tourism has created a serious shortage of agricultural labor and has tremendously inflated wage rates is in the upper Imja Khola during hay-cutting season. This coincides with a time when Pangboche people can easily obtain work with expeditions that have completed their climb and need porters to help them haul their gear out of Khumbu. To compete with the wages offered to porters, Pangboche families must pay men double the usual agricultural day wage for hay cutting. In autumn 1990 this was eighty rupees per day.

19. Some lodge-owning families have also redoubled their efforts to buy more land so that they can also expand their potato production and cut down on the amount of potatoes they must obtain for lodge use either through deals with individual farmers or at the weekly market.

20. Similar complaints about lowland wage earners' lack of knowledge and sense of responsibility is also said to limit their usefulness as herders and to lead to careless and destructive fuel-wood gathering.

21. Field fertility, for example, could be lowered by an increasing depletion of key nutrients, including trace minerals. It may also be that fields are less well fertilized in some places where forest floor leaves and needles are less abundant now than in earlier periods.

22. At Nyeshe near Nauje there are many terraces that were used until only about ten to fifteen years ago along with other terraces that have been abandoned for a considerable time. Reasons for the abandoning of these fields in-

cluded poor harvests and increasing problems with wildlife depredations (mainly by Himalayan tahr). No one indicated that the problem was a lack of labor. Some land was converted to buckwheat production for several years before being abandoned. This suggests that additional potato production was no longer required, possibly reflecting larger yellow potato harvests. Buckwheat would have been especially difficult to guard against tahr and livestock depredation. The abandonment of wheat production at Tashilung in the middle of the twentieth century was also precipitated by problems with tahr. The great increase in the number of tahr in some parts of Khumbu, including the Nauje area, during the late 1980s and early 1990s concerns many farmers, and some believe that Sagarmatha National Park will need to take action. The fact that some fields at Nyeshe have been reclaimed during the last few years by families with little land in Nauje itself, despite the risks there of crop damage from livestock, testifies to the current degree of Sherpa interest in expanding crop production.

23. In some other areas of Nepal, however, such as Langtang and the Takshindo area adjacent to Shorung, the establishment of cheese factories with Swiss aid has had an effect on local pastoralism. This cheese is primarily destined for tourist consumption in Kathmandu. Some Takshindo cheese is brought to Khumbu and sold to lodges and Nauje shops.

24. One Khumjung family, for instance, kept about eight urang zopkio about fifty years ago when they were heavily engaged in salt trading, and a number of other families there owned a few urang zopkio that were used during winter trading trips south. Some urang zopkio were kept by a few Thamicho and Nauje families for similar purposes in more recent times.

25. Very few urang zopkio, however, are kept by Phurtse or Pangboche families. Only four Phurtse families had urang zopkio in the autumn of 1990 and among them owned only nine head of stock.

26. The 1984 figures may have included calves, which would have diminished the percentage of urang zopkio given that only three-year and older urang zopkio are imported.

27. Tourism has increased the local demand for milk and cheese. This demand, as well as the greater part of the increased Sherpa demand for more dairy products (especially butter but also some types of cheese such as *shomar*), are met, however, by dairy products imported from outside of Khumbu. Most milk consumed in the region, and almost the only milk that any tourist ever tastes, is foreign powdered milk sold in the weekly market. Butter and other dairy products sold in Nauje and consumed by tourists and local residents alike are imported from Shorung, Kulung, and Pharak.

28. This did not mean, as Fürer-Haimendorf has assumed, that enough fuel wood had to be stored at base camp for the needs of the army of porters as well (1984:58), since they were dismissed and returned home after the establishment of the camp.

29. The felling of trees for this purpose ceased after the establishment of Sagarmatha National Park.

30. Showers are offered by most Nauje lodges and also by some lodges at the Tengboche monastery and even by a few lodges above the forest line at the

settlements of Pheriche, Dingboche, and Chukkung. Some Nauje lodges provide more than fifteen showers a day to tourists during the peak of the season. Four Nauje lodges have electric water heating. The others heat water on a wood stove. By mixing hot water with cold in nearly equal proportions and keeping the amount of water issued per shower to thirty-five liters, a load of birch will heat enough water for eight to ten showers.

31. The increasing pressure for timber was instead shifted to forests outside the park in Pharak. The unfortunate impact of Khumbu timber demand on Pharak forests offers dramatic testimony of what change the national park policy may have averted in Khumbu itself. In effect the shift of impact created by national park policy is very similar to that which took place under local Sherpa forest-protection systems when forest-use pressure was shunted to areas outside of rani ban and lama's forests—only in this case all of Khumbu is administered as protected forest.

32. It was not considered possible to forbid Sherpas or other Nepalis from making cookfires for their own use. This meant that expeditions and groups with large numbers of porters still contributed to increased demand on forest and woodland along the main trails and on the high-altitude juniper that supplied base camps. In spring 1991, however, a sign at the entry station to the park advised tourists that their porters were also forbidden to use fuel wood inside the national park.

33. The development of a small hydroelectric facility in Nauje in 1983 has had very little impact on local fuel-wood use. Under the best operating conditions the plant produces only enough electricity to power hotplates and immersion coils for water heating in four of the village's lodges from early morning until 6:00 P.M. No other village households have any electric power whatsoever for cooking (see Stevens 1989 for more detail).

34. This is reflected, for example, in the increasing importation of rice and its greater role now in Khumbu diets (especially in Nauje, Khumjung, and Kunde). The role of grain grown outside Khumbu in general has also increased and has made possible the shift of much land from buckwheat to potatoes. But cropland is not being abandoned as a result of this process, and the increased consumption of rice does not represent a fundamental shift in Sherpa food tastes. Potatoes and barley are still the preferred foods in Khumbu.

35. That some of these beams were used in building lodges at Pheriche is not the critical point. The key was the undermining of the local ability to enforce traditional rules.

36. There seems to be no link, however, between overgrazing and what Blower and some later reports described as severe gullying near Kunde, Thami Teng, and in the upper Dudh Kosi valley. These instead probably are the result of the geomorphological processes of debris and earth flows, wind ablation, landslides, and slumps, whose origins are unconnected with grazing pressure. The landslides in the upper Dudh Kosi valley probably reflect tectonic activity (Vuichard 1986), whereas the major landscape feature just to the east of Kunde village that had been assumed to be an erosive gully is a long-standing (and still active) debris flow generated high on the walls of Khumbila, and the supposed gully near Thami Teng is probably a feature of wind ablation.

37. Erosion rates may be higher, however, during the occasional years when heavy rains occur. In some years very heavy mid-summer and autumn rains occur, the yerchu and tenju rains which are of concern to farmers. There are also sometimes very intense rainfalls in May and early June before the onset of the main monsoon season. On several occasions in the mid-1980s rainfall was intense enough and runoff heavy enough that the trails in Nauje ran like creeks. They carried enough silt that the small hydroelectric plant down slope had to be closed for fear that the filtration system would be inadequate to prevent damage to the generators. In the late 1960s a heavy autumn rain flooded several Nauje houses and a number of people spent the night seeking refuge beside the village shrine, and on September 18, 1987 I recorded a rainfall in Nauje of more than 100 millimeters in a twenty-four-hour period. The erosive power of these unusual events is not known. It may be substantial in years such as 1985 when a relatively dry winter and spring retards spring grass growth and heavy late-May and early-June rains fall on relatively bare slopes. The role of wind erosion also needs to be evaluated, and it may be considerable in the spring before the growth of new ground cover.

38. Erosion levels, moreover, would undoubtedly have been far higher still if he had been able to measure wind erosion as well.

39. It is unclear how these figures of forest cover loss were reached, or how so much deforestation could take place on the snowy slopes of Mount Everest, or even at its base, an area that is a thousand meters above the forest line.

40. The comparison of several pairs of photographs from 1962 and 1984 that Byers carried out (1987b:207–208, 213–214, 217–218) and a pair from 1950 and 1989 made by J. Fisher (1990:144–145) suggest that relatively little deforestation has occurred in these particular sites. These photographs also offer support for Sherpa oral history accounts of localized vegetation change, including the clearing of juniper at Shyangboche in the 1960s and early 1970s, the thinning of juniper above Khumjung after 1965, tree felling between Nauje and Sonasa during the 1960s and 1970s, and pre-1950 forest use on the slope west of the Tengboche monastery.

41. These mules are currently mainly employed in transporting material for the building of the hydroelectric project near Thami Og and the rebuilding of the Tengboche monastery. When these projects are completed in the next year or two the mules may well be shifted to the tourist trade. Sherpas note that mules carry 100-kilogram loads rather than the 60-kilogram loads that urang zopkio do, and that they travel twice as fast. Mules, for example, routinely make the trip between Nauje and Lukla in a single day. It may be that Sherpas will decide that mules can be even more profitable for transport businesses than zopkio or yak. This could have an enormous impact on regional pastoralism as well as significantly change the character of the Khumbu rural landscape.

Conclusions

1. It should be reemphasized that Khumbu is not a homogeneous society and that within individual settlements a range of lifestyles are followed by different

households. This complexity increases when viewed diachronically, for both the demographic cycle and broader historical changes related to intensification, migration, and political economy widen the range of subsistence strategies followed by a given family or community. Yet a number of common themes characterize Khumbu Sherpa subsistence strategies: diversification that combines agropastoralism with trade or income from tourism; multialtitudinal resource use; specialization in high-altitude varieties of crops and livestock and in particular in potatoes, buckwheat, and yak and yak-cattle crossbreeds; seasonal movement between multiple dwellings; and localized transhumance.

2. Examples of household decisions to choose taste over yield in crop selection include not cultivating yellow potatoes and development potatoes and growing barley rather than buckwheat at Dingboche even though it requires considerably more labor for manuring and irrigating.

3. Not all villages, however, were able to respond to declining resource availability by developing community resource-management systems. The main villages of the Bhote Kosi valley (other than Nauje), for example, failed to establish controls over tree felling in nearby areas despite what must have been a very evident decline in the number of trees suitable for house beams. And in the early twentieth century establishment of the rani ban outside intervention in the form of orders from Kathmandu was critical in creating a climate in which local leaders were able to mobilize cooperation to establish new conservation measures based on local perception of environmental conditions and resource needs.

4. This leaves open the possibility that some indigenous peoples may decide to follow ways of life that may one day force them to adjust their land-use practices to different environmental conditions that they themselves have precipitated. Indeed, many indigenous peoples may have radically reshaped local ecosystems intentionally through swidden cultivating, clearing forest for farm and pasture land, changing forest and grassland composition through the use of fire, introducing species, depleting some forms of wildlife and plants through hunting and gathering for consumption, security, and trade, changing the courses of streams and the contours of slopes, and even encouraging or acquiescing in ecological degradation in one place in order to create environments that are more useful for subsistence activities in other places. An example of this would be a lack of concern with erosion from upper slopes in the knowledge that this would only increase alluvium and fertility in down-valley farmlands, a process which appears to have been important in early New Guinea (Blaikie and Brookfield 1987) and perhaps elsewhere in the Pacific.

5. The degree to which those Khumbu Sherpas who have migrated to Kathmandu will be able (or interested) in maintaining some aspects of their identity—or even whether what they perceive as central to their identities as Sherpas is the same as that perceived in Khumbu communities—is a very different question. James Fisher (1990) suggests that in Kathmandu language, to note just one important aspect of culture, is not being maintained and being transmitted to the children. It is interesting, however, the degree to which adult migrants to Kathmandu attempt to maintain some aspects of their former way of life. They

often do not sell their houses in Khumbu, either in the hope of some day returning after they have made their fortunes or, as is also common, so that they have a home to return to each year for at least a few months. Many Kathmandu-based Sherpas spend the summer monsoon months in Khumbu, the season when the great regional festivals of Dumje, Yerchang, and Pangyi take place and the time of weddings and other occasions for social interaction and celebration. In Kathmandu the emigrants try to maintain some sense of community, celebrate religious festivals and life-cycle rites, and even (in the case of the women) wear a modified form of traditional dress. But clearly there are important changes, among which is the sacrifice of the roles of homeland and traditional subsistence lifestyles as a part of their identity. Here the urban dwellers may be developing a generational gap with their parents for whom being Khumbu Sherpa continues to mean living in Khumbu on the land and practicing forms of agriculture and pastoralism that reach far back into the Sherpa experience.

Bibliography

Acharya, Harihar Prasad
    1989   "Jirel Property Arrangements and the Management of Forest and Pasture Resources in Highland Nepal." *Development Anthropology Network* (Bulletin of the Institute for Development Anthropology) 7(2):16–24.
    1990   "Processes of Forest and Pasture Management in a Jirel Community of Highland Nepal." Ph.D. dissertation, Cornell University.

Adams, Vincanne
    1989   "Healing Buddhas and Mountain Guides: The Production of Self within Society through Medication." Ph.D. dissertation, Medical Anthropology, University of California, Berkeley, and University of California, San Francisco.

Allan, Nigel J. R.
    1986   "Accessibility and Altitudinal Zonation Models of Mountains." *Mountain Research and Development* 6:185–194.

Allan, Nigel J. R., G. W. Knapp, and C. Stadel, eds.
    1988   *Human Impact on Mountains.* Totowa, N. J.: Rowman and Littlefield.

Arnold, J. E. M., and J. Gabriel Campbell
    1986   "Collective Management of Hill Forests in Nepal: The Community Forest Development Project." In *Proceedings of the Conference on Common Property Resource Management,* 425–454. Washington, D.C.: National Academy Press.

Aziz, Barbara N.
    1978   *Tibetan Frontier Families: Reflections of Three Generations from D'ing-ri.* New Delhi: Vikas Publishing House.

Bajracharya, Deepak
    1983   "Deforestation in the Food/Fuel Context: Historical and Political

507

Perspectives from Nepal." *Mountain Research and Development* 3(3):227–240.

Barth, Frederich
1956   "Ecologic Relationships of Ethnic Groups in Swat, North Pakistan." *American Anthropologist* 58:1079–1089.

Basham, A. L.
1985   *The Wonder That Was India.* 3d rev. ed. London: Sidgwick and Jackson.

Begley, S., and R. Moreau
1987   "Mount Everest is a Junk Pile." *Newsweek,* November 16.

Bennett, John
1976   *The Ecological Transition.* London: Pergamon Press.

Bernbaum, Edwin
1990   *Sacred Mountains of the World.* San Francisco: Sierra Club.

Berreman, Gerald D.
1960   "Culture Variability and Drift in the Himalayan Hills." *American Anthropologist* 62(5):774–794.

1963*a*   *Hindus of the Himalayas.* Berkeley and Los Angeles: University of California Press.

1963*b*   "Peoples and Cultures of the Himalayas." *Asian Survey* 3(6):289–304.

Bhatt, D. D.
1977   *Natural History and Economic Botany of Nepal.* rev. ed. New Delhi: Orient Longman Ltd.

Bishop, Barry C.
1990   *Karnali Under Stress.* Chicago: Department of Geography Research Publications.

Bishop, Naomi
1989   "From Zomo to Yak: Change in a Sherpa Village." *Human Ecology* 17(2):177–204.

Bjønness, Inger Marie
1980*a*   "Animal Husbandry and Grazing, a Conservation and Management Problem in Sagarmatha (Mt. Everest) National Park, Nepal." *Norsk Geografisk Tidsskrift* 34(2):59–76.

1980*b*   "Ecological Conflicts and Economic Dependency on Tourist Trekking in Sagarmatha (Mt. Everest) National Park, Nepal: An Alternative Approach to Park Planning." *Norsk Geografisk Tidsskrift* 34(3):119–138.

1983   "External Economic Dependency and Changing Human Adjustment to Marginal Environment in the High Himalaya, Nepal." *Mountain Research and Development* 3(3):263–272.

1986   "Mountain Hazard Perception and Risk-avoiding Strategies among the Sherpas of Khumbu Himal." *Mountain Research and Development* 6(4):277–292.

Blaikie, Piers, and Harold Brookfield
1987   *Land Degradation and Society.* London: Methuen.

Blower, John H.
1971   "Proposed National Park in Khumbu District." Unpublished report to His Majesty's Government's Secretary of Forests.

Boserup, Esther
  1965   *The Conditions of Agricultural Growth.* Chicago: Aldine.
Bremer-Kamp, Cherie
  1987   *Living on the Edge.* Layton, Utah: Peregrine Smith Books.
Brookfield, Harold C.
  1964   "Questions on the Human Frontiers of Geography." *Economic Geography* 40(4):283–303.
  1973   "Introduction: Explaining or Understanding? The Study of Adaptation and Change." In *The Pacific in Transition: Geographical Perspectives on Adaptation and Change,* ed. Harold C. Brookfield, 3–23. New York: St. Martin's Press.
Brower, Barbara A.
  1987   "Livestock and Landscape: The Sherpa Pastoral System in Sagarmatha (Mt. Everest) National Park, Nepal." Ph.D. dissertation, Department of Geography, University of California, Berkeley.
Brush, Stephen
  1976   "Cultural Adaptations to Mountain Ecosystems: An Introduction." *Human Ecology* 4(2):125–133.
  1977   *Mountain, Field, and Family: The Economy and Human Ecology of an Andean Valley.* Philadelphia: University of Pennsylvania Press.
  1987   "Diversity and Change in Andean Agriculture." In *Lands at Risk in the Third World: Local Level Perspectives,* ed. P. D. Little et al., 271–289. Boulder, Colo.: Westview Press.
Byers, Alton
  1986   "A Geomorphic Study of Man-induced Soil Erosion in the Sagarmatha (Mt. Everest) National Park, Khumbu, Nepal." *Mountain Research and Development* 6(1):83–87.
  1987a   "An Assessment of Landscape Change in the Khumbu Region of Nepal." *Mountain Research and Development* 7(1):77–80.
  1987b   "A Geoecological Study of Landscape Change and Man-accelerated Soil Loss: The Case of Sagarmatha (Mt. Everest) National Park, Khumbu, Nepal." Ph.D. dissertation, Department of Geography, University of Colorado, Boulder.
  1987c   "Landscape Change and Man-accelerated Soil Loss: The Case of the Sagarmatha (Mt. Everest) National Park, Khumbu, Nepal." *Mountain Research and Development* 7(3):209–216.
Cameron, I. C.
  1984   *Mountains of the Gods.* London: Century Publishing.
Carter, H. A.
  1985   "Classification of the Himalaya." *The American Alpine Journal* 27(59):109–141.
Casimir, Michael V., and Aparna Rao
  1985   "Vertical Control in the Western Himalaya: Some Notes on the Pastoral Ecology of the Nomadic Bakrwal of Jammu and Kashmir." *Mountain Research and Development* 5(3):221–232.
Chayanov, A. V.
  1966   *The Theory of Peasant Economy.* Homewood, Ill.: Richard D. Irwin for the American Economic Association.

Clarke, G. E.
1980*a*   "A Helambu History." *Journal of the Nepal Research Centre* 4(1):
          1–38.
1980*b*   "Lama and Tamang in Yolmo." In *Tibetan Studies in Honour of
          Hugh Richardson,* ed. by Michael Aris and Aung San Suu Kyi, 79–
          86. Warminster, England: Aris and Phillips.
Clarke, W. C.
1971   *Place and People: An Ecology of a New Guinea Community.* Berke-
          ley, Los Angeles, London: University of California Press.
Cole, John and Eric R. Wolf
1974   *The Hidden Frontier: Ecology and Ethnicity in an Alpine Valley.*
          New York: Academic Press.
Conklin, Harold C.
1954   "An Ethnoecological Approach to Shifting Agriculture." *Transi-
          tions* ser. 2(17):133–142.
1957   *Hanunoo Agriculture.* Rome: Food and Agriculture Organization.
Croft, Peter
1976   "Sagarmatha (Mt. Everest) National Park Draft Management Plan."
          Unpublished manuscript. Sagarmatha National Park Headquarters,
          Namche Bazar, Nepal.
De Kadt, E.
1976   *Tourism: Passport to Development?* New York: Oxford University
          Press.
Dittert, R., G. Chevally, and R. Lambert
1954   *Forerunners to Everest: The Story of the Two Swiss Expeditions of
          1952.* New York: Harper and Brothers.
Dutt, B. L.
1964   "Late Blight of Potato in India II, Occurrence and Severity." *India
          Potato Journal* 6(2):70–86.
Ekvall, Robert
1968   *Fields on the Hoof: Nexus of Tibetan Nomad Pastoralism.* New
          York: Holt, Rinehart, Winston.
English, Richard
1982   "Gorkhali and Kiranti: Political Economy in the Eastern Hills of
          Nepal." Ph.D. dissertation, New School for Social Research.
1985   "Himalayan State Formation and the Impact of British Rule in the
          Nineteenth Century." *Mountain Research and Development* 5(1):
          61–78.
Fisher, James F.
1986*a*   "Tourists and Sherpas." *Contributions to Nepalese Studies* 14(1):
          37–62.
1986*b*   *Trans-Himalayan Traders: Economy, Society, and Culture in North-
          west Nepal.* Berkeley, Los Angeles, Oxford: University of Califor-
          nia Press.
1990   *Sherpas: Reflections on Change in Himalayan Nepal.* Berkeley, Los
          Angeles, Oxford: University of California Press.

Fisher, R. J.
1989   *Indigenous Systems of Common Property Forest Management in Nepal.* East-West Environment and Policy Institute Working Paper 18. Honolulu: East-West Center.

Fox, J. M.
1983   "Managing Public Lands in a Subsistence Economy: The Perspective from a Nepali Village." Ph.D. dissertation, University of Wisconsin, Madison.

Fricke, Thomas
1986   *Himalayan Households: Tamang Demography and Domestic Processes.* Ann Arbor: UMI Research Press.
1989   "Human Ecology in the Himalaya." *Human Ecology* 131–145.

Funke, F.
1969   *Religiöses Leben der sherpa, khumbu himal.* Innsbruck: Universitätsverlag Wagner.

Fürer-Haimendorf, Christoph von
1964   *The Sherpas of Nepal: Buddhist Highlanders.* London: John Murray.
1974   *Contributions to the Anthropology of Nepal.* Warminster, England: Aris and Phillips.
1975   *Himalayan Traders: Life in Highland Nepal.* London: John Murray.
1979   *The Sherpas of Nepal: Buddhist Highlanders.* 3d ed. London: East-West Publications.
1981   *Asian Highland Societies in Anthropological Perspective.* New Delhi: Sterling Publishers.
1984   *The Sherpas Transformed: Social Change in a Buddhist Society of Nepal.* New Delhi: Sterling Publishers.

Garratt, Keith J.
1981   *Sagarmatha National Park Management Plan.* Wellington, New Zealand: Department of Lands and Survey.

Gilmour, Donald A.
1989   *Forest Resources and Indigenous Management in Nepal.* Working Paper No. 17. Honolulu: East-West Center Environment and Policy Institute.

Gilmour, Donald A., and Robert J. Fisher.
1991   *Villagers, Forests, and Foresters: The Philosophy, Process, and Practice of Community Forestry in Nepal.* Kathmandu, Nepal: Sahayogi Press.

Goldstein, Melvyn
1974   "Tibetan-speaking Agro-pastoralists of Limi: A Cultural Ecological Overview of High Altitude Adaptation in the Northwest Himalaya." *Objets et Mondes* 14:259–268.
1981   "High Altitude Tibetan Populations in the Remote Himalaya: Social Transformation and its Demographic, Economic, and Ecological Consequences." *Mountain Research and Development* 1(1): 5–18.

Goldstein, Melvyn, and Cynthia M. Beall
1990   *Nomads of Western Tibet: The Survival of a Way of Life.* Berkeley, Los Angeles, Oxford: University of California Press.

Goldstein, Melvyn, and Donald Messerschmidt
1980   "The Significance of Latitudinality in Himalayan Mountain Ecosystems." *Human Ecology* 88(2):117–134.

Goldstein, Melvyn, Cynthia M. Beall, and Richard P. Cincotta
1990   "Traditional Nomadic Pastoralism and Ecological Conservation on Tibet's Northern Plateau." *National Geographic Research* 6(2): 139–156.

Grenard, F.
1974   *Tibet: The Country and its Inhabitants.* Reprint. Delhi: Cosmo Publications.

Grossman, Lawrence
1977   "Man-Environment Relationships in Anthropology and Geography." *Annals of the Association of American Geographers* 67(1): 126–144.

1984   *Peasants, Subsistence Ecology, and Development in the Highlands of Papua New Guinea.* Princeton: Princeton University Press.

Guibaut, A.
1987   *Tibetan Venture.* Reprint 1947 ed. New York: Oxford University Press.

Guillet, David
1983   "Toward a Cultural Ecology of Mountains: The Central Andes and the Himalayas Compared." *Current Anthropology* 24(5):561–574.

Gurung, Harka
1980   *Vignettes of Nepal.* Kathmandu: Sajha Prakashan.

1983   *Maps of Nepal: Inventory and Evaluation.* Bangkok: White Orchid Press.

Hagen, Toni
1980   *Nepal: The Kingdom in the Himalayas.* New Delhi: Oxford and IBH Publishing Company.

Hagen, Toni, et al.
1963   *Mount Everest: Formation, Population and Exploration of the Everest Region.* London: Oxford University Press.

Hardie, Norman
1957   *In Highest Nepal: Our Life Among the Sherpas.* London: Allen and Unwin, Ltd.

Hardie, Norman, et al.
1987   "Nepal-New Zealand Project of Forest Management in Khumbu-Pharak." Unpublished report to the Himalayan Trust and Volunteer Service Abroad.

Hardin, Garrett
1968   "The Tragedy of the Commons." *Science* 162:1243–1248.

Hillary, Sir Edmund
1982   "Preserving a Mountain Heritage." *National Geographic* 161(6): 696–702.

Hinrichsen, Don, et al.
1983   "Saving Sagarmatha." *Ambio* 11(5):203–205.

Holmberg, David H.
1989   *Order in Paradox: Myth, Ritual, and Exchange among Nepal's Tamang.* Ithaca, N.Y.: Cornell University Press.

Hooker, Sir Joseph Dalton
1969   *Himalayan Journals: Notes of a Naturalist.* Reprint. New Delhi: Today and Tomorrow's Printers and Publishers.

Houston, Charles S.
1987   "Deforestation in Solu Khumbu." *Mountain Research and Development* 7(1):76.

Hunt, John
1954   "Triumph on Everest: Siege and Assault." *National Geographic* 106(1):1–44.

Hutt, M.
1986   "Diversity and Change in the Languages of Highland Nepal." *Contributions to Nepalese Studies* 14(1):1–24.

Inoue, J.
1976   "Climate of Khumbu Himal." *Seppyo* 38:66–73.

Ives, Jack D., and Bruno Messerli
1989   *The Himalayan Dilemma: Reconciling Development and Conservation.* London and New York: Routledge.

Jackson, John A.
1955   *More Than Mountains.* London: Harrap.

James, Preston E.
1972   *All Possible Worlds: A History of Geographical Ideas.* Indianapolis and New York: Odyssey Press.

Jeffries, B. E.
1982   "Sagarmatha National Park: The Impact of Tourism in the Himalayas." *Ambio* 11(5):274–281.

Jerstad, L.
1969   *Mani-Rimdu: A Sherpa Dance Drama.* Seattle: University of Washington Press.

Joshi, D. P.
1982   "The Climate of Namche Bazar." *Mountain Research and Development* 2(4):399–403.

Karan, Pradyumna P., and Cotton Mather
1985   "Tourism and Environment in the Mount Everest Region." *Geographical Review* 75(1):93–95.
1987   "Population Characteristics of the Himalayan Region." *Mountain Research and Development* 7(3):271–274.

Khanal, Prakash
1988   "Super Potato!" *Himal* 1(2):27–28.

Kirkpatrick, William
1975   *An Account of the Kingdom of Nepaul, Being the Substance of Observations Made During a Mission to that Country in the Year 1793.* Reprint 1811 ed. New Delhi: Asian Publications Services.

Kunwar, Ramesh Raj
1989   *Fire of Himal: An Anthropological Study of the Sherpas of Nepal Himalayan Region.* New Delhi: Nirala Publications.
Langer, R. H. M., and G. D. Hill
1982   *Agricultural Plants.* Cambridge: Cambridge University Press.
Li, S., ed.
1986   *Gannan Zangzu Zizhizhou Gaikuang* (Basic facts on Gannan Tibetan autonomous region). Lanzhou, Gansu: Gansu People's Publishing House.
Limberg, W.
1982   *Untersuchungen über Besiedlung, Landbesitz und Feldbau in Solu-Khumbu (Mount Everestgebiet, Ostnepal), Khumbu Himal.* Innsbruck: Universitätsverlag Wagner.
Little, Peter D., and Michael M. Horowitz, eds.
1987   *Land at Risk in the Third World.* Boulder, Colo.: Westview Press.
Lucas, P. H. C., N. D. Hardie, and R. A. C. Hodder
1974   *Report of the New Zealand Mission on Sagarmatha (Mt. Everest) National Park.* Wellington: Ministry of Foreign Affairs.
MacDonald, Alexander W.
1980   "The Coming of Buddhism to the Sherpa Area of Nepal." *Acta Orientalia* 34(1–3):139–146.
1987   *Essays on the Ethnology of Nepal and South Asia.* Kathmandu: Ratna Pustak Bhandar.
McDougal, Charles
1979   *The Kulunge Rai.* Kathmandu: Ratna Pustak Bhandar.
MacFarlane, Alan
1976   *Resources and Population: A Study of the Gurungs of Nepal.* Cambridge: Cambridge University Press.
McNeely, Jeffrey
1985   "Man and Nature in the Himalaya: What Can Be Done to Ensure that Both Can Prosper." Paper presented at the International Workshop on the Management of National Parks and Protected Areas of the Hindu Kush-Himalaya, May 6–11, Kathmandu.
March, K.
1977   "Of People and Naks: The Meaning of High-altitude Herding among Contemporary Solu Sherpas." *Contributions to Nepalese Studies* 4(2):83–97.
Mason, Kenneth
1955   *Abode of Snow: A History of Himalayan Exploration and Mountaineering.* London: Rupert Hart-Davis.
Masuda, Shozo, Izumi Shimada, and Craig Morris, eds.
1985   *Andean Ecology and Civilization.* Tokyo: University of Tokyo Press.
Mayer, Enrique
1985   "Production Zones." In *Andean Ecology and Civilization,* ed. S. Masuda, I. Shimada, and C. Morris. Tokyo: University of Tokyo Press.

Messerschmidt, Donald A.
1986   "People and Resources in Nepal: Customary Resource Manage-
            ment Systems of the Upper Kali Gandaki." In *Proceedings of the
            Conference on Common Property Resource Management,* 455–480.
            Washington D.C.: National Academy Press.
1987   "Conservation and Society in Nepal: Traditional Forest Manage-
            ment and Innovative Development." In *Land at Risk in the Third
            World,* ed. Pctcr D. Little and Michael M. Horowitz, 373–397.
            Boulder, Colo.: Westview Press.
Metz, John J.
1989   "A Framework for Classifying Subsistence Production Types of Ne-
            pal." *Human Ecology* 17(2):147–176.
1990   "Conservation Practices at an Upper-elevation Village of West Ne-
            pal." *Mountain Research and Development* 10(1):7–15.
Mishra, Hemanta R.
1973   *Conservation in Khumbu: The Proposed Mt. Everest National Park.
            A Preliminary Report.* Kathmandu, Nepal: Department of National
            Parks and Wildlife Conservation.
Moench, Marcus H.
1985   "Resource Utilization and Degradation: An Integrated Analysis of
            Biomass Utilization Patterns in a Garhwal Hill Village, Northern
            Uttar Pradesh, India." Master's thesis, University of California,
            Berkeley, Energy and Resources Group.
Moran, Emilio F.
1979   *Human Adaptability: An Introduction to Ecological Anthropology.*
            North Scituate, Mass.: Duxbury Press.
Moran, Emilio F., ed.
1984   *The Ecosystem Concept in Anthropology.* AAAS Selected Sympo-
            sium 92. Boulder, Colo.: Westview Press.
Muller, F.
1959   "Eight Months of Glacier and Soil Research in the Everest Re-
            gion." In *The Mountain World 1958/59,* 191–208. London: George
            Allen and Unwin.
Murra, John V.
1972   "El Control Vertical de un Máximo de Pisos Ecológicos en la
            Economía de las Sociedades Andinas." In *Inigo Ortiz de Zuniga,
            Visita de la Provincia de León de Huanuco en 1562,* vol. 2, ed. J. V.
            Murra, 429–476. Huanuco, Peru: Universidad Nacional Hermilio
            Valdizan.
1985a   " 'El Archipielago Vertical' Revisited." In *Andean Ecology and Civi-
            lization,* ed. S. Masuda, I. Shimada, and C. Morris, 3–13. Tokyo:
            University of Tokyo Press.
1985b   "The Limits and Limitations of the 'Vertical Archipelago' in the
            Andes." In *Andean Ecology and Civilization,* ed. S. Masuda,
            I. Shimada, and C. Morris, 15–20. Tokyo: University of Tokyo
            Press.

Naylor, R.
1970   "Colombo Plan Assignment in Nepal." Unpublished report to the New Zealand Forest Service.

Nepal. Central Bureau of Statistics.
1984   *Population Census 1981.* Kathmandu: His Majesty's Government of Nepal National Planning Commission Secretariat.

Nepal. Ministry of Tourism.
1985   *Annual Statistical Report 1984.* Kathmandu: His Majesty's Government of Nepal, Department of Tourism.

[1989]   *Annual Statistical Report 1988.* Kathmandu: His Majesty's Government of Nepal, Department of Tourism.

Netting, Robert McC.
1971   "Of Men and Meadows: Strategies of Alpine Land Use." *Anthropological Quarterly* 38:132–144.

1981   *Balancing on an Alp: Ecological Change and Continuity in a Swiss Mountain Community.* Cambridge: Cambridge University Press.

1984   "Reflections on an Alpine Village as Ecosystem." In *The Ecosystem Concept in Anthropology,* ed. Emilio F. Moran, 225–235. Boulder, Colo.: Westview Press.

Nietschmann, Bernard Q.
1973   *Between Land and Water: The Subsistence Ecology of the Miskito Indians, Eastern Nicaragua.* New York: Seminar Press.

Oppitz, Michael
1968   *Geschichte und Sozialordnung der Sherpa, Khumbu Himal.* Innsbruck: Universitätsverlag Wagner.

1974   "Myths and Facts: Reconsidering Some Data Concerning the Clan History of the Sherpa." In *Contributions to the Anthropology of Nepal,* ed. C. von Fürer-Haimendorf, 232–243. Warminster, England: Aris and Phillips.

Orlove, Benjamin
1980   "Ecological Anthropology." *Annual Review of Anthropology* 9: 235–273.

Orlove, Benjamin, and David W. Guillet
1985   "Theoretical and Methodological Considerations on the Study of Mountain Peoples: Reflections on the Idea of Subsistence Type and the Role of History in Human Ecology." *Mountain Research and Development* 5(1):3–18.

Ortner, Sherry B.
1978   *Sherpas Through Their Rituals.* Cambridge: Cambridge University Press.

1989   *High Religion: A Cultural and Political History of Sherpa Buddhism.* Princeton, N. J.: Princeton University Press.

Palmieri, Richard
1976   "Domestication and Exploitation of Livestock in the Nepal Himalaya and Tibet: An Ecological, Functional, and Cultural Historical Study of Yaks and Yak Hybrids in Society, Economy, and Culture."

Ph.D. dissertation, Department of Geography, University of California, Davis.

Pant, S. D.
1935   *The Social Economy of the Himalayans.* London: Allen and Unwin.

Parker, Anne
1989   "The Meanings of "Sherpa," an Evolving Social Category." *Himalayan Research Bulletin* 9(3):11–14.

Paul, R.
1979   "Dumje: Paradox and Resolution in Sherpa Ritual Symbolism." *American Ethnologist* 6(2):274–304.
1982   *The Tibetan Symbolic World: Psychoanalytic Explorations.* Chicago: University of Chicago Press.

Pawson, Ivan G., D. D. Stanford, and V. A. Adams
1984   "Effects of Modernization on the Khumbu Region of Nepal: Changes in Population Structure, 1970–1982." *Mountain Research and Development* 4(1):73–81.

Pawson, Ivan G. et al.
1984   "Growth of Tourism in Nepal's Everest Region: Impact on the Physical Environment and Structure of Human Settlements." *Mountain Research and Development* 4(3):237–246.

Perry, R.
1981   *Mountain Wildlife.* Harrisburg, Pa.: Stackpole Books.

Pradhan, Kumar
1991   *The Gorkha Conquests: The Process and Consequences of the Unification of Nepal, with Particular Reference to Eastern Nepal.* Calcutta: Oxford University Press.

Price, L. W.
1981   *Mountains and Man.* Berkeley, Los Angeles, London: University of California Press.

Rappaport, R. A.
1968   *Pigs for the Ancestors.* New Haven, Conn.: Yale University Press.

Rauber, Hanna
1980   "The Humli-Khyampas of Far Western Nepal: A Study in Ethnogenesis." *Contributions to Nepalese Studies* 8(1):57–79.
1982   "Humli-Khyampas and the Indian Salt Trade: Changing Economy of Nomadic Traders in Far West Nepal." In *Contemporary Nomadic and Pastoral Peoples: Asia and the North,* Studies in Third World Societies 18, ed. Philip Carl Salzman, 141–176. Williamsburg, Va.: Department of Anthropology, College of William and Mary.

Rawat, Indra Singh
1973   *Indian Explorers of the Nineteenth Century: An Account of Explorations in the Himalayas, Tibet, Mongolia, and Central Asia.* New Delhi: Ministry of Information and Broadcasting, Government of India.

Rhoades, Robert E., and S. I. Thompson
1975   "Adaptive Strategies in Alpine Environments: Beyond Ecological Particularism." *American Ethnologist* 2(3):535–551.

Rowell, Galen
    1980   *Many People Come, Looking, Looking.* Seattle: The Mountaineers.
Sacherer, Janice
    1975   "The Sherpas of Rolwaling: Human Adaptation to a Harsh Moun-
            tain Environment." *Objets et Mondes* 14(4):317–324.
    1981   "The Recent Social and Economic Impact of Tourism on a Remote
            Sherpa Community." In *Asian Highland Societies in Anthropologi-
            cal Perspective,* ed. C. von Fürer-Haimendorf, 157–167. Atlantic
            Highlands, N.J.: Humanities Press.
Sagant, Philippe
    1976   *Le Paysan Limbu: Sa Maison et ses Champs.* Paris: Mouton.
Sahlins, Marshall
    1972   *Stone Age Economics.* Chicago: Aldine.
Salaman, R. N.
    1985   *The History and Social Influence of the Potato.* Cambridge: Cam-
            bridge University Press.
Salomon, Frank
    1985   "The Dynamic Potential of the Complementarity Concept." In
            Andean Ecology and Civilization, ed. S. Masuda, I. Shimada, and
            C. Morris, 511–531. Tokyo: University of Tokyo Press.
Sandberg, G.
    1987   *The Exploration of Tibet.* Delhi: Cosmo Publications.
Sauer, Carl O.
    1925   "The Morphology of Landscape." *University of California Publica-
            tions in Geography* 2(2):19–53.
    1956   "The Agency of Man on the Earth." In *Man's Role in Changing the
            Face of the Earth,* ed. W. L. Thomas, 46–49. Chicago: University of
            Chicago Press.
Schaffer, E. H.
    1963   *The Golden Peaches of Samarkand: A Study of T'ang Exotics.*
            Berkeley and Los Angeles: University of California Press.
Schneider, Erwin
    1963   "Forward to the Map of the Mount Everest Area." In *Mount
            Everest: Formation, Population and Exploration of the Everest Re-
            gion,* ed. Toni Hagen *et al.,* 182–191. London: Oxford University
            Press.
Schrader, Heiko
    1988   *Trading Patterns in the Nepal Himalayas.* Saarbrücken and Fort Lau-
            derdale, Fla.: Verlag breitenbach.
Schweinfurth, V.
    1983   "Man's Impact on Vegetation and Landscape in the Himalayas." In
            Man's Impact on Vegetation, ed. W. Holzer, M. Werger, and I.
            Ikusima, 297–309. The Hague: Dr. W. Junk Publishers.
Seddon, David
    1987   *Nepal: A State of Poverty.* New Delhi: Vikas Publishing House.
Sherpa, Nima Wangchu
    1979   "A Report on Firewood Use in Sagarmatha National Park, Khumbu

Region, Nepal." Unpublished report to His Majesty's Government's Department of National Parks and Wildlife Conservation.

Shimada, Izumi
    1985   "Introduction." In *Andean Ecology and Civilization,* ed. S. Masuda, I. Shimada, and C. Morris, xi–xxxii. Tokyo: University of Tokyo Press.

Slusser, M. S.
    1982   *Nepal Mandala: A Cultural Study of the Kathmandu Valley.* Princeton, N. J.: Princeton University Press.

Smith, Valene, ed.
    1989   *Hosts and Guests: The Anthropology of Tourism.* 2d rev. ed. Philadephia: University of Pennsylvania Press.

Speechly, H. T.
    1976   *Proposal for Forest Management in Sagarmatha National Park.* Unpublished report to His Majesty's Government's Department of National Parks and Wildlife Conservation.

Stevens, Stanley F.
    1983   "Tourism and Change in Khumbu." Bachelor's thesis, Department of Geography, University of California, Berkeley.
    1986*a*   "Inhabited National Parks: Indigenous Peoples in Protected Landscapes." Canberra: Australia National University, Centre for Resource and Environmental Studies. East Kimberley Impact Assessment Project, Working Paper 10.
    1986*b*   "Sherpa Forests: Forest Use, Protection, and Destruction in the Mount Everest Region of Nepal." Paper presented to the First Annual California Universities' Conference on South Asia, University of California, Berkeley.
    1988*a*   "Sacred and Profaned Himalayas." *Natural History* 97(1):26–35.
    1988*b*   "Tourism Development in Nepal." *Kroeber Anthropological Society Papers* 67–68:67–80.
    1989   "Sherpa Settlement and Subsistence: Cultural Ecology and History in Highland Nepal." Ph.D. dissertation, Department of Geography, University of California, Berkeley.

Steward, J.
    1936   "The Economic and Social Basis of Primitive Bands." In *Essays in Anthropology Presented to A. L. Kroeber,* ed. R. Lowie, 331–347. Berkeley and Los Angeles: University of California Press.
    1955   *Theory of Culture Change.* Urbana: University of Illinois Press.

Stiller, Ludwig F.
    1973   *The Rise of the House of Gorkha: A Study in the Unification of Nepal 1768–1816.* Patna, India: The Patna Jesuit Society.

Teschke, G.
    1977   *Anthropologie der Sherpa, Khumbu Himal.* Innsbruck: Universitätsverlag Wagner.

Thomas, W. L., Jr.
    1956   *Man's Role in Changing the Face of the Earth.* Chicago: University of Chicago Press.

Thompson, Michael, and Michael Warburton
    1985   "Uncertainty on a Himalayan Scale." *Mountain Research and Development* 5(2):115–135.
Tilman, H. W.
    1952   *Nepal Himalaya.* Cambridge: Cambridge University Press.
Troll, Carl
    1966   "The Cordilleras of the Tropical Americas, Aspects of Climactic, Phytogeographical and Agrarian Ecology." *Colloquium Geographicum* 9, Bonn: Ferd Dummlers Verlag.
    1988   "Comparative Geography of High Mountains of the World in the View of Landscape Ecology: A Development of Three and a Half Decades of Research and Organization." In *Human Impact on Mountains,* ed. N. J. R. Allan, G. W. Knapp, and C. Stadel. Totowa, N. J.: Rowman and Littlefield.
Unsworth, W.
    1981   *Everest.* Middlesex, England: Penguin Books.
Vansina, J.
    1985   *Oral Tradition as History.* Madison: University of Wisconsin Press.
Vuichard, D.
    1986   "Geological and Petrographical Investigations for the Mountain Hazards Mapping Project, Khumbu Himal, Nepal." *Mountain Research and Development* 6(1):41–52.
Yamamoto, Noria
    1985   "The Ecological Complementarity of Agro-pastoralism: Some Comments." In *Andean Ecology and Civilization,* ed. S. Masuda, I. Shimada, and C. Morris, 85–99. Tokyo: University of Tokyo Press.
Yan, Z., ed.
    1986   *Qinghaisheng Qing* (The situation of Qinghai province). Xining, Qinghai: Qinghai People's Publishing House.
Zhang, F. X., D. T. Xue, and Z. X. Tian
    1987   *Qinghai Xumu* (Qinghai pastoralism). Xining, Qinghai: Qinghai People's Publishing House.
Zimmerman, M., M. Bichsel, and H. Kienholz
    1986   "Mountain Hazards Mapping in the Khumbu Himal, Nepal, with Prototype Map Scale 1:50,000." *Mountain Research and Development* 6(1):29–40.
Zurick, D. N.
    1988   "Resource Needs and Land Stress in Rapti Zone, Nepal." *The Professional Geographer* 40(4):428–443.
    1990   "Traditional Knowledge and Conservation as a Basis for Development in a West Nepal Village." *Mountain Research and Development* 10(1):23–33.

Index

Note: Sherpa personal names are alphabetized by the first name, not by the surname Sherpa.

Designer: U.C. Press Staff
Compositor: Huron Valley Graphics
Text: 10/13 Sabon
Display: Sabon
Printer: Malloy Lithographing, Inc.
Binder: John H. Dekker & Sons